IGCSE **Physics**

Endorsed by
University of Cambridge International Examinations

IGCSE Physics

Second Edition

Tom Duncan and Heather Kennett

HODDER EDUCATION
AN HACHETTE UK COMPANY

To F.D.B.

Although every effort has been made to ensure that website addresses are correct at time of going to press, Hodder Education cannot be held responsible for the content of any website mentioned in this book. It is sometimes possible to find a relocated web page by typing in the address of the home page for a website in the URL window of your browser.

Hachette UK's policy is to use papers that are natural, renewable and recyclable products and made from wood grown in sustainable forests. The logging and manufacturing processes are expected to conform to the environmental regulations of the country of origin.

Orders: please contact Bookpoint Ltd, 130 Milton Park, Abingdon, Oxon OX14 4SB. Telephone: (44) 01235 827720. Fax: (44) 01235 400454. Lines are open 9.00–5.00, Monday to Saturday, with a 24-hour message answering service. Visit our website at www.hoddereducation.com

First published in 2002 by
Hodder Education, an Hachette UK Company,
338 Euston Road
London NW1 3BH

This second edition published 2009

Impression number 8 7 6 5
Year 2014 2013 2012

Cover photo © TEK Image/Science Photo Library
Illustrations by Fakenham Prepress Solutions, Fakenham, Norfolk NR21 8NN and Wearset
Typeset in 11.5/13pt Bembo by Fakenham Prepress Solutions
Printed and bound in Dubai.

A catalogue record for this title is available from the British Library

ISBN 978 0 340 98187 0

Contents

Preface

IGCSE Physics Second Edition aims to provide an up-to-date and comprehensive coverage of the Core and Extended curriculum in Physics specified in the current University of Cambridge International Examinations IGCSE syllabus.

Within each chapter, sections marked ■ contain Core material (e.g. p.1), those marked □ contain Supplement material (e.g. p.16) and sections marked [F] contain material for further reading (e.g. p.30).

Some restructuring of the book has occurred and material no longer in the latest syllabus has been removed. The following topics have been modified, extended or brought up-to-date:

- **Waves and sound** – a new experiment to measure the speed of sound in air is described; seismic waves are introduced.
- **Forces and resources** – in addition to forces, pressure and energy transfer, this section now includes the chapter on energy sources.
- **Matter and measurements** – an experiment to measure the time period of a simple pendulum is included.
- **Heat and energy** – discussion of the kinetic theory of matter has been moved to this section. New experiments on determining the specific latent heats for ice and steam are described. Treatments of condensation and solidification are expanded.
- **Electricity** – electronics has been integrated into this section with two new chapters: Electronic systems and Digital electronics. Analogue and digital electronics, logic gates and control systems are now covered and additional examples of the use of relays in switching circuits provided.
- **Electromagnetism** – discussion of magnetisation and demagnetisation has been extended.
- **Electrons and atoms** – the nuclear energy section has been expanded to include the fusion reactions occurring in the Sun.
- **Checklists** – all are aligned more closely with the syllabus requirements.
- **Questions** – new questions from recent exam papers have been included in the *Additional questions* (after each group of related topics) and in the *Revision questions* (at the end of the book, for quick comprehensive revision before exams). More difficult questions or those on Supplement material are marked with a blue stripe down the left-hand side.

The accompanying **Revision CD-ROM** provides invaluable exam preparation and practice. It has two sections:

- *Revision Support* includes sample examination papers and answers.
- *Test Your Knowledge* includes interactive tests that cover both the Core and Extended curriculum. These are organised by syllabus topic.

The authors would like to thank Mike Folland for his comprehensive review of the IGCSE syllabus and his helpful comments and suggestions for changes to the text, and Brian Kennett for his valuable advice.

T.D. and H.K.

Acknowledgement is made to the following examining boards for permission to reproduce questions (answers given being the sole responsibility of the authors):

AQA, incorporating:
NEAB (Northern Examinations and Assessment Board)
SEG (Southern Examining Group)
EDEXCEL: London (London Examinations)
Edexcel Ltd accepts no responsibility whatsoever for the accuracy or method of working in the answers given
OCR (Oxford, Cambridge and RSA Examinations)
UCLES (Cambridge International Examinations, part of the University of Cambridge Local Examinations Syndicate)

IGCSE Physics syllabus and textbook matching grid

Key for syllabus levels
IGCSE Core curriculum material
IGCSE Extended curriculum material

Notes:

- Definitions are given during and at the end of appropriate chapters.

IGCSE Physics syllabus topic	Cross-reference to book	Cross-reference to CD-ROM
1 General physics		
1.1 Length and time	pp.44–48, pp.93–94, pp.47–49	Test 1 Test 2
1.2 Speed, velocity and acceleration	pp.92–94, pp.96–97, p.101, pp.92–94, p.96, pp.100–102, p.107	Test 3 Test 4
1.3 Mass and weight	p.47, pp.54–55, p.104, p.106	
1.4 Density	pp.51–52, p.52	
1.5 Forces (a) Effects of forces (b) Turning effect (c) Conditions for equilibrium (d) Centre of mass (e) Scalars and vectors	 pp.54–55, p.68, pp.104–105, p.55, p.105, pp.113–115 pp.60–61, pp.60–61 p.62 pp.64–66 pp.69–70	Tests 5–6 Tests 7–8
1.6 Energy, work and power (a) Energy (b) Energy resources (c) Work (d) Power	 pp.71–72, p.74, p.79, p.110, p.109 pp.76–78, p.259, p.74, p.259 p.73, p.73 p.73, p.73	Tests 5–6 Tests 7–8
1.7 Pressure	pp.83–84, p.86, p.83, p.85	
2 Thermal physics		
2.1 Simple kinetic molecular model of matter (a) States of matter (b) Molecular model (c) Evaporation (d) Pressure changes	 p.123 pp.122–124, p.127, p.134, pp.136–137, p.122–123 pp.144–145, p.144 pp.135–136, p.135	Tests 9–10 Test 11
2.2 Thermal properties (a) Thermal expansion of solids, liquids and gases (b) Measurement of temperature (c) Thermal capacity (d) Melting and boiling	 pp.129–130, p.133–134, pp.129–130 pp.126–127, pp.126–127 p.127, p.139, p.140 pp.142–145, p.142–144	
2.3 Transfer of thermal energy (a) Conduction (b) Convection (c) Radiation (d) Consequences of energy transfer	 p.148, p.149 p.150 p.153, p.153 p.149, p.151, p.154	

IGCSE Physics syllabus topic	Cross-reference to book	Cross-reference to CD-ROM
3 Properties of waves, including light and sound		
3.1 General wave properties	pp.26–29, p.32, p.36, p.27, pp.29–30	
3.2 Light (a) Reflection of light (b) Refraction of light (c) Thin converging lens (d) Dispersion of light (e) Electromagnetic spectrum	 p.5, pp.8–9, pp.5–6 pp.11–12, p.15, p.12, p.16 pp.18–20, pp.20–21 p.13 pp.32–34, pp.32–33	Tests 12–13 Test 14
3.3 Sound	pp.36–38, p.37	
4 Electricity and magnetism		
4.1 Simple phenomena of magnetism		Tests 15–16
4.2 Electrical quantities (a) Electric charge (b) Current (c) Electro–motive force (d) Potential difference (e) Resistance (f) Electrical energy	 p.160, p.161, p.164, pp.161–162, p.164–165, p.167 pp.166–167, p.167, pp.224–225 p.171, pp.170–171 pp.170–172, pp.224–225 pp.174–175, p.178, p.178 p.183	Tests 17–18
4.3 Electrical circuits (a) Circuit diagrams (b) Series and parallel circuits (c) Action and use of circuit components (d) Digital electronics	 p.167, p.175, p.185, p.192, p.215, p.217, p.233, pp.193–197 pp.167–168, pp.176–177, p.168, p.172, p.177 p.186, pp.175–176, p.178, pp.180–182, pp.191–193, p.176, pp.192–197 pp.198–200	Test 19 Test 20
4.4 Dangers of electricity	pp.185–189	
4.5 Electromagnetic effects (a) Electromagnetic induction (b) a.c. generator (c) Transformer (d) The magnetic effect of a current (e) Force on a current-carrying conductor (f) d.c.motor	 p.227, p.228 p.228 pp.233–234, p.233–234 pp.214–217, p.214–215 p.220, p.221, p.241, p.248 p.221, p.221	Test 21 Test 22
4.6 Cathode ray oscilloscopes (a) Cathode rays (b) Simple treatment of cathode-ray oscilloscope	 pp.240–241 pp.242–243	
5 Atomic physics		
5.1 Radioactivity (a) Detection of radioactivity (b) Characteristics of the three types of emission (c) Radioactive decay (d) Half-life (e) Safety precautions	 p.246 p.247, p.250, pp.247–248 p.249, p.257 p.249, p.252 pp.250–251	Tests 23–24 Tests 25–26
5.2 The nuclear atom (a) Atomic model (b) Nucleus (c) Isotopes	 p.161, p.255, p.258, p.255 p.256 p.250, p.256, p.259	

Physics and technology

Physicists explore the Universe. Their investigations range from particles that are smaller than atoms to stars that are millions and millions of kilometres away, Figures 1a, b.

Figure 1a This image, produced by a scanning tunnelling microscope, shows an aggregate of gold just three atoms thick on a graphite substrate. Individual graphite (carbon) atoms are shown as green

As well as having to find the **facts** by observation and experiment, physicists also must try to discover the **laws** that summarize (often as mathematical equations) these facts. Sense has then to be made of the laws by thinking up and testing **theories** (thought-models) to explain the laws. The reward, apart from a satisfied curiosity, is a better understanding of the physical world. Engineers and technologists use physics to solve **practical problems** for the benefit of people, though, in solving them, social, environmental and other problems may arise.

In this book we will study the behaviour of **matter** (the stuff things are made of) and the different kinds of **energy** (such as light, sound, heat, electricity). We will also consider the applications of physics in the home, in transport, medicine, research, industry, energy production and electronics. Figures 2a, b, c and d show some examples.

Mathematics is an essential tool of physics and a 'reference section' of some of the basic mathematics is given at the end of the book along with suggested procedures for solving physics problems.

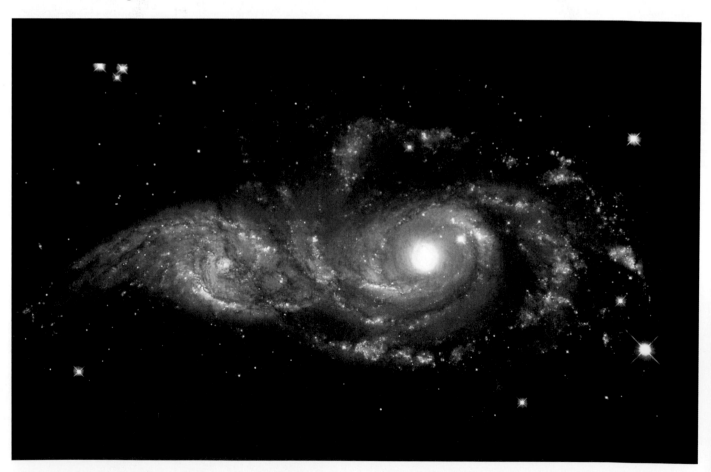

Figure 1b The many millions of stars in the Universe, of which the Sun is just one, are grouped in huge galaxies. This photograph of two interacting spiral galaxies was taken with the Hubble Space Telescope. This orbiting telescope is enabling astronomers to tackle one of the most fundamental questions in science, i.e. the age and scale of the Universe, by giving much more detailed information about individual stars than is possible with ground-based telescopes

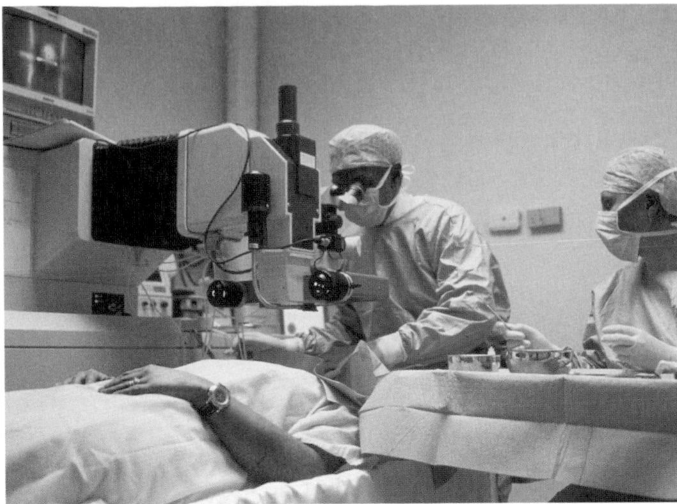

Figure 2a The modern technology of laser surgery enables very delicate operations to be performed. Here the surgeon is removing thin sheets of tissue from the surface of the patient's cornea, in order to alter its shape and correct severe short-sightedness

Figure 2b Mobile phones provide us with the convenience of instant communication wherever we are – but does the electromagnetic radiation they use pose a hidden risk to our health?

Figure 2c The manned exploration of space is such an expensive operation that international co-operation is seen as the way forward. This is the International Space Station, built module-by-module in orbit around the Earth. It is operated as a joint venture by the USA and Russia

Figure 2d In the search for alternative energy sources, 'wind farms' of 20 to 100 wind turbines have been set up in suitable locations, such as this one in North Wales, to generate at least enough electricity for the local community

Scientific enquiry

During your course you will have to carry out a few experiments and investigations aimed at encouraging you to develop some of the **skills** and **abilities** that scientists use to solve real-life problems.

Simple experiments may be designed to measure, for example, the temperature of a liquid or the electric current in a circuit. Longer investigations may be designed to establish or verify a relationship between two or more physical quantities.

Investigations may arise from the topic you are currently studying in class, or your teacher may provide you with suggestions to choose from, or you may have your own ideas. However an investigation arises, it will probably require at least one hour of laboratory time, but often longer and will involve you in the following four aspects.

1 **Planning** how you are going to set about finding answers to the questions the problem poses. Making predictions and hypotheses (informed guesses) may help you to focus on what is required at this stage.
2 **Obtaining** the necessary experimental data safely and accurately. You will have to decide on the equipment needed, what observations and measurements have to be made and what variable quantities need to be manipulated. Do not dismantle the equipment until you have completed your analysis and are sure you do not need to repeat any of the measurements!
3 **Presenting** and **interpreting** the evidence in a way that enables any relationships between quantities to be established.
4 **Considering** and **evaluating** the evidence by drawing conclusions, assessing the reliability of data and making comparisons with what was expected.

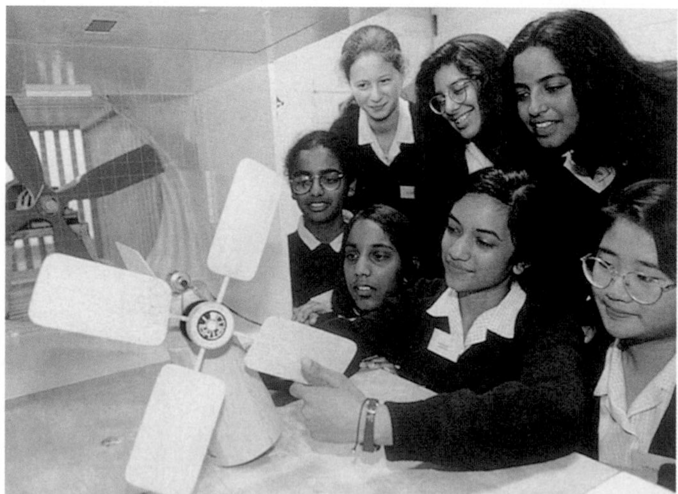

Figure 3 Girls from Copthall School, London, with their winning entry for a contest to investigate, design and build the most efficient, elegant and cost-effective windmill

A **written report** of the investigation would normally be made. This should include:

- the **aim** of the work.
- a list of all items of **apparatus** used and a record of the smallest division of the scale of each measuring device. For example, the smallest division on a metre rule is 1 mm. The scale of the rule can be read to the nearest mm. So when used to measure a length of 100 mm (0.1 m), the length is measured to the nearest 1 mm, the degree of accuracy of the measurement being 1 part in 100. When used to measure 10 mm (0.01 m), the degree of accuracy of the measurement is 1 part in 10. A thermometer is calibrated in degrees Celsius and may be read to the nearest 1 °C. A temperature may be measured to the nearest 1 °C. So when used to measure a temperature of 20 °C, the degree of accuracy is 1 part in 20 (this is 5 parts in 100).
- details of **procedures**, observations and measurements made. A clearly labelled *diagram* will be helpful here; any difficulties encountered or precautions taken to achieve accuracy should be mentioned.
- presentation of **results** and **calculations**. If several measurements of a quantity are made, draw up a *table* in which to record your results. Use the column headings, or start of rows, to *name* the measurement and state its *unit*; for example 'Mass of load/kg'. *Repeat* the measurement of each observation; record each value in your table, then calculate an average value. Numerical values should be given to the number of *significant figures* appropriate to the measuring device (see Chapter 10).

  If you decide to make a *graph* of your results you will need at least eight data points taken over as large a range as possible; be sure to *label* each axis of a graph with the name and unit of the quantity being plotted (see Chapter 20).
- **conclusions** which can be drawn from the evidence. These can take the form of a numerical value (and unit), the statement of a known law, a relationship between two quantities, or a statement related to the aim of the experiment (sometimes experiments do not achieve the intended objective).
- an **evaluation** and discussion of the findings which should include

  (i) a comparison with expected outcomes,

  (ii) a comment on the reliability of the readings, especially in relation to the scale of the measuring apparatus,

  (iii) a reference to any apparatus that was unsuitable for the experiment,

(iv) a comment on any graph drawn, its shape and whether the graph points lie on the line,
(v) a comment on any trend in the readings, usually shown by the graph,
(vi) how the experiment might be modified to give more reliable results, for example in an electrical experiment by using an ammeter with a more appropriate scale.

Suggestions for investigations

Investigations which extend the practical work or theory covered in some chapters are listed below. The section *Further experimental investigations* on p. 291 details how you can carry out some of these investigations.

1 Pitch of a note from a vibrating wire (Chapter 9).
2 Stretching of a rubber band (Chapter 12 and *Further experimental investigations*, p. 291).
3 Stretching of a copper wire – **wear safety glasses** (Chapter 12).
4 Toppling (*Further experimental investigations*, p. 291).
5 Friction – factors affecting (Chapter 15).
6 Energy values from burning fuel, e.g. a firelighter (Chapter 16).
7 Model wind turbine design (Chapter 17).
8 Speed of a bicycle and its stopping distance (Chapter 23).
9 Circular motion using a bung on a string (Chapter 24).
10 Heat loss using different insulating materials (Chapter 31).
11 Cooling and evaporation (*Further experimental investigations*, p. 291).
12 Variation of the resistance of a thermistor with temperature (Chapter 36).
13 Variation of the resistance of a wire with length (*Further experimental investigations*, p. 292).
14 Heating effect of an electric current (Chapter 38).
15 Strength of an electromagnet (Chapter 42).
16 Efficiency of an electric motor (Chapter 43).

Ideas and evidence in science

You may find that in some of the investigations you perform in the school laboratory you do not interpret your data in the same way as your friends do; perhaps you will argue with them as to the best way to explain your results and try to convince them that your interpretation is right. Scientific controversy frequently arises through people interpreting evidence differently.

Observations of the heavens led the ancient Greek philosophers to believe that the Earth was at the centre of the planetary system, but a complex system of rotation was needed to match observations of the apparent movement of the planets across the sky. In 1543 Nicolaus Copernicus made the radical suggestion that all the planets revolved not around the Earth but around the Sun. (His book '*On the Revolutions of the Celestial Spheres*' gave us the modern usage of the word 'revolution'.) It took time for his ideas to gain acceptance. The careful astronomical observations of planetary motion documented by Tycho Brahe were studied by Johannes Kepler, who realised that the data could be explained if the planets moved in elliptical paths (not circular) with the Sun at one focus. Galileo's observations of the moons of Jupiter with the newly invented telescope led him to support this 'Copernican view' and to be imprisoned by the Catholic Church in 1633 for disseminating heretical views. About 50 years later, Isaac Newton introduced the idea of gravity and was able to explain the motion of all bodies, whether on Earth or in the heavens (Chapter 22), which led to full acceptance of the Copernican model. Newton's mechanics were refined further at the beginning of the 20th century when Einstein developed his theories of relativity; even today, data from the Hubble Space Telescope is providing new evidence which confirms Einstein's ideas.

Many other scientific theories have had to wait for new data, technological inventions, or time and the right social and intellectual climate for them to become accepted. In the field of health and medicine, for example, because cancer takes a long time to develop it took several years before people recognized that X-rays and radioactive materials could be dangerous (Chapter 48).

At the beginning of the 20th century scientists were trying to reconcile the wave theory and the particle theory of light (Chapter 47) by means of the new ideas of quantum mechanics.

Today we are collecting evidence on possible health risks from microwaves used in mobile phone networks. The cheapness and popularity of mobile phones may make the public and the manufacturers reluctant to accept adverse findings, even if risks are made widely known in the press and on television. Although scientists can provide evidence and evaluation of that evidence, there may still be room for controversy and a reluctance to accept scientific findings, particularly if there are vested social or economic interests to contend with. This is most clearly shown today in the issue of global warming.

Light and sight

1 Light rays

Sources of light	*Shadows*	*Practical work*
Rays and beams	*Speed of light*	The pinhole camera.

Sources of light

You can see an object only if light from it enters your eyes. Some objects such as the Sun, electric lamps and candles make their own light. We call these **luminous** sources.

Most things you see do not make their own light but reflect it from a luminous source. They are **non-luminous** objects. This page, you and the Moon are examples. Figure 1.1 shows some others.

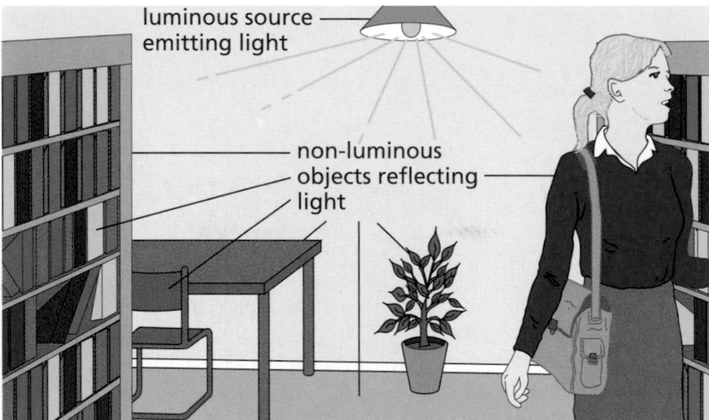

Figure 1.1 Luminous and non-luminous objects

Luminous sources radiate light when their atoms become 'excited' as a result of receiving energy. In a light bulb, for example, the energy comes from electricity. The 'excited' atoms give off their light haphazardly in most luminous sources.

A light source that works differently is the **laser**, invented in 1960. In it the 'excited' atoms act together and emit a narrow, very bright beam of light. The laser has a host of applications. It is used in industry to cut through plate metal, in scanners to read the bar code at shop and library check-outs, in CD players, in optical fibre telecommunication systems, in delicate medical operations on the eye or inner ear, for example, Figure 1.2, in printing, and in surveying and range-finding.

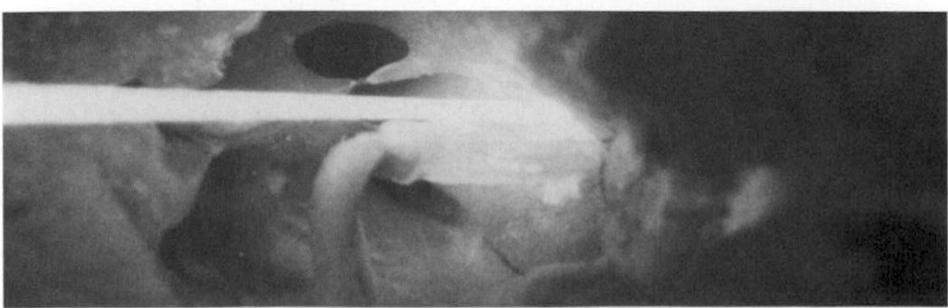

Figure 1.2 Laser surgery in the inner ear

Rays and beams

Sunbeams streaming through trees, Figure 1.3, and light from a cinema projector on its way to the screen both suggest that **light travels in straight lines**. The beams are visible because dust particles in the air reflect light into our eyes.

The direction of the path in which light is travelling is called a **ray** and is represented in diagrams by a straight line with an arrow on it. A **beam** is a stream of light and is shown by a number of rays; it may be parallel, diverging (spreading out) or converging (getting narrower), Figure 1.4.

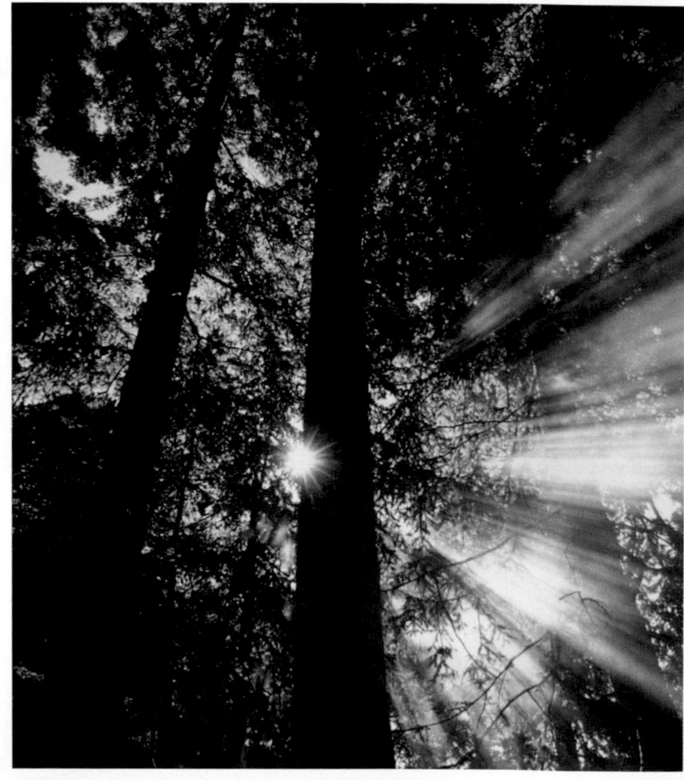

Figure 1.3 Light travels in straight lines

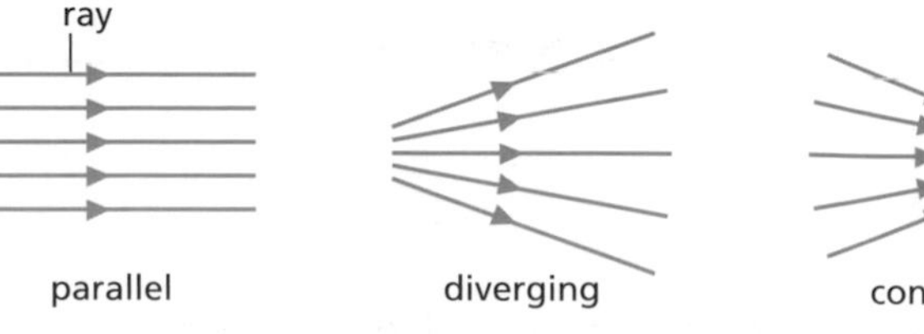

Figure 1.4 Beams of light

Practical work

The pinhole camera

A simple pinhole camera is shown in Figure 1.5a. Make a small pinhole in the centre of the black paper. Half-darken the room. Hold the box at arm's length so that the pinhole end is nearer to and about 1 metre from a luminous object, e.g. a carbon filament lamp or a candle. Look at the **image** on the screen (an image is a likeness of an object and need not be an exact copy).

Can you see *three* ways in which the image differs from the object? What is the effect of moving the camera closer to the object?

Make the pinhole larger. What happens to the

(i) brightness,
(ii) sharpness,
(iii) size of the image?

Make several small pinholes round the large hole, Figure 1.5b, and view the image again.

The forming of an image is shown in Figure 1.6.

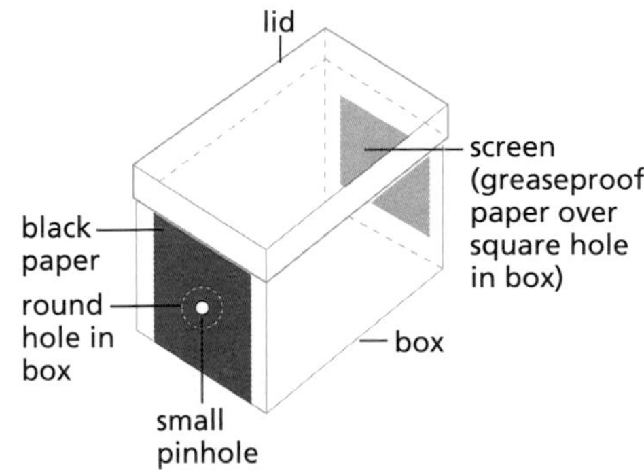

a A pinhole camera

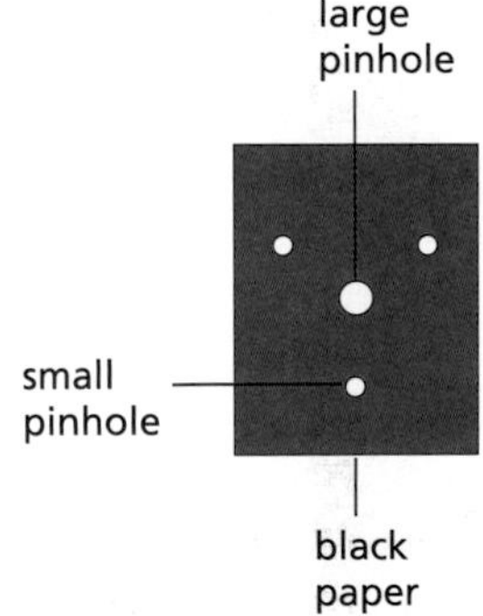

b

Figure 1.5

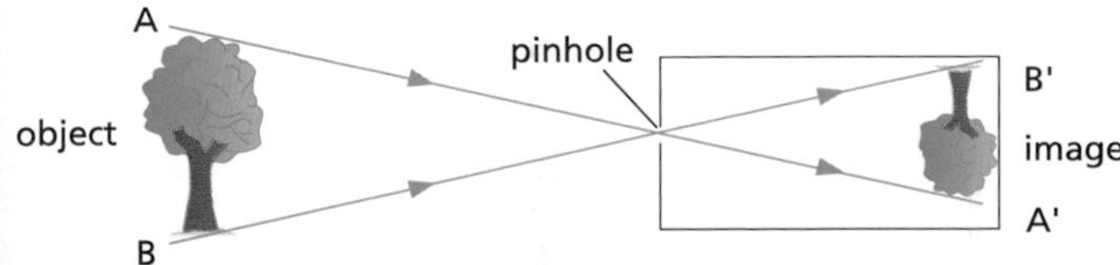

Figure 1.6 Forming an image in a pinhole camera

Shadows

Shadows are formed for two reasons. First, because some objects, which are said to be **opaque**, do not allow light to pass through them. Second, because light travels in straight lines. The sharpness of the shadow depends on the size of the light source. A very small source of light, called a **point** source, gives a sharp shadow which is equally dark all over. This may be shown as in Figure 1.7a where the small hole in the card acts as a point source.

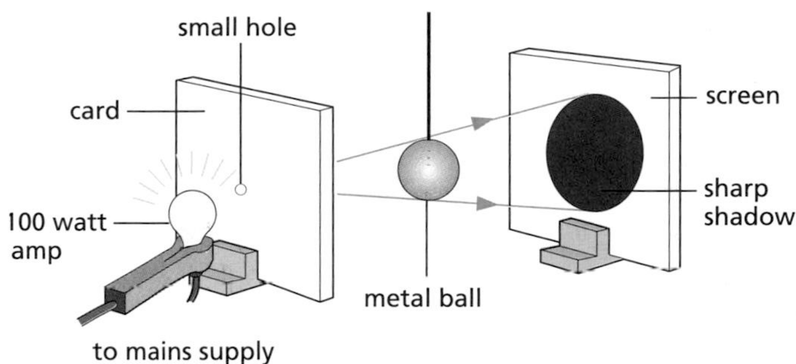

a With a point source

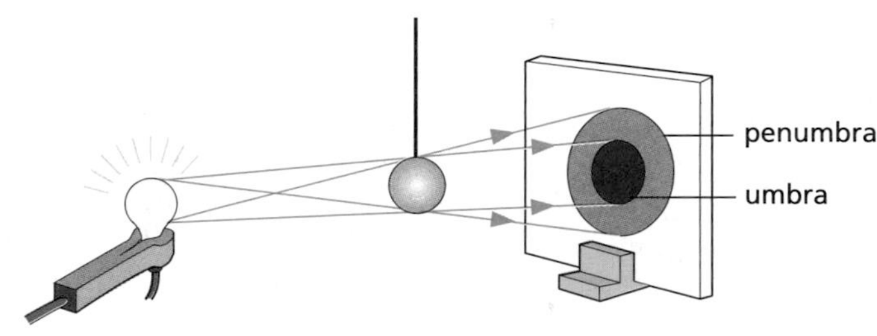

b With an extended source

Figure 1.7 Forming a shadow

If the card is removed the lamp acts as a large or **extended** source, Figure 1.7b. The shadow is then larger and has a central dark region, the **umbra**, surrounded by a ring of partial shadow, the **penumbra**. You can see by the rays that some light reaches the penumbra but none reaches the umbra.

Speed of light

Proof that light travels very much faster than sound is provided by a thunderstorm. The flash of lightning is seen before the thunder is heard. The length of the time lapse is greater the further the observer is from the storm.

The speed of light has a definite value; light does not travel instantaneously from one point to another but takes a certain, very small time. Its speed is about 1 million times greater than that of sound.

Questions

1 How would the size and brightness of the image formed by a pinhole camera change if the camera were made longer?

2 What changes would occur in the image if the single pinhole in a camera were replaced by
a four pinholes close together,
b a hole 1 cm wide?

3 In Figure 1.8 the completely dark region on the screen is

A PQ **B** PR **C** QR **D** QS **E** RS

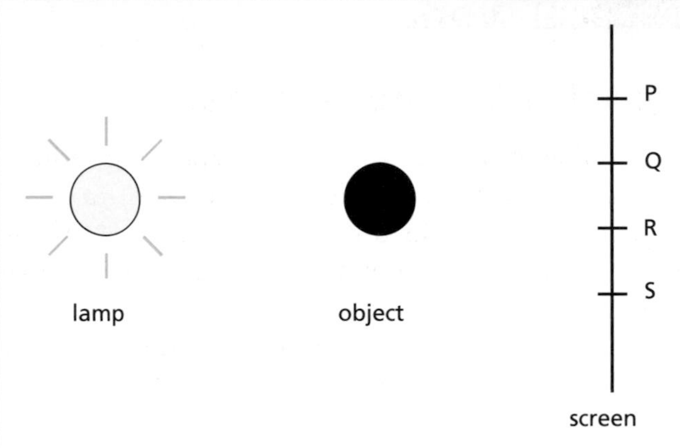

Figure 1.8

4 When watching a distant firework display do you see the cascade of lights before or after hearing the associated bang? Explain your answer.

Checklist

After studying this chapter you should be able to

- give examples of effects which show that light travels in a straight line,
- explain the operation of a pinhole camera and draw ray diagrams to show the result of varying the object distance or the length of the camera,
- draw diagrams to show how shadows are formed using point and extended sources, and use the terms **umbra** and **penumbra**,
- recall that light travels much faster than sound.

2 Reflection of light

Law of reflection	***Practical work***
Periscope	Reflection by a plane mirror.
Regular and diffuse reflection	

If we know how light behaves when it is reflected we can use a mirror to change the direction in which the light is travelling. This happens when a mirror is placed at the entrance of a concealed drive to give warning of approaching traffic.

An ordinary mirror is made by depositing a thin layer of silver on one side of a piece of glass and protecting it with paint. The silver – at the *back* of the glass – acts as the reflecting surface.

Law of reflection

Terms used in connection with reflection are shown in Figure 2.1. The perpendicular to the mirror at the point where the incident ray strikes it is called the **normal**. Note that the angle of incidence i is the angle between the incident ray and the normal; similarly the angle of reflection r is the angle between the reflected ray and the normal.

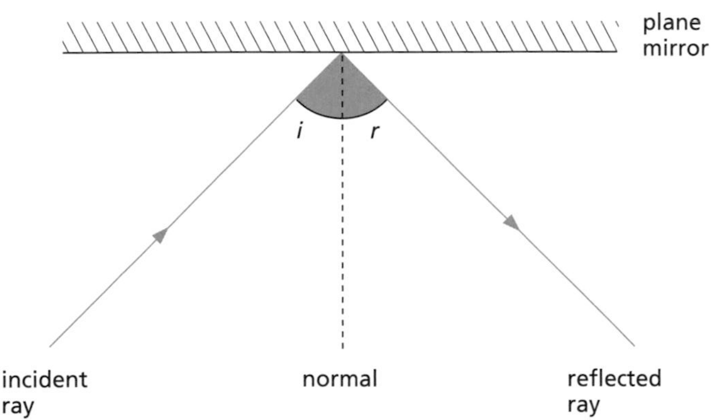

Figure 2.1 Reflection of light by a plane mirror

The law of reflection states:

The angle of incidence equals the angle of reflection.

The incident ray, the reflected ray and the normal all lie in the same plane. (*This means that they can all be drawn on a flat sheet of paper.*)

Practical work

Reflection by a plane mirror

Draw a line AOB on a sheet of paper and using a protractor mark angles on it. Measure them from the perpendicular ON, which is at right angles to AOB. Set up a plane (flat) mirror with its reflecting surface on AOB.

Shine a narrow ray of light along, say, the 30° line onto the mirror, Figure 2.2.

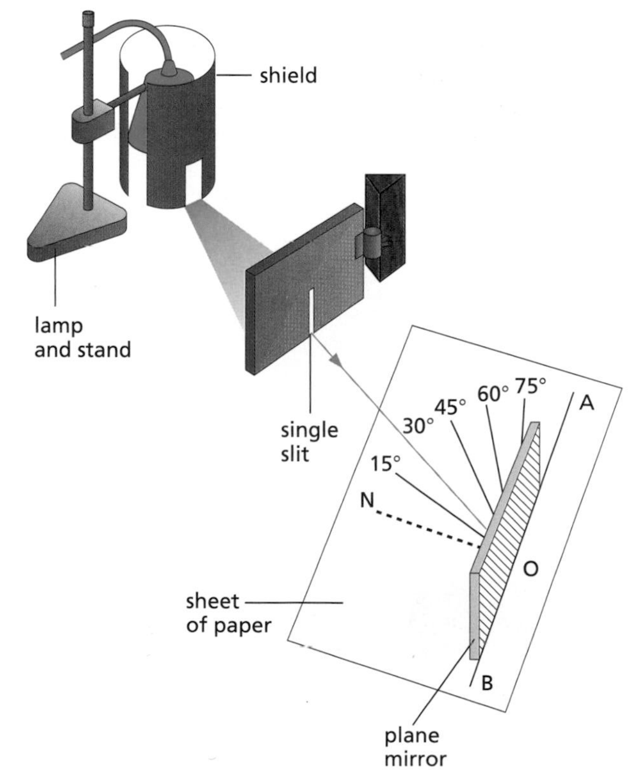

Figure 2.2

Mark the position of the reflected ray, remove the mirror and measure the angle between the reflected ray and ON. Repeat for rays at other angles. What can you conclude?

Periscope

A simple periscope consists of a tube containing two plane mirrors, fixed parallel to and facing one another. Each makes an angle of 45° with the line joining them, Figure 2.3. Light from the object is turned through 90° at each reflection and an observer is able to see over a crowd, for example, Figure 2.4, or over the top of an obstacle.

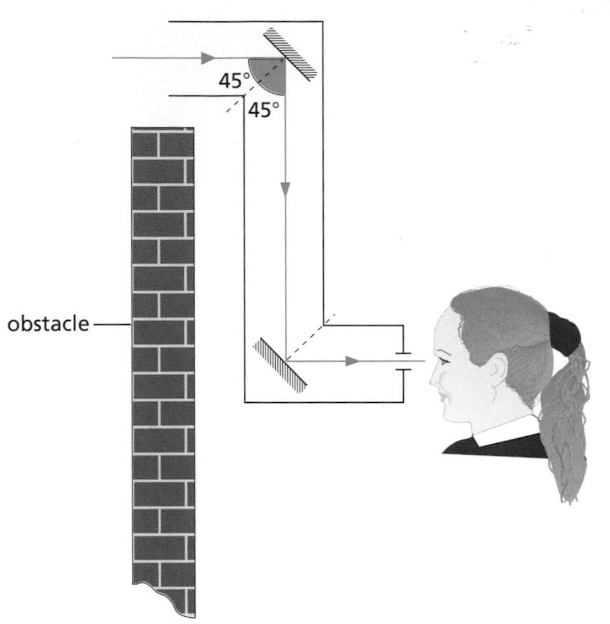

Figure 2.3 Action of a simple periscope

Figure 2.4 Periscopes being used by people in a crowd

In more elaborate periscopes like those used in submarines, prisms replace mirrors (see Chapter 5).

Make your own periscope from a long, narrow cardboard box measuring about 40 cm × 5 cm × 5 cm (e.g. one in which aluminium cooking foil or cling film is sold), two plane mirrors (7.5 cm × 5 cm) and sticky tape. When you have got it to work, make modifications that turn it into a 'See-back-o-scope' which lets you see what is behind you.

Regular and diffuse reflection

If a parallel beam of light falls on a plane mirror it is reflected as a parallel beam, Figure 2.5a, and **regular** reflection occurs. Most surfaces, however, reflect light irregularly and the rays in an incident parallel beam are reflected in many directions, Figure 2.5b.

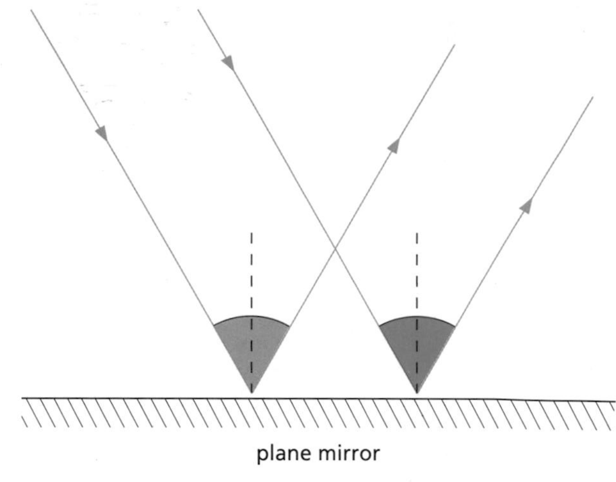

a Regular reflection

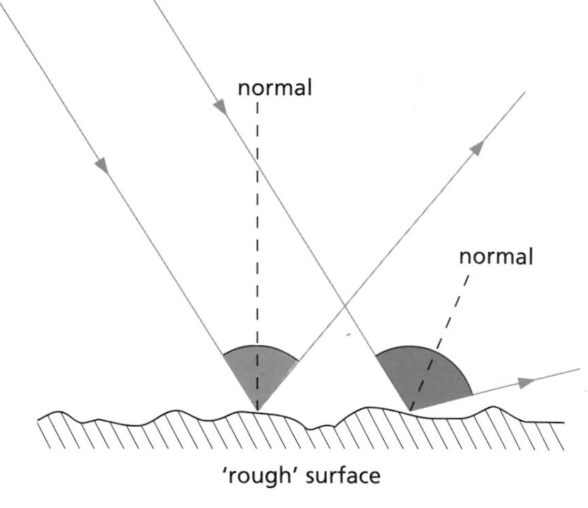

b Diffuse reflection

Figure 2.5

Irregular or **diffuse** reflection is due to the surface of the object not being perfectly smooth like a mirror. At each point on the surface the laws of reflection are obeyed but the angle of incidence and so the angle of reflection varies from point to point. The reflected rays are scattered haphazardly. Most objects, being rough, are seen by diffuse reflection.

Questions

1 Figure 2.6 shows a ray of light PQ striking a mirror AB. The mirror AB and the mirror CD are at right angles to each other. QN is a normal to the mirror AB.

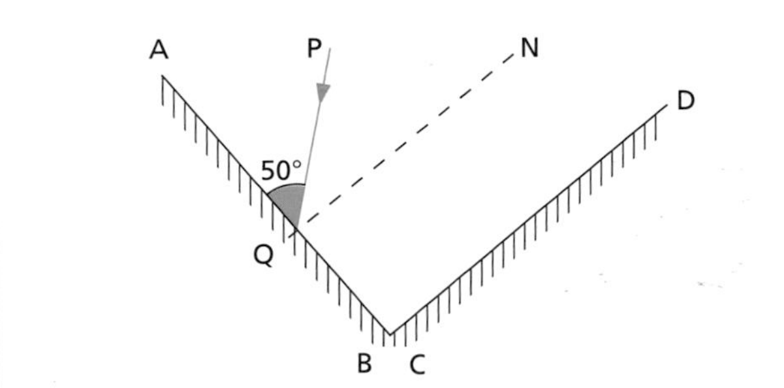

Figure 2.6

- **a** What is the value of the angle of incidence of the ray PQ on the mirror AB?
- **b** Copy the diagram and continue the ray PQ to show the path it takes after reflection at both mirrors.
- **c** What are the values of the angle of reflection at AB, the angle of incidence at CD and the angle of reflection at CD?
- **d** What do you notice about the path of the ray PQ and the final reflected ray?

2 A ray of light strikes a plane mirror at an angle of incidence of 60°, is reflected from the mirror and then strikes a second plane mirror placed so that the angle between the mirrors is 45°. The angle of reflection at the second mirror, in degrees, is

A 15 **B** 25 **C** 45 **D** 65 **E** 75

3 A person stands in front of a mirror, Figure 2.7. How much of the mirror is used to see from eye to toes?

Figure 2.7

Checklist

After studying this chapter you should be able to

- state the law of reflection and use it to solve problems,
- describe an experiment to show that the angle of incidence equals the angle of reflection,
- draw a ray diagram to show how a periscope works.

3 Plane mirrors

Real and virtual images	*Properties of the image*	*Practical work*
Lateral inversion	*Kaleidoscope*	Position of the image.

When you look into a plane mirror on the wall of a room you see an image of the room behind the mirror; it is as if there were another room. Restaurants sometimes have a large mirror on one wall just to make them look larger. You may be able to say how much larger after the next experiment.

The position of the image formed by a mirror depends on the position of the object.

Practical work

Position of the image

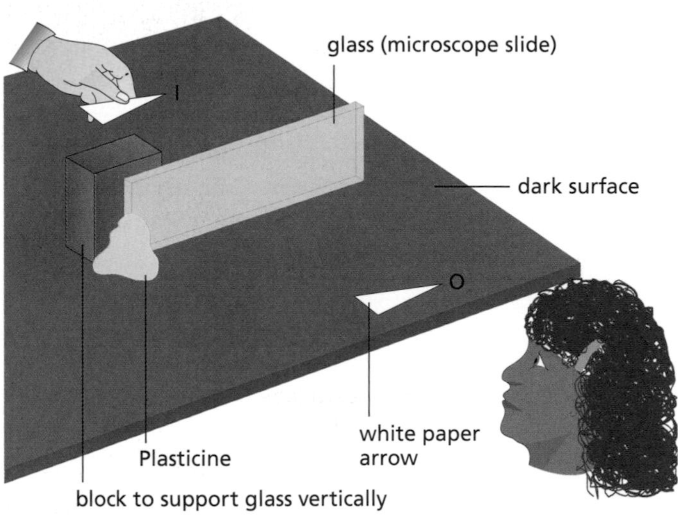

Figure 3.1

Support a piece of thin glass on the bench, as in Figure 3.1. It must be *vertical* (at 90° to the bench). Place a small paper arrow, O, about 10 cm from the glass. The glass acts as a poor mirror and an image of O will be seen in it; the darker the bench top, the brighter the image will be.

Lay another identical arrow, I, on the bench behind the glass; move it until it coincides with the image of O. How do the sizes of O and its image compare? Imagine a line joining them. What can you say about it? Measure the distances of the points of O and I from the glass along the line joining them. How do they compare? Try placing O at other distances.

Real and virtual images

A **real** image is one which can be produced on a screen (as in a pinhole camera) and is formed by rays that actually pass through it.

A **virtual** image cannot be formed on a screen and is produced by rays which seem to come from it but do not pass through it. The image in a plane mirror is virtual. Rays from a point on an object are reflected at the mirror and appear to come from the point behind the mirror where the eye imagines the rays intersect when produced backwards, Figure 3.2. IA and IB are construction lines and are shown by broken lines.

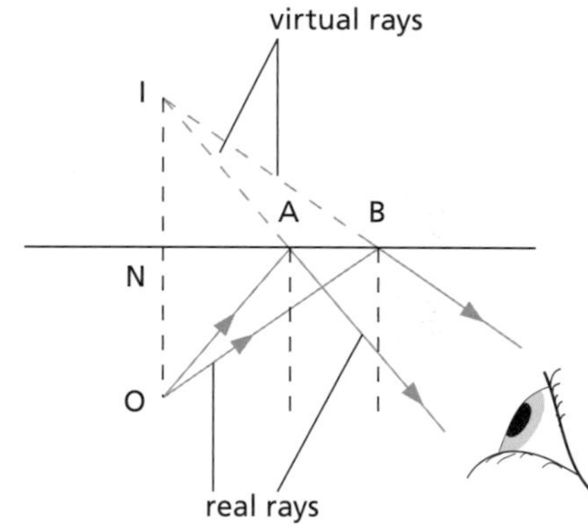

Figure 3.2 A plane mirror forms a virtual image

Lateral inversion

If you close your left eye, your image in a plane mirror seems to close the right eye. In a mirror image, left and right are interchanged and the image appears to be **laterally inverted**. The effect occurs whenever an image is formed by one reflection and is very evident if print is viewed in a mirror, Figure 3.3. What happens if two reflections occur, as in a periscope?

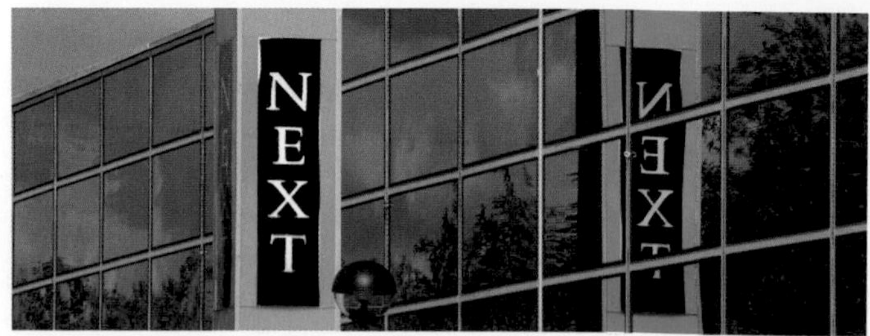

Figure 3.3 The image in a plane mirror is laterally inverted

Properties of the image

The image in a plane mirror is

(i) as far behind the mirror as the object is in front and the line joining the object and image is perpendicular to the mirror,
(ii) the same size as the object,
(iii) virtual,
(iv) laterally inverted.

Kaleidoscope

Figure 3.4 A kaleidoscope produces patterns using the images formed by two plane mirrors – the patterns change as the object moves

To see how a kaleidoscope works, draw on a sheet of paper two lines at right angles to one another. Using different coloured pens or pencils draw a design between them, Figure 3.5a. Place a small mirror along each line and look into the mirrors, Figure 3.5b. You will see three reflections which join up to give a circular pattern. If you make the angle between the mirrors smaller, more reflections appear but you always get a complete design.

In a kaleidoscope the two mirrors are usually kept at the same angle (about 60°) and different designs are made by hundreds of tiny coloured beads which can be moved around between the mirrors.

Now make a kaleidoscope using a cardboard tube (e.g. from half a kitchen roll), some thin card, grease-proof paper, clear sticky tape, small pieces of different coloured cellophane and two mirrors (10 cm × 3 cm) or a single plastic mirror (10 cm × 6 cm) bent to form two mirrors at 60° to each other, as shown in Figure 3.5c.

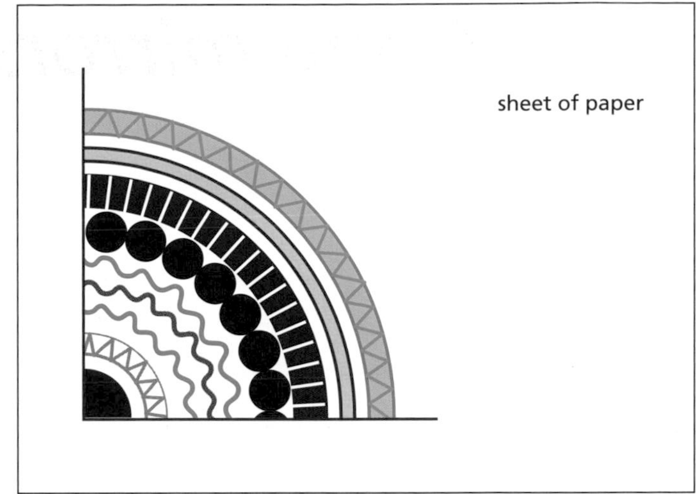

a

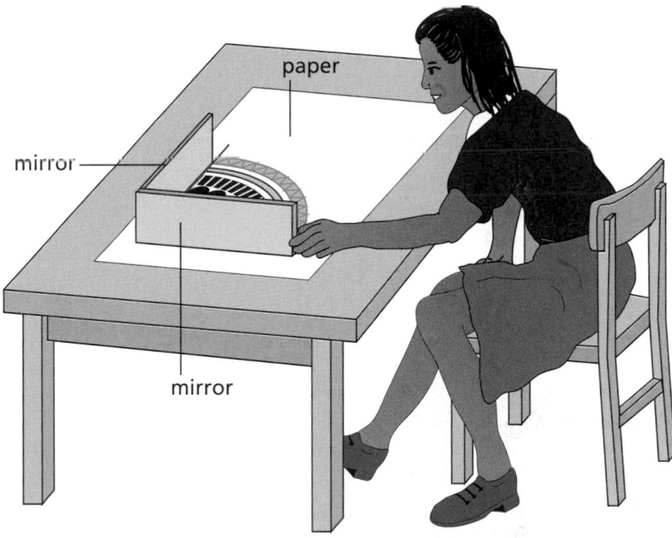

b

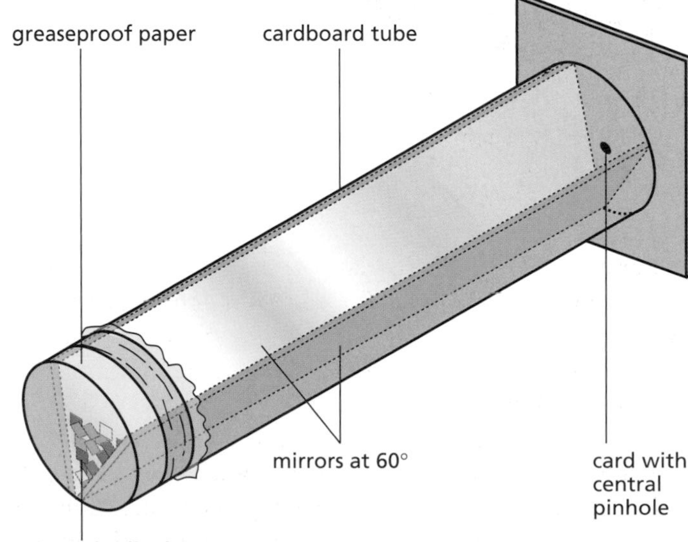

c

Figure 3.5

Questions

1 In Figure 3.6 at which of the points **A** to **E** will the observer see the image of the object in the plane mirror?

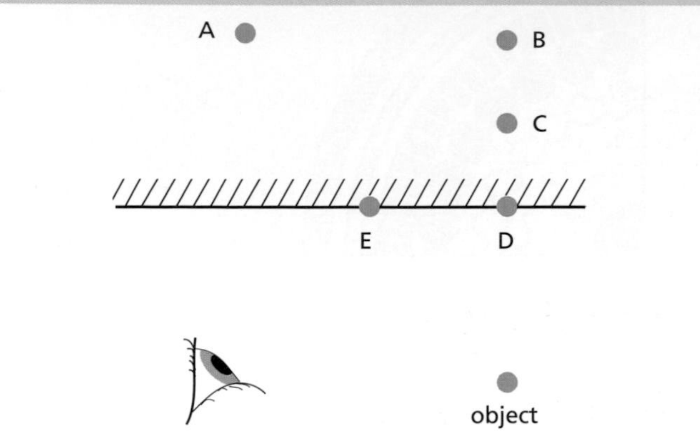

Figure 3.6

2 Figure 3.7 shows the image in a plane mirror of a clock. The correct time is

A 2.25 **B** 2.35 **C** 6.45 **D** 9.25

Figure 3.7

3 A girl stands 5 m away from a large plane mirror. How far must she walk to be 2 m away from her image?

Checklist

After studying this chapter you should be able to

- describe an experiment to show that the image in a plane mirror is as far behind the mirror as the object is in front, and that the line joining the object and image is at right angles to the mirror,
- draw a diagram to explain the formation of a virtual image by a plane mirror,
- explain the apparent lateral inversion of an image in a plane mirror,
- □ explain how a kaleidoscope works.

4 Refraction of light

If you place a coin in an empty mug and move back until you *just* cannot see it, the result is surprising if someone *gently* pours in water. Try it.

Although light travels in straight lines in one transparent material, such as air, if it passes into a different material, such as water, it changes direction at the boundary between the two, i.e. it is bent. The **bending of light** when it passes from one material (called a medium) to another is called **refraction**. It causes effects like the coin trick.

Facts about refraction

(i) A ray of light is bent **towards** the normal when it enters an optically denser medium at an angle (e.g. from air to glass), i.e. the angle of refraction r is less than the angle of incidence i, Figure 4.1a.
(ii) A ray of light is bent **away from** the normal when it enters an optically less dense medium (e.g. from glass to air).
(iii) A ray emerging from a parallel-sided block is **parallel** to the ray entering, but is displaced sideways.
(iv) A ray travelling along the normal is **not refracted**, Figure 4.1b.

Note 'Optically denser' means having a greater refraction effect; the actual density may or may not be greater.

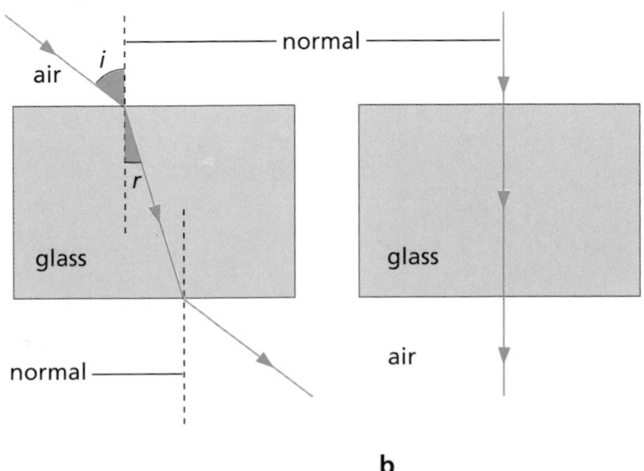

Figure 4.1 Refraction of light in glass

Practical work

Refraction in glass

Shine a ray of light at an angle on to a glass block (which has its lower face painted white or frosted), as in Figure 4.2. Draw the outline ABCD of the block on the sheet of paper under it. Mark the positions of the various rays in air and in glass.

Remove the block and draw the normals on the paper at the points where the ray enters AB (see Figure 4.2) and where it leaves CD.

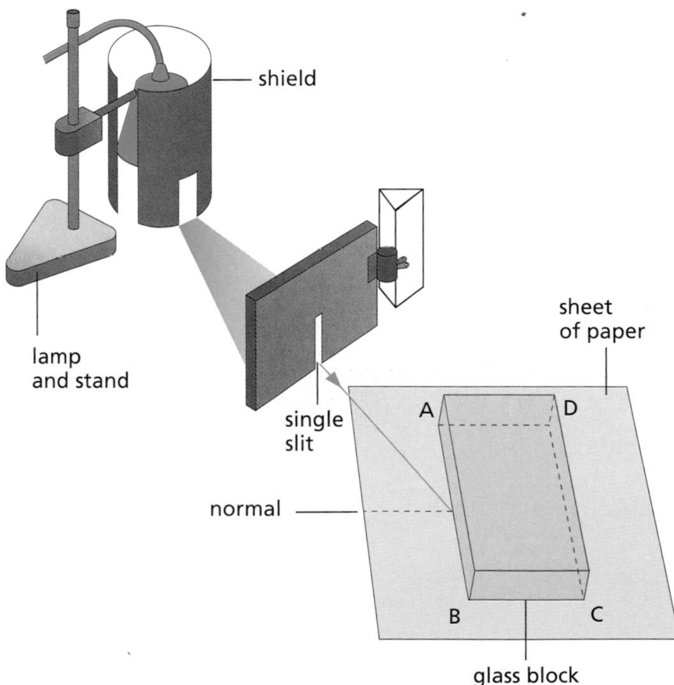

Figure 4.2

What *two* things happen to the light falling on AB? When the ray enters the glass at AB is it bent towards or away from the part of the normal in the block? How is it bent at CD? What can you say about the direction of the ray falling on AB and the direction of the ray leaving CD?

What happens if the ray hits AB at right angles?

Real and apparent depth

Rays of light from a point O on the bottom of a pool are refracted away from the normal at the water surface because they are passing into an optically less dense medium, i.e. air, Figure 4.3. On entering the eye they appear to come from a point I *above* O; I is the virtual image of O formed by refraction. The apparent depth of the pool is less than its real depth.

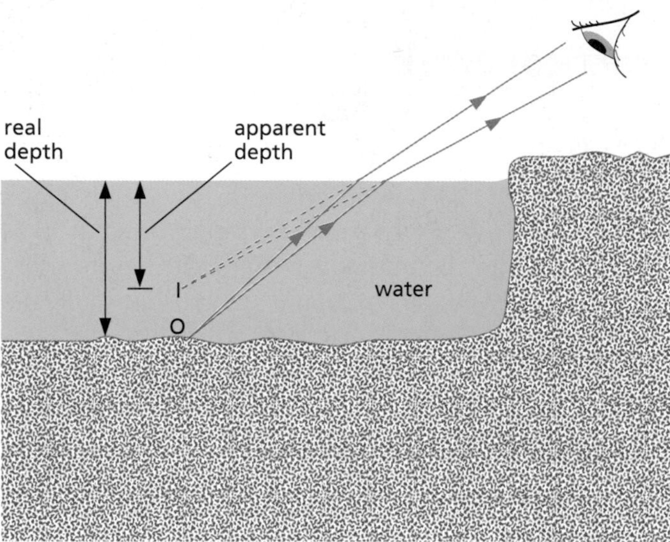

Figure 4.3 A pool of water appears shallower than it is

Figure 4.4 A pencil put in water seems to bend at the water surface. Why?

Refractive index

Light is refracted because its speed changes when it enters another medium. An analogy helps to explain why.

Suppose three people A, B, C are marching in line, with hands linked, on a good road surface. If they approach marshy ground at an angle, Figure 4.5a, A is slowed down first, followed by B and then C. This causes the whole line to swing round and change its direction of motion.

In air (and a vacuum) light travels at 300 000 km/s (3×10^8 m/s), in glass its speed falls to 200 000 km/s (2×10^8 m/s), Figure 4.5b. The **refractive index** n of the medium, i.e. glass, is defined by the equation

$$\text{refractive index } n = \frac{\text{speed of light in air (or a vacuum)}}{\text{speed of light in medium}}$$

$$\therefore \quad n = \frac{300\,000 \text{ km/s}}{200\,000 \text{ km/s}} = \frac{3}{2}$$

Experiments also show that

$$n = \frac{\text{sine of angle of incidence}}{\text{sine of angle of refraction}}$$

$$= \frac{\sin i}{\sin r} \quad \text{(see Figure 4.1a)}$$

The more light is slowed down when it enters a medium from air, the greater is the refractive index of the medium and the more it is bent.

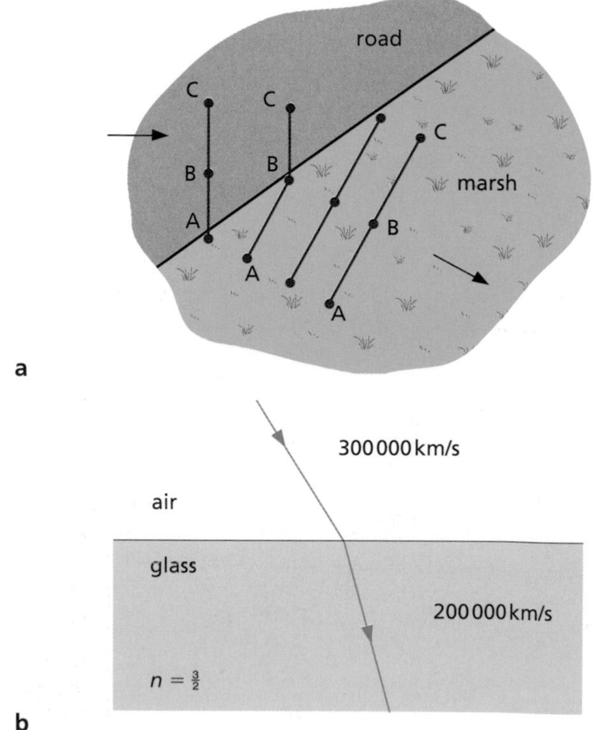

Figure 4.5

Refraction by a prism

In a triangular glass prism, Figure 4.6a, the deviation (bending) of a ray due to refraction at the first surface is added to the deviation at the second surface, Figure 4.6b. The deviations do not cancel out as in a parallel-sided block where the emergent ray, although displaced, is parallel to the incident ray.

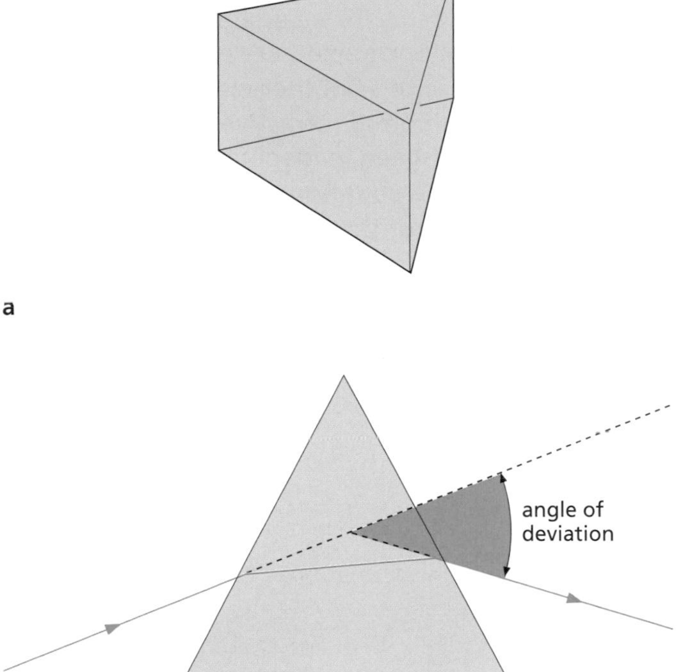

Figure 4.6

Figure 4.7b White light shining through cut crystal can produce several spectra

Dispersion

When sunlight (white light) falls on a triangular glass prism, Figure 4.7a, a band of colours called a **spectrum** is obtained, Figure 4.7b. The effect is termed **dispersion** and arises because white light is a mixture of many colours which the prism separates because the refractive index of glass is different for each colour (it is greatest for violet light).

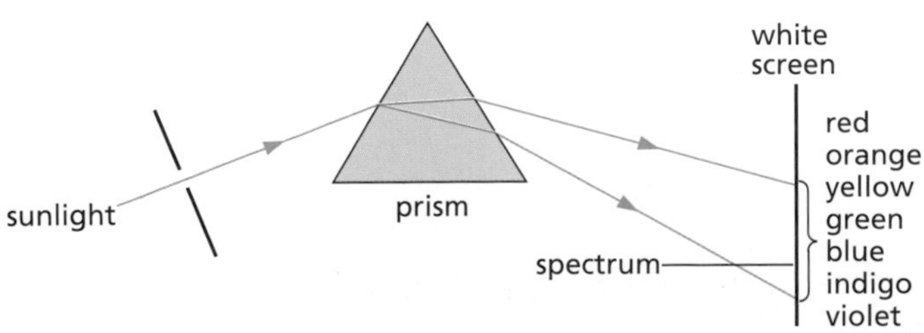

Figure 4.7a Forming a spectrum with a prism

Questions

1 Figure 4.8 shows a ray of light entering a rectangular block of glass.
   a Copy the diagram and draw the normal at the point of entry.
   b Sketch the approximate path of the ray through the block and out of the other side.

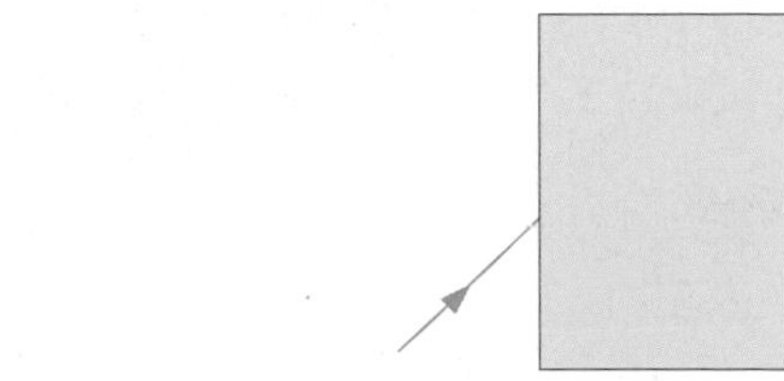

Figure 4.8

2 Draw two rays from a point on a fish in a stream to show where someone on the bank will see the fish. Where must the person aim to spear the fish?

3 What is the speed of light in a medium of refractive index 6/5 if its speed in air is 300 000 km/s?

Continued

4 Figure 4.9 shows a ray of light OP striking a glass prism and then passing through it. Which of the rays **A** to **D** is the correct representation of the emerging ray?

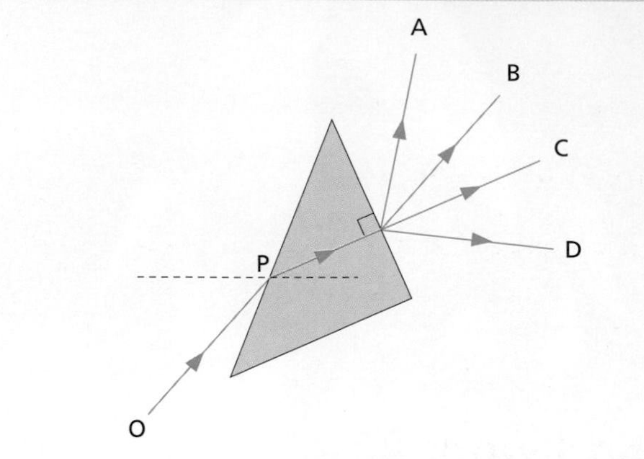

Figure 4.9

5 A beam of white light strikes the face of a prism. Copy Figure 4.10 and draw the path taken by red and blue rays of light as they pass through the prism and on to the screen AB.

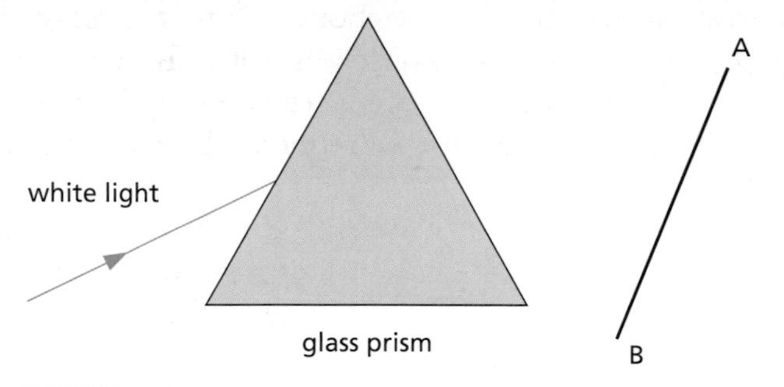

Figure 4.10

Checklist

After studying this chapter you should be able to

- state what the term **refraction** means,
- give examples of effects that show light can be refracted,
- describe an experiment to study refraction,
- draw diagrams of the passage of light rays through rectangular blocks and recall that lateral displacement occurs for a parallel-sided block,
- recall that light is refracted because it changes speed when it enters another medium,
- □ recall the definition of refractive index as n = speed in air/speed in medium,
- □ recall and use the equation $n = \sin i/\sin r$
- draw a diagram for the passage of a light ray through a prism,
- explain the terms **spectrum** and **dispersion**,
- describe how a prism is used to produce a spectrum from white light.

5 Total internal reflection

Critical angle
Multiple images in a mirror
Totally reflecting prisms
Light pipes and optical fibres
Practical work
Critical angle of glass.

Critical angle

When light passes at small angles of incidence from an optically dense to a less dense medium, e.g. from glass to air, there is a strong refracted ray and a weak ray reflected back into the denser medium, Figure 5.1a. Increasing the angle of incidence increases the angle of refraction.

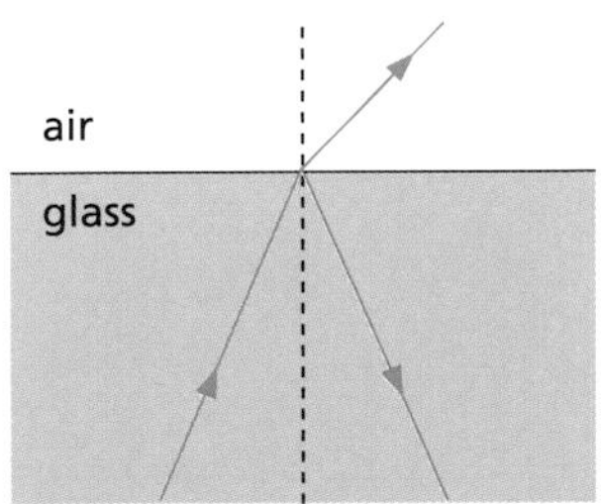

a

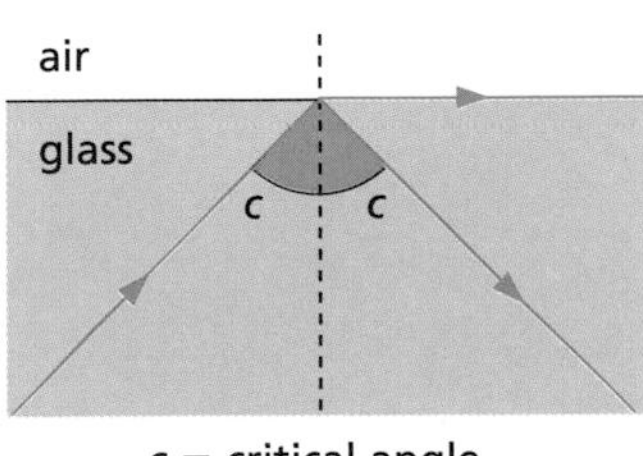

b

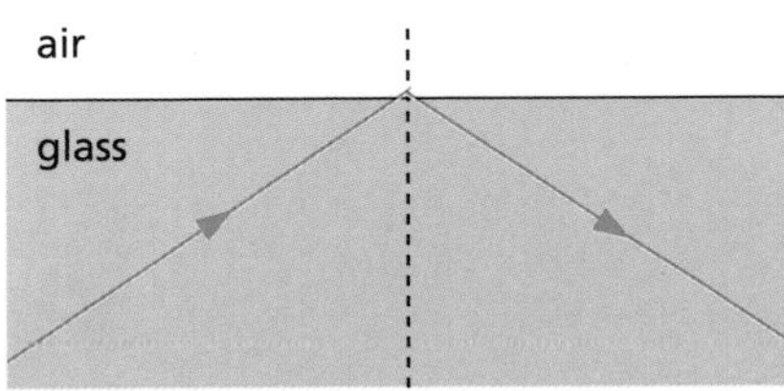

c

Figure 5.1

At a certain angle of incidence, called the **critical angle** *c*, the angle of refraction is 90°, Figure 5.1b. For angles of incidence greater than *c*, the refracted ray disappears and *all* the incident light is reflected inside the denser medium, Figure 5.1c. The light does not cross the boundary and is said to undergo **total internal reflection**.

Practical work

Critical angle of glass

Place a semicircular glass block on a sheet of paper, Figure 5.2, and draw the outline LOMN where O is the centre and ON the normal at O to LOM. Direct a narrow ray (at an angle of about 30° to the normal) *along a radius towards* O. The ray is not refracted at the curved surface. Why? Note the refracted ray in the air beyond LOM and also the weak internally reflected ray in the glass.

Slowly rotate the paper so that the angle of incidence increases until total internal reflection *just* occurs. Mark the incident ray. Measure the angle of incidence; it equals the critical angle.

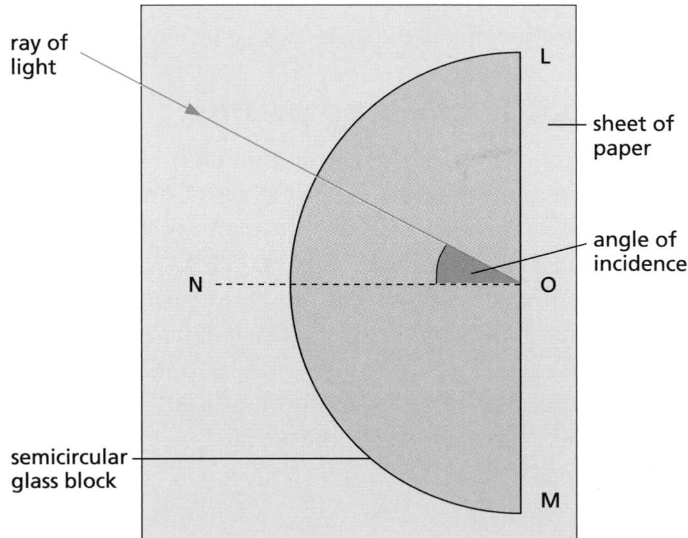

Figure 5.2

Multiple images in a mirror

An ordinary mirror silvered at the back forms several images of one object, due to multiple reflection inside the glass, Figures 5.3a and b. These blur the main image I (which is formed by one reflection at the silvering), especially if the glass is thick. The problem is absent in front-silvered mirrors but they are easily damaged.

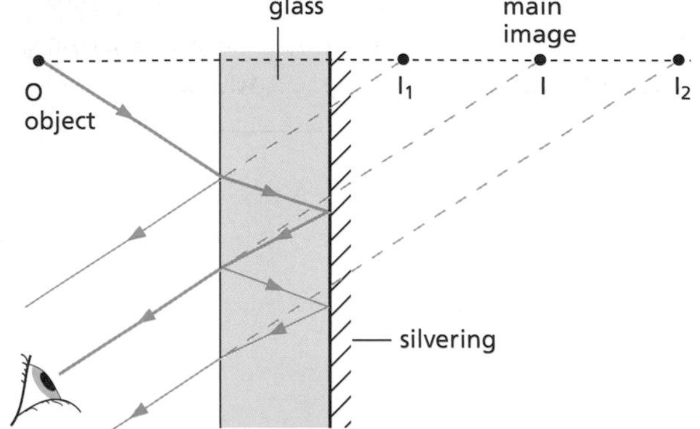

Figure 5.3a Multiple reflections in a mirror

Figure 5.3b The multiple images cause blurring

Totally reflecting prisms

The defects of mirrors are overcome if 45° right-angled glass prisms are used. The critical angle of ordinary glass is about 42° and a ray falling normally on face PQ of such a prism, Figure 5.4a, hits face PR at 45°. Total internal reflection occurs and the ray is turned through 90°. Totally reflecting prisms replace mirrors in good periscopes.

Light can also be reflected through 180° by a prism, Figure 5.4b; this happens in binoculars.

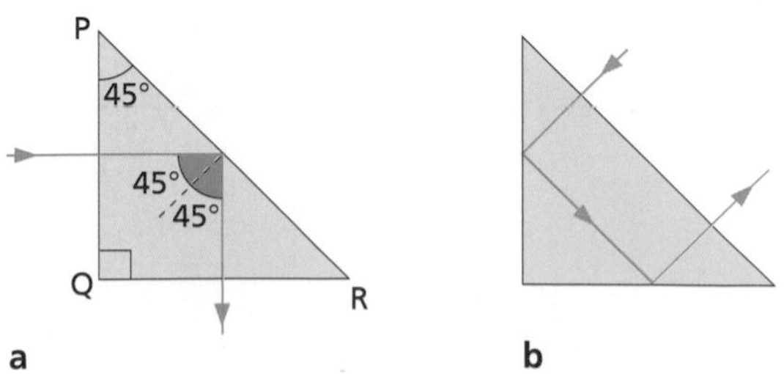

Figure 5.4 Reflection of light by a prism

Light pipes and optical fibres

Light can be trapped by total internal reflection inside a bent glass rod and 'piped' along a curved path, Figure 5.5. A single, very thin glass fibre behaves in the same way. If several thousand such fibres are taped together a flexible light pipe is obtained that can be used, for example, by doctors as an 'endoscope', Figure 5.6a, to obtain an image of an internal organ in the body, Figure 5.6b, or by engineers to light up some awkward spot for inspection. The latest telephone 'cables' are optical (very pure glass) fibres carrying information as pulses of laser light.

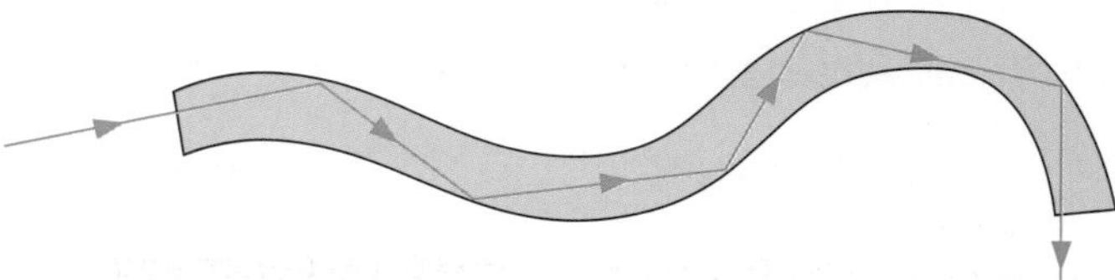

Figure 5.5 Light travels through a curved glass rod or fibre by total internal reflection

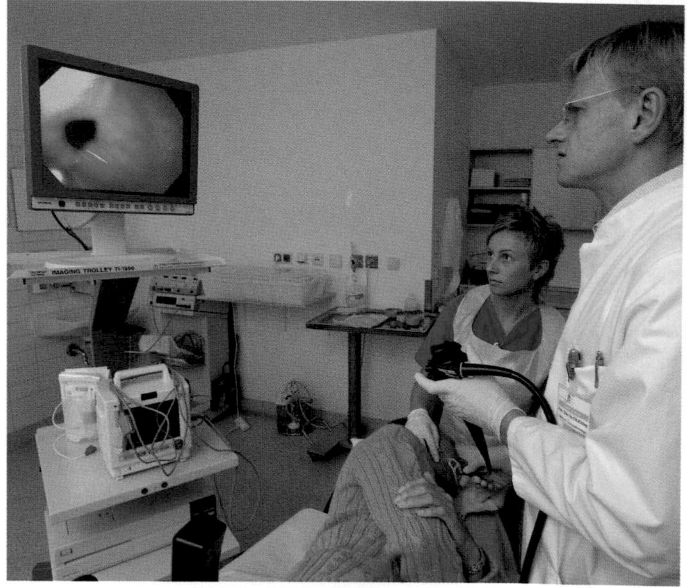

Figure 5.6a Endoscope in use

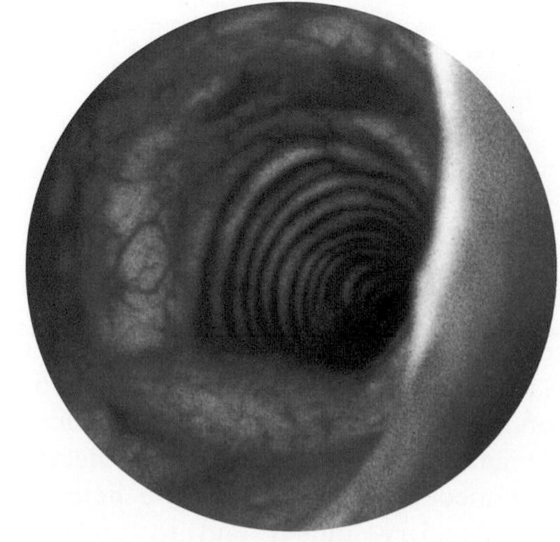

Figure 5.6b Trachea (windpipe) viewed by an endoscope

Questions

1 Figure 5.7 shows rays of light in a semicircular glass block.

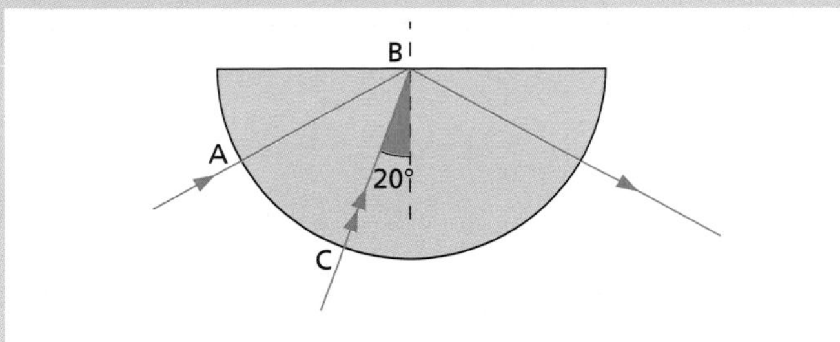

Figure 5.7

- **a** Explain why the ray entering the glass at A is not bent.
- **b** Explain why the ray AB is reflected at B and not refracted.
- **c** Ray CB does not stop at B. Copy the diagram and draw its approximate path after it leaves B.

2 Copy Figures 5.8a and b and complete the paths of the rays through the glass prisms.

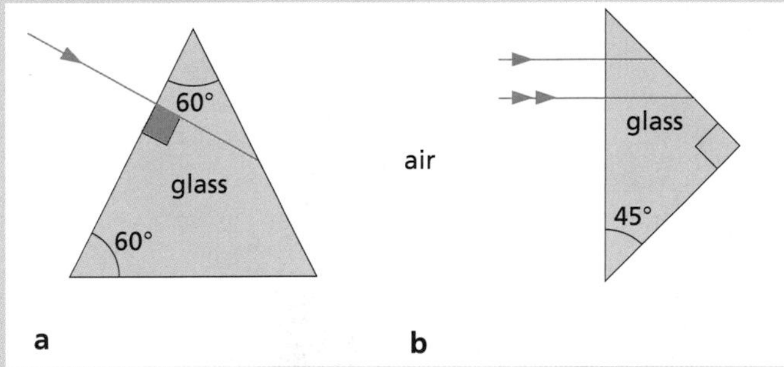

Figure 5.8

3 Name two instruments that use prisms to reflect light.

Checklist

After studying this chapter you should be able to

- explain with the aid of diagrams what is meant by **critical angle** and **total internal reflection,**
- describe an experiment to find the critical angle of glass or Perspex,
- ☐ draw diagrams to show the action of totally reflecting prisms in periscopes and binoculars,
- ☐ explain the action of optical fibres.

6 Lenses

Converging and diverging lenses
Principal focus
Ray diagrams
Magnification
Power of a lens
Magnifying glass
Spectacles
Practical work
f of a converging lens.
Images formed by a converging lens.

Converging and diverging lenses

Lenses are used in optical instruments such as cameras, spectacles, microscopes and telescopes; they often have spherical surfaces and there are two types. A **converging** (or convex) lens is thickest in the centre and bends light inwards, Figure 6.1a. You may have used one as a magnifying glass, Figure 6.2a, or as a burning glass. A **diverging** (or concave) lens is thinnest in the centre and spreads light out, Figure 6.1b; it always gives a diminished image, Figure 6.2b.

The centre of a lens is its **optical centre** C; the line through C at right angles to the lens is the **principal axis**.

The action of a lens can be understood by treating it as a number of prisms (most with the tip removed), each of which bends the ray towards its base, as in Figure 6.1c and d. The centre acts as a parallel-sided block.

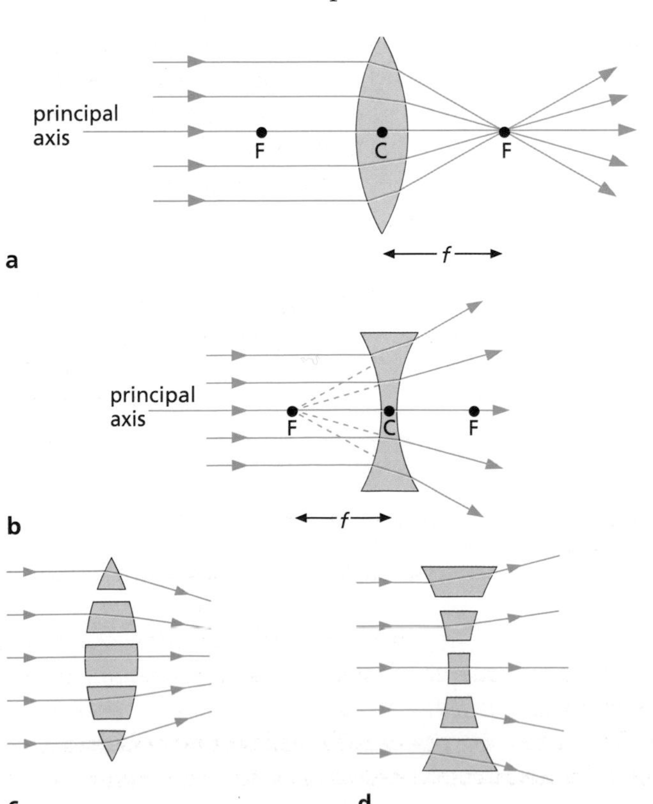

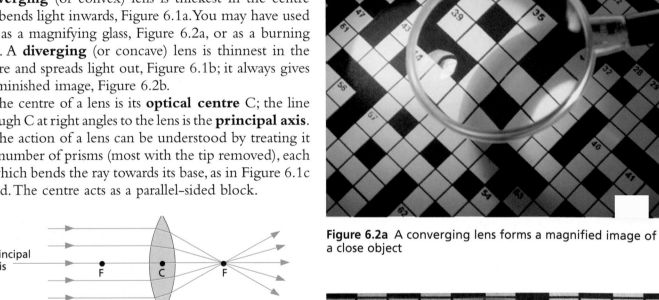

Figure 6.1 Action of lenses on a parallel beam of light

Figure 6.2a A converging lens forms a magnified image of a close object

Figure 6.2b A diverging lens always forms a diminished image

Principal focus

When a beam of light parallel to the principal axis passes through a converging lens it is refracted so as to converge to a point on the axis called the **principal focus** F. It is a real focus. A diverging lens has a virtual principal focus behind the lens, from which the refracted beam seems to diverge.

Since light can fall on both faces of a lens it has two principal foci, one on each side, equidistant from C. The distance CF is the **focal length** *f* of the lens (see Figure 6.1a); it is an important property of a lens. The more curved the lens faces are, the smaller is *f* and the more powerful is the lens.

Practical work

f of a converging lens

We use the fact that rays from a **point** on a very distant object, i.e. at infinity, are nearly parallel, Figure 6.3a.

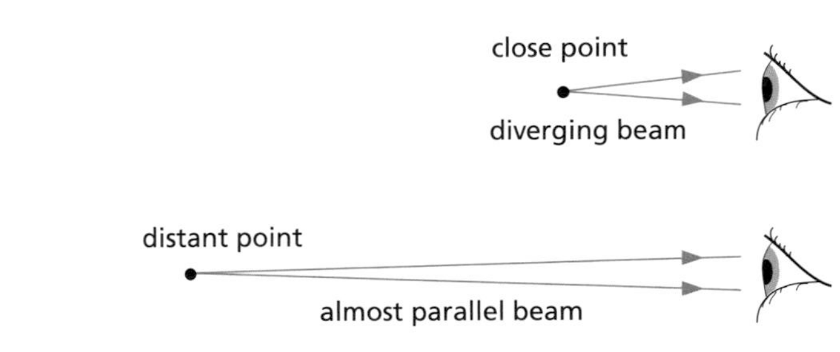

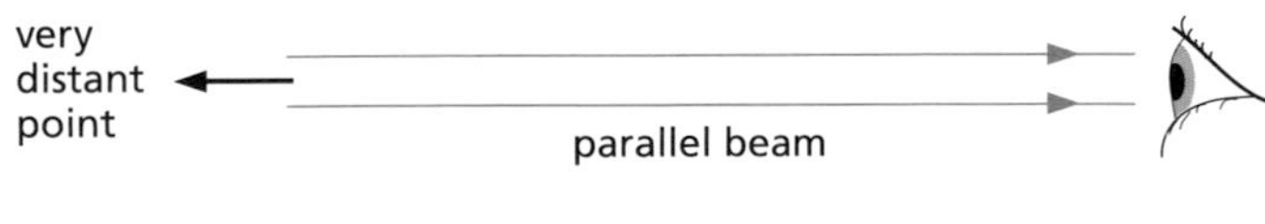

a

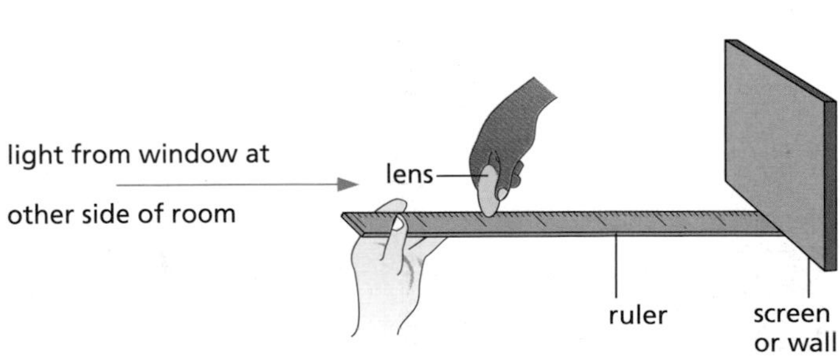

b

Figure 6.3

Move the lens, arranged as in Figure 6.3b, until a **sharp** image of a window at the other side of the room is obtained on the screen. The distance between the lens and the screen is *f*, roughly. Why?

Practical work

Images formed by a converging lens

In the formation of images by lenses, two important points on the principal axis are F and 2F; 2F is at a distance of twice the focal length from C.

First find the focal length of the lens by the 'distant object method' just described, then fix the lens upright with Plasticine at the centre of a metre rule. Place small pieces of Plasticine at the points F and 2F on both sides of the lens, as in Figure 6.4.

Place a small light source, e.g. a torch bulb, as the object supported on the rule beyond 2F and move a white card, on the other side of the lens from the light, until a sharp image is obtained on the card.

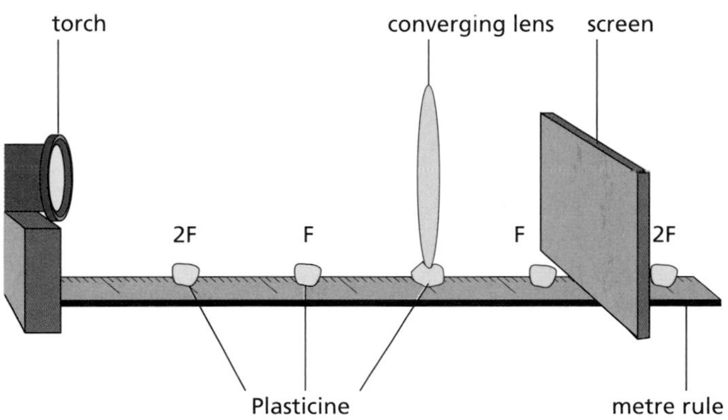

Figure 6.4

Note and record, in a table like the one below, the image position as 'beyond 2F', 'between 2F and F' or 'between F and lens'. Also note whether the image is 'larger' or 'smaller' than the actual bulb or 'same size' and if it is 'upright' or 'inverted'. Now repeat with the light at 2F, then between 2F and F.

Object position	Image position	Larger, smaller or same size?	Upright or inverted?
Beyond 2F			
At 2F			
Between 2F and F			
Between F and lens			

So far all the images have been real since they can be obtained on a screen. When the light is between F and the lens, the image is **virtual** and is seen by *looking through the lens* at the light. Do this. Is the virtual image larger or smaller than the object? Is it upright or inverted? Record your findings in your table.

Ray diagrams

Information about the images formed by a lens can be obtained by drawing *two* of the following rays.

1 A ray parallel to the principal axis which is refracted through the principal focus F.
2 A ray through the optical centre C which is undeviated for a thin lens.
3 A ray through the principal focus F which is refracted parallel to the principal axis.

In diagrams a thin lens is represented by a straight line at which all the refraction is considered to occur.

In each ray diagram in Figure 6.5 two rays are drawn from the top A of an object OA and where they intersect after refraction gives the top B of the image IB. The foot I of each image is on the axis since ray OC passes through the lens undeviated. In part d the broken rays, and the image, are virtual.

F Magnification

The **linear magnification** m is given by

$$m = \frac{\text{height of image}}{\text{height of object}}$$

It can be shown that in all cases

$$m = \frac{\text{distance of image from lens}}{\text{distance of object from lens}}$$

F Power of a lens

The shorter the focal length of a lens, the stronger it is, i.e. the more it converges or diverges a beam of light. We define the **power** of a lens P to be 1/focal length of the lens, where the focal length is measured in metres:

$$P = \frac{1}{f}$$

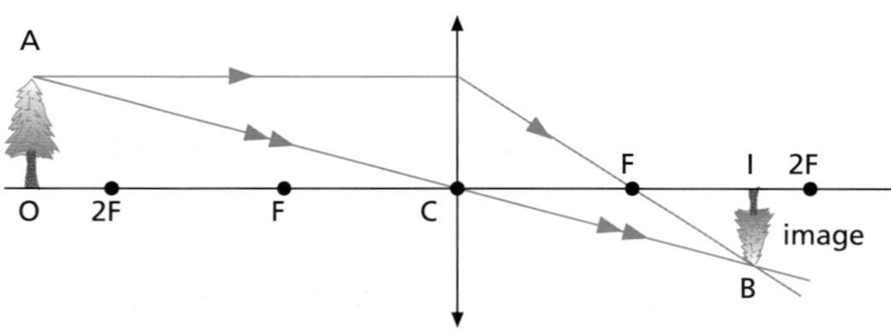

Image is between F and 2F, real, inverted, smaller

a Object beyond 2F

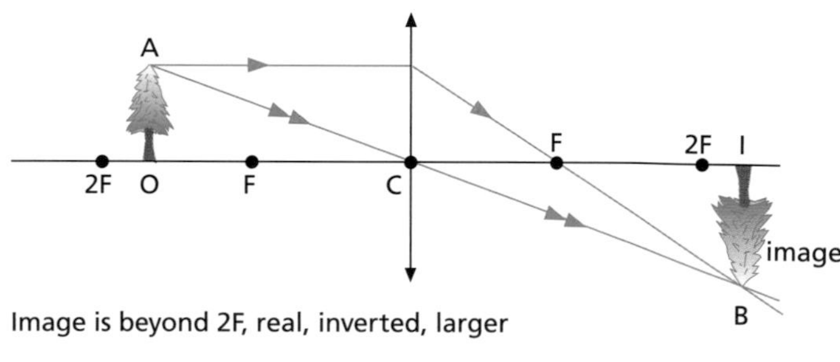

Image is beyond 2F, real, inverted, larger

b Object between 2F and F

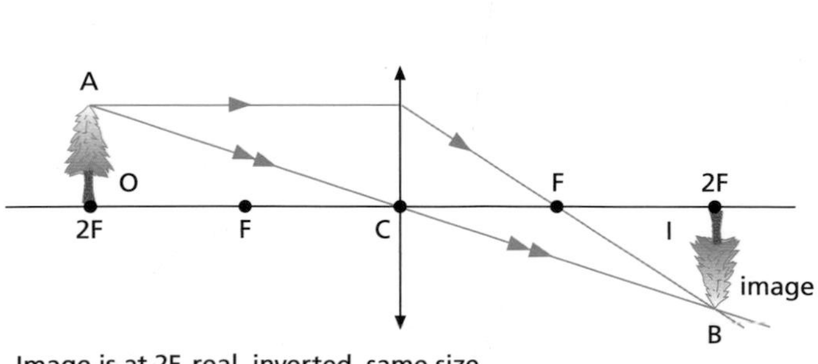

Image is at 2F, real, inverted, same size

c Object at 2F

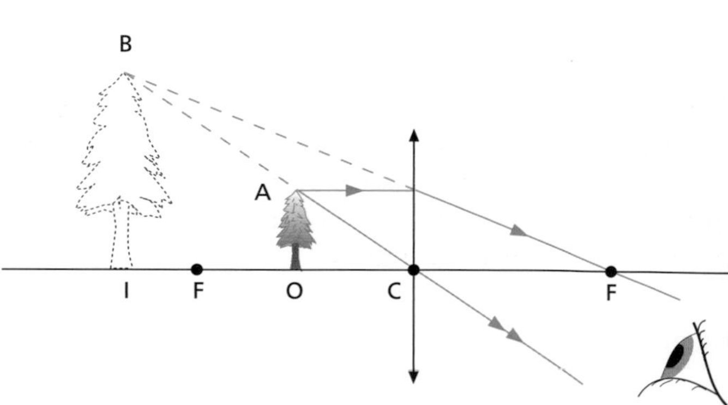

Image is behind object, virtual, erect, larger

d Object between F and C

Figure 6.5 Ray diagrams for a converging lens

Magnifying glass

The apparent size of an object depends on its actual size and on its distance from the eye. The sleepers on a railway track are all the same length but those nearby seem longer. This is because they enclose a larger angle at your eye than more distant ones: their image on the retina is larger, so making them appear bigger.

A converging lens gives an enlarged, upright virtual image of an object placed inside its principal focus F, Figure 6.6a. It acts as a magnifying glass since the angle β made at the eye by the image, formed at the near point, is greater than the angle α made by the object when it is viewed directly at the near point without the magnifying glass, Figure 6.6b.

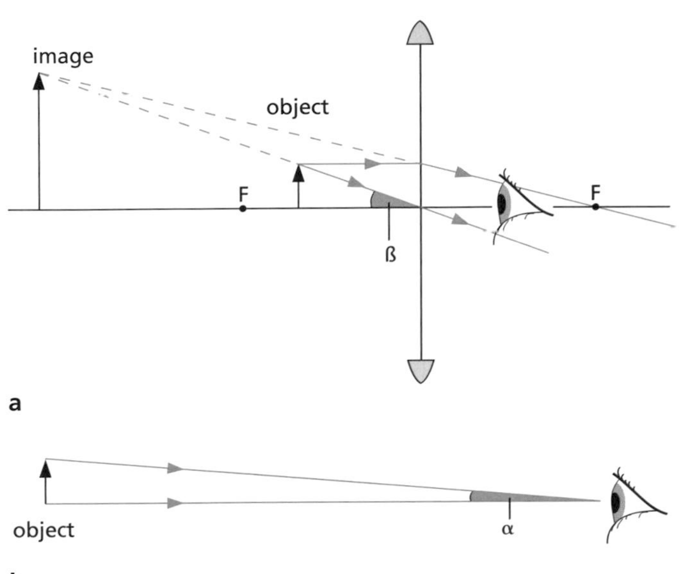

Figure 6.6 Magnification by a converging lens: $\beta > \alpha$

The fatter (more curved) a converging lens is, the shorter its focal length and the more it magnifies. Too much curvature however distorts the image.

Spectacles

From the ray diagrams shown in Figure 6.5 (page 20) we would expect that the converging lens in the eye will form a real **inverted** image on the retina as shown in Figure 6.7. Since an object normally appears upright, the brain must invert the image.

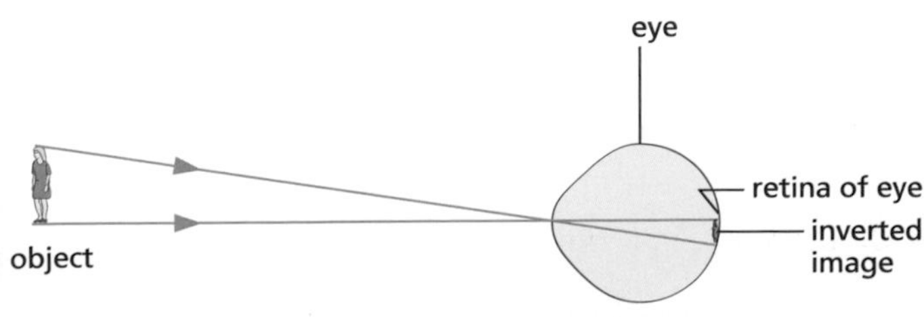

Figure 6.7 Inverted image on the retina

The average adult eye can focus objects comfortably from about 25 cm (the **near point**) to infinity (the **far point**). Your near point may be less than 25 cm; it gets farther away with age.

a) Short sight

A short-sighted person sees near objects clearly but distant objects appear blurred. The image of a distant object is formed in front of the retina because the eyeball is too long or because the eye lens cannot be made thin enough, Figure 6.8a. The defect is corrected by a diverging spectacle lens (or contact lens) which diverges the light before it enters the eye, to give an image on the retina, Figure 6.8b.

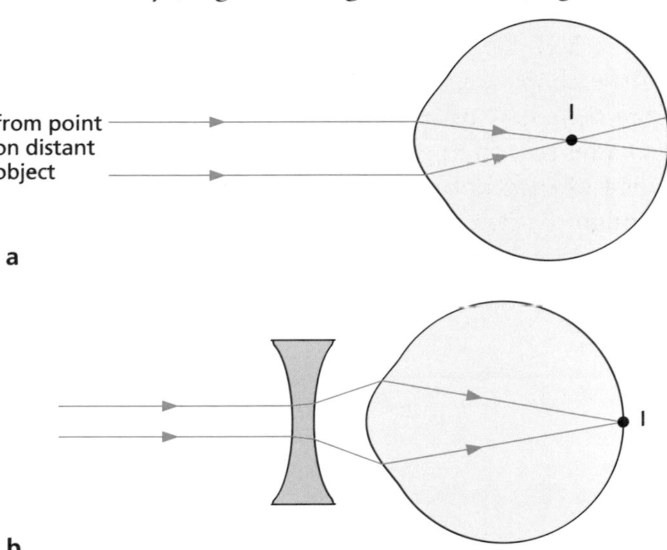

Figure 6.8 Short sight and its correction by a diverging lens

b) Long sight

A long-sighted person sees distant objects clearly but close objects appear blurred. The image of a near object is focused behind the retina because the eyeball is too short or because the eye lens cannot be made thick enough, Figure 6.9a. A converging spectacle lens (or contact lens) corrects the defect, Figure 6.9b.

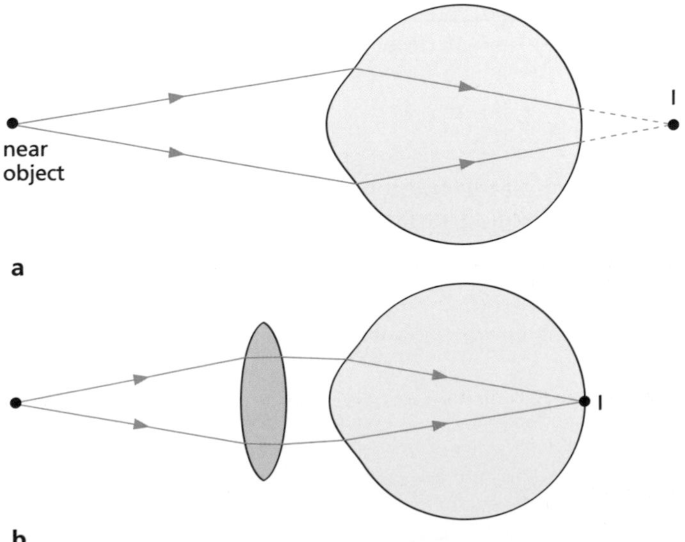

Figure 6.9 Long sight and its correction by a converging lens

Questions

1 A small torch bulb is placed at the focal point of a converging lens. When the bulb is switched on, does the lens produce a convergent, divergent or parallel beam of light?

2 a What kind of lens is shown in Figure 6.10?

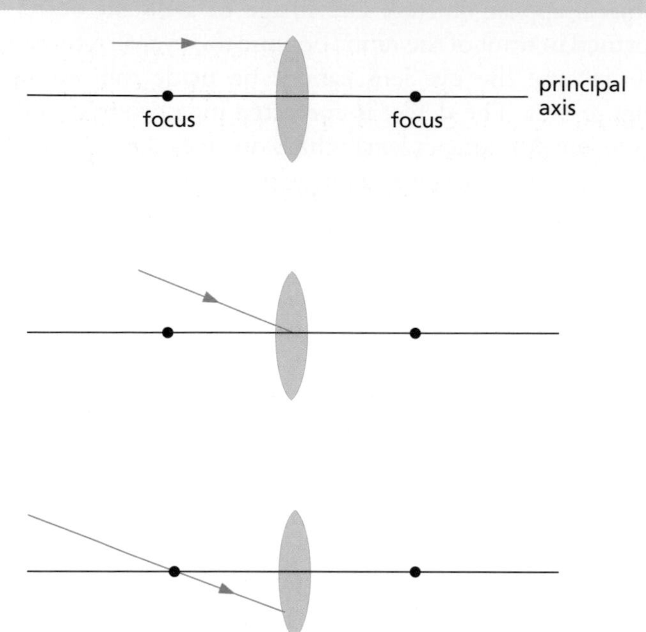

Figure 6.10

b Copy the diagrams and complete them to show the path of the light after passing through the lens.

c Figure 6.11 shows an object AB 6 cm high placed 18 cm in front of a lens of focal length 6 cm. Draw the diagram to scale and by tracing the paths of rays from A find the position and size of the image formed.

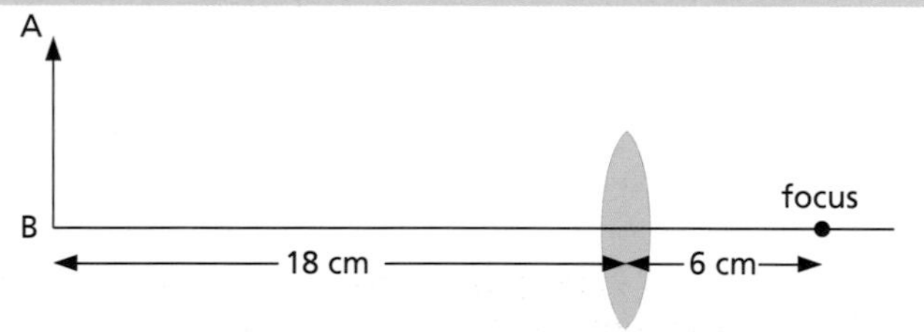

Figure 6.11

3 Where must the object be placed for the image formed by a converging lens to be
a real, inverted and smaller than the object,
b real, inverted and same size as the object,
c real, inverted and larger than the object,
d virtual, upright and larger than the object?

4 Figure 6.12 shows a camera focused on an object in the middle distance. Should the lens be moved towards or away from the film so that the image of a more distant object is in focus?

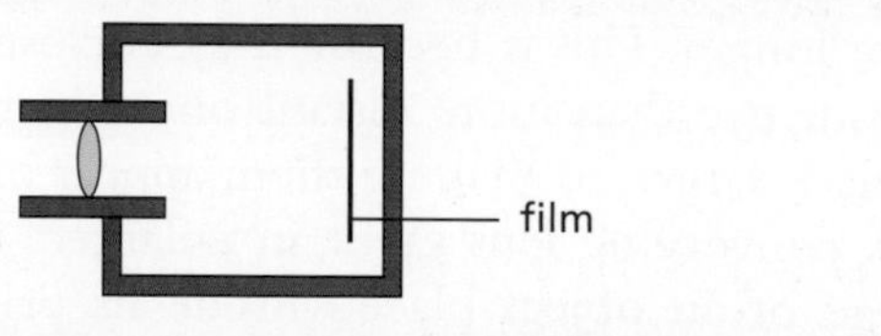

Figure 6.12

5 a Three converging lenses are available having focal lengths of 4 cm, 40 cm and 4 m respectively. Which one would you choose as a magnifying glass?

b An object 2 cm high is viewed through a converging lens of focal length 8 cm. The object is 4 cm from the lens. By means of a ray diagram find the position, nature and magnification of the image.

Checklist

After studying this chapter you should be able to

- explain the action of a lens in terms of refraction by a number of small prisms,
- draw diagrams showing the effects of a converging lens on a beam of parallel rays,
- recall the meaning of **optical centre, principal axis, principal focus** and **focal length,**
- describe an experiment to measure the focal length of a converging lens,
- draw ray diagrams to show formation of a real image by a converging lens,
- draw scale diagrams to solve problems on converging lenses,
- □ draw ray diagrams to show formation of a virtual image by a single lens,
- □ show how a single lens is used as a magnifying glass.

Light and sight
Additional questions

Light rays; reflection of light; plane mirrors; refraction of light; total internal reflection; lenses

1 The image in a plane mirror is

A upright, real and larger
B upright, virtual and the same size
C inverted, real and smaller
D inverted, virtual and the same size
E inverted, real and larger.

2 In this question, drawing should be done carefully.

A ray of light is shown striking mirror 1 at point X.

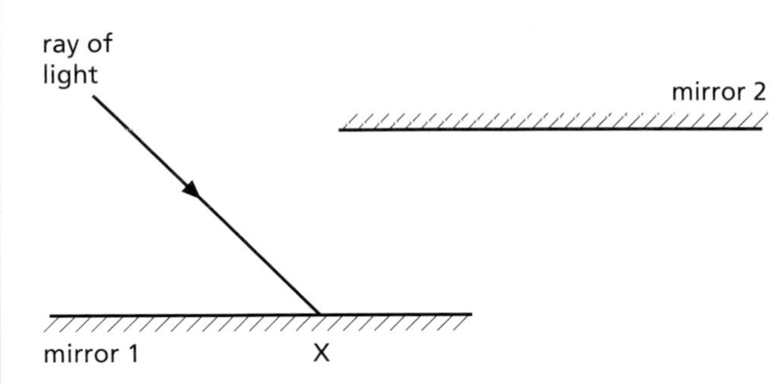

a (i) Draw the normal at X.
(ii) Draw the ray reflected from mirror 1.
(iii) Mark the angle of incidence using the letter *i* and the angle of reflection using the letter *r*.

b Mirror 2 is parallel to mirror 1. The reflected ray from mirror 1 strikes mirror 2.

Compare the direction of the ray reflected from mirror 2 with the incident ray at X. You may do a further construction if you wish. Complete the sentence below.

The reflected ray from mirror 2 is . . .

(*UCLES IGCSE Physics Core, May 2003*)

3 The diagram shows the view from above of a triangular object on one side of a vertical mirror.

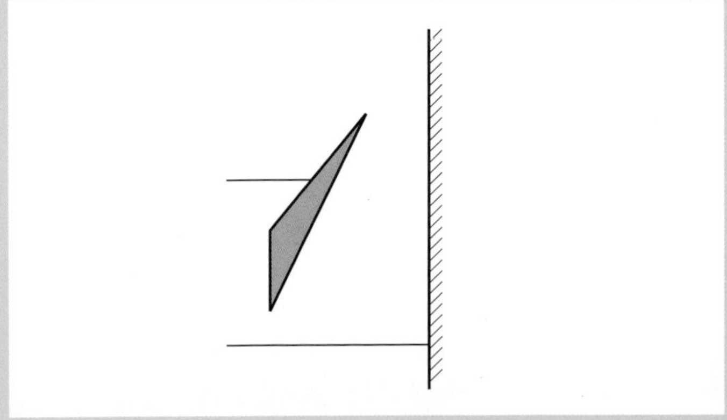

Copy the diagram and carefully draw the image formed by the mirror.

(*UCLES IGCSE Physics Core, May 2001*)

4 Light travels up through a pond of water of critical angle 49°. What happens at the surface if the angle of incidence is **a** 30°, **b** 60°?

5 Which diagram shows the ray of light refracted correctly?

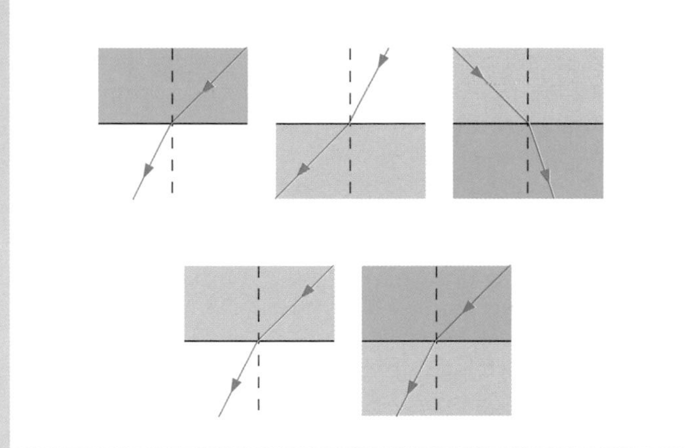

6 Which diagram shows the correct path of the ray through the prism?

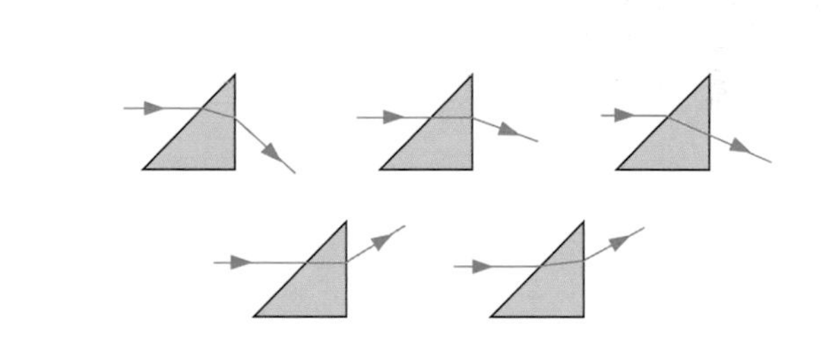

7 a The filament of a lamp is placed at the principal focus of a lens, as shown below.

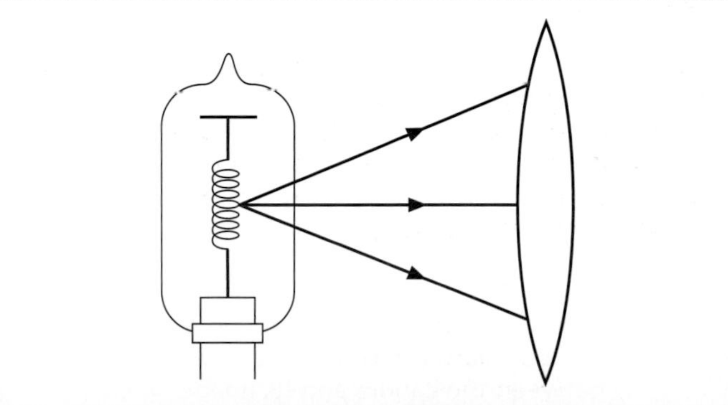

Continue the three rays through the lens and out into the air on the right of the lens.

continued

b The lens below has a focal length of 2.0 cm

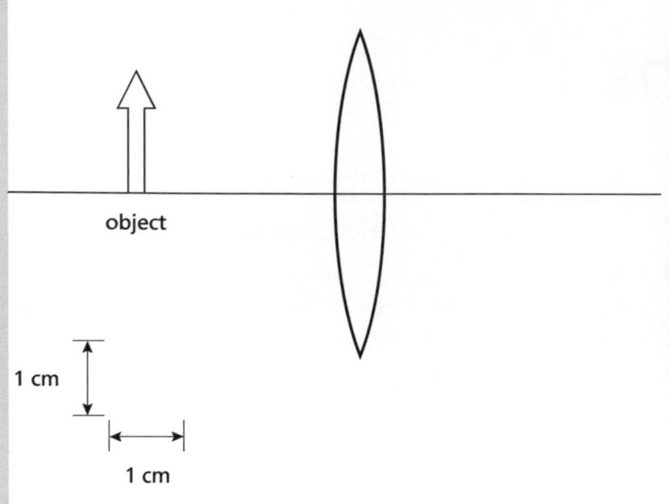

(i) Mark and label the positions of the principal focus on the left of the lens and the principal focus on the right of the lens.
(ii) Carefully draw a ray from the top of the object, parallel to the axis, through the lens and continue it until it reaches the edge of the squared area.
(iii) Carefully draw a ray from the top of the object, which travels parallel to the axis after it has passed through the lens.
(iv) Draw and label the image.
(*UCLES IGCSE Physics Core, Nov 2005*)

8 An illuminated object is placed at a distance of 20 cm from a converging lens of focal length 15 cm. The image obtained on a screen is

A upright and magnified
B upright and the same size
C inverted and magnified
D inverted and the same size
E inverted and diminished.

9 The diagram shows a candle 7 cm in front of a plane mirror.

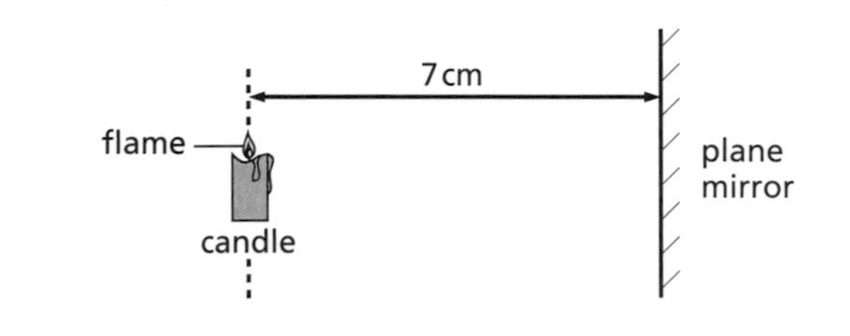

a By copying the diagram and accurately drawing the paths of *two* rays from the top of the flame, locate the image of the flame in the mirror.
b (i) Write down the distance between the candle and its image.
(ii) The candle is moved 2 cm towards the mirror. Write down the distance between the new position of the candle and its image.
(iii) Calculate the *change* in the distance between the candle and its image when the candle has been moved 2 cm.
c List three properties of an image formed in a plane mirror.
(*UCLES IGCSE Physical Science Extended, Nov 2000*)

10 The diagram shows some apparatus in use in an experiment to find the critical angle for blue light.

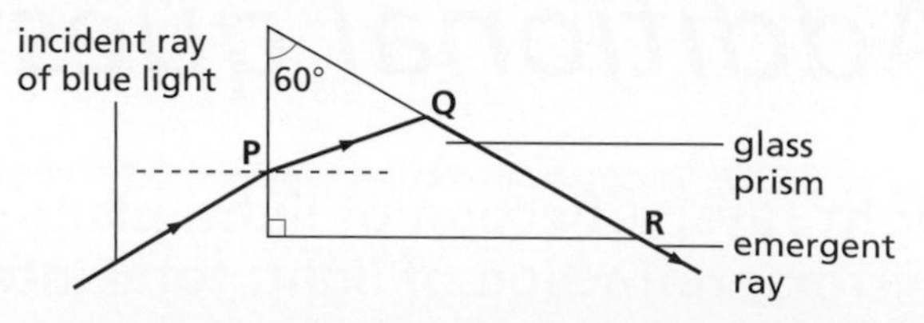

The ray hits the prism at point **P**, then crosses the prism to point **Q**. Part of the ray emerges along the surface **QR** as shown.

a (i) By using measurements taken from the diagram, find the critical angle of the glass for blue light.
(ii) Use your value to explain how total internal reflection of blue light could be made to occur at point **Q**.
b Using measured angles on the diagram, calculate the refractive index of the glass for blue light.
(*UCLES IGCSE Physics Extended, May 2001*)

11 The figure shows wavefronts of light crossing the edge of a glass block from air into glass.

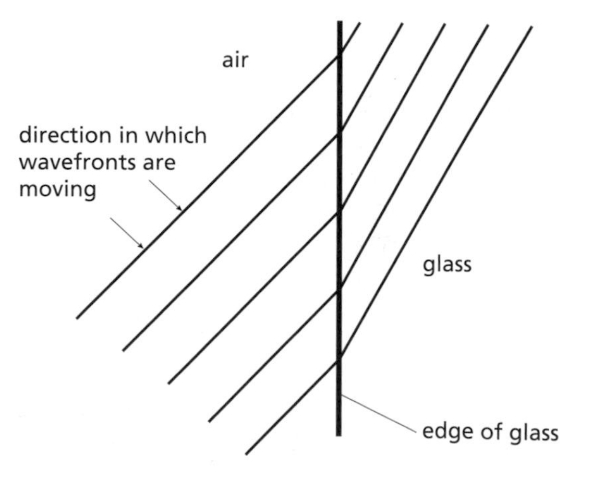

a (i) Draw in an incident ray, a normal and a refracted ray that meet at the same point on the edge of the glass block.
(ii) Label the angle of incidence and the angle of refraction.
(iii) Measure the two angles and record their values.
b Calculate the refractive index of the glass.
(*UCLES IGCSE Physics Extended, May 2003*)

12 An object is placed 10 cm in front of a lens, A; the details of the image are given below. The process is repeated for a different lens, B.

Lens A Real, inverted, magnified and at a great distance.
Lens B Real, inverted and same size as the object.

Estimate the focal length of each lens and state whether it is converging or diverging.

Waves and sound

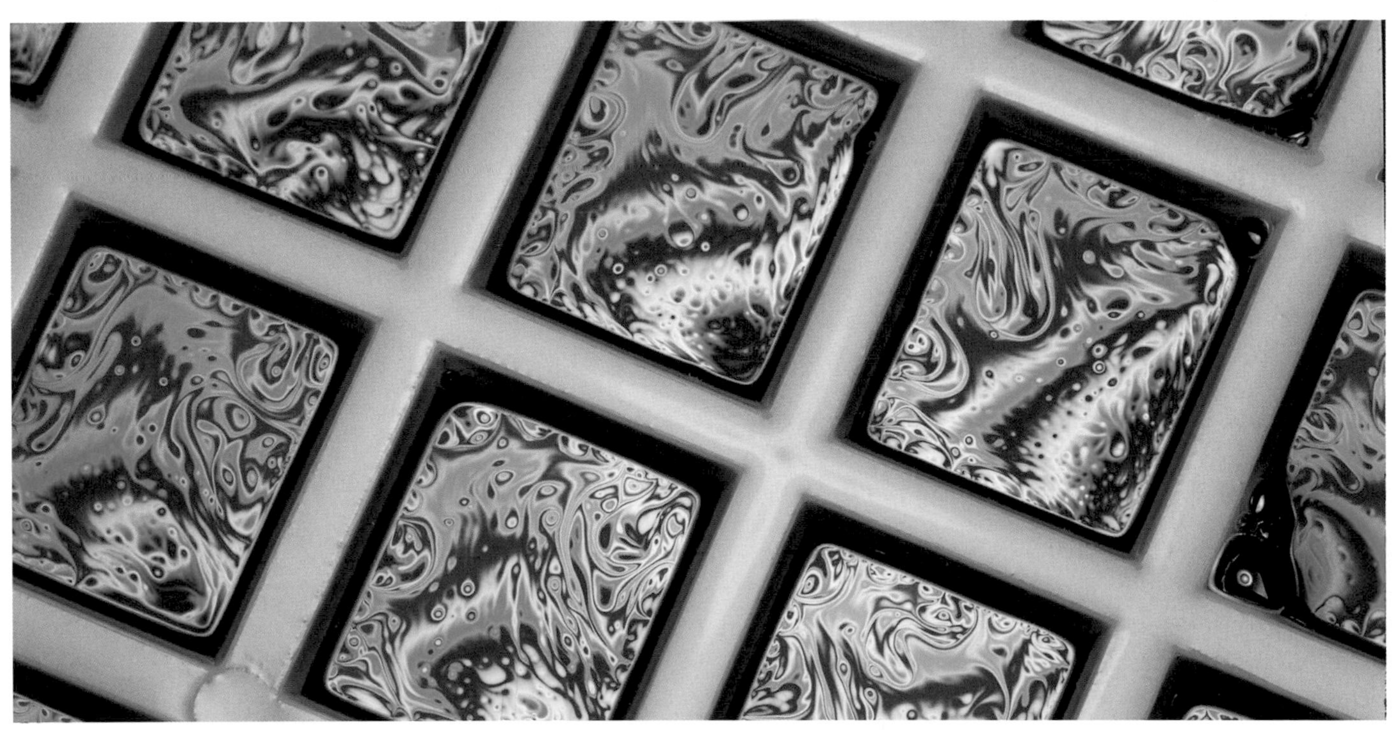

7 Mechanical waves

Types of wave
Describing waves
The wave equation
Wavefronts and rays
Reflection
Refraction
Diffraction
Wave theory
Interference
Polarization
Practical work
The ripple tank.

Types of wave

Several kinds of wave occur in physics. **Mechanical waves** are produced by a disturbance, e.g. a vibrating object, in a material medium and are transmitted by the particles of the medium vibrating to and fro. Such waves can be seen or felt and include waves on a rope or spring, water waves and sound waves in air or in other materials.

A **progressive** or travelling wave is a disturbance which carries energy from one place to another without transferring matter. There are two types, **transverse** and **longitudinal**. Longitudinal waves are dealt with in Chapter 9.

In a transverse wave, the direction of the disturbance is at **right angles** to the direction of travel of the wave. One can be sent along a rope (or a spring) by fixing one end and moving the other rapidly up and down, Figure 7.1. The disturbance generated by the hand is passed on from one part of the rope to the next which performs the same motion but slightly later. The humps and hollows of the wave travel along the rope as each part of the rope vibrates transversely about its undisturbed position.

Water waves are transverse waves.

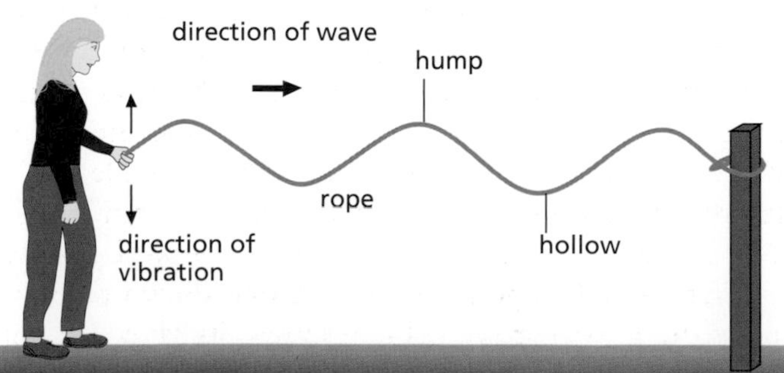

Figure 7.1 A transverse wave

Describing waves

Terms used to describe waves can be explained with the aid of a **displacement–distance** graph, Figure 7.2. It shows, for a certain instant of time, the distance moved (sideways from their undisturbed positions) by the parts of the medium vibrating at different distances from the cause of the wave.

a) Wavelength

The wavelength of a wave, represented by the Greek letter λ (lambda), is the distance between successive crests.

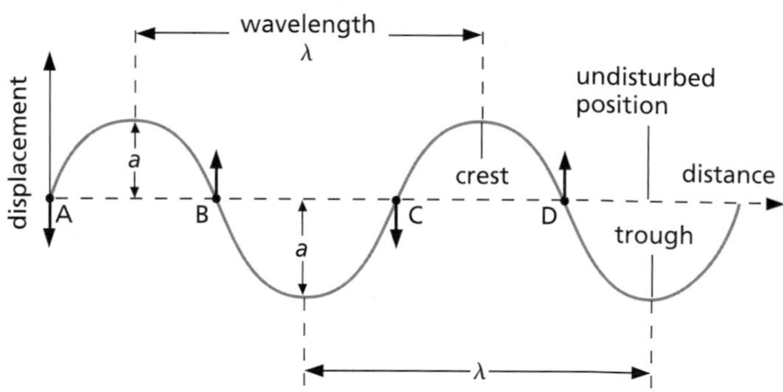

Figure 7.2 Displacement–distance graph for a wave at a particular instant

b) Frequency

The frequency f is the number of complete waves generated per second. If the end of a rope is moved up and down twice in a second, two waves are produced in this time. The frequency of the wave is 2 vibrations per second or 2 **hertz** (2 Hz; the hertz being the unit of frequency) which is the same as the frequency of the movement of the end of the rope. That is, the frequencies of the wave and its source are equal.

The frequency of a wave is also the number of crests passing a chosen point per second.

c) Speed

The speed v of the wave is the distance moved in the direction of travel of the wave by a crest or any point on the wave in 1 second.

d) Amplitude

The amplitude a is the height of a crest or the depth of a trough measured from the undisturbed position of what is carrying the wave, e.g. a rope.

e) Phase

The short arrows at A, B, C, D on Figure 7.2 show the directions of vibration of the parts of the rope at these points. The parts at A and C have the same speed in the same direction and are **in phase**. At B and D the parts are also in phase but they are **out of phase** with those at A and C because their directions of vibration are opposite.

The wave equation

The faster the end of a rope is vibrated, the shorter the wavelength of the wave produced. That is, the higher the frequency of a wave, the smaller its wavelength. There is a useful connection between f, λ and v, which is true for all types of wave.

Suppose waves of wavelength $\lambda = 20$ cm travel on a long rope and three crests pass a certain point every second. The frequency $f = 3$ Hz. If Figure 7.3 represents this wave motion then if crest A is at P at a particular time, 1 second later it will be at Q, a distance from P of three wavelengths, i.e. $3 \times 20 = 60$ cm. The speed of the wave is $v = 60$ cm per second (60 cm/s), obtained by multiplying f by λ. Hence

speed of wave = frequency × wavelength

or $$v = f\lambda$$

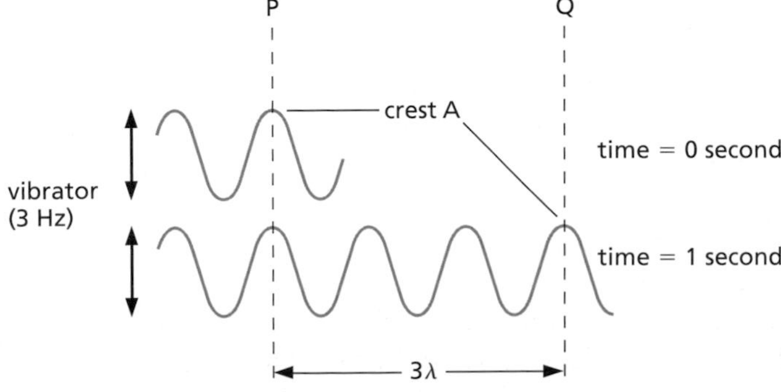

Figure 7.3

Practical work

The ripple tank

The behaviour of water waves can be studied in a ripple tank. It consists of a transparent tray containing water, having a light source above and a white screen below to receive the wave images, Figure 7.4.

Pulses (i.e. short bursts) of ripples are obtained by dipping a finger in the water for circular ones and a ruler for straight ones. **Continuous ripples** are generated by an electric motor and a bar which gives straight ripples if it just touches the water or circular ripples if the bar is raised and a small ball fitted to it.

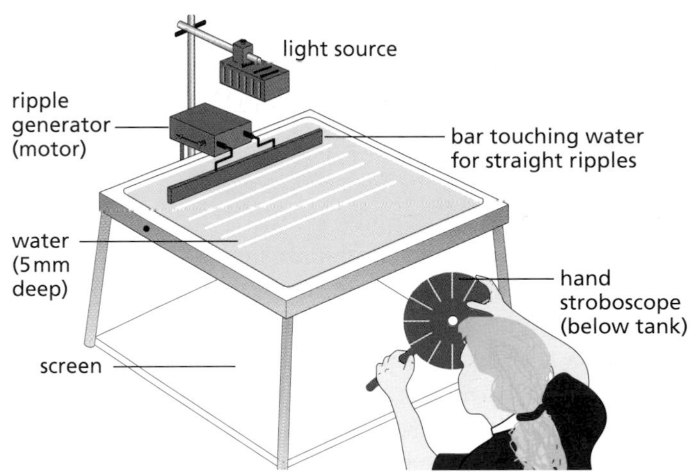

Figure 7.4 A ripple tank

Continuous ripples are studied more easily if they are *apparently* stopped ('frozen') by viewing the screen through a disc with equally spaced slits, which can be spun by hand, i.e. a stroboscope. If the disc speed is such that the waves have advanced one wavelength each time a slit passes your eye, they appear at rest.

Wavefronts and rays

In two dimensions, a **wavefront** is a line on which the disturbance has the same phase at all points; the **crests of waves** in a ripple tank can be thought of as wavefronts. A vibrating source produces a succession of wavefronts, all of the same shape. In a ripple tank, straight wavefronts are produced by a vibrating bar (a line source) and circular wavefronts are produced by a vibrating ball (a point source). A line drawn at right angles to a wavefront, which shows its direction of travel, is called a **ray**. Straight wavefronts and the corresponding rays are shown in Figure 7.5; circular wavefronts can be seen in Figure 7.9.

Reflection

In Figure 7.5 **straight** water waves are falling on a metal strip placed in a ripple tank at an angle of 60°, i.e. the angle i between the direction of travel of the waves and the normal to the strip is 60°, as is the angle between the wavefront and the strip. The wavefronts are represented by straight lines and can be thought of as the crests of the waves. They are at right angles to the direction of travel, i.e. to the rays. The angle of reflection r is 60°. Incidence at other angles shows that **the angles of reflection and incidence are always equal**.

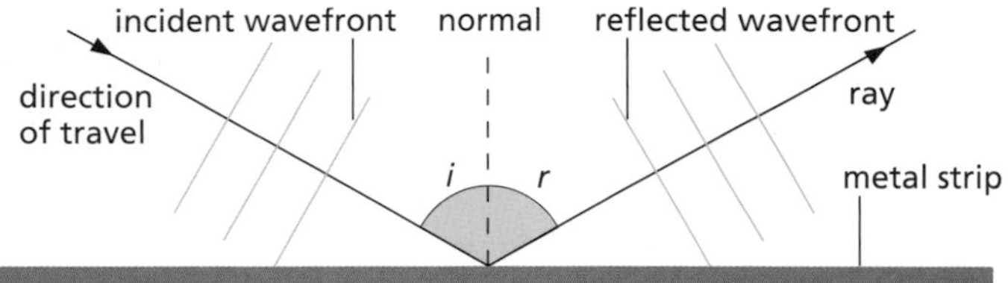

Figure 7.5 Reflection of waves

Refraction

If a glass plate is placed in a ripple tank so that the water over it is about 1 mm deep but 5 mm elsewhere, continuous straight waves in the shallow region are found to have a shorter wavelength than those in the deeper parts, i.e. the wavefronts are closer together, Figure 7.6. Both sets of waves have the frequency of the vibrating bar and since $v = f\lambda$, if λ has decreased so has v, since f is fixed. Hence **waves travel more slowly in shallow water**.

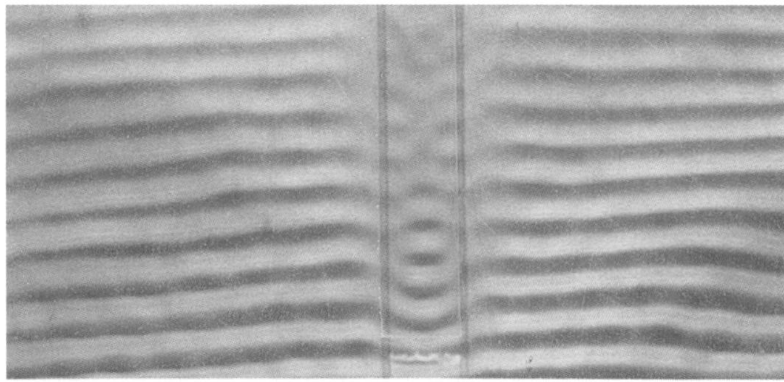

Figure 7.6 Waves in shallower water have a shorter wavelength

When the plate is at an angle to the waves, Figure 7.7a, their direction of travel in the shallow region is bent towards the normal, Figure 7.7b, i.e. refraction occurs. We saw earlier (Chapter 4) that light is refracted because its speed (and wavelength) changes (but not its frequency) when it enters another medium. The refraction of water waves for the same reason seems to suggest that light may be a kind of wave motion.

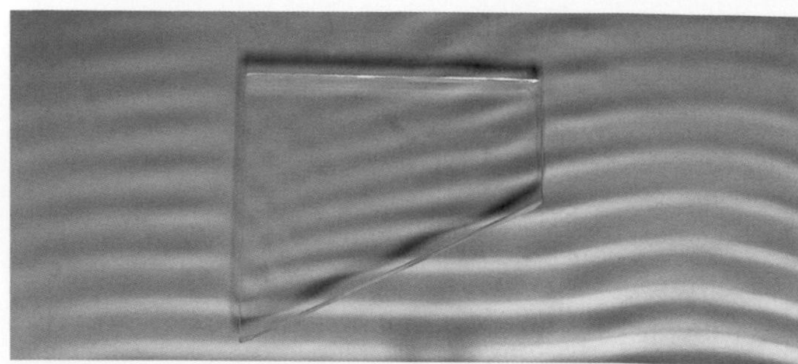

Figure 7.7a Waves are refracted at the boundary between deep and shallow regions

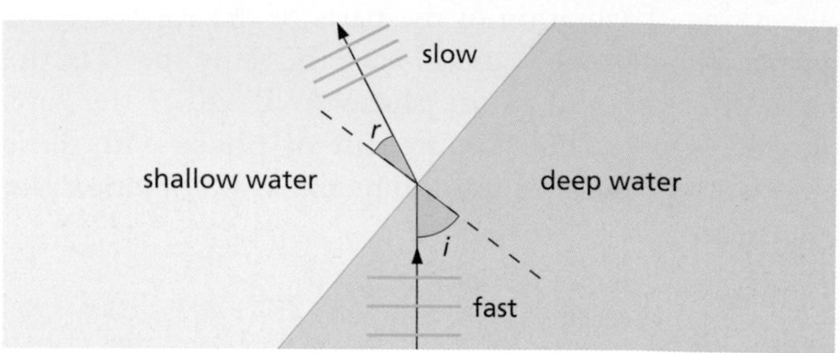

Figure 7.7b The direction of travel is bent towards the normal in the shallow region

Diffraction

In Figures 7.8a and b, straight water waves in a ripple tank are falling on gaps formed by obstacles. In photograph a the gap width is about the same as the wavelength of the waves (1 cm); those passing through are circular and spread out in all directions. In photograph b the gap is wide (10 cm) compared with the wavelength and the waves continue straight on; some spreading occurs but it is less obvious.

The spreading of waves at the edges of obstacles is called **diffraction**; when designing harbours, engineers use models like that in Figure 7.9 to study it.

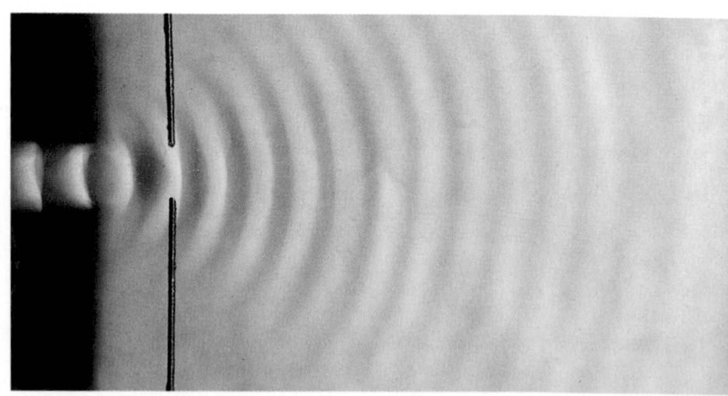

Figure 7.8a Spreading of waves after passing through a narrow gap

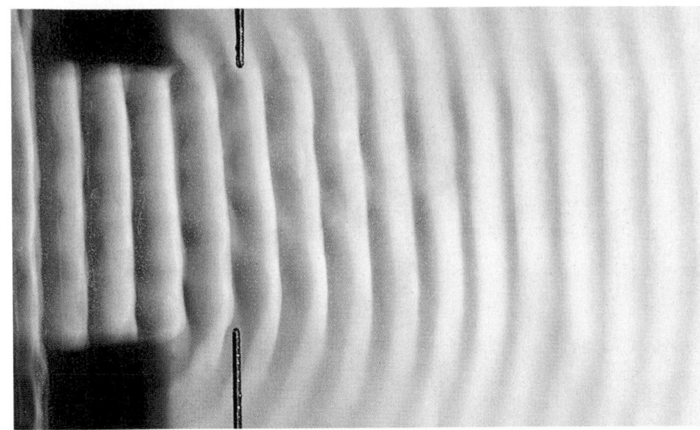

Figure 7.8b Spreading of waves after passing through a wide gap

Figure 7.9 Model of a harbour used to study wave behaviour

Wave theory

If the position of a wavefront is known at one instant, its position at a later time can be found using **Huygens' construction**. Each point on the wavefront is considered to be a source of **secondary** spherical wavelets which spread out at the wave speed; the new wavefront is the surface that touches all the wavelets (in the forward direction). A simple example is shown in Figure 7.10; the straight wavefront AB is travelling from left to right with speed v. At a time t later, the spherical wavelets from AB will be a distance vt from the secondary sources and the new surface which touches all the wavelets is the straight wavefront CD.

Wave theory can be used to explain reflection, refraction and diffraction effects.

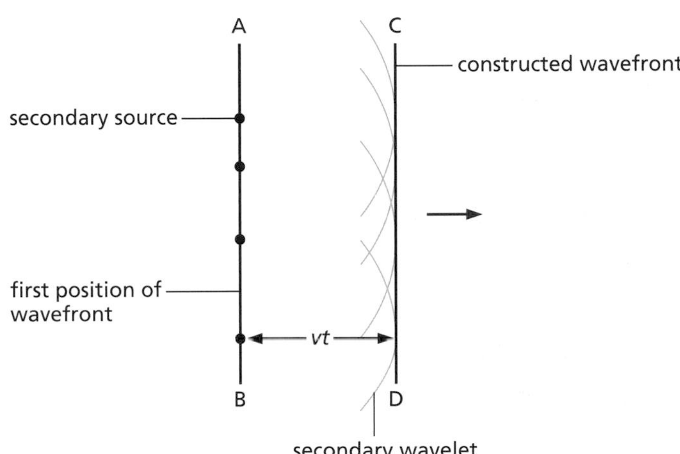

Figure 7.10 Huygens' construction for a straight wavefront

a) Reflection

Figure 7.11 shows a straight wavefront AB incident at an angle i on a reflecting surface; the wavefront has just reached the surface at A. The position of the wavefront a little later, when B reaches the reflecting surface, can be found using Huygens' construction. A circle of radius BB′ is drawn about A; the reflected wavefront is then A′B′, the tangent to the wavelet from B′. Measurements of the angle of incidence i and the angle of reflection r show that they are equal.

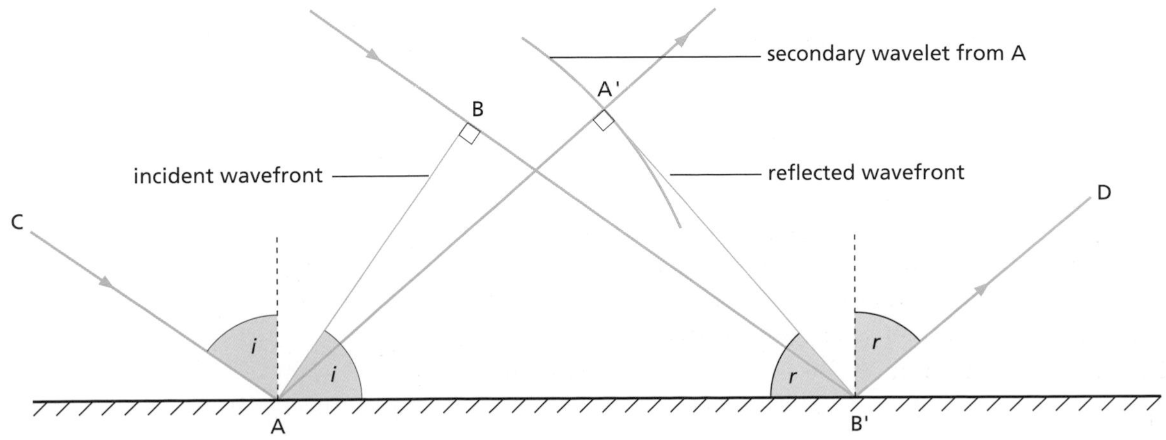

Figure 7.11 Reflection of a straight wavefront

b) Refraction

Huygens' construction can also be used to find the position of a wavefront when it enters a second medium in which the speed of travel of the wave is different from in the first (see Figure 7.7b).

In Figure 7.12, A, on the straight wavefront AB, has just reached the boundary between two media. When B reaches the boundary the secondary wavelet from A will have moved on to A′. If the wave travels more slowly in the second medium AA′ < BB′. The new wavefront is then A′B′, the tangent to the wavelet from B′; it is clear that the direction of travel of the wave has changed – refraction has occurred.

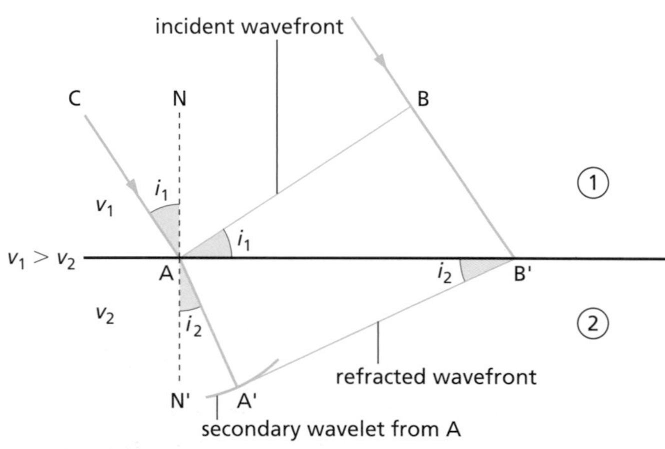

Figure 7.12 Refraction of a straight wavefront

c) Diffraction

Diffraction effects, such as those shown in Figure 7.8, can be explained if secondary wavelets arise from point sources on the unrestricted part of the wavefronts in the gaps.

F *Interference*

When two sets of continuous **circular** waves cross in a ripple tank, a pattern like that in Figure 7.13 is obtained.

At points where a crest from one source, S_1, arrives at the same time as a crest from the other source, S_2, a bigger crest is formed and the waves are said to be **in phase**. At points where a crest and a trough arrive together, they cancel out (if their amplitudes are equal); the waves are exactly **out of phase** (due to travelling different distances from S_1 and S_2) and the water is undisturbed; in Figure 7.13 the blurred lines radiating from between S_1 and S_2 join such points.

Figure 7.13 Interference of circular waves

Interference or **superposition** is the combination of waves to give a larger or a smaller wave, Figure 7.14.

Study this effect with two ball 'dippers' about 3 cm apart on the bar of a ripple tank. Also observe the effect of changing (i) the frequency and (ii) the separation of the 'dippers'; use a stroboscope when necessary. You will find that if the frequency is increased, i.e. the wavelength decreased, the blurred lines are closer together. Increasing the separation has the same effect.

Similar patterns are obtained if straight waves fall on two small gaps: interference occurs between the sets of emerging (circular) diffracted waves.

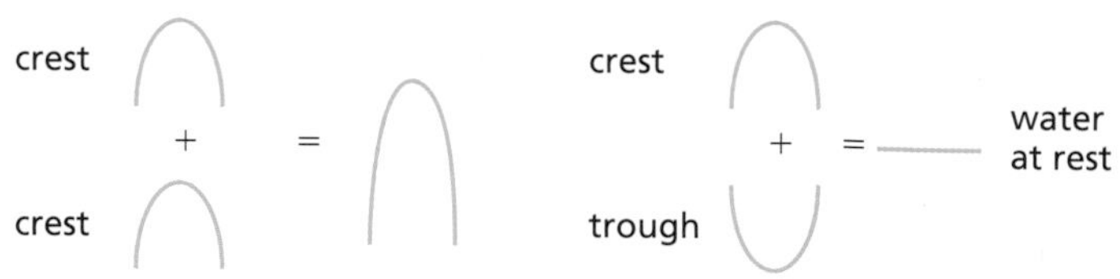

Figure 7.14

F *Polarization*

This effect occurs only with transverse waves. It can be shown by fixing a rope at one end, D in Figure 7.15, and passing it through two slits B and C. If end A is vibrated in all directions (as shown by the short arrowed lines), vibrations of the rope occur in every plane and transverse waves travel towards B.

At B only waves due to vibrations in a vertical plane can emerge from the vertical slit. The wave between B and C is said to be **plane polarized** (in the vertical plane containing the slit at B). By contrast the waves between A and B are unpolarized. If the slit at C is vertical, the wave travels on, but if it is horizontal as shown, the wave is stopped and the slits are said to be 'crossed'.

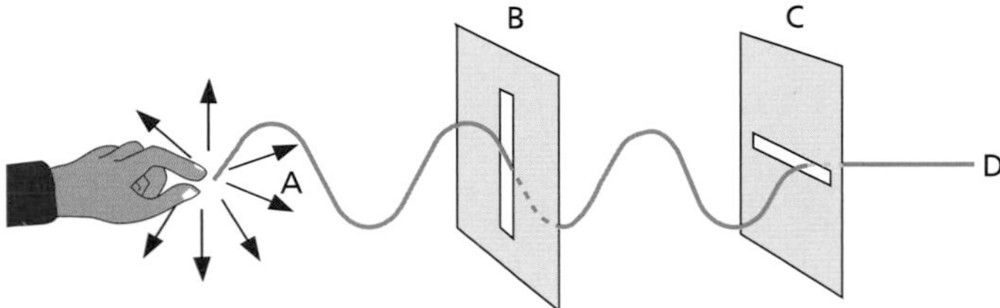

Figure 7.15 Polarizing waves on a rope

Questions

1 The lines in Figure 7.16 are crests of straight ripples.
   a What is the wavelength of the ripples?
   b If 5 seconds ago ripple A occupied the position now occupied by ripple F, what is the frequency of the ripples?
   c What is the speed of the ripples?

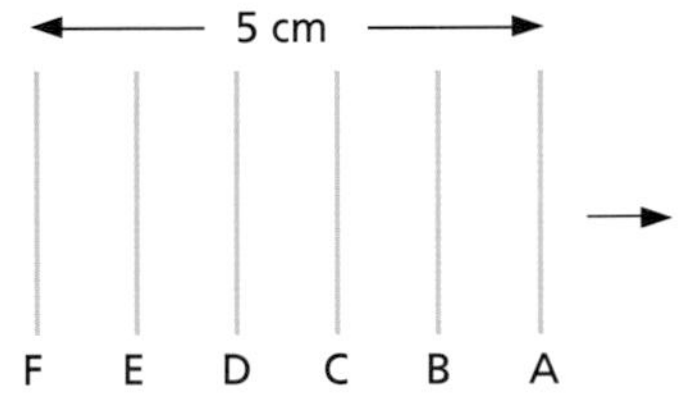

Figure 7.16

2 During the refraction of a wave which *two* of the following properties change?

   A the speed
   B the frequency
   C the wavelength

3 One side of a ripple tank ABCD is raised slightly, Figure 7.17, and a ripple started at P by a finger. After 1 second the shape of the ripple is as shown.
   a Why is it not circular?
   b Which side of the tank has been raised?

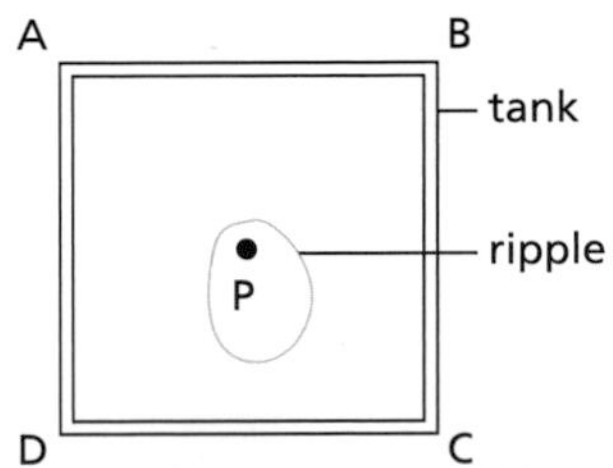

Figure 7.17

4 Figure 7.18 gives a full-scale representation of the water in a ripple tank 1 second after the vibrator was started. The coloured lines represent crests.
   a What is represented at A at this instant?
   b Estimate
      (i) the wavelength,
      (ii) the speed of the waves, and
      (iii) the frequency of the vibrator.
   c Sketch a suitable attachment which could have been vibrated up and down to produce this wave pattern.

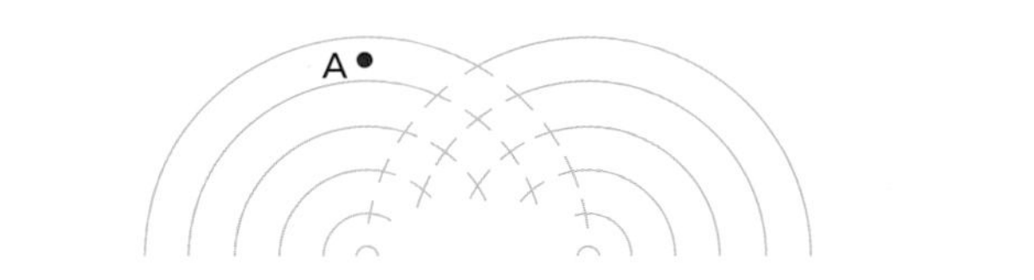

Figure 7.18

5 Copy Figure 7.19 and show on it what happens to the waves as they pass through the gap, if the water is much shallower on the right-hand side than on the left.

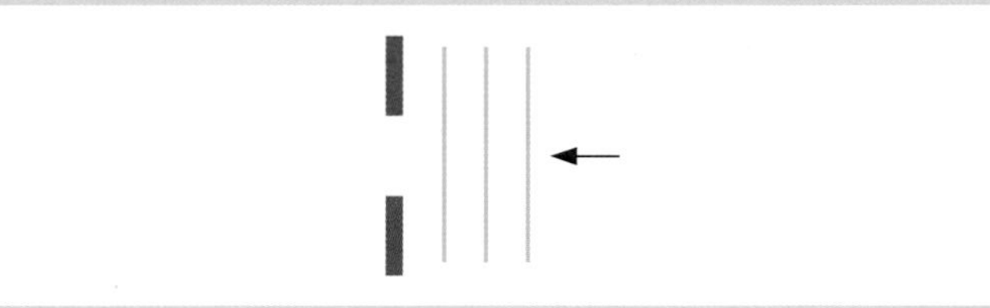

Figure 7.19

Checklist

After studying this chapter you should be able to

- describe the production of pulses and progressive transverse waves on ropes, springs and ripple tanks,
- recall the meaning of **wavelength**, **frequency**, **speed**, and **amplitude**,
- represent a transverse wave on a displacement–distance graph and extract information from it,
- recall the wave equation $v = f\lambda$ and use it to solve problems,
- use the term wavefront,
- recall that the angle of reflection equals the angle of incidence and draw a diagram for the reflection of straight wavefronts at a plane surface,
- describe experiments to show reflection and refraction of waves,
- recall that refraction at a straight boundary is due to change of wave speed but *not* of frequency,
- draw a diagram for the refraction of straight wavefronts at a straight boundary,
- explain the term **diffraction**,
- describe experiments to show diffraction of waves,
- draw diagrams for the diffraction of straight wavefronts at single slits of different widths,
- predict the effect of changing the wavelength or the size of the gap on diffraction of waves at a single slit,
- explain reflection, refraction and diffraction in terms of wave theory.

8 Electromagnetic radiation

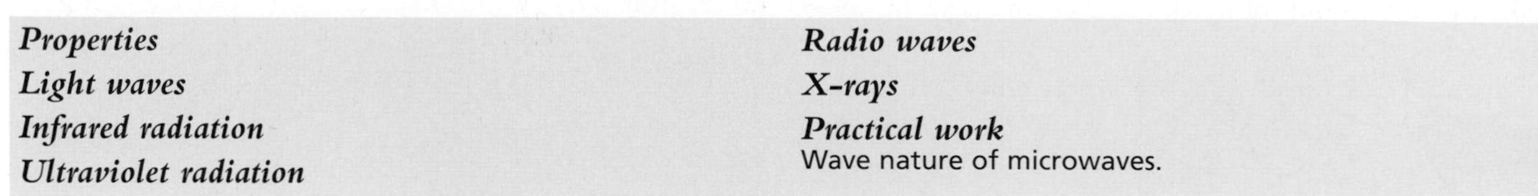

Properties
Light waves
Infrared radiation
Ultraviolet radiation
Radio waves
X-rays
Practical work
Wave nature of microwaves.

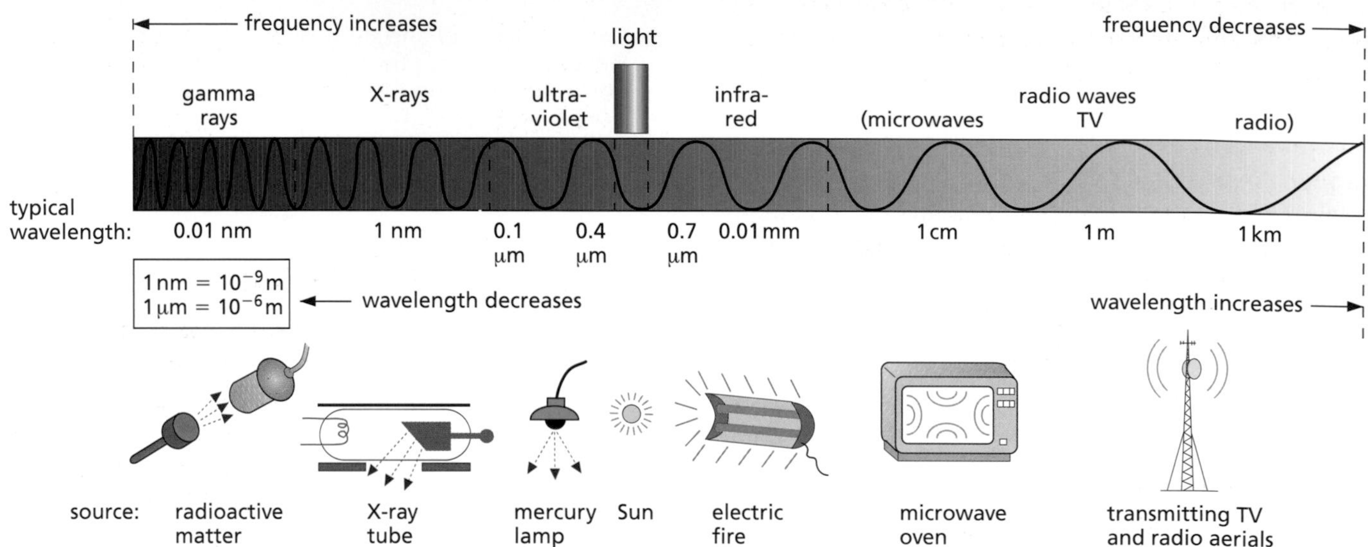

Figure 8.1 The electromagnetic spectrum and sources of each type of radiation

Light is one member of the family of electromagnetic radiation, which forms a continuous spectrum beyond both ends of the visible (light) spectrum, Figure 8.1. While each type of radiation has a different source, all result from electrons in atoms undergoing an energy change and all have certain properties in common.

Properties

1 All types of electromagnetic radiation **travel through a vacuum at 300 000 km/s** (3×10^8 m/s), i.e. with the speed of light.

2 They **exhibit interference, diffraction and polarization**, which suggests they have a transverse wave nature.

3 They **obey the wave equation** $v = f\lambda$ where v is the speed of light, f is the frequency of the waves and λ is the wavelength. Since v is constant for a particular medium, it follows that large f means small λ.

4 They **carry energy from one place to another and can be absorbed by matter to cause heating and other effects**. The higher the frequency and the smaller the wavelength of the radiation, the greater is the energy carried, i.e. gamma rays are more 'energetic' than radio waves. This is shown by the **photoelectric effect** in which electrons are ejected from metal surfaces when electromagnetic waves fall on them. As the frequency of the waves increases so too does the speed (and energy) with which electrons are emitted.

Because of its electrical origin, its ability to travel in a vacuum (e.g. from the Sun to the Earth) and its wave-like properties (i.e. **2** above), electromagnetic radiation is regarded as a **progressive transverse wave**. The wave is a combination of travelling electric and magnetic fields. The fields vary in value and are directed at right angles to each other and to the direction of travel of the wave, as shown by the representation in Figure 8.2.

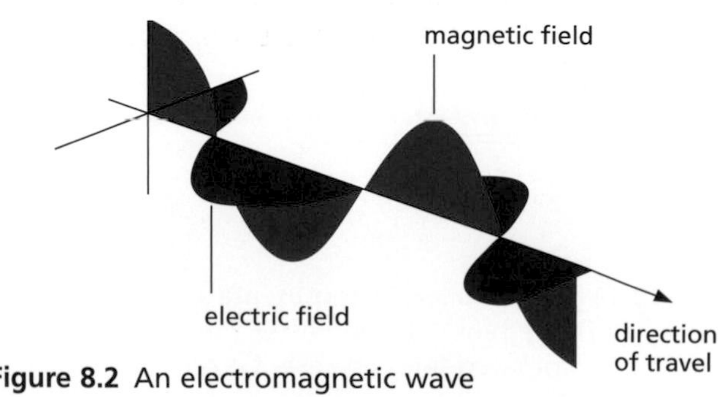

Figure 8.2 An electromagnetic wave

Light waves

Red light has the longest wavelength, which is about 0.000 7 mm (7×10^{-7} m = 0.7 μm), while violet light has the shortest wavelength of about 0.000 4 mm (4×10^{-7} m = 0.4 μm). Colours between these in the spectrum of white light have intermediate values. Light of one colour and so of one wavelength is called **monochromatic** light.

Since $v = f\lambda$ for all waves including light, it follows that red light has a lower frequency, f, than violet light since (i) the wavelength, λ, of red light is greater and (ii) all colours travel with the same speed, v, of 3×10^8 m/s in air (strictly in a vacuum). It is the **frequency** of light which decides its colour, rather than its wavelength which is different in different media, as is its speed (Chapter 4).

Different frequencies of light travel at different speeds through a transparent medium and so are refracted by different amounts. This explains dispersion (Chapter 4), i.e. why the refractive index of a material depends on the wavelength of the light.

The **amplitude** of a light (or any other) wave is greater the higher the **intensity** of the source, i.e. the brighter it is in the case of light.

Infrared radiation

Our bodies detect infrared radiation (IR) by its heating effect on the skin. It is sometimes called 'radiant heat' or 'heat radiation'.

Anything which is hot but not glowing, i.e. below 500 °C, emits IR alone. At about 500 °C a body becomes red-hot and emits red light as well as IR – the heating element of an electric fire, a toaster or a grill are examples. At about 1500 °C things such as lamp filaments are white-hot and radiate IR and white light, i.e. all the colours of the visible spectrum.

Infrared is also detected by special temperature-sensitive photographic films which allow pictures to be taken in the dark. Infrared sensors are used on satellites and aircraft for weather forecasting, monitoring of land use, Figure 8.3, assessing heat loss from buildings, and locating victims of earthquakes.

Infrared lamps are used to dry the paint on cars during manufacture and in the treatment of muscular complaints. Remote control keypads for televisions contain a small infrared transmitter for interacting with the programmes.

Figure 8.3 Infrared aerial photograph of Washington DC

Ultraviolet radiation

Ultraviolet (UV) rays have shorter wavelengths than light. They cause sun-tan and produce vitamins in the skin but can penetrate deeper causing skin cancer. Dark skin is able to absorb more UV, so reducing the amount reaching deeper tissues. Exposure to the harmful UV rays present in sunlight can be reduced by wearing protective clothing such as a hat or using sunscreen lotion.

Ultraviolet causes fluorescent paints and clothes washed in some detergents to fluoresce, Figure 8.4. They glow by re-radiating as light the energy they absorb as UV. This effect may be used to verify 'invisible' signatures on bank documents.

Figure 8.4 White clothes fluorescing in a club

A UV lamp used for scientific or medical purposes contains mercury vapour and this emits UV when an electric current passes through it. Fluorescent tubes also contain mercury vapour and their inner surfaces are coated with special powders called phosphors which radiate light.

Radio waves

Radio waves have the longest wavelengths in the electromagnetic spectrum. They are radiated from aerials and used to 'carry' sound, pictures and other information over long distances.

a) Long, medium and short waves (wavelengths of 2 km to 10 m)

These diffract round obstacles so can be received when hills etc. are in their way, Figure 8.5a. They are also reflected by layers of electrically charged particles in the upper atmosphere (the **ionosphere**), which makes long-distance radio reception possible, Figure 8.5b.

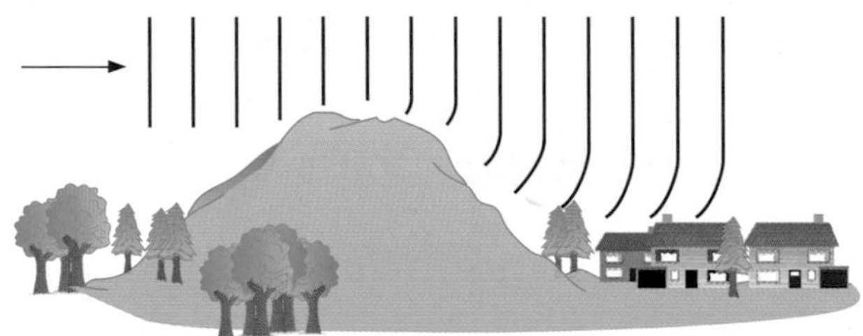

a Diffraction of radio waves

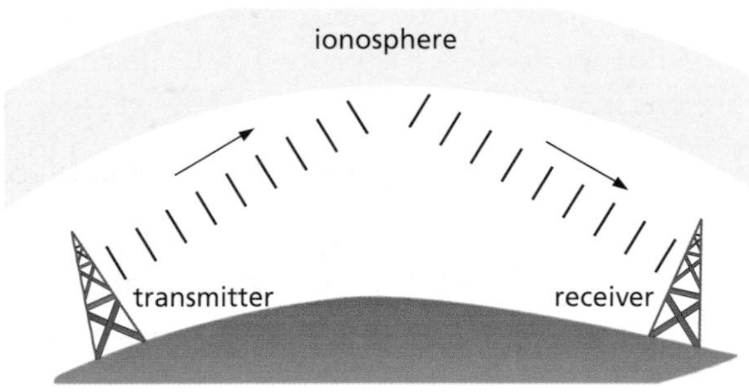

b Reflection of radio waves

Figure 8.5

b) VHF (very high frequency) and UHF (ultra high frequency) waves (wavelengths of 10 m to 10 cm)

These shorter wavelength radio waves need a clear, straight-line path to the receiver. They are not reflected by the ionosphere. They are used for local radio and for television.

c) Microwaves (wavelengths of a few cm)

These are used for international telecommunications and television relay via geostationary satellites and for mobile phone networks via microwave aerial towers and low-orbit satellites (Chapter 24). The microwave signals are transmitted through the ionosphere by dish aerials, amplified by the satellite and sent back to a dish aerial in another part of the world.

Microwaves are also used for **radar** detection of ships and aircraft, and in police speed traps.

Microwaves can be used for cooking since they cause water molecules in the moisture of the food to vibrate vigorously at the frequency of the microwaves. As a result, heating occurs inside the food which cooks itself.

Living cells can be damaged or killed by the heat produced when microwaves are absorbed by water in the cells. There is some debate at present as to whether their use in mobile phones is harmful; 'hands-free' mode, where separate earphones are used, may be safer.

Practical work

Wave nature of microwaves

The 3 cm microwave transmitter and receiver shown in Figure 8.6 can be used. The three metal plates can be set up for double-slit interference with 'slits' about 3 cm wide; the reading on the meter connected to the horn receiver rises and falls as it is moved across behind the plates.

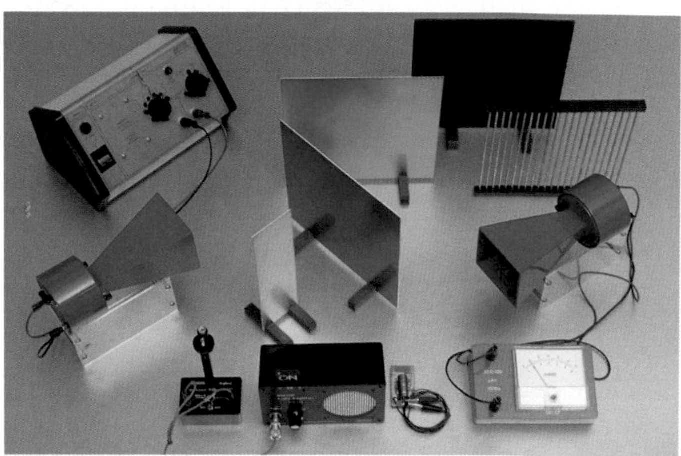

Figure 8.6

Diffraction can be shown similarly using the two wide metal plates to form a single slit.

If the grid of vertical metal wires is placed in front of the transmitter, the signal is absorbed but transmission occurs when the wires are horizontal, showing that the microwaves are vertically polarized. This can also be shown by rotating the receiver through 90° in a vertical plane from the maximum signal position, when the signal decreases to a minimum.

X-rays

These are produced when high-speed electrons are stopped by a metal target in an X-ray tube. X-rays have smaller wavelengths than UV.

They are absorbed to some extent by living cells but can penetrate some solid objects and affect a photographic film. With materials like bones, teeth and metals which they do not pass through easily, shadow pictures can be taken, like that in Figure 8.7 of someone shaving. In industry X-ray photography is used to inspect welded joints.

X-ray machines need to be shielded with lead since normal body cells can be killed by high doses and made cancerous by lower doses.

Gamma rays (Chapter 48) are more penetrating and dangerous than X-rays. They are used to kill cancer cells, and also harmful bacteria in food and on surgical instruments.

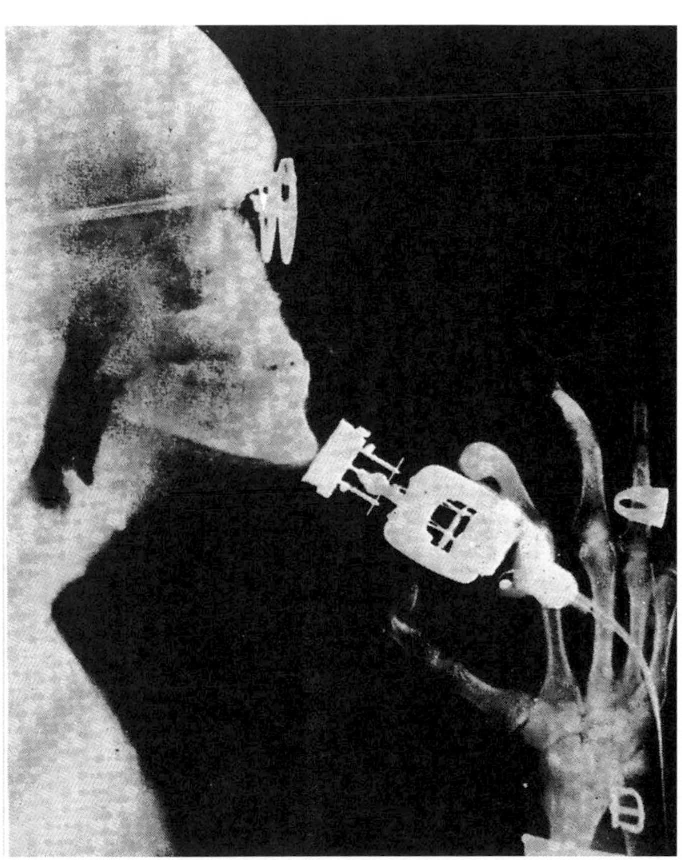

Figure 8.7 X-rays cannot penetrate bone and metal

Questions

1 Give the approximate wavelength in micrometres (μm) of
- **a** red light,
- **b** violet light.

2 Which of the following types of radiation has
- **a** the longest wavelength,
- **b** the highest frequency?

  - **A** UV
  - **B** radio waves
  - **C** light
  - **D** X-rays
  - **E** IR

3 Name one type of electromagnetic radiation which
- **a** causes sun-tan,
- **b** is used for satellite communication,
- **c** is used to sterilize surgical instruments,
- **d** is used in TV remote control,
- **e** is used to cook food,
- **f** is used to detect a break in a bone.

4 A VHF radio station transmits on a frequency of 100 MHz (1 MHz = 10^6 Hz). If the speed of radio waves is 3×10^8 m/s,
- **a** what is the wavelength of the waves,
- **b** how long does the transmission take to travel 60 km?

Checklist

After studying this chapter you should be able to

- recall the types of electromagnetic radiation,
- recall that all electromagnetic waves have the same speed in space and are progressive transverse waves,
- recall that the colour of light depends on its frequency, that red light has a lower frequency (but longer wavelength) than blue light and that all colours travel at the same speed in air,
- use the term monochromatic,
- distinguish between **infrared radiation, ultraviolet radiation, radio waves** and **X-rays** in terms of their wavelengths, properties and uses,
- be aware of the harmful effects of different types of electromagnetic radiation and of how exposure to them can be reduced.

9 Sound waves

Origin and transmission of sound	*Reflection and echoes*	*Ultrasonics*
Longitudinal waves	*Speed of sound*	*Seismic waves*
Limits of audibility	*Musical notes*	*Practical work* Speed of sound in air.

Origin and transmission of sound

Sources of sound all have some part which **vibrates**. A guitar has strings, Figure 9.1, a drum has a stretched skin and the human voice has vocal cords. The sound travels through the air to our ears and we hear it. That the air is necessary may be shown by pumping out a glass jar containing a ringing electric bell, Figure 9.2; the sound disappears though the striker can still be seen hitting the gong. Evidently sound cannot travel in a vacuum as light can. Other materials, including solids and liquids, transmit sound.

Sound also gives interference and diffraction effects. Because of this and its other properties, we believe it is a form of energy (as the damage from supersonic booms shows) which travels as a progressive wave, but of a type called **longitudinal**.

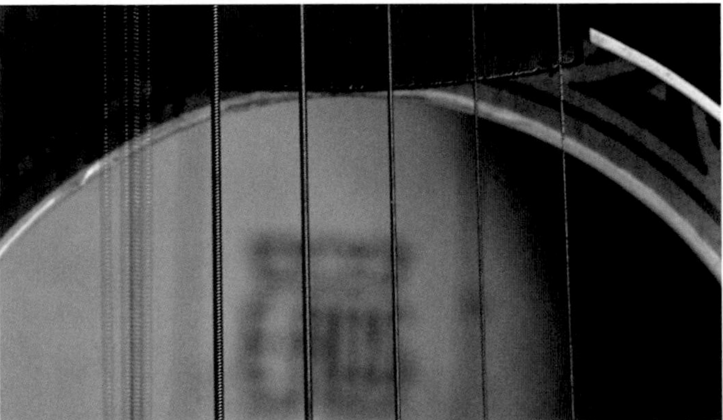

Figure 9.1 A guitar string vibrating. The sound waves produced are amplified when they pass through the circular hole into the guitar's sound box

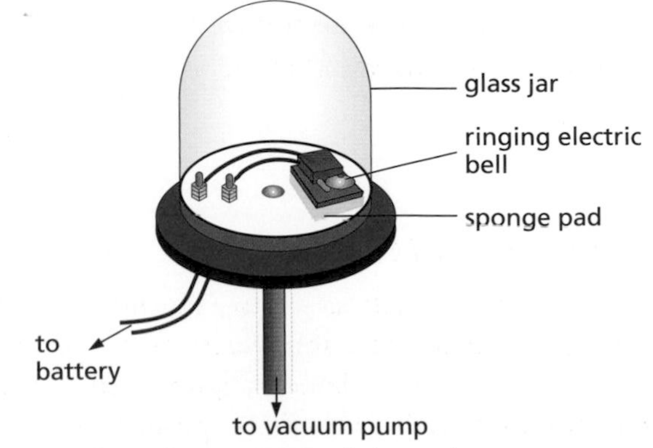

Figure 9.2 Sound cannot travel through a vacuum

Longitudinal waves

a) Waves on a spring

In a progressive longitudinal wave the particles of the transmitting medium vibrate to and fro along the same line as that in which the wave is travelling and not at right angles to it as in a transverse wave. A longitudinal wave can be sent along a spring, stretched out on the bench and fixed at one end, if the free end is repeatedly pushed and pulled sharply. Compressions C (where the coils are closer together) and rarefactions R (where the coils are farther apart), Figure 9.3, travel along the spring.

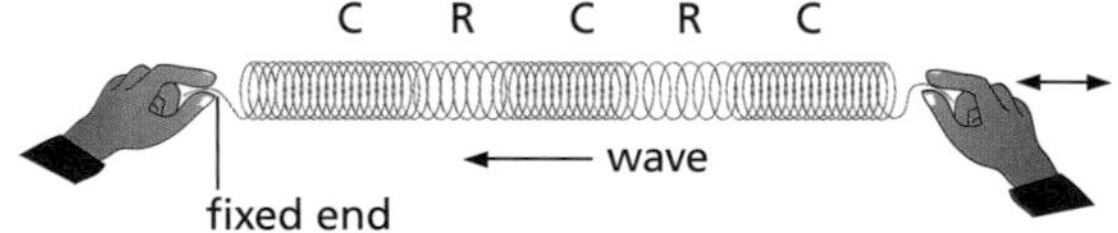

Figure 9.3 A longitudinal wave

b) Sound waves

A sound wave, produced for example by a loudspeaker, consists of a train of compressions ('squashes') and rarefactions ('stretches') in the air, Figure 9.4.

The speaker has a cone which is made to vibrate in and out by an electric current. When the cone moves out, the air in front is compressed; when it moves in, the air is rarefied (goes 'thinner'). The wave progresses through the air but the air as a whole does not move. The air particles (molecules) vibrate backwards and forwards a little as the wave passes. When the wave enters your ear the compressions and rarefactions cause small, rapid pressure changes on the eardrum and you experience the sensation of sound.

The number of compressions produced per second is the frequency f of the sound wave (and equals the frequency of the vibrating loudspeaker cone); the distance between successive compressions is the wavelength λ. As for transverse waves the speed is $v = f\lambda$.

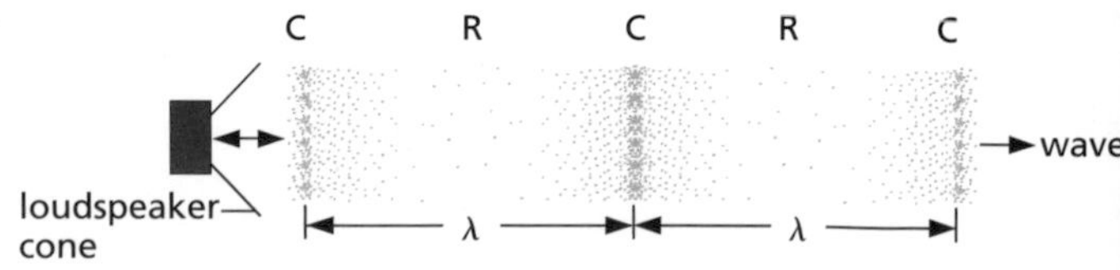

Figure 9.4 Sound travels as a longitudinal wave

Reflection and echoes

Sound waves are reflected well from hard, flat surfaces such as walls or cliffs and obey the same laws of reflection as light. The reflected sound forms an **echo**.

If the reflecting surface is nearer than 15 m from the source of sound, the echo joins up with the original sound which then seems to be prolonged. This is called **reverberation**. Some is desirable in a concert hall to stop it sounding 'dead', but too much causes 'confusion'. Modern concert halls are designed for the optimum amount of reverberation. Seats and some wall surfaces are covered with sound-absorbing material.

Speed of sound

The speed of sound depends on the material through which it is passing. It is greater in solids than in liquids or gases because the molecules in a solid are closer together than in a liquid or a gas. Some values are given in Table 9.1.

Table 9.1 Speed of sound in different materials

Material	air (0 °C)	water	concrete	steel
Speed/ metres/second	330	1400	5000	6000

In air the speed **increases with temperature** and at high altitudes, where the temperature is lower, it is less than at sea-level. Changes of atmospheric pressure do not affect it.

An estimate of the speed of sound can be made directly if you stand about 100 metres from a high wall or building and clap your hands. Echoes are produced. When the clapping rate is such that each clap coincides with the echo of the previous one, the sound has travelled to the wall and back in the time between two claps, i.e. one interval. By timing 30 intervals with a stopwatch, the time t for one interval can be found. Also, knowing the distance d to the wall, a rough value is obtained from

$$\text{speed of sound in air} = \frac{2d}{t}$$

The speed of sound in air can be found directly by measuring the time t taken for a sound to travel past two microphones separated by a distance d:

$$\text{speed of sound in air} = \frac{\text{distance travelled by the sound}}{\text{time taken}}$$

$$= \frac{d}{t}$$

Limits of audibility

Humans hear only sounds with frequencies from about 20 Hz to 20 000 Hz. These are the **limits of audibility**; the upper limit decreases with age.

Practical work

Speed of sound in air

Set two microphones about a metre apart, and attach one to the 'start' terminal and the other to the 'stop' terminal of a digital timer, as shown in Figure 9.5. The timer should have millisecond accuracy. Measure and record the distance d between the centres of the microphones with a metre ruler. With the small hammer and metal plate to one side of the 'start' microphone, produce a sharp sound. When the sound reaches the 'start' microphone, the timer should start; when it reaches the 'stop' microphone, the timer should stop. The time displayed is then the time taken for the sound to travel the distance d. Record the time and then reset the timer; repeat the experiment a few times and work out an *average* value for t.

$$\text{Calculate the speed of sound in air} = \frac{d}{t}$$

How does your value compare with that given in Table 9.1?

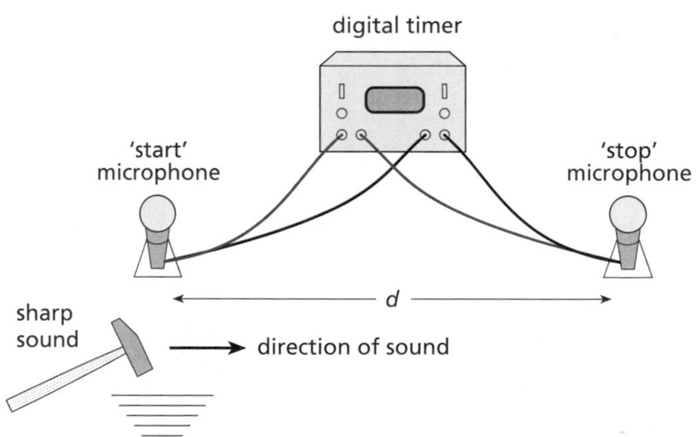

Figure 9.5 Measuring the speed of sound

Musical notes

Irregular vibrations such as those of motor engines cause **noise**; regular vibrations such as occur in the instruments of a brass band, Figure 9.6, produce **musical notes** which have three properties – pitch, loudness and quality.

Figure 9.6 Musical instruments produce regular sound vibrations

a) Pitch

The pitch of a note depends on the frequency of the sound wave reaching the ear, i.e. on the frequency of the source of sound. A high-pitched note has a high frequency and a short wavelength. The frequency of middle C is 256 vibrations per second or 256 Hz and that of upper C is 512 Hz. Notes are an **octave** apart if the frequency of one is twice that of the other. Pitch is like colour in light; both depend on the frequency.

Notes of known frequency can be produced in the laboratory by a signal generator supplying alternating electric current (a.c.) to a loudspeaker. The cone of the speaker vibrates at the frequency of the a.c. which can be varied and read off a scale on the generator. A set of tuning forks with frequencies marked on them can also be used. A tuning fork, Figure 9.7, has two steel prongs which vibrate when struck; the prongs move in and out *together*, generating compressions and rarefactions.

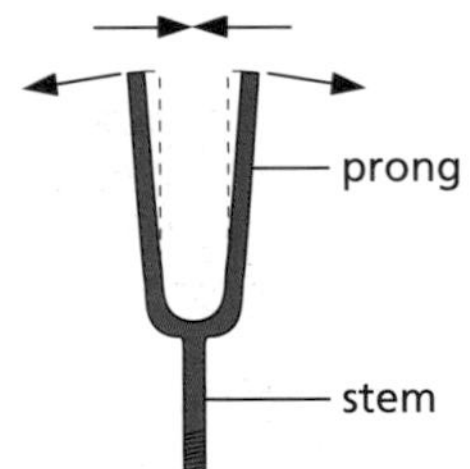

Figure 9.7 A tuning fork

b) Loudness

A note becomes louder when more sound energy enters our ears per second than before. This will happen when the source is vibrating with a larger amplitude. If a violin string is bowed more strongly, its amplitude of vibration increases as does that of the resulting sound wave and the note heard is louder because morc energy has been used to produce it.

c) Quality

The same note on different instruments sounds different; we say the notes differ in **quality** or **timbre**. The difference arises because no instrument (except a tuning fork and a signal generator) emits a 'pure' note, i.e. of one frequency. Notes consist of a main or **fundamental** frequency mixed with others, called **overtones**, which are usually weaker and have frequencies that are exact multiples of the fundamental. The number and strength of the overtones decides the quality of a note. A violin has more and stronger higher overtones than a piano. Overtones of 256 Hz (middle C) are 512 Hz, 768 Hz and so on.

The **waveform** of a note played near a microphone connected to a cathode ray oscilloscope (CRO, see Chapter 47) can be displayed on the CRO screen. Those for the *same* note on three instruments are given in Figure 9.8. Their different shapes show that while they have the same fundamental frequency, their quality differs. The 'pure' note of a tuning fork has a **sine** waveform and is the simplest kind of sound wave.

Note Although the waveform on the CRO screen is transverse it represents a longitudinal sound wave.

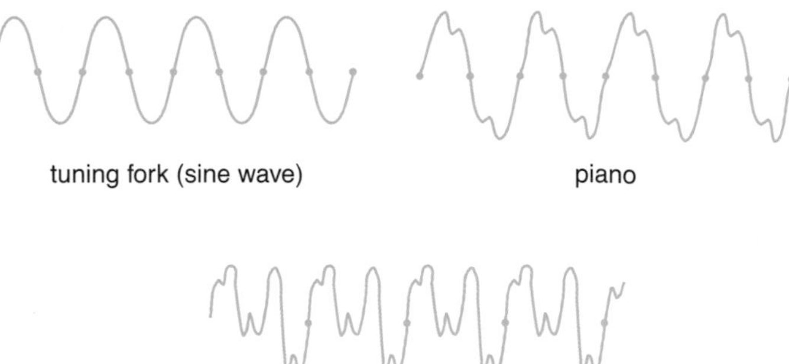

Figure 9.8 Notes of the same frequency (pitch) but different quality

F *Ultrasonics*

Sound waves with frequencies above 20 kHz are called **ultrasonic** waves; their frequency is too high to be detected by the human ear but they can be detected electronically and displayed on a CRO.

a) Quartz crystal oscillators

Ultrasonic waves are produced by a quartz crystal which is made to vibrate electrically at the required frequency; they are emitted in a narrow beam in the direction in which the crystal oscillates. An ultrasonic receiver also consists of a quartz crystal but it works in reverse, i.e. when it is set into vibration by ultrasonic waves it generates an electrical signal which is then amplified. The same quartz crystal can act as both a transmitter and a receiver.

b) Ultrasonic echo techniques

Ultrasonic waves are partially or totally reflected from surfaces at which the density of the medium changes; this property is exploited in techniques such as the non-destructive testing of materials, sonar and medical ultrasound imaging. A bat emitting ultrasonic waves can judge the distance away of an object from the time taken by the reflected wave or 'echo' to return.

Ships with **sonar** can determine the depth of a shoal of fish or the sea-bed, Figure 9.9, in the same way; motion sensors (Chapter 19) also work on this principle.

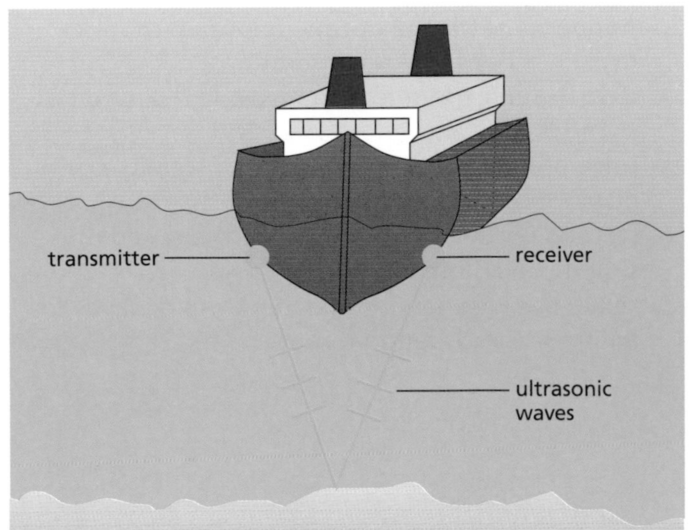

Figure 9.9 A ship using sonar

In **medical ultrasound imaging**, used in pre-natal clinics to monitor the health and sometimes to determine the sex of an unborn baby, an ultrasonic transmitter/receiver is scanned over the mother's abdomen and a detailed image of the fetus is built up, Figure 9.10. Reflection of the ultrasonic pulses occurs from boundaries of soft tissue, in addition to bone, so images of internal organs not seen by using X-rays can be obtained. Less detail of bone structure is seen than with X-rays, as the wavelength of ultrasonic waves is larger, typically about 1 mm, but ultrasound has no harmful effects on human tissue.

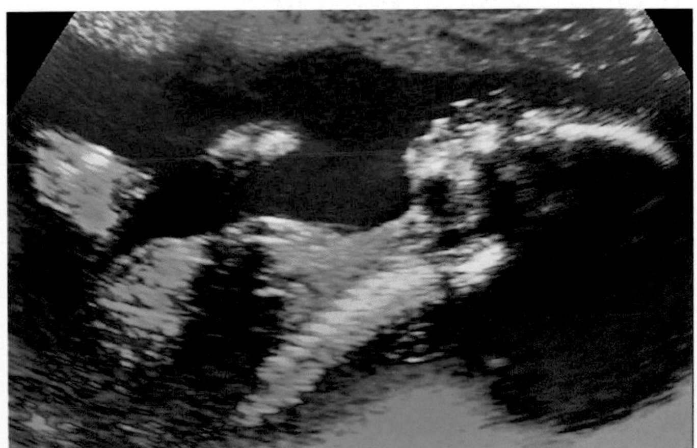

Figure 9.10 Checking the development of a fetus using ultrasound imaging

c) Other uses

Ultrasound can also be used in ultrasonic drills to cut holes of any shape or size in hard materials such as glass and steel. Jewellery, or more mundane objects such as street lamp covers, can be cleaned by immersion in a tank of solvent which has an ultrasonic vibrator in the base.

Seismic waves

Earthquakes produce both longitudinal waves (P-waves) and transverse waves (S-waves) that are known as **seismic** waves. These travel through the Earth at speeds of up to 13 000 m/s. When seismic waves pass under buildings, severe structural damage may occur. If the earthquake occurs under the sea, the seismic energy can be transmitted to the water and produce **tsunami** waves that may travel for very large distances across the ocean. As a tsunami wave approaches shallow coastal waters, it slows down (see Chapter 7) and its amplitude increases, which can lead to massive coastal destruction. This happened in Sri Lanka, see Figure 9.11, and Thailand after the great 2004 Sumatra–Andaman earthquake. The time of arrival of a tsunami wave can be predicted if its speed of travel and the distance from the epicentre of the earthquake are known; it took about 2 hours for tsunami waves to cross the ocean to Sri Lanka from Indonesia. A similar time was needed for the tsunami waves to travel less far to Thailand. This was because the route was through shallower water and the waves travelled more slowly. If an early-warning system had been in place, many lives could have been saved.

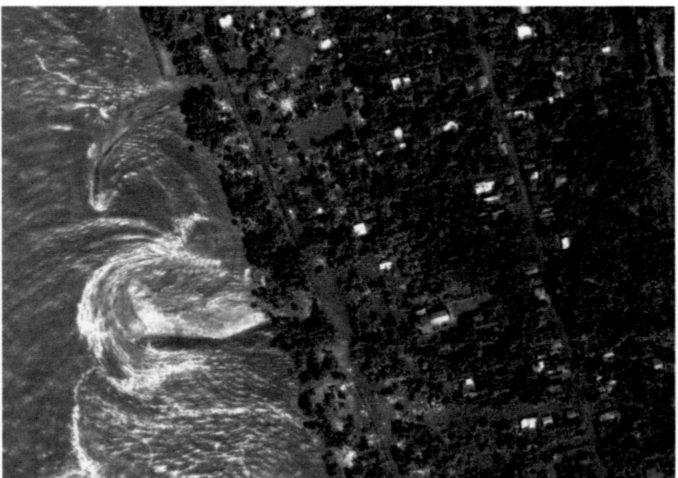

Figure 9.11 This satellite image shows the tsunami that hit the south-western coast of Sri Lanka on 26 December 2004 as it pulled back out to sea having caused utter devastation in coastal areas

Questions

1 If 5 seconds elapse between a lightning flash and the clap of thunder how far away is the storm? Speed of sound = 330 m/s.

2 a A girl stands 160 m away from a high wall and claps her hands at a steady rate so that each clap coincides with the echo of the one before. If she makes 60 claps in 1 minute, what value does this give for the speed of sound?
b If she moves 40 m closer to the wall she finds the clapping rate has to be 80 per minute. What value do these measurements give for the speed of sound?
c If she moves again and finds the clapping rate becomes 30 per minute, how far is she from the wall if the speed of sound is the value you found in **a**?

3 a What properties of sound suggest it is a wave motion?
b How does a progressive transverse wave differ from a longitudinal one? Which type is sound?

4 a A sound wave in air is made up of compressions and rarefactions.
(i) State what is meant by a *compression*.
(ii) State what is meant by a *rarefaction*.
b The distance between two consecutive rarefactions in a sound wave is 2.5 m. The speed of sound in air is 330 m/s. Calculate the frequency of this sound wave.
c A person makes a loud sound and hears the echo of this sound 1.2 s later. Calculate how far the person is from the object causing the echo. Assume that the speed of sound is 330 m/s.
(*UCLES IGCSE Physics Extended, May/June 2000*)

5 a Draw the waveform of
(i) a loud, low-pitched note, and
(ii) a soft, high-pitched note.
b If the speed of sound is 340 m/s what is the wavelength of a note of frequency
(i) 340 Hz,
(ii) 170 Hz?

Checklist

After studying this chapter you should be able to

- recall that sound is produced by vibrations,
- describe an experiment to show that sound is not transmitted through a vacuum,
- describe how sound travels in a medium as progressive longitudinal waves,
- recall the limits of audibility (i.e. the range of frequencies) for the normal human ear,
- explain echoes and reverberation,
- describe an experiment to measure the speed of sound in air,
- solve problems using the speed of sound, e.g. distance of thundercloud,
- state the order of magnitude of the speed of sound in air, liquid and solids,
- relate the loudness and pitch of sound waves to amplitude and frequency.

Waves and sound
Additional questions

Mechanical waves

1 The lines in the diagram are the crests of straight ripples produced in a ripple tank by a wave generator.

a What is the wavelength of the ripples if there are 6 complete waves in a distance of 30 cm?

b If the wave generator makes 4 vibrations per second what is the frequency of the ripples?

c What is the equation connecting the wavelength (λ), the frequency (f) and the speed (v) of the ripples?

d Calculate the speed of the ripples.

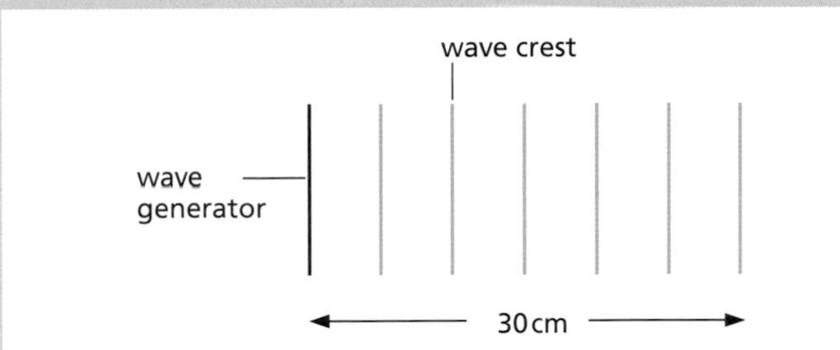

2 The diagrams **a** and **b** show ripples approaching metal plates with gaps in them. Gap **a** is narrow compared with the wavelength and gap **b** is wide compared with the wavelength.

Copy the diagrams and draw the shapes of the ripples after they have passed through the gaps.

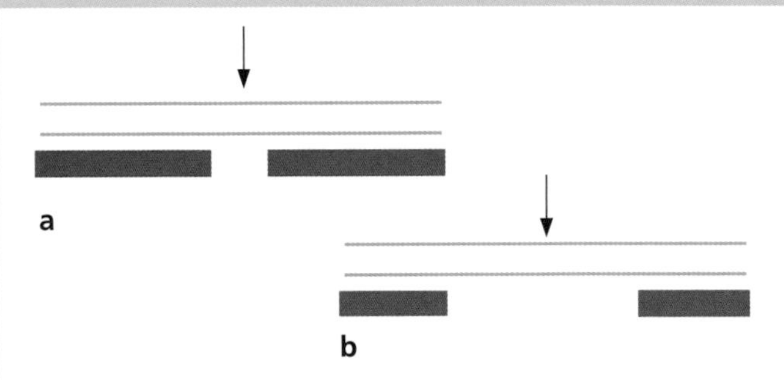

3 The blue curve shows the position of a water wave travelling to the right. X, Y and Z are corks floating on the surface and the black dotted line is the undisturbed water surface. As the wave moves forward, state whether (i) X, (ii) Y, (iii) Z, moves up or down or stays where it is.

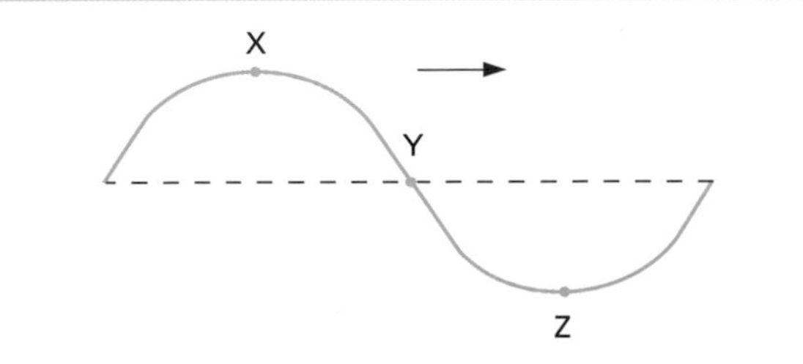

Electromagnetic radiation

4 In the diagram light waves are incident on an air–glass boundary. Some are reflected and some are refracted in the glass. One of the following is the same for the incident wave and the refracted wave. Which?

A speed **B** wavelength **C** direction
D brightness **E** frequency

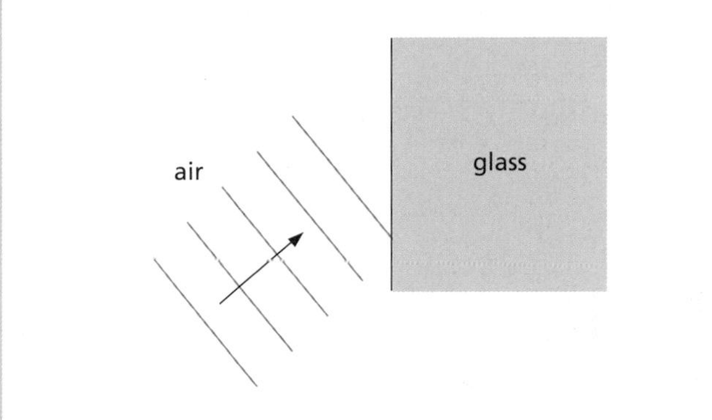

5 The diagram represents a wave.

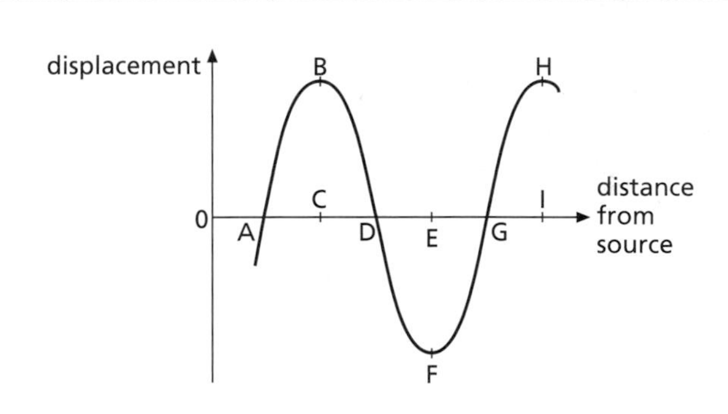

a Making use of the letters on the diagram, state which distances you would measure to find
(i) the wavelength of the wave: measure between .... and ....
(ii) the amplitude of the wave: measure between .... and ....

b What is meant by the *frequency* of the wave?

c One complete wave takes 0.2 s to generate. Calculate the frequency of the wave.

(*UCLES IGCSE Physics Core, May 2001*)

Continued

6 The diagram shows a ray of light, PQRS, passing along a simple optical fibre.

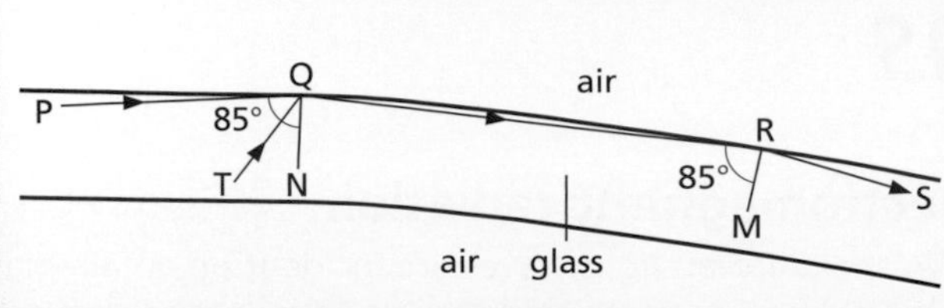

a Calculate the angle between the ray PQ and the ray RS.
b Explain why the ray PQ does not leave the fibre at Q.
c Another ray TQ also strikes the surface at Q. The refractive index of the glass is 1.50.
(i) Calculate the critical angle for this glass.
(ii) Explain why the ray TQ leaves the fibre.
d The light waves travelling towards Q are monochromatic and have a frequency of 4×10^{14} Hz and a wavelength of 5×10^{-7} m.
(i) What is meant by *monochromatic*?
(ii) Calculate the speed of these waves in the glass.
(iii) Waves travelling along TQ pass into the air. The refractive index of the glass is 1.50. Write down an expression from which the speed of the light waves in air could be found.

(*UCLES IGCSE Physics Extended, Nov 1999*)

Sound waves

7 An echo-sounder in a trawler receives an echo from a shoal of fish 0.4 s after it was sent. If the speed of sound in water is 1500 m/s, how deep is the shoal?

A 150 m B 300 m C 600 m D 7500 m
E 10 000 m

8 The figure below shows an unlabelled diagram which a teacher draws to represent a sound wave in air.

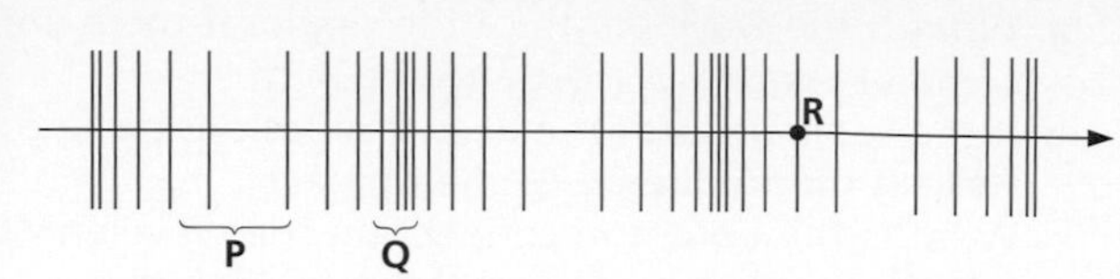

a What label should be put on the line with the arrow?
b (i) What does the uneven spacing of the lines show?
(ii) What is being shown at **P**?
(iii) What is being shown at **Q**?
c Describe the motion of an air particle at **R**.
d From the diagram, measure the wavelength of the sound wave.

(*UCLES IGCSE Physics Extended, May 2001*)

Matter and measurements

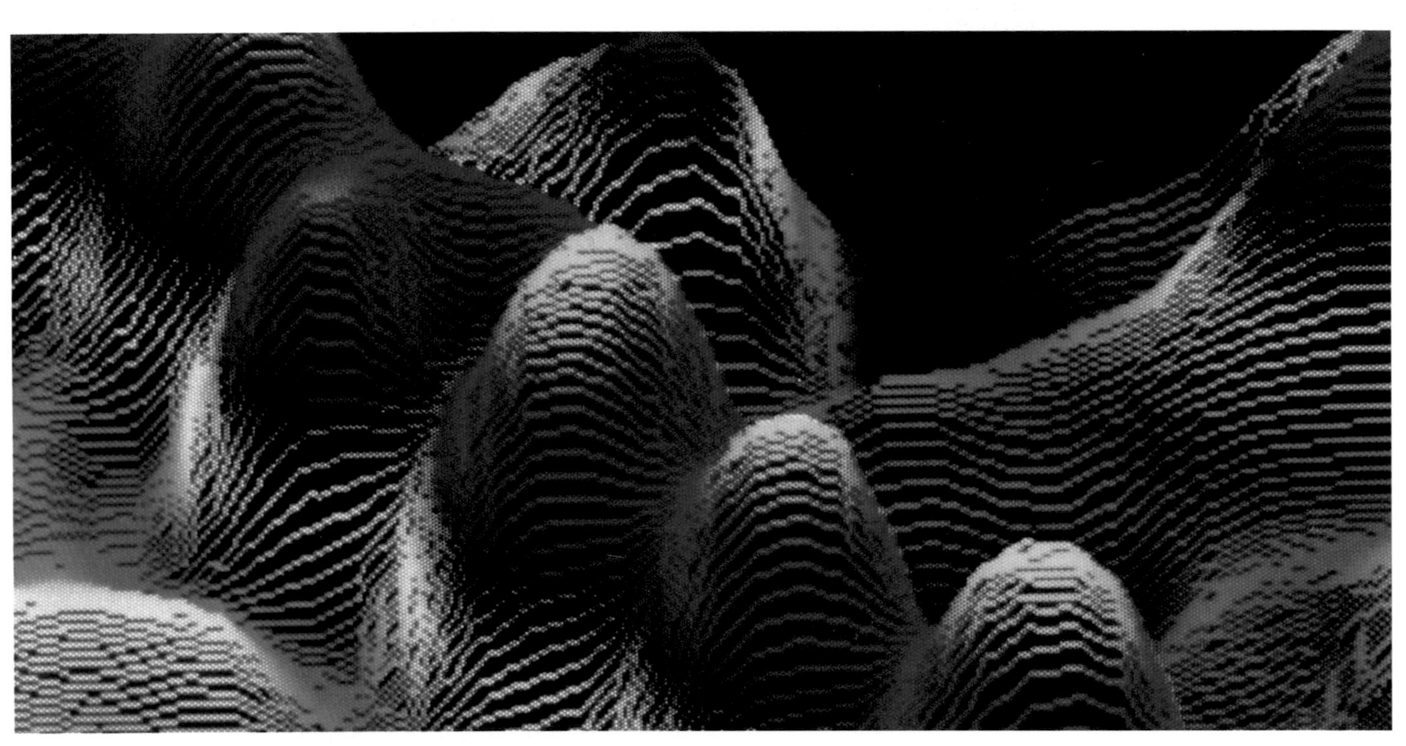

10 Measurements

Units and basic quantities
Powers of ten shorthand
Length
Significant figures
Area
Volume
Mass
Time
Systematic errors
Vernier scales and micrometers
Practical work
Period of a simple pendulum.

Units and basic quantities

Before a measurement can be made, a standard or **unit** must be chosen. The size of the quantity to be measured is then found with an instrument having a scale marked in the unit.

Three basic quantities we measure in physics are **length**, **mass** and **time**. Units for other quantities are based on them. The SI (Système International d'Unités) system is a set of metric units now used in many countries. It is a decimal system in which units are divided or multiplied by 10 to give smaller or larger units.

Figure 10.1 Measuring instruments on the flight deck of the space shuttle Discovery provide the crew with information about the performance of the shuttle, here during re-entry

Powers of ten shorthand

This is a neat way of writing numbers especially if they are large or small. It works like this:

$$
\begin{aligned}
4000 &= 4 \times 10 \times 10 \times 10 &&= 4 \times 10^3 \\
400 &= 4 \times 10 \times 10 &&= 4 \times 10^2 \\
40 &= 4 \times 10 &&= 4 \times 10^1 \\
4 &= 4 \times 1 &&= 4 \times 10^0 \\
0.4 &= 4/10 \quad = 4/10^1 &&= 4 \times 10^{-1} \\
0.04 &= 4/100 \quad = 4/10^2 &&= 4 \times 10^{-2} \\
0.004 &= 4/1000 \quad = 4/10^3 &&= 4 \times 10^{-3}
\end{aligned}
$$

The small figures 1, 2, 3, etc., are called **powers of ten**. The power gives the number of times the number has to be multiplied by 10 if the power is greater than 0 or divided by 10 if the power is less than 0. Note that 1 is written as 10^0.

This way of writing numbers is called **standard notation**.

Length

The unit of length is the **metre** (m) and is the distance travelled by light in a vacuum during a specific time interval. At one time it was the distance between two marks on a certain metal bar. Submultiples are:

$$
\begin{aligned}
1 \text{ centimetre (cm)} &= 10^{-2}\text{ m} \\
1 \text{ millimetre (mm)} &= 10^{-3}\text{ m} \\
1 \text{ micrometre } (\mu\text{m}) &= 10^{-6}\text{ m} \\
1 \text{ nanometre (nm)} &= 10^{-9}\text{ m}
\end{aligned}
$$

A multiple for large distances is

$$1 \text{ kilometre (km)} = 10^3 \text{ m } (\tfrac{5}{8} \text{ mile approx.})$$

Many length measurements are made with rulers; the correct way to read one is shown in Figure 10.2. The reading is 76 mm or 7.6 cm. Your eye must be right over the mark on the scale or the thickness of the ruler causes parallax errors.

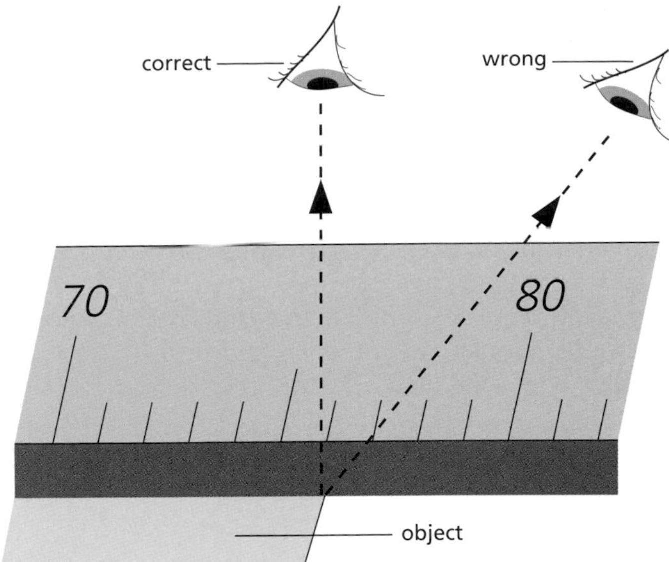

Figure 10.2 The correct way to measure with a ruler

Significant figures

Every measurement of a quantity is an attempt to find its true value and is subject to errors arising from limitations of the apparatus and the experimenter. The number of figures, called **significant** figures, given for a measurement indicates how accurate we think it is and more figures should not be given than is justified.

For example, a value of 4.5 for a measurement has two significant figures; 0.0385 has three significant figures, 3 being the most significant and 5 the least, i.e. it is the one we are least sure about since it might be 4 or it might be 6. Perhaps it had to be estimated by the experimenter because the reading was between two marks on a scale.

When doing a calculation your answer should have the same number of significant figures as the measurements used in the calculation. For example, if your calculator gave an answer of 3.4185062, this would be written as 3.4 if the measurements had two significant figures. It would be written as 3.42 for three significant figures. Note that in deciding the least significant figure you look at the next figure. If it is less than 5 you leave the least significant figure as it is (hence 3.41 becomes 3.4) but if it equals or is greater than 5 you increase the least significant figure by 1 (hence 3.418 becomes 3.42).

If a number is expressed in standard notation, the number of significant figures is the number of digits before the power of ten. For example, 2.73×10^3 has three significant figures.

Area

The area of the square in Figure 10.3a with sides 1 cm long is 1 square centimetre (1 cm^2). In Figure 10.3b the rectangle measures 4 cm by 3 cm and has an area of $4 \times 3 = 12$ cm^2 since it has the same area as twelve squares each of area 1 cm^2. The **area of a square** or **rectangle** is given by

$$\text{area} = \text{length} \times \text{breadth}$$

The SI unit of area is the square metre (m^2) which is the area of a square with sides 1 m long. Note that

$$1\text{ cm}^2 = \frac{1}{100}\text{ m} \times \frac{1}{100}\text{ m} = \frac{1}{10\,000}\text{ m}^2 = 10^{-4}\text{ m}^2$$

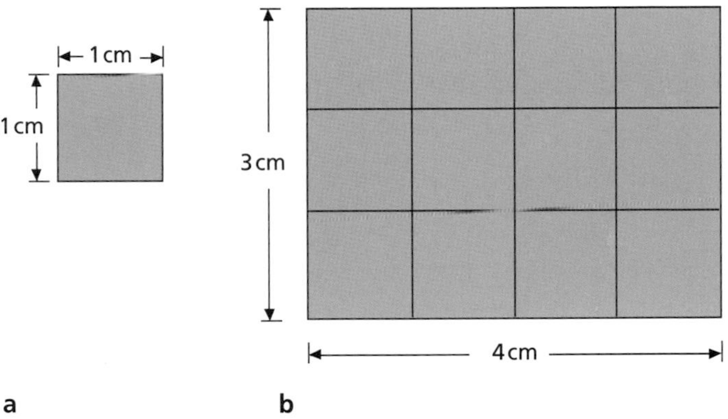

Figure 10.3

Sometimes we need to know the **area of a triangle** (Chapter 20). It is given by

$$\text{area of triangle} = \tfrac{1}{2} \times \text{base} \times \text{height}$$

For example in Figure 10.4

$$\text{area } \triangle ABC = \tfrac{1}{2} \times AB \times AC$$
$$= \tfrac{1}{2} \times 4\text{ cm} \times 6\text{ cm} = 12\text{ cm}^2$$

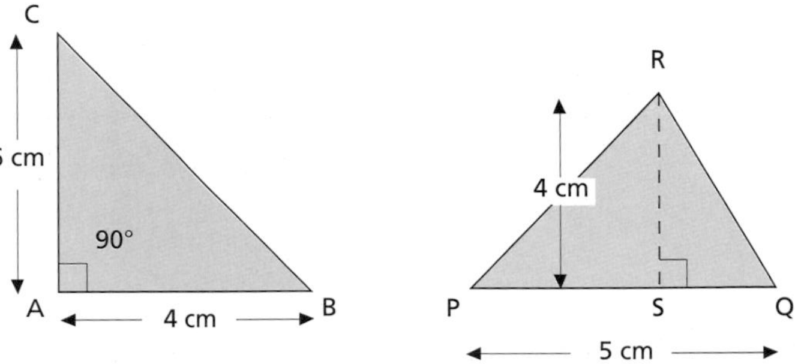

Figure 10.4

and

$$\text{area } \triangle PQR = \tfrac{1}{2} \times PQ \times SR$$
$$= \tfrac{1}{2} \times 5\text{ cm} \times 4\text{ cm} = 10\text{ cm}^2$$

The **area of a circle** of radius r is πr^2 where $\pi = 22/7$ or 3.14; its **circumference** is $2\pi r$.

Volume

Volume is the amount of space occupied. The unit of volume is the **cubic metre** (m^3) but as this is rather large, for most purposes the **cubic centimetre** (cm^3) is used. The volume of a cube with 1 cm edges is 1 cm^3. Note that

$$1\ cm^3 = \frac{1}{100}\ m \times \frac{1}{100}\ m \times \frac{1}{100}\ m$$

$$= \frac{1}{1\,000\,000}\ m^3 = 10^{-6}\ m^3$$

For a regularly shaped object such as a rectangular block, Figure 10.5 shows that

$$\text{volume} = \text{length} \times \text{breadth} \times \text{height}$$

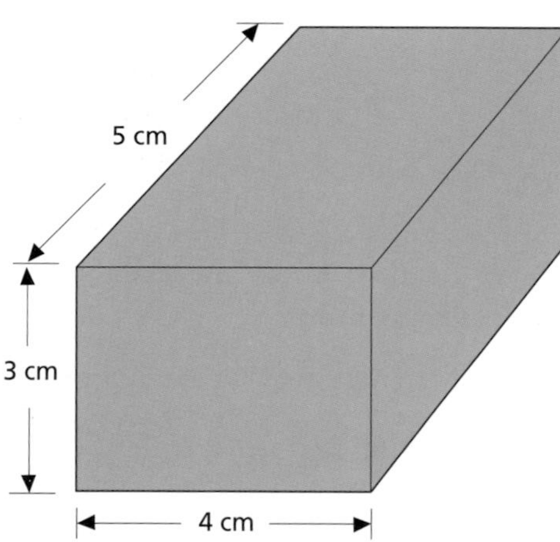

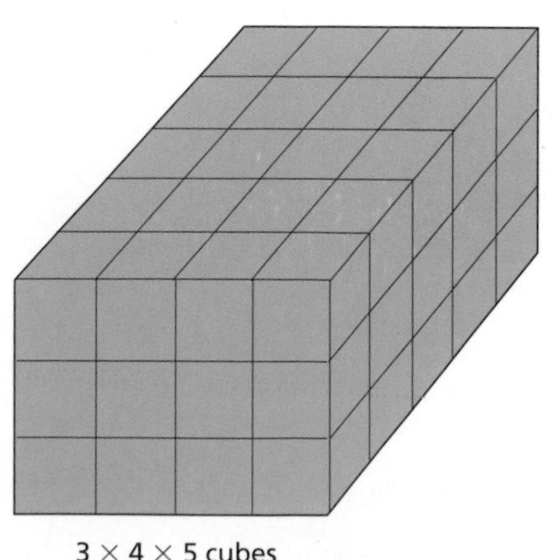

3 × 4 × 5 cubes

Figure 10.5

The **volume of a sphere** of radius r is $\frac{4}{3}\pi r^3$ and that of a **cylinder** of radius r and height h is $\pi r^2 h$.

The **volume of a liquid** may be obtained by pouring it into a measuring cylinder, Figure 10.6a. A known volume can be run off accurately from a burette, Figure 10.6b. When making a reading both vessels must be upright and your eye must be level with the bottom of the curved liquid surface, i.e. the meniscus. The meniscus formed by mercury is curved oppositely to that of other liquids and the top is read.

Liquid volumes are also expressed in litres (l); 1 litre = 1000 cm^3. One millilitre (1 ml) = 1 cm^3.

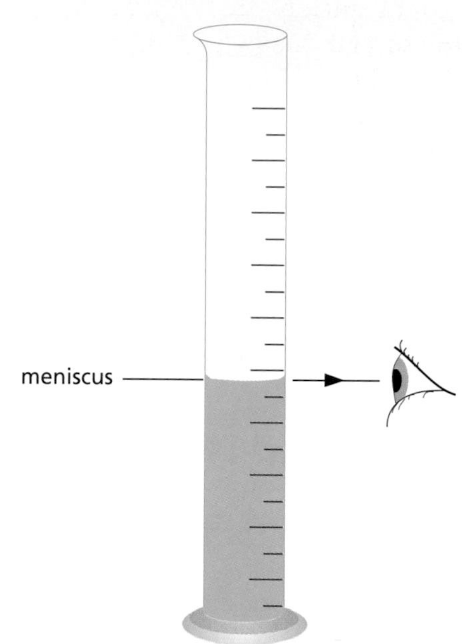

a A measuring cylinder

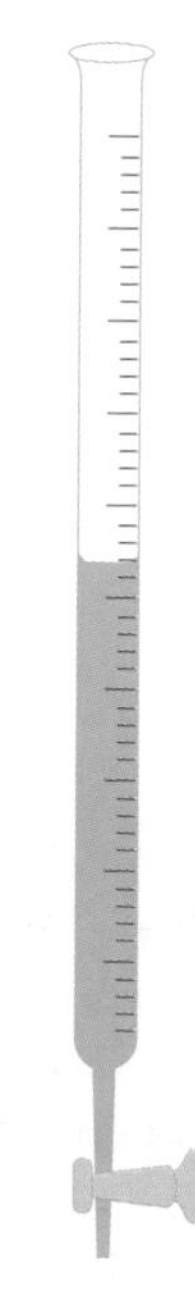

b A burette

Figure 10.6

Mass

The mass of an object is the measure of the amount of matter in it. The unit of mass is the **kilogram** (kg) and is the mass of a piece of platinum–iridium alloy at the Office of Weights and Measures in Paris. The gram (g) is one-thousandth of a kilogram.

$$1 \text{ g} = 1/1000 \text{ kg} = 10^{-3} \text{ kg} = 0.001 \text{ kg}$$

The term **weight** is often used when mass is really meant. In science the two ideas are distinct and have different units as we shall see later. The confusion is not helped by the fact that mass is found on a balance by a process we unfortunately call 'weighing'!

There are several kinds of balance. In the **beam balance** the unknown mass in one pan is balanced against known masses in the other pan. In the **lever balance** a system of levers acts against the mass when it is placed in the pan. A direct reading is obtained from the position on a scale of a pointer joined to the lever system. A digital **top-pan balance** is shown in Figure 10.7.

Figure 10.7 A digital top-pan balance

Time

The unit of time is the **second** (s) which used to be based on the length of a day, this being the time for the Earth to revolve once on its axis. However, days are not all of exactly the same duration and the second is now defined as the time interval for a certain number of energy changes to occur in the caesium atom.

Time-measuring devices rely on some kind of constantly repeating oscillations. In traditional clocks and watches a small wheel (the balance wheel) oscillates to and fro; in digital clocks and watches the oscillations are produced by a tiny quartz crystal. A swinging pendulum controls a pendulum clock.

To measure an interval of time in an experiment, first choose a timer that is accurate enough for the task. A stopwatch is adequate for finding the period in seconds of a pendulum, see Figure 10.8, but to measure the speed of sound (Chapter 9), a clock that can time in milliseconds is needed. To measure very short time intervals, a digital clock that can be triggered to start and stop by an electronic signal from a microphone, photogate or mechanical switch is useful. Tickertape timers or dataloggers are often used to record short time intervals in motion experiments (Chapter 19).

Accuracy can be improved by measuring longer time intervals. Several oscillations (rather than just one) are timed to find the period of a pendulum. 'Tenticks' (rather than 'ticks') are used in tickertape timers.

Practical work

Period of a simple pendulum

In this investigation you have to make time measurements using a stopwatch or clock.

Attach a small metal ball (called a bob) to a piece of string, and suspend it as shown in Figure 10.8. Pull the bob a small distance to one side, and then release it so that it oscillates to and fro through a small angle.

Find the time for the bob to make several complete oscillations; one oscillation is from A to O to B to O to A, Figure 10.8. Repeat the timing a few times for the same number of oscillations and work out the average. The time for one oscillation is the **period** T. What is it for your system? The **frequency** f of the oscillations is the number of complete oscillations per second and equals $1/T$. Calculate f.

How does the amplitude of the oscillations, Figure 10.8, change with time?

Investigate the effect on T of (i) a longer string, (ii) a heavier bob. A motion sensor connected to a datalogger and computer (Chapter 19) could be used instead of a stopwatch for these investigations.

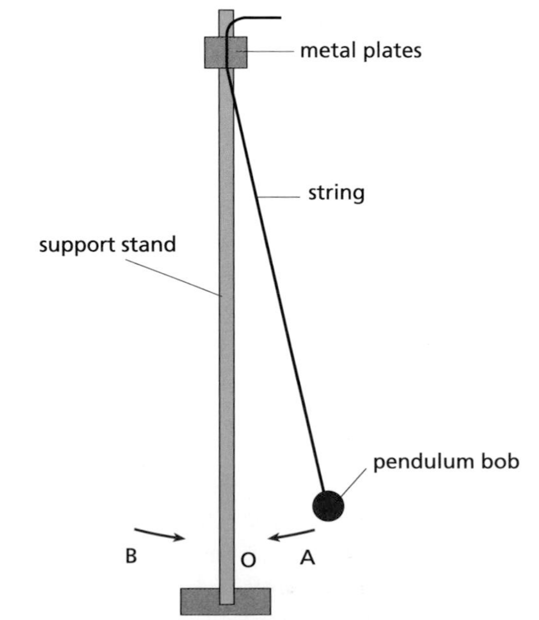

Figure 10.8

Systematic errors

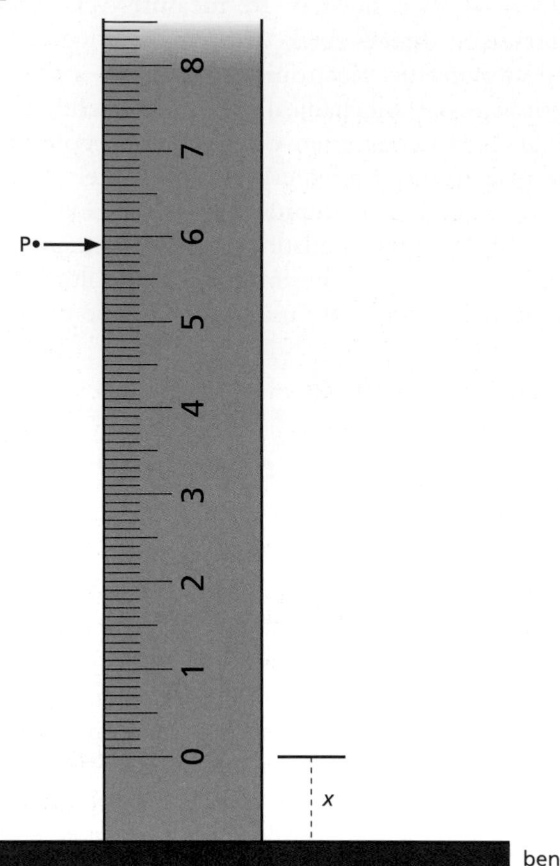

Figure 10.9 Introducing a systematic error into a measurement

Figure 10.9 shows a part of a rule used to measure the height of a point P above the bench. The rule chosen has a space before the zero of the scale. This is shown as the length x. The height of the point P is given by the scale reading added to the value of x. The equation for the height is

$$\text{height} = \text{scale reading} + x$$
$$\text{height} = 5.9 + x$$

By itself the scale reading is not equal to the height. It is too small by the value of x.

This type of error is known as a **systematic error**. The error is introduced by the system. A half-metre rule has the zero at the *end of the rule* and so can be used without introducing a systematic error.

When using a rule to determine a height, the rule must be held so that it is vertical. If the rule is at an angle to the vertical a systematic error is introduced.

Vernier scales and micrometers

Lengths can be measured with a ruler to an accuracy of about 1 mm. Some investigations may need a more accurate measurement of length, which can be achieved by using vernier calipers (Figure 10.10) or micrometer screw gauge.

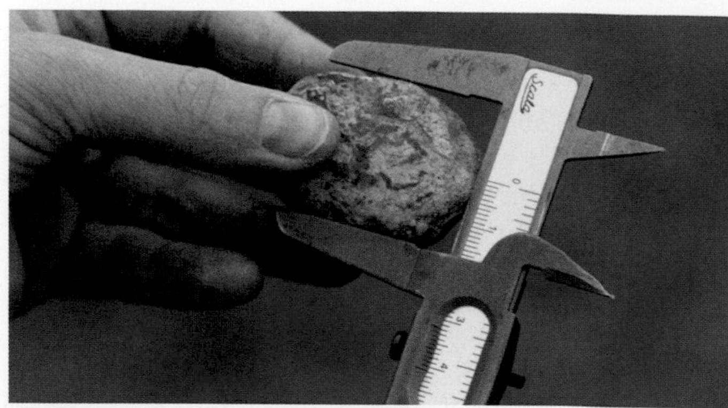

Figure 10.10 Vernier calipers in use

a) Vernier scale

The calipers shown above use a **vernier scale**. The simplest type enables a length to be measured to 0.01 cm. It is a small sliding scale which is 9 mm long but divided into 10 equal divisions, Figure 10.11a.

$$\begin{aligned} 1 \text{ vernier division} &= \frac{9}{10} \text{ mm} \\ &= 0.9 \text{ mm} \\ &= 0.09 \text{ cm} \end{aligned}$$

One end of the length to be measured is made to coincide with the zero of the millimetre scale and the other end with the zero of the vernier scale. The length of the object in Figure 10.11b is between 1.3 cm and 1.4 cm. The reading to the second place of decimals is obtained by finding the vernier mark which is exactly opposite (or nearest to) a mark on the millimetre scale. In this case it is the 6th mark and the length is 1.36 cm, since

$$\begin{aligned} \text{OA} &= \text{OB} - \text{AB} \\ \therefore \quad \text{OA} &= (1.90 \text{ cm}) - (6 \text{ vernier divisions}) \\ &= 1.90 \text{ cm} - 6(0.09) \text{ cm} \\ &= (1.90 - 0.54) \text{ cm} \\ &= 1.36 \text{ cm} \end{aligned}$$

Vernier scales are also used on barometers, travelling microscopes and spectrometers.

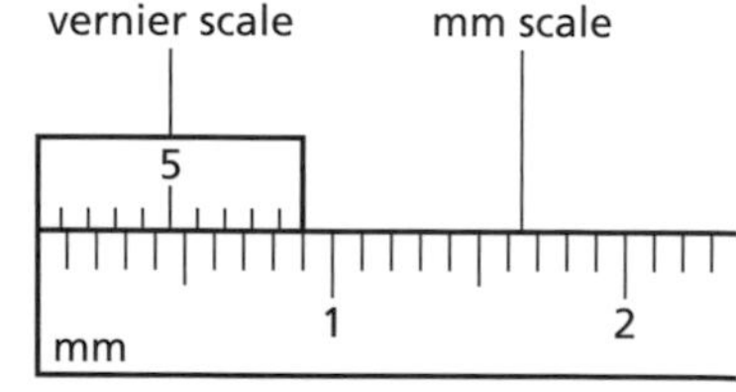

a

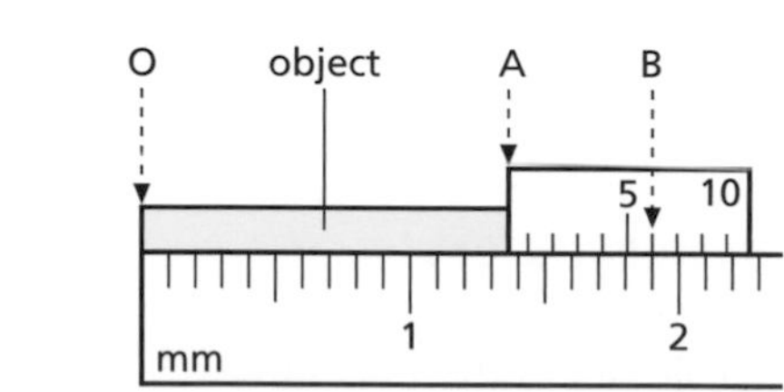

b

Figure 10.11 Vernier scale

b) Micrometer screw gauge

This measures very small objects to 0.001 cm. One revolution of the drum opens the accurately flat, parallel jaws by 1 division on the scale on the shaft of the gauge; this is usually $\frac{1}{2}$ mm, i.e. 0.05 cm. If the drum has a scale of 50 divisions round it, then rotation of the drum by 1 division opens the jaws by 0.05/50 = 0.001 cm, Figure 10.12. A friction clutch ensures that the jaws exert the same force when the object is gripped.

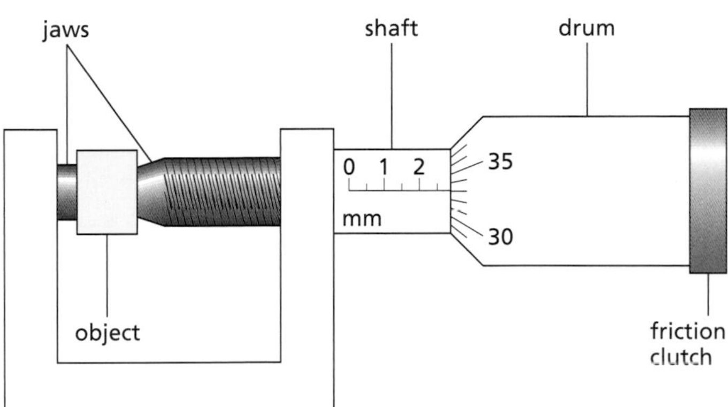

Figure 10.12 Micrometer screw gauge

The object shown in Figure 10.12 has a length of

2.5 mm on the shaft scale +
33 divisions on the drum scale
= 0.25 cm + 33(0.001) cm
= 0.283 cm

Before making a measurement, check to ensure that the reading is zero when the jaws are closed. Otherwise the zero error must be allowed for when the reading is taken.

Questions

1 How many millimetres are there in
a 1 cm, b 4 cm, c 0.5 cm, d 6.7 cm, e 1 m?

2 What are these lengths in metres:
a 300 cm, b 550 cm, c 870 cm, d 43 cm, e 100 mm?

3 a Write the following as powers of ten with one figure before the decimal point:

100 000 3500 428 000 000 504 27 056

b Write out the following in full:

10^3 2×10^6 6.9×10^4 1.34×10^2 10^9

4 a Write these fractions as powers of ten:

1/1000 7/100 000 1/10 000 000 3/60 000

b Express the following decimals as powers of ten with one figure before the decimal point:

0.5 0.084 0.000 36 0.001 04

5 The pages of a book are numbered 1 to 200 and each leaf is 0.10 mm thick. If each cover is 0.20 mm thick, what is the thickness of the book?

6 How many significant figures are there in a length measurement of:
a 2.5 cm, b 5.32 cm, c 7.180 cm, d 0.042 cm?

7 A rectangular block measures 4.1 cm by 2.8 cm by 2.1 cm. Calculate its volume giving your answer to an appropriate number of significant figures.

8 A metal block measures 10 cm × 2 cm × 2 cm. What is its volume? How many blocks each 2 cm × 2 cm × 2 cm have the same total volume?

9 How many blocks of ice cream each 10 cm × 10 cm × 4 cm can be stored in the compartment of a freezer measuring 40 cm × 40 cm × 20 cm?

10 A Perspex container has a 6 cm square base and contains water to a height of 7 cm, Figure 10.13.
a What is the volume of the water?
b A stone is lowered into the water so as to be completely covered and the water rises to a height of 9 cm. What is the volume of the stone?

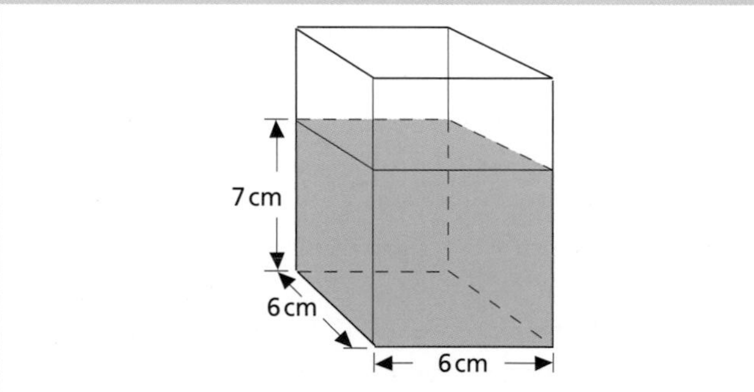

Figure 10.13

Continued

11 What are the readings on the vernier scales in Figures 10.14a and b?

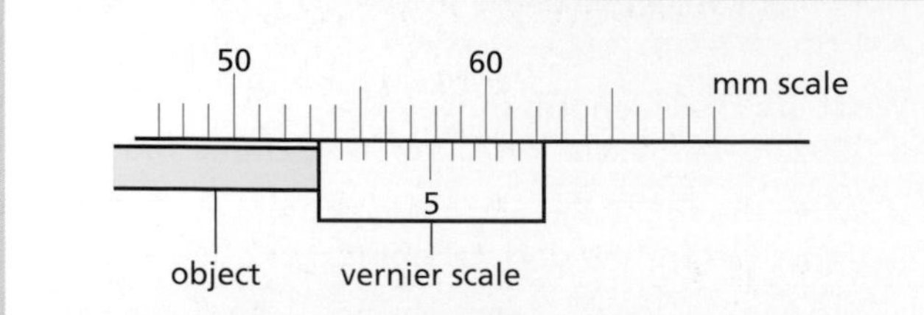

a

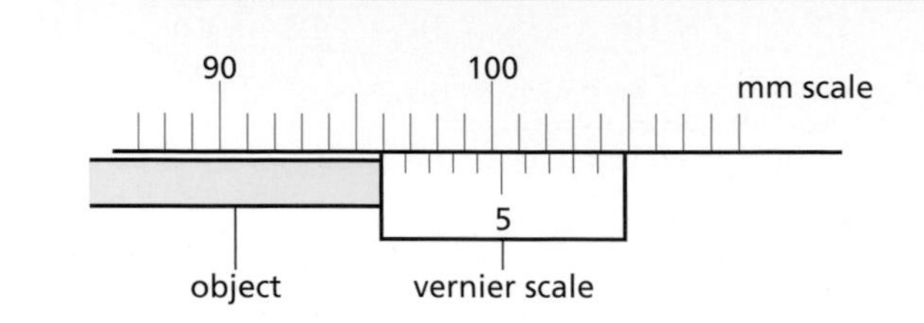

b

Figure 10.14

12 What are the readings on the micrometer screw gauges in Figures 10.15a and b?

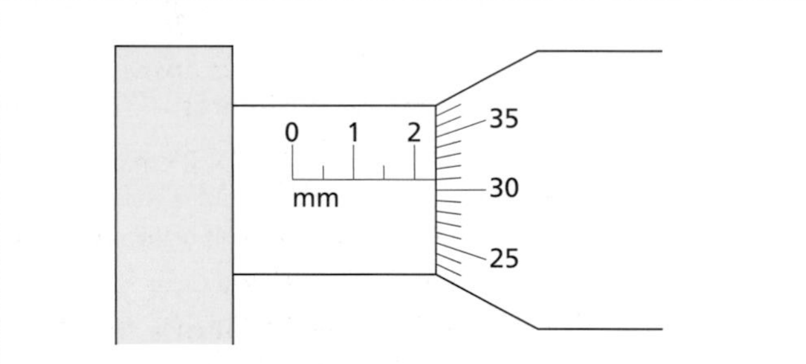

a

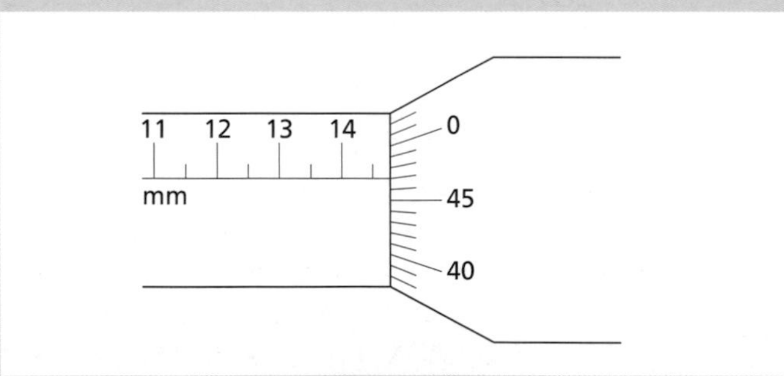

b

Figure 10.15

Checklist

After studying this chapter you should be able to

- recall three basic quantities in physics,
- write a number in powers of ten (standard notation),
- recall the unit of length and the meaning of the prefixes **kilo, centi, milli, micro, nano,**
- use a ruler to measure length so as to minimize errors,
- give a result to an appropriate number of significant figures,
- measure areas of squares, rectangles, triangles and circles,
- measure the volume of regular solids and of liquids,
- recall the unit of mass and how mass is measured,
- recall the unit of time and how time is measured,
- describe the use of clocks and devices for measuring an interval of time,
- describe an experiment to find the period of a pendulum,
- understand how a systematic error may be introduced when measuring,
- take measurements with vernier calipers and a micrometer screw gauge.

11 Density

Calculations
Simple density measurements
Floating and sinking

In everyday language, lead is said to be 'heavier' than wood. By this it is meant that a certain volume of lead is heavier than the same volume of wood. In science such comparisons are made by using the term **density**. This is the **mass per unit volume** of a substance and is calculated from

$$\text{density} = \frac{\text{mass}}{\text{volume}}$$

The density of lead is 11 grams per cubic centimetre (11 g/cm^3) and this means that a piece of lead of volume 1 cm^3 has mass 11 g. A volume of 5 cm^3 of lead would have mass 55 g. Knowing the density of a substance the mass of *any* volume can be calculated. This enables engineers to work out the weight of a structure if they know from the plans the volumes of the materials to be used and their densities. Strong enough foundations can then be made.

The SI unit of density is the **kilogram per cubic metre**. To convert a density from g/cm^3, normally the most suitable unit for the size of sample we use, to kg/m^3, we multiply by 10^3. For example the density of water is 1.0 g/cm^3 or 1.0×10^3 kg/m^3.

The approximate densities of some common substances are given in Table 11.1.

Table 11.1 Densities of some common substances

Solids	Density/ g/cm^3	Liquids	Density/ g/cm^3
aluminium	2.7	paraffin	0.80
copper	8.9	petrol	0.80
iron	7.9	pure water	1.0
gold	19.3	mercury	13.6
glass	2.5	**Gases**	kg/m^3
wood (teak)	0.80	air	1.3
ice	0.92	hydrogen	0.09
polythene	0.90	carbon dioxide	2.0

Calculations

Using the symbols ρ (rho) for density, m for mass and V for volume, the expression for density is

$$\rho = \frac{m}{V}$$

Rearranging the expression gives

$$m = V \times \rho \qquad \text{and} \qquad V = \frac{m}{\rho}$$

These are useful if ρ is known and m or V have to be calculated. If you do not see how they are obtained refer to the *Mathematics for physics* section on p. 287. The triangle in Figure 11.1 is an aid to remembering them. If you cover the quantity you want to know with a finger, e.g. m, it equals what you can still see, i.e. $\rho \times V$. To find V, cover V and you get $V = m/\rho$.

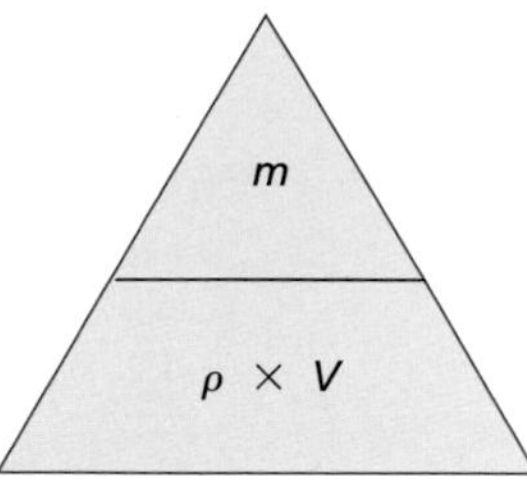

Figure 11.1

Worked example

Taking the density of copper as 9 g/cm^3, find **a** the mass of 5 cm^3 and **b** the volume of 63 g.

a $\rho = 9$ g/cm^3, $V = 5$ cm^3 and m is to be found.

$$\therefore \quad m = V \times \rho = 5\ \text{cm}^3 \times 9\ \text{g/cm}^3 = 45\ \text{g}$$

b $\rho = 9$ g/cm^3, $m = 63$ g and V is to be found.

$$\therefore \quad V = \frac{m}{\rho} = \frac{63\ \text{g}}{9\ \text{g/cm}^3} = 7\ \text{cm}^3$$

Simple density measurements

If the mass m and volume V of a substance are known its density can be found from $\rho = m/V$.

a) Regularly shaped solid

The mass is found on a balance and the volume by measuring its dimensions with a ruler.

b) Irregularly shaped solid, e.g. pebble or glass stopper

The mass of the solid is found on a balance. Its volume is measured by one of the methods shown in Figures 11.2a and b. In diagram **a** the volume is the difference between the first and second readings. In diagram **b** it is the volume of water collected in the measuring cylinder.

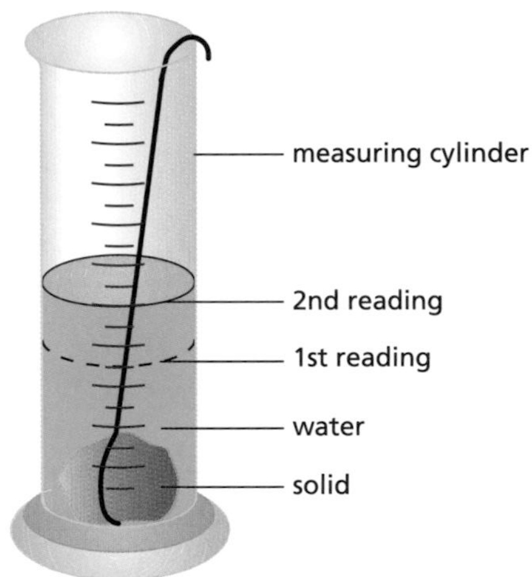

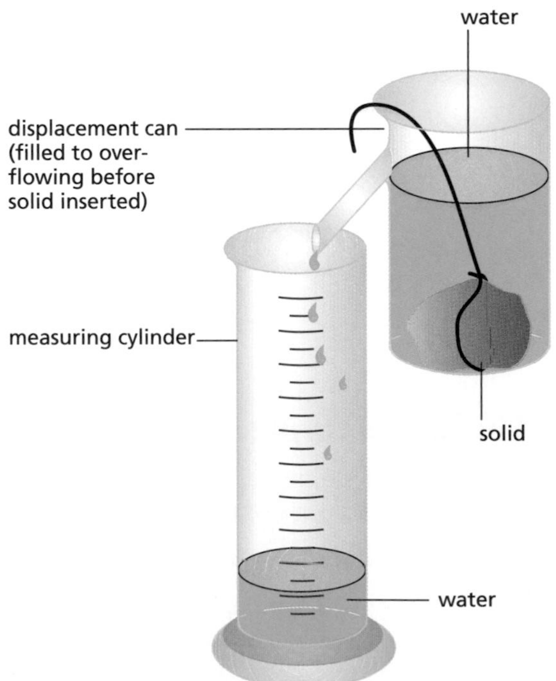

Figure 11.2 Measuring the volume of an irregular solid

c) Liquid

The mass of an empty beaker is found on a balance. A known volume of the liquid is transferred from a burette or a measuring cylinder into the beaker. The mass of the beaker plus liquid is found and the mass of liquid is obtained by subtraction.

d) Air

Using a balance the mass of a 500 cm^3 round-bottomed flask full of air is found and then after removing the air with a vacuum pump; the difference gives the mass of air in the flask. The volume of air is found by filling the flask with water and pouring it into a measuring cylinder.

F Floating and sinking

An object sinks in a liquid of lower density than its own; otherwise it floats, partly or wholly submerged. For example, a piece of glass of density 2.5 g/cm^3 sinks in water (density 1.0 g/cm^3) but floats in mercury (density 13.6 g/cm^3). An iron nail sinks in water but an iron ship floats because its average density is less than that of water.

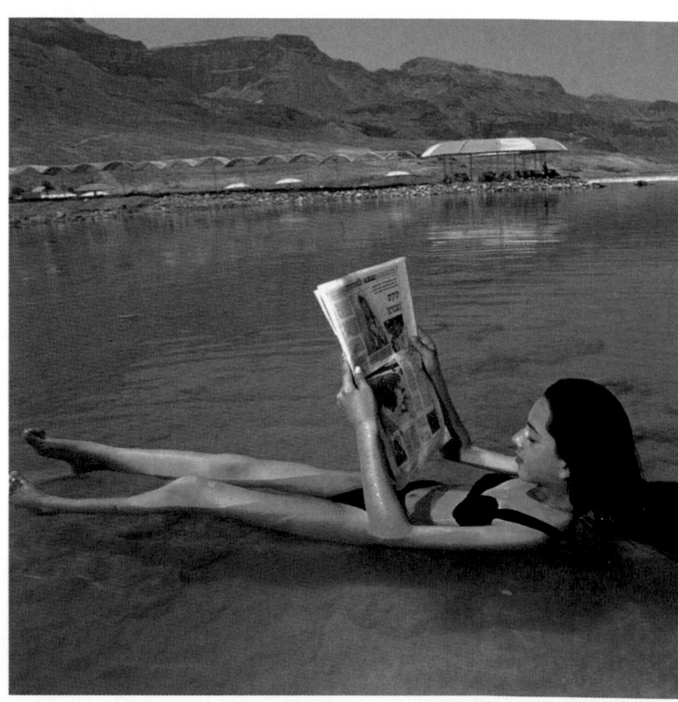

Figure 11.3 Why is it easy to float in the Dead Sea?

Questions

1 a If the density of wood is 0.5 g/cm^3 what is the mass of
(i) 1 cm^3,
(ii) 2 cm^3,
(iii) 10 cm^3?
b What is the density of a substance of
(i) mass 100 g and volume 10 cm^3,
(ii) volume 3 m^3 and mass 9 kg?
c The density of gold is 19 g/cm^3. Find the volume of
(i) 38 g,
(ii) 95 g of gold.

2 A piece of steel has a volume of 12 cm^3 and a mass of 96 g. What is its density in
a g/cm^3,
b kg/m^3?

3 What is the mass of 5 m^3 of cement of density 3000 kg/m^3?

4 What is the mass of air in a room measuring 10 m $\times$ 5.0 m $\times$ 2.0 m if the density of air is 1.3 kg/m^3?

5 When a golf ball is lowered into a measuring cylinder of water, the water level rises by 30 cm^3 when the ball is completely submerged. If the ball weighs 33 g in air, find its density.

Checklist

After studying this chapter you should be able to

- define **density** and perform calculations using $\rho = m/V$,
- describe experiments to measure the density of solids, liquids and air.

12 Weight and stretching

Force
Weight
The newton
Hooke's law
Practical work
Stretching a spring.

Force

A **force** is a push or a pull. It can cause a body at rest to move, or if the body is already moving it can change its speed or direction of motion. It can also change its shape or size.

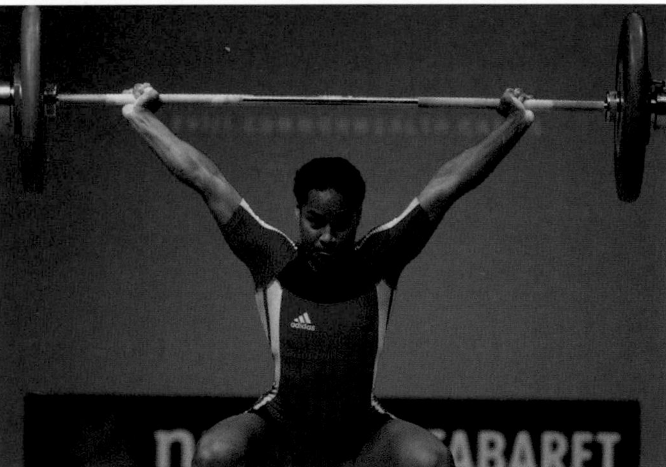

Figure 12.1 A weightlifter in action exerts first a pull and then a push

Weight

We all constantly experience the force of gravity, i.e. the pull of the Earth. It causes an unsupported body to fall from rest to the ground.

The weight of a body is the force of gravity on it.

The nearer a body is to the centre of the Earth the more the Earth attracts it. Since the Earth is not a perfect sphere but is flatter at the poles, the weight of a body varies over the Earth's surface. It is greater at the poles than at the equator.

Gravity is a force which can act through space, i.e. there does not need to be contact between the Earth and the object on which it acts as there does when we push or pull something. Other action-at-a-distance forces which, like gravity, decrease with distance are:

(i) **magnetic** forces between magnets, and
(ii) **electric** forces between electric charges.

The newton

The unit of force is the **newton** (N). It will be defined later (Chapter 22); the definition is based on the change of speed a force can produce in a body. Weight is a force and therefore should be measured in newtons.

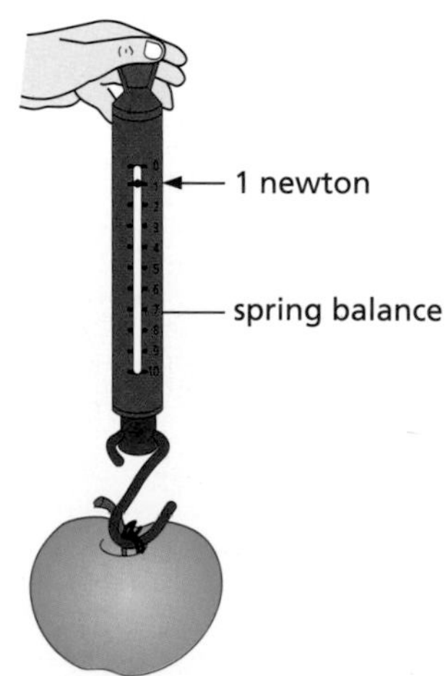

Figure 12.2 The weight of an average-sized apple is about 1 newton

The weight of a body can be measured by hanging it on a **spring balance** marked in newtons, Figure 12.2, and letting the pull of gravity stretch the spring in the balance. The greater the pull, the more the spring stretches. On most of the Earth's surface:

The weight of a body of mass 1 kg is 9.8 N.

Often this is taken as 10 N. A mass of 2 kg has a weight of 20 N and so on. The mass of a body is the same wherever it is and, unlike weight, does not depend on the presence of the Earth.

Practical work

Stretching a spring

Arrange a steel spring as in Figure 12.3. Read the scale opposite the bottom of the hanger. Add 100 g loads one at a time (thereby increasing the stretching force by steps of 1 N) and take the readings after each one. Enter the readings in a table for loads up to 500 g. Note that it is usual to give the name of the quantity or its symbol/unit at the head of columns (or rows) in data tables.

Stretching force/N	Scale reading/mm	Total extension/mm

Do the results suggest any rule about how the spring behaves when it is stretched?

Sometimes it is easier to discover laws by displaying the results on a graph. Do this on graph paper by plotting **stretching force** readings along the x-axis (horizontal axis) and **total extension** readings along the y-axis (vertical axis). Every pair of readings will give a point; mark them by small crosses and draw a smooth line through them. What is its shape?

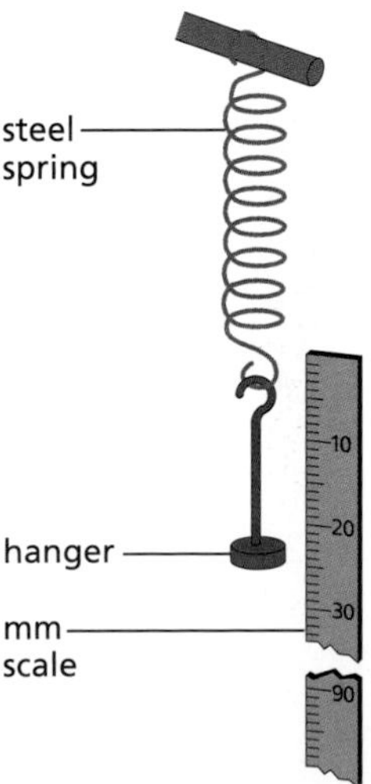

Figure 12.3

Hooke's law

Springs were investigated by Robert Hooke nearly 350 years ago. He found that the extension was proportional to the stretching force so long as the spring was not permanently stretched. This means that doubling the force doubles the extension, trebling the force trebles the extension and so on. Using the sign for proportionality we can write **Hooke's law** as

extension $\propto$ stretching force

It is true only if the **elastic limit** or limit of proportionality of the spring is not exceeded. The spring returns to its original length when the force is removed.

The graph of Figure 12.4 is for a spring stretched beyond its elastic limit E. OE is a straight line passing through the origin O and is graphical proof that Hooke's law holds over this range. If the force for point A on the graph is applied to the spring, the proportionality limit is passed and on removing the force some of the extension (OS) remains. Over which part of the graph does a spring balance work?

The **force constant** k of a spring is the force needed to cause unit extension, i.e. 1 m. If a force F produces extension x then

$$k = \frac{F}{x}$$

Rearranging the equation gives

$$F = kx$$

This is the usual way of writing Hooke's law in symbols.

Hooke's law also holds when a force is applied to a straight metal wire or an elastic band, as long as they are not permanently stretched. Force–extension graphs similar to Figure 12.4 are obtained. You should label each axis of your graph with the name of the quantity or its symbol/unit as shown.

For a rubber band, a small force causes a large extension.

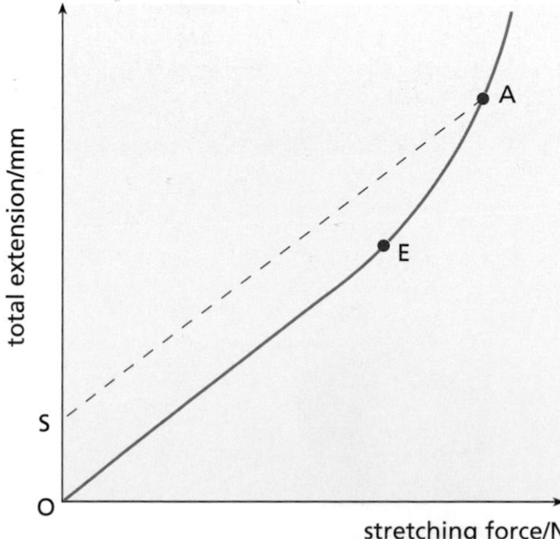

Figure 12.4

Worked example

A spring is stretched 10 mm (0.01 m) by a weight of 2.0 N. Calculate **a** the force constant k and **b** the weight W of an object that causes an extension of 80 mm (0.08 m).

a $k = \frac{F}{x} = \frac{2.0\ \text{N}}{0.01\ \text{m}} = 200\ \text{N/m}$

b W = stretching force F

$= k \times x$

$= 200\ \text{N/m} \times 0.08\ \text{m}$

$= 16\ \text{N}$

Questions

1 A body of mass 1 kg has weight 10 N at a certain place. What is the weight of
a 100 g,
b 5 kg,
c 50 g?

2 The force of gravity on the Moon is said to be one-sixth of that on the Earth. What would a mass of 12 kg weigh
a on the Earth, and
b on the Moon?

3 What is the force constant of a spring which is stretched
a 2 mm by a force of 4 N,
b 4 cm by a mass of 200 g?

4 The spring in Figure 12.5 stretches from 10 cm to 22 cm when a force of 4 N is applied. If it obeys Hooke's law, its total length in cm when a force of 6 N is applied is

A 28 **B** 42 **C** 50 **D** 56 **E** 100

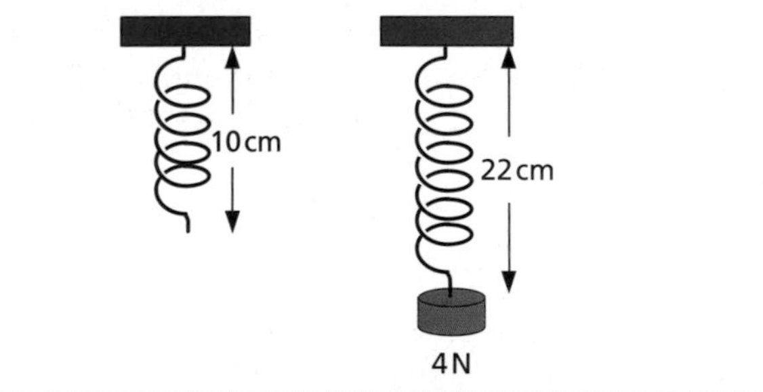

Figure 12.5

Checklist

After studying this chapter you should be able to

- recall that a force can cause a change in the motion, size or shape of a body,
- recall that the weight of a body is the force of gravity on it,
- recall the unit of force and how force is measured,
- describe an experiment to study the relation between force and extension for springs,
- draw conclusions from force–extension graphs,
- recall Hooke's law and solve problems using it,
- recognise the significance of the term **limit of proportionality**.

Matter and measurements

Additional questions

Measurements; density

1 a Name the basic units of: length, mass, time.

b What is the difference between two measurements with values of 3.4 and 3.42?

c Write expressions for
(i) the area of a circle,
(ii) the volume of a sphere,
(iii) the volume of a cylinder.

2 a An unopened bottle of olive oil has a mass of 0.97 kg. The empty bottle has a mass of 0.51 kg. Calculate the mass of the olive oil.

b The olive oil is poured into three 250 cm³ measuring cylinders. The first two cylinders are filled to the 250 cm³ mark. The third is shown below.

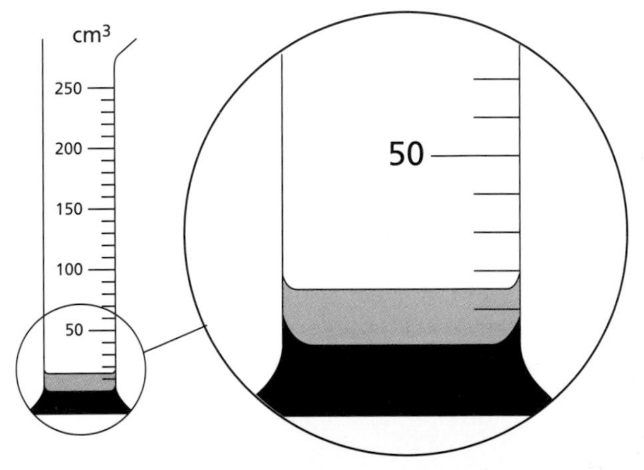

(i) What is the volume of the olive oil in the third measuring cylinder?
(ii) Calculate the volume of the olive oil in the unopened bottle.
(iii) Calculate the density of the olive oil. Express your answer to 2 significant figures.

(UCLES IGCSE Physics Core, May 2003)

3 A machine operator is making metal cylinders. The factory inspector wants to check whether the machine operator is working fast enough.

a He tells the operator to start working when the clock on the wall of the factory shows the time in the diagram below.
What time is this?

A 3.01 **B** 1.03 **C** 3.05 **D** 5.03

b The operator is told to stop when the clock shows the time in the diagram below.
What time is this?

A 3.07 **B** 7.03 **C** 3.35 **D** 4.35

c How long did the test take?

d During this time, the operator makes 5 cylinders. What is the average time to make one cylinder?

(UCLES IGCSE Physics Core, May 2001)

Weight and stretching

4 The springs in the diagram are identical. If the extension produced in **a** is 4 cm, what are the extensions in **b** and **c**?

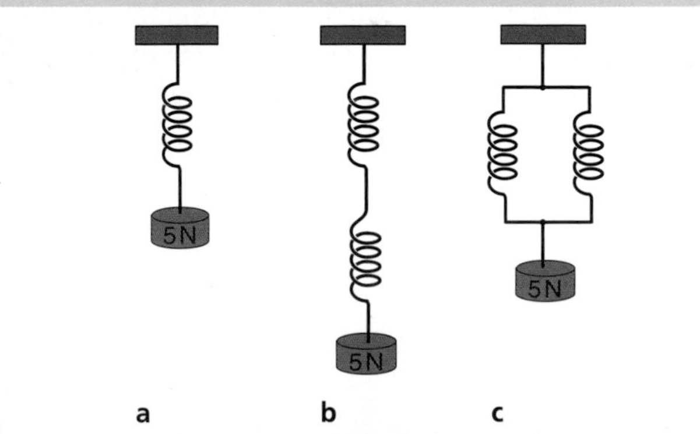

continued

5 The length of a spring is measured when various loads from 1.0 N to 6.0 N are hanging from it. A graph of the results is shown below.

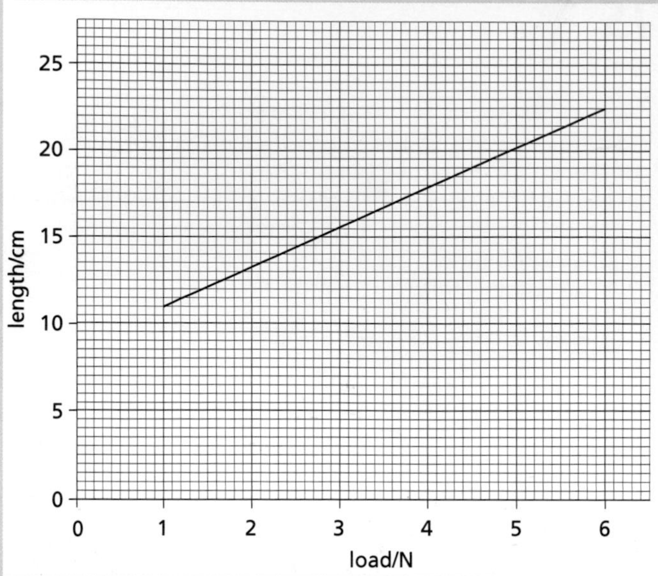

Use the graph to find

a the length of the spring with no load attached,
b the length of the spring with 4.5 N attached,
c the extension caused by a 4.5 N load.

(*UCLES IGCSE Physics Core, Nov 2005*)

6 The graph shows how the extension of a wire varies with the force applied to it.

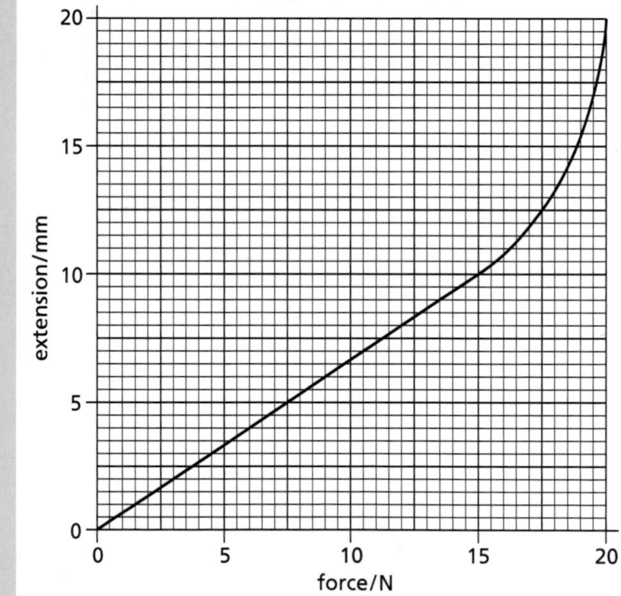

a Describe in words how the extension varies with the force applied to the wire.
b Copy the graph and mark the limit of proportionality with an **X**.
c (i) Write down the extension corresponding to a force of 9 N.
(ii) *If* the extension was proportional to the force for *all* forces, find the extension that a force of 19 N would produce. Show all your working.

(*UCLES IGCSE Physical Science Extended, June 2000*)

Forces and resources

13 Moments and levers

Moment of a force	*Conditions for equilibrium*
Balancing a beam	*Practical work*
Levers	Law of moments.

Moment of a force

The handle on a door is at the outside edge so that it opens and closes easily. A much larger force would be needed if the handle were near the hinge. Similarly it is easier to loosen a nut with a long spanner than with a short one.

The **turning effect** or **moment of a force** is a measure of the turning effect of a force. It depends on both the size of the force and how far it is applied from the pivot or **fulcrum**. It is measured by multiplying the force by the **perpendicular distance** of the line of action of the force from the fulcrum. The unit is the **newton metre** (N m).

moment of a force = force × perpendicular distance of the line of action of the force from fulcrum

In Figure 13.1a a force F acts on a gate at its edge, and in b it acts at the centre.

In a:

moment of F about O = 5 N × 3 m = 15 N m

In b:

moment of F about O = 5 N × 1.5 m = 7.5 N m

The turning effect of F is greater in the first case; this agrees with the fact that a gate opens most easily when pushed or pulled at the edge furthest from the hinge.

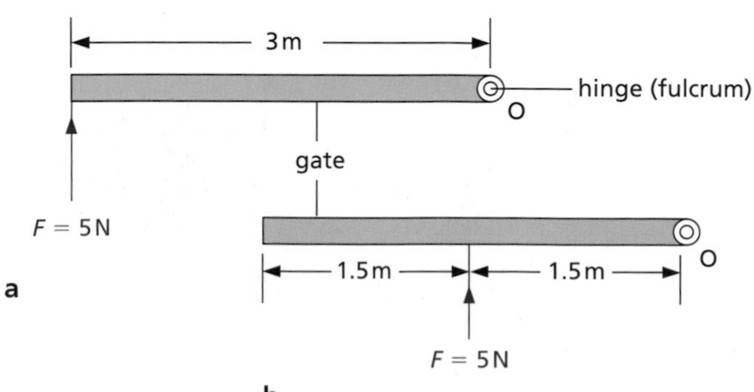

Figure 13.1

Balancing a beam

To balance a beam about a pivot, Figure 13.2, the weights must be moved so that the clockwise turning effect equals the anticlockwise turning effect and the net moment on the beam becomes zero. If the beam tends to swing clockwise, m_1 can be moved further from the pivot to increase its turning effect; alternatively m_2 can be moved nearer to the pivot to reduce its turning effect. What adjustment would you make to the position of m_2 to balance the beam if it is tending to swing anticlockwise?

Practical work

Law of moments

Balance a half-metre ruler at its centre, adding Plasticine to one side or the other until it is horizontal.

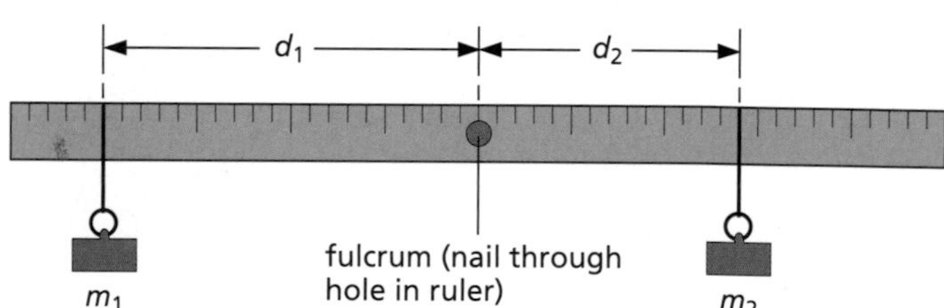

Figure 13.2

Hang unequal loads m_1 and m_2 from either side of the fulcrum and alter their distances d_1 and d_2 from the centre until the ruler is again balanced, Figure 13.2. Forces F_1 and F_2 are exerted by gravity on m_1 and m_2 and so on the ruler; the force on 100 g is 1 N. Record the results in a table and repeat for other loads and distances.

m_1 /g	F_1 /N	d_1 /cm	$F_1 \times d_1$ /N cm	m_2 /g	F_2 /N	d_2 /cm	$F_2 \times d_2$ /N cm

F_1 is trying to turn the ruler anticlockwise and $F_1 \times d_1$ is its moment. F_2 is trying to cause clockwise turning and its moment is $F_2 \times d_2$. When the ruler is balanced or, as we say, **in equilibrium**, the results should show that the anticlockwise moment $F_1 \times d_1$ equals the clockwise moment $F_2 \times d_2$.

The **law of moments** (also called the **law of the lever**) is stated as follows.

> When a body is in equilibrium the sum of the clockwise moments about any point equals the sum of the anticlockwise moments about the same point.

There is no *net* moment on a body which is in equilibrium.

Worked example

The see-saw in Figure 13.3 balances when Shani of weight 320 N is at A, Tom of weight 540 N is at B and Harry of weight W is at C. Find W.

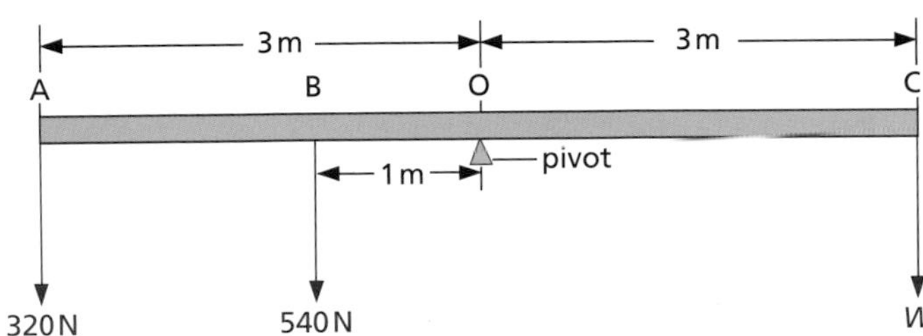

Figure 13.3

Taking moments about the fulcrum O:

$$\begin{aligned}\text{anticlockwise moment} &= 320\text{ N} \times 3\text{ m} + 540\text{ N} \times 1\text{ m}\\ &= 960\text{ N m} + 540\text{ N m}\\ &= 1500\text{ N m}\end{aligned}$$

$$\text{clockwise moment} = W \times 3\text{ m}$$

By the law of moments,

$$\text{clockwise moments} = \text{anticlockwise moments}$$

$$\therefore \quad W \times 3\text{ m} = 1500\text{ N m}$$

$$\therefore \quad W = \frac{1500\text{ N m}}{3\text{ m}} = 500\text{ N}$$

Levers

A lever is any device which can turn about a pivot. In a working lever a force called the **effort** is used to overcome a resisting force called the **load**. The pivotal point is called the **fulcrum**.

When we use a crowbar to move a heavy boulder, Figure 13.4, our hands apply the effort at one end of the bar and the load is the force exerted by the boulder on the other end. If distances from the fulcrum O are as shown and the load is 1000 N (i.e. the part of the weight of the boulder supported by the crowbar), the effort can be calculated from the law of moments. As the boulder *just begins* to move we can say, taking moments about O, that

$$\text{clockwise moment} = \text{anticlockwise moment}$$

$$\text{effort} \times 200\text{ cm} = 1000\text{ N} \times 10\text{ cm}$$

$$\text{effort} = \frac{10\,000\text{ N cm}}{200\text{ cm}} = 50\text{ N}$$

Examples of other levers are shown in Figure 13.5.

How does the effort compare with the load for scissors and a spanner in Figures 13.5c and d?

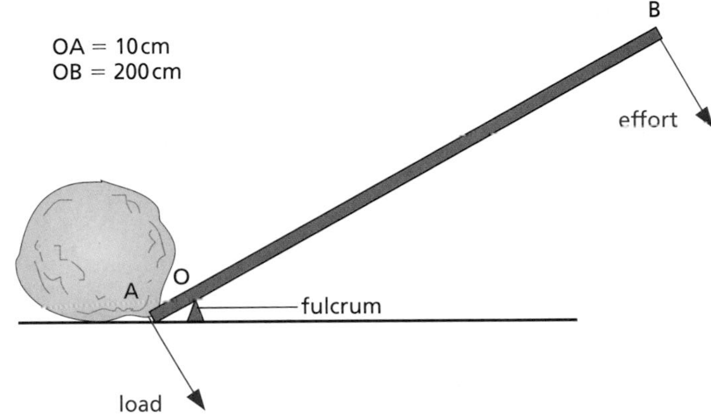

Figure 13.4 Crowbar

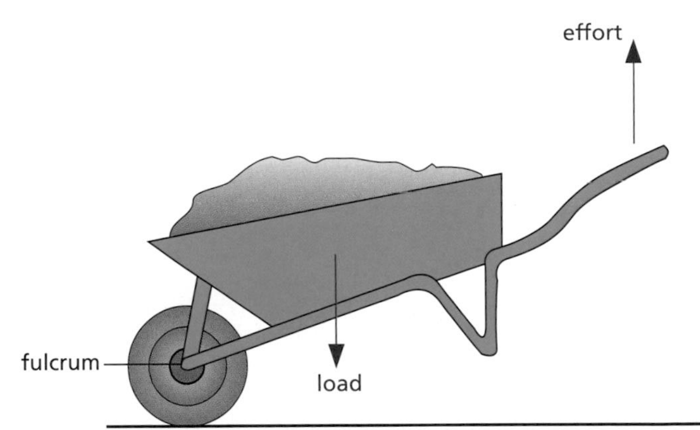

Figure 13.5a Other levers: Wheelbarrow

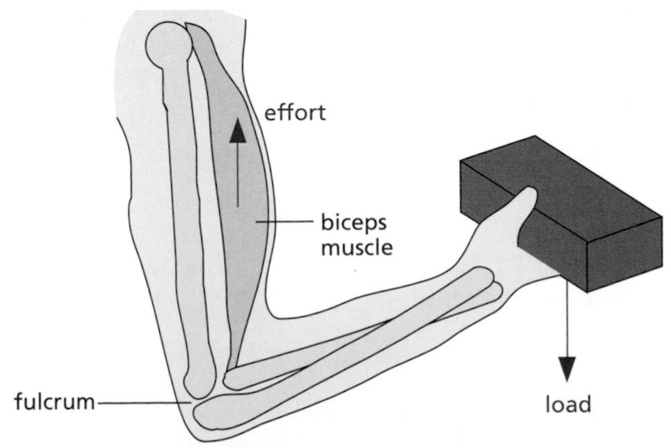

Figure 13.5b Other levers: Forearm

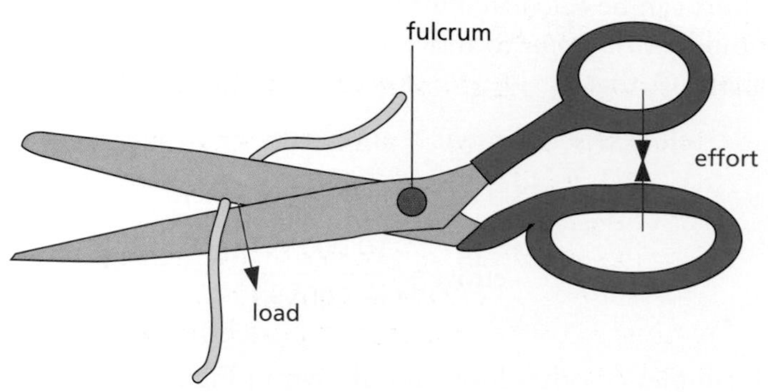

Figure 13.5c Other levers: Scissors

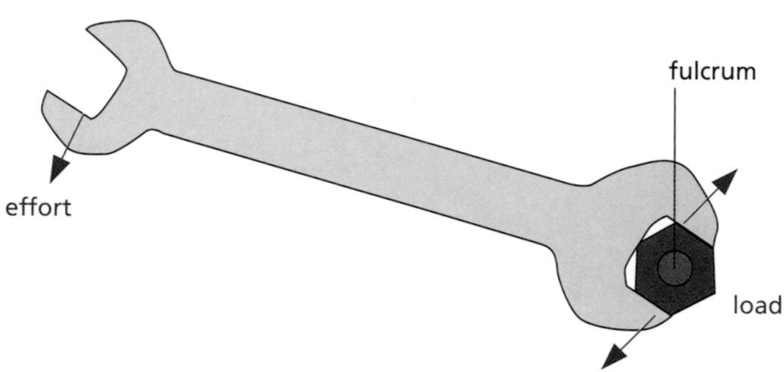

Figure 13.5d Other levers: Spanner

Conditions for equilibrium

Sometimes a number of parallel forces act on a body so that it is in equilibrium. We can then say:

(i) The sum of the forces in one direction equals the sum of the forces in the opposite direction.
(ii) The law of moments must apply.

A body is in equilibrium when there is no resultant force and no resultant turning effect acting on it.

As an example consider a heavy plank resting on two trestles, Figure 13.6. In the next chapter we will see that the whole weight of the plank (400 N) may be taken to act vertically downwards at its centre, O. If P and Q are the upward forces exerted by the trestles on the plank (called **reactions**) then we have from (i)

$$P + Q = 400 \text{ N} \qquad (1)$$

Moments can be taken about any point but if we take them about C the moment due to Q is zero.

$$\text{clockwise moment} = P \times 5 \text{ m}$$

$$\text{anticlockwise moments} = 400 \text{ N} \times 2 \text{ m} = 800 \text{ N m}$$

Since the plank is in equilibrium we have from (ii)

$$P \times 5 \text{ m} = 800 \text{ N m}$$

$$\therefore \qquad P = \frac{800 \text{ N m}}{5 \text{ m}} = 160 \text{ N}$$

From (1), $\qquad Q = 240 \text{ N}$

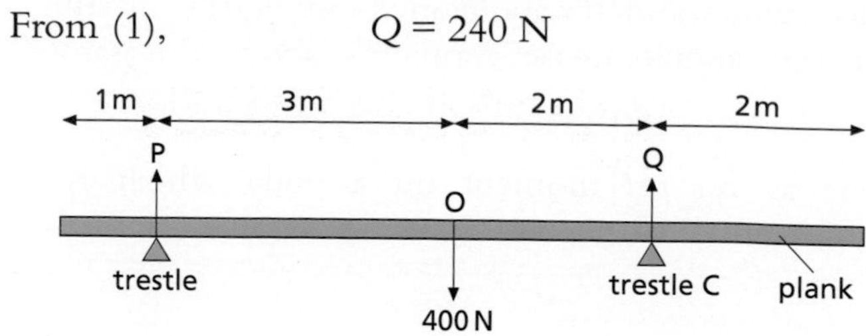

Figure 13.6

Questions

1 The metre rule in Figure 13.7 is pivoted at its centre. If it balances, the mass of M is given by the equation

A $M + 50 = 40 + 100$
B $M \times 40 = 100 \times 50$
C $M/50 = 100/40$
D $M/50 = 40/100$
E $M \times 50 = 100 \times 40$

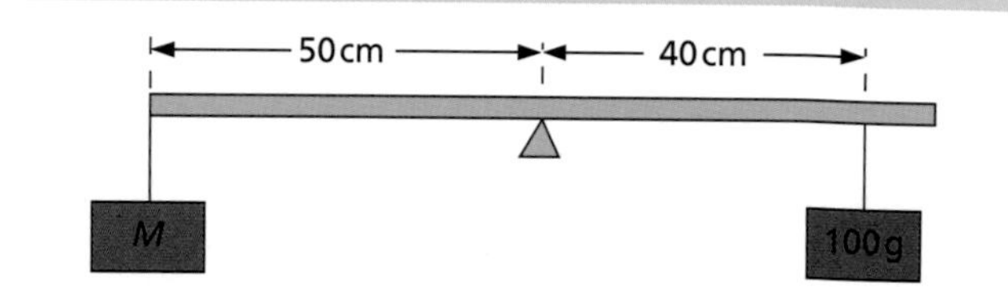

Figure 13.7

2 Figure 13.8 shows three positions of the pedal on a bicycle which has a crank 0.20 m long. If the cyclist exerts the same vertically downward push of 25 N with his foot, in which case, A, B or C, is the turning effect
(i) $25 \times 0.2 = 5$ N m,
(ii) 0,
(iii) between 0 and 5 N m?
Explain your answers.

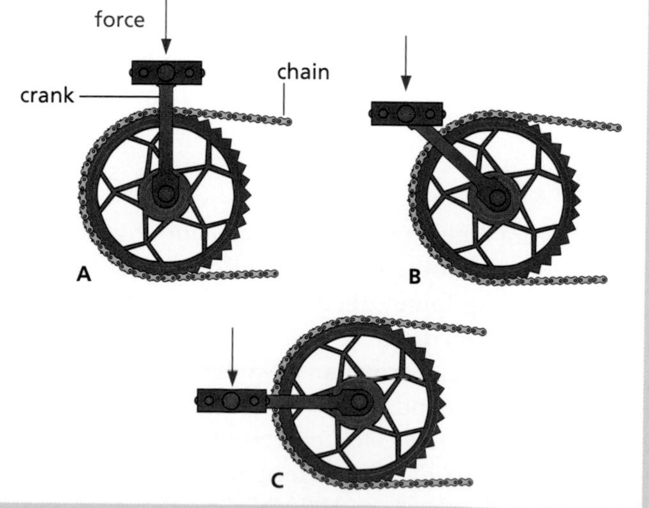

Figure 13.8

3 Figure 13.9 shows a shelf supported by a hinge at one end and a rope at the other.

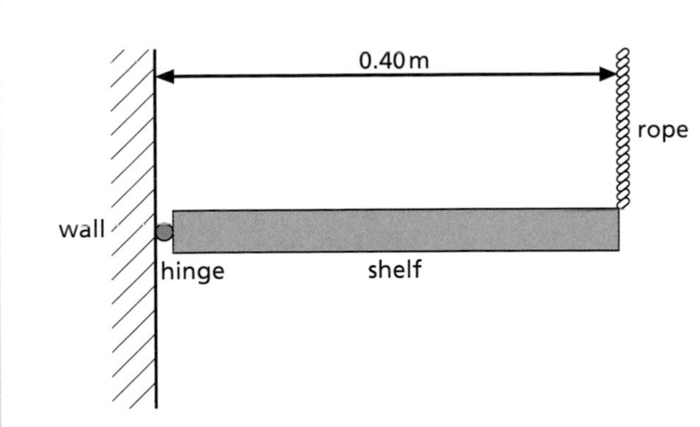

Figure 13.9

a (i) The mass of the shelf is 0.15 kg. Calculate its weight, showing all your working. (g = 10 N/kg)
(ii) The shelf's weight acts through its centre. Copy the diagram and draw an arrow to show the direction of the weight.
(iii) On your diagram, show the direction of a second vertical force acting on the shelf.
(iv) By taking moments about the hinge, calculate a value for the force in (iii), showing all your working.

b When books are placed on the shelf, the vertical forces change.
(i) Explain how the vertical forces change.
(ii) Suggest *two* factors which determine if the shelf can take the extra weight without collapsing.
(*UCLES IGCSE Physical Science Extended, Nov 1999*)

Checklist

After studying this chapter you should be able to

- define the **moment** of a force about a point,
- describe qualitatively the balancing of a beam about a pivot,
- ☐ describe an experiment to verify that there is no net moment on a body in equilibrium,
- ☐ state the law of moments and use it to solve problems,
- explain the action of common tools and devices as levers,
- state the conditions for equilibrium when parallel forces act on a body.

14 Centres of mass

Toppling
Stability
Balancing tricks and toys
Practical work
Centre of mass using a plumb line.

A body behaves as if its whole mass were concentrated at one point, called its **centre of mass** or **centre of gravity**, even though the Earth attracts every part of it. The body's weight can be considered to act at this point. The centre of mass of a uniform ruler is at its centre and when supported there it can be balanced, Figure 14.1a. If it is supported at any other point it topples because the moment of its weight W about the point of support is not zero, Figure 14.1b.

Your centre of mass is near the centre of your body and the vertical line from it to the floor must be within the area enclosed by your feet or you will fall over. You can test this by standing with one arm and the side of one foot pressed against a wall, Figure 14.2. Now try to raise the other leg sideways.

A tight-rope walker has to keep his centre of mass exactly above the rope. Some carry a long pole to help them to balance, Figure 14.3. The combined weight of the walker and pole is then spread out more and if the walker begins to topple to one side he moves the pole to the other side.

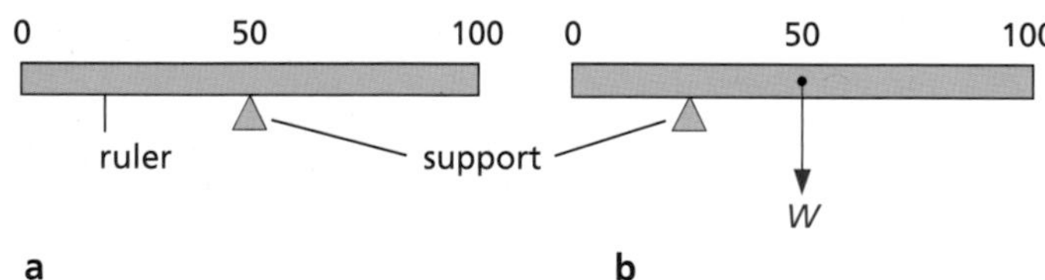

Figure 14.1

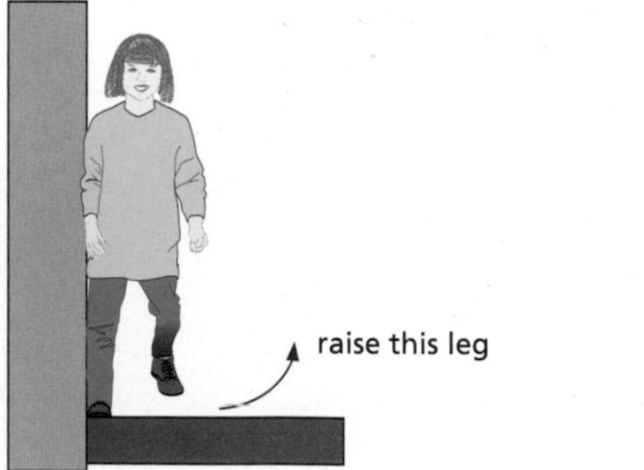

Figure 14.2 Can you do this without falling over?

The centre of mass of a regularly shaped body of the same density throughout is at its centre. In other cases it can be found by experiment.

Figure 14.3 A tight-rope walker using a long pole

Practical work

Centre of mass using a plumb line

Suppose we have to find the centre of mass of an irregularly shaped lamina (a thin sheet) of cardboard.

Make a hole A in the lamina and hang it so that it can *swing freely* on a nail clamped in a stand. It will come to rest with its centre of mass vertically below A. To locate the vertical line through A tie a plumb line (a thread and a weight) to the nail, Figure 14.4, and mark its position AB on the lamina. The centre of mass lies on AB.

Hang the lamina from another position C and mark the plumb line position CD. The centre of mass lies on CD and must be at the point of intersection of AB and CD. Check this by hanging the lamina from a third hole. Also try balancing it at its centre of mass on the tip of your forefinger.

Devise a method using a plumb line for finding the centre of mass of a tripod.

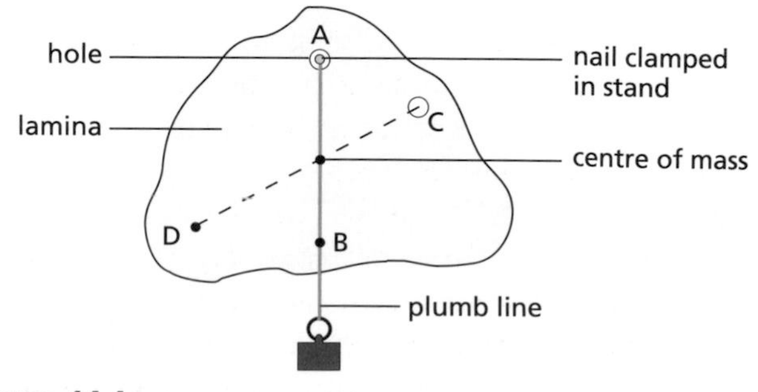

Figure 14.4

Toppling

The position of the centre of mass of a body affects whether or not it topples over easily. This is important in the design of such things as tall vehicles (which tend to overturn when rounding a corner), racing cars, reading lamps and even drinking glasses.

A body topples when the vertical line through its centre of mass falls outside its base, Figure 14.5a. Otherwise it remains stable, Figure 14.5b.

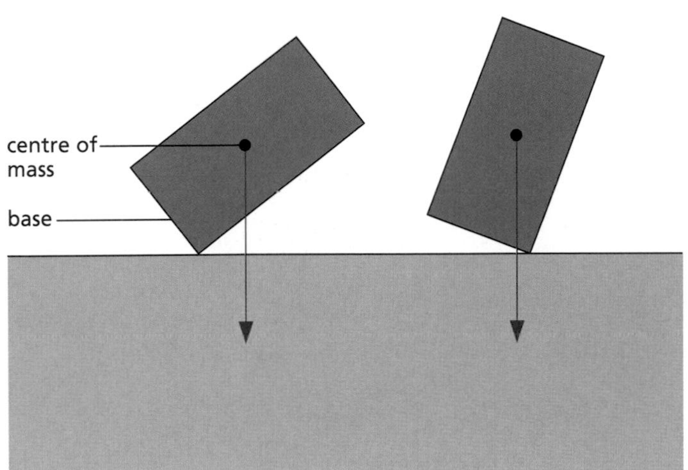

a Topples **b** Stable

Figure 14.5

Toppling can be investigated by placing an empty can on a plank (with a rough surface to prevent slipping) which is slowly tilted. The angle of tilt is noted when the can falls over. This is repeated with a mass of 1 kg in the can. How does this affect the position of the centre of mass? The same procedure is followed with a second can of the same height as the first but of greater width. It will be found that the second can with the mass in it can be tilted through the greater angle.

The stability of a body is therefore increased by

(i) lowering its centre of mass, and
(ii) increasing the area of its base.

In Figure 14.6a the centre of mass of a tractor is being found. It is necessary to do this when testing a new design since tractors are often driven over sloping surfaces and any tendency to overturn must be discovered.

The stability of double-decker buses is being tested in Figure 14.6b. When the top deck only is fully laden with passengers (represented by sand bags in the test), it must not topple if tilted through an angle of 28°.

Racing cars have a low centre of mass and a wide wheel base for maximum stability.

Figure 14.6a A tractor under test to find its centre of mass

Figure 14.6b A double-decker bus being tilted to test its stability

Stability

Three terms are used in connection with stability.

a) Stable equilibrium

A body is in 'stable equilibrium' if when slightly displaced and then released it returns to its previous position. The ball at the bottom of the dish in Figure 14.7a is an example. Its centre of mass rises when it is displaced. It rolls back because its weight has a moment about the point of contact.

b) Unstable equilibrium

A body is in 'unstable equilibrium' if it moves farther away from its previous position when slightly displaced and released. The ball in Figure 14.7b behaves in this way. Its centre of mass falls when it is displaced slightly because there is a moment which increases the displacement. Similarly in Figure 14.1a (p. 64) the balanced ruler is in unstable equilibrium.

c) Neutral equilibrium

A body is in 'neutral equilibrium' if it stays in its new position when displaced, Figure 14.7c. Its centre of mass does not rise or fall because there is no moment to increase or decrease the displacement.

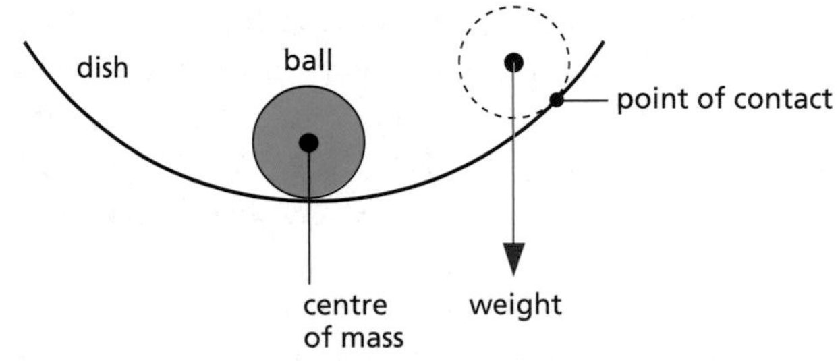

a Stable

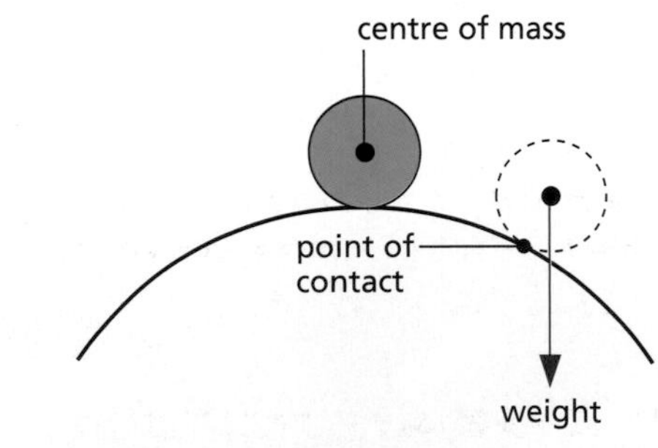

b Unstable

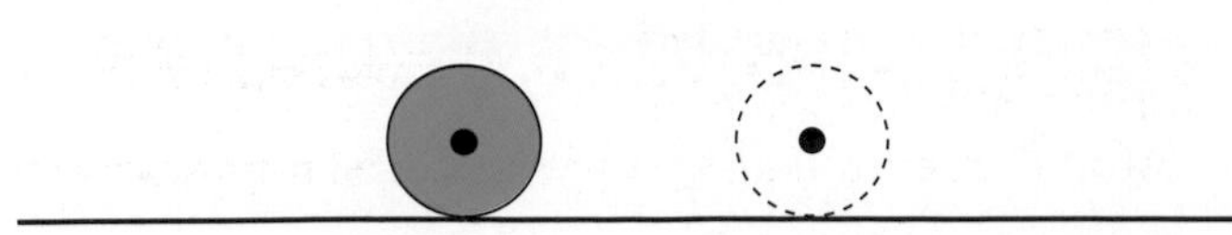

c Neutral

Figure 14.7 States of equilibrium

Balancing tricks and toys

Some tricks that you can try or toys you can make are shown in Figure 14.8. In each case the centre of mass is vertically below the point of support and equilibrium is stable.

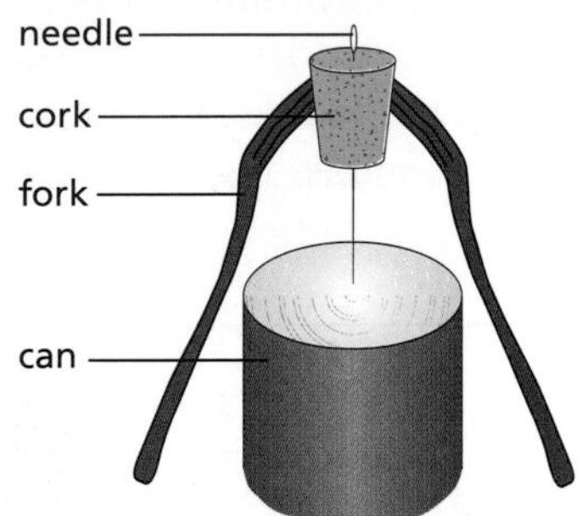

a Balancing a needle on its point

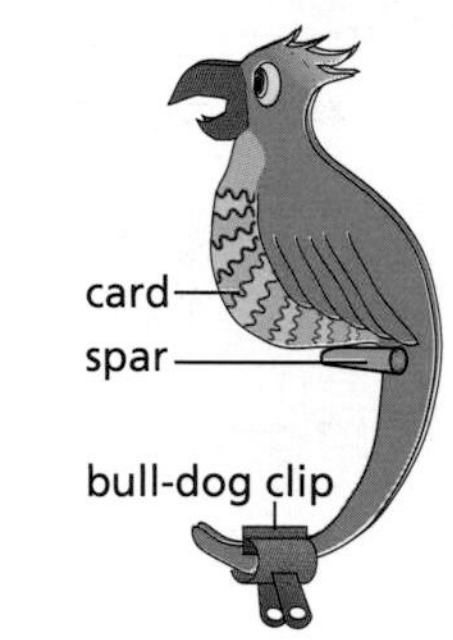

b The perched parrot

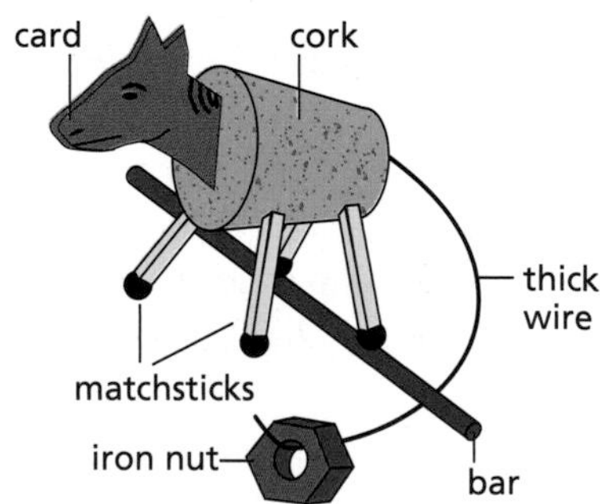

c A rocking horse

Figure 14.8 Balancing tricks

A self-righting toy, Figure 14.9, has a heavy base and, when tilted, the weight acting through the centre of mass has a moment about the point of contact. This restores it to the upright position.

Figure 14.9 A self-righting toy

Questions

1 When the cardboard shape in Figure 14.10 is freely hung from A, line AX is vertical.

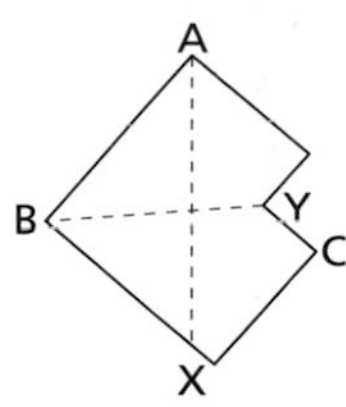

Figure 14.10

When it is freely hung from B, line BY is vertical.

a Copy the diagram and mark the position of the centre of mass of the shape, using a clear dot (•).

b Draw a line through C which would be vertical if the shape were to be freely hung from C.

(*UCLES IGCSE Physics Core, Nov 1999*)

2 Figure 14.11 shows a Bunsen burner in three different positions. State in which position it is in

a stable equilibrium,

b unstable equilibrium,

c neutral equilibrium.

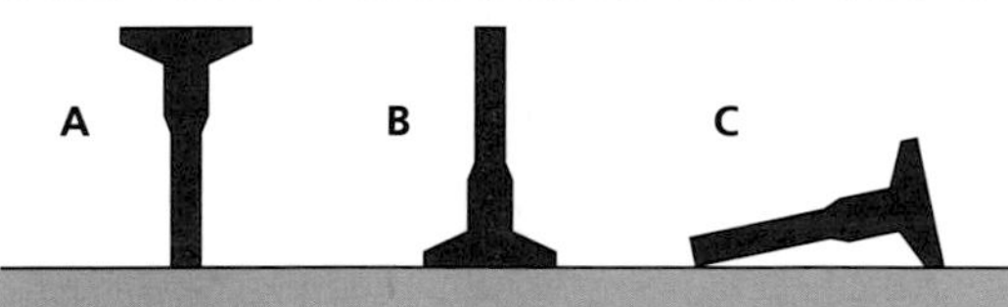

Figure 14.11

3 The weight of the uniform bar in Figure 14.12 is 10 N. Does it balance, tip to the right or tip to the left?

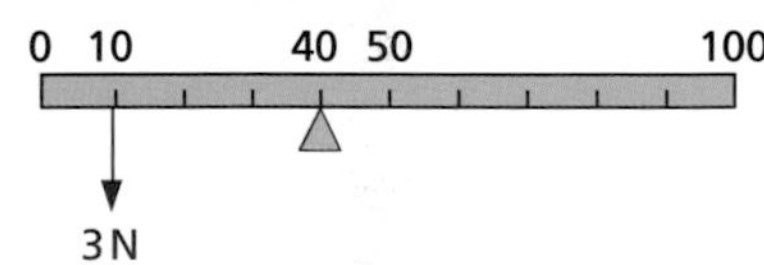

Figure 14.12

4 Figure 14.13 shows apparatus for investigating moments of forces.

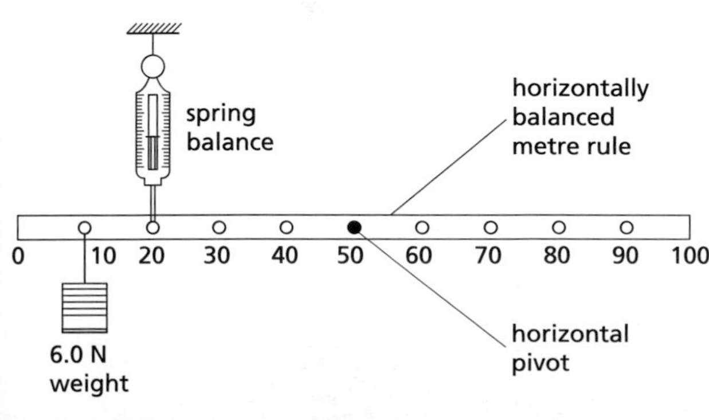

Figure 14.13

The uniform metre rule shown in Figure 14.13 is in equilibrium.

a Write down two conditions for the metre rule to be in equilibrium.

b Show that the value of the reading on the spring balance is 8.0 N.

c The weight of the uniform metre rule is 1.5 N. Calculate the force exerted by the pivot on the metre rule.

(*UCLES IGCSE Physics Extended, Nov 2005*)

5 The following paragraphs appeared in a newspaper.

JEEP FAILS GOVERNMENT TEST

A car manufacturer confirmed yesterday that one of its four-wheel drive mini-jeeps rolled over at 38 mph during stability tests conducted by the Government.

Testing has been halted until safety cages can be fitted to seven makes of mini-jeeps which the Department of Transport has agreed to test after repeated complaints from the Consumers' Association. The Association claims its own tests show that the narrow track, short-wheelbase vehicles are prone to rolling over. The Government's tests highlight how passengers raise the centre of mass. All seven vehicles passed the test unladen, although one raised two wheels.

a Write down *two* factors mentioned in the newspaper article which affect the stability of vehicles.

b Figure 14.14 shows a tilted vehicle.

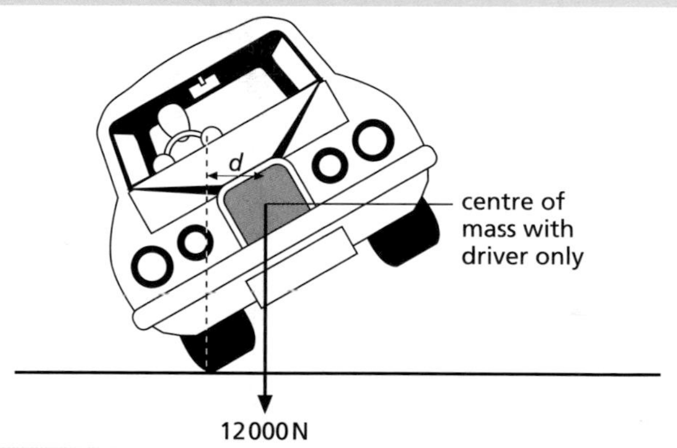

Figure 14.14

The distance *d* shown in the diagram is 50 cm. Calculate the moment of the force about the point of contact with the road.

c Explain how passengers make the vehicles more likely to roll over (less stable). You may use diagrams if you wish.

(*NEAB Higher, June 1998*)

Checklist

After studying this chapter you should be able to

- recall that an object behaves as if its whole mass were concentrated at its centre of mass,
- recall that an object's weight acts through the centre of mass (or centre of gravity),
- describe an experiment to find the centre of mass of an object,
- connect the stability of an object to the position of its centre of mass.

15 Adding forces

Forces and resultants
Examples of addition of forces
Vectors and scalars
Friction
Practical work
Parallelogram law.

Forces and resultants

Force has both magnitude (size) and direction. It is represented in diagrams by a straight line with an arrow to show its direction of action.

Usually more than one force acts on an object. As a simple example, an object resting on a table is pulled downwards by its weight W and pushed upwards by a force R due to the table supporting it, Figure 15.1. Since the object is at rest, the forces must balance, i.e. $R = W$.

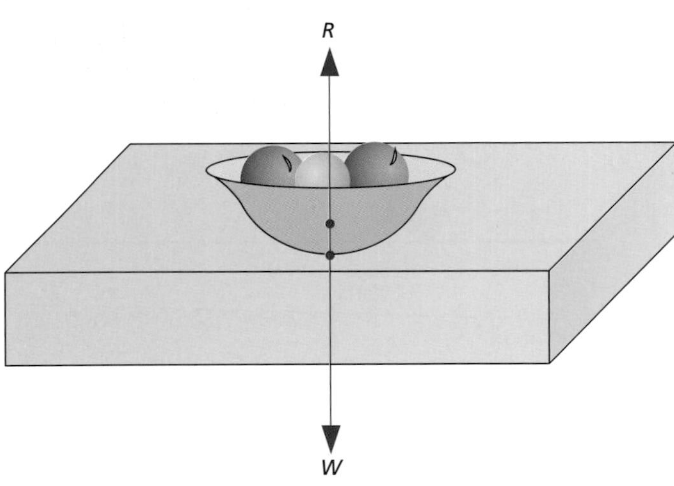

Figure 15.1

Figure 15.2 The design of an off-shore oil platform requires an understanding of the combination of many forces

In structures such as a giant oil platform, Figure 15.2, two or more forces may act at the same point. It is then often useful for the design engineer to know the value of the single force, i.e. the **resultant**, which has exactly the same effect as these forces. If the forces act in the same straight line the resultant is found by simple addition or subtraction as shown in Figure 15.3, but if they do not they are added by the 'parallelogram law'.

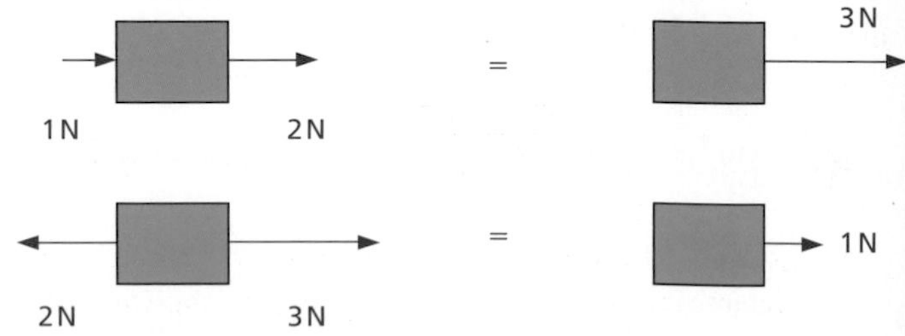

Figure 15.3 The resultant of forces acting in the same straight line is found by addition or subtraction

Practical work

Parallelogram law

Arrange the apparatus as in Figure 15.4a with a sheet of paper behind it on a vertical board. We have to find the resultant of forces P and Q.

Read the values of P and Q from the spring balances. Mark on the paper the directions of P, Q and W as shown by the strings. Remove the paper and, using a scale of 1 cm to represent 1 N, draw OA, OB and OD to represent the three forces P, Q and W which act at O, Figure 15.4b. (W = weight of the 1 kg mass = 9.8 N, therefore OD = 9.8 cm.)

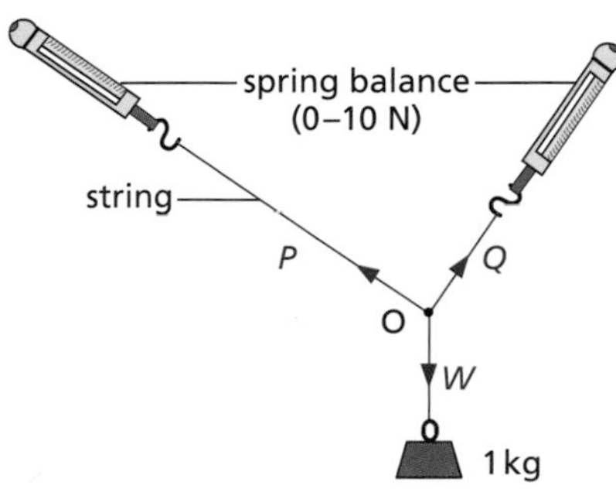

a

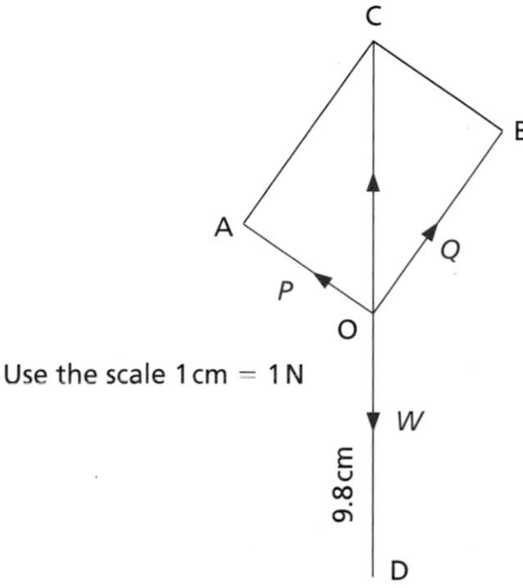

b

Figure 15.4 Finding a resultant by the parallelogram law

P and Q together are balanced by W and so their resultant must be a force equal and opposite to W.

Complete the parallelogram OACB. Measure the diagonal OC; if it is equal in size (i.e. 9.8 cm) and opposite in direction to W then it represents the resultant of P and Q.

The **parallelogram law** for adding two forces is:

> If two forces acting at a point are represented in size and direction by the sides of a parallelogram drawn from the point, their resultant is represented in size and direction by the diagonal of the parallelogram drawn from the point.

Worked example

Find the resultant of two forces of 4.0 N and 5.0 N acting at an angle of 45° to each other.

Using a scale of 1.0 cm = 1.0 N, draw parallelogram ABDC with AB = 5.0 cm, AC = 4.0 N and angle CAB = 45°, Figure 15.5. By the parallelogram law, the diagonal AD represents the resultant in magnitude and direction; it measures 8.3 cm and angle BAD = 20°.

∴ Resultant is a force of 8.3 N acting at an angle of 20° to the force of 5.0 N.

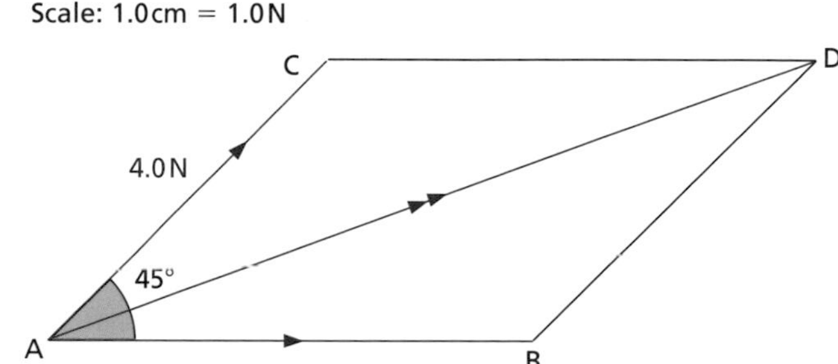

Figure 15.5

Examples of addition of forces

1 Two people carrying a heavy bucket. The weight of the bucket is balanced by the resultant F of F_1 and F_2, Figure 15.6a.

2 Two tugs pulling a ship. The resultant of T_1 and T_2 is forwards, Figure 15.6b, and so the ship moves forwards (as long as the resultant is greater than the resistance to motion of the sea and the wind).

a

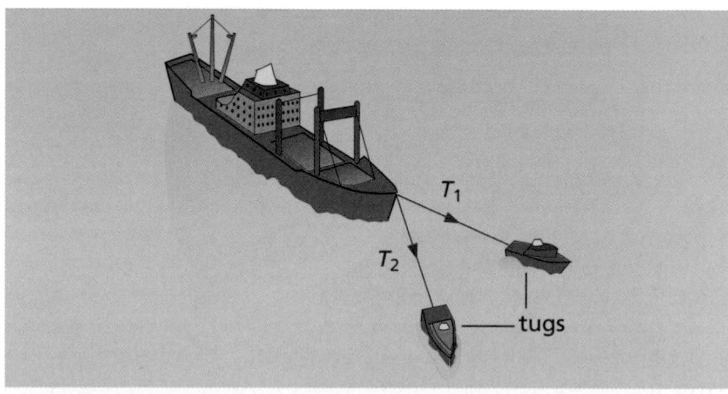

b

Figure 15.6

Vectors and scalars

A **vector** quantity is one such as force which is described completely only if both its size (magnitude) and direction are stated. It is not enough to say, for example, a force of 10 N, but rather a force of 10 N acting vertically downwards.

A vector can be represented by a straight line whose length represents the magnitude of the quantity and whose direction gives its line of action. An arrow on the line shows which way along the line it acts.

A **scalar** quantity has magnitude only. Mass is a scalar and is completely described when its value is known. Scalars are added by ordinary arithmetic; vectors are added geometrically, taking account of their directions as well as their magnitudes.

F *Friction*

Friction is the force that opposes one surface moving, or trying to move, over another. It can be a help or a hindrance. We could not walk if it did not exist between the soles of our shoes and the ground. Our feet would slip backwards, as they tend to if we walk on ice. On the other hand, engineers try to reduce friction to a minimum in the moving parts of machinery by using lubricating oils and ball-bearings.

When a gradually increasing force P is applied through a spring balance to a block on a table, Figure 15.7, it does not move at first. This is because an equally increasing but opposing frictional force F acts where the block and table touch. At any instant P and F are equal and opposite.

If P is increased further the block eventually moves; as it does so F has its maximum value, called **starting** or **static friction**. When it is moving at a steady speed the balance reading is slightly less than that for starting friction. **Sliding** or **dynamic friction** is therefore less than starting or static friction.

A mass on the block increases the force pressing the surfaces together and increases friction.

When work is done against friction, the temperatures of the bodies in contact rise (as you can test by rubbing your hands together); mechanical energy is being changed into heat energy (see Chapter 16).

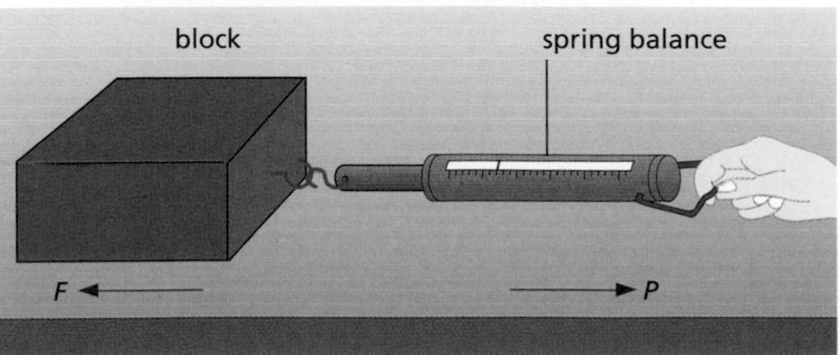

Figure 15.7 Friction opposes motion between surfaces in contact

Questions

1 Jo, Daniel and Helen are pulling a metal ring. Jo pulls with a force of 100 N in one direction and Daniel with a force of 140 N in the opposite direction. If the ring does not move, what force does Helen exert if she pulls in the same direction as Jo?

2 A boy drags a suitcase along the ground with a force of 100 N. If the frictional force opposing the motion of the suitcase is 50 N, what is the resultant forward force on the suitcase?

3 A picture is supported by two vertical strings; if the weight of the picture is 50 N what is the force exerted by each string?

4 Using a scale of 1 cm to represent 10 N find the size and direction of the resultant of forces of 30 N and 40 N acting at right angles to each other.

Checklist

After studying this chapter you should be able to

- combine forces acting along the same straight line to find their resultant,
- add vectors graphically to determine a resultant,
- distinguish between vectors and scalars and give examples of each.

16 Energy transfer

Forms of energy	*Energy of food*
Energy transfers	*Combustion of fuels*
Energy measurements	*Practical work* Measuring power.
Energy conservation	

Energy is a theme that pervades all branches of science. It links a wide range of phenomena and enables us to explain them. It exists in different forms and when something happens, it is likely to be due to energy being transferred from one form to another. **Energy transfer** is needed to enable people, computers, machines and other devices to work and to enable processes and changes to occur. For example, in Figure 16.1, the water skier can only be pulled along by the boat if there is energy transfer in its engine from the burning petrol to its rotating propeller.

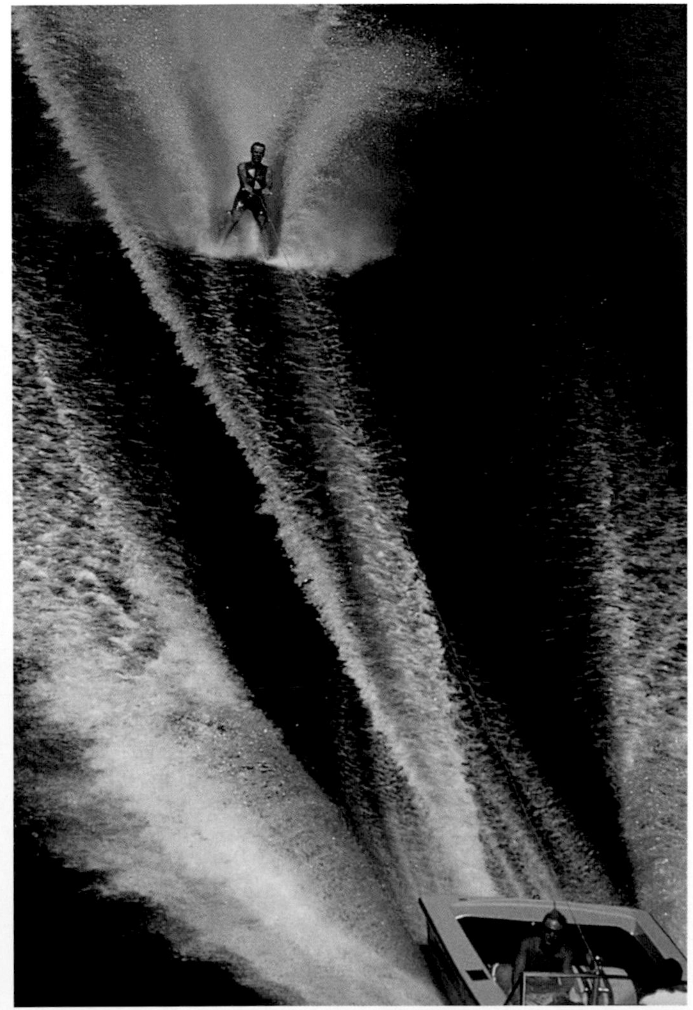

Figure 16.1 Energy transfer in action

Forms of energy

a) Chemical energy

Food and fuels, like oil, gas, coal and wood, are concentrated stores of chemical energy (see Chapter 17). The energy of food is released by chemical reactions in our bodies, and during the transfer to other forms we are able to do useful jobs. Fuels cause energy transfers when they are burnt in an engine or a boiler. Batteries are compact sources of chemical energy, which in use is transferred to electrical energy.

b) Potential energy (p.e.)

This is the energy a body has because of its position or condition. A body above the Earth's surface, like water in a mountain reservoir, has p.e. stored in the form of **gravitational potential energy**. Work has to be done to compress or stretch a spring or elastic material and energy is transferred to p.e.; the p.e. is stored in the form of **strain energy** (or elastic potential energy). If the catapult in Figure 16.3c were released, the strain energy would be transferred to the projectile.

c) Kinetic energy (k.e.)

Any moving body has k.e. and the faster it moves the more k.e. it has. As a hammer drives a nail into a piece of wood, there is a transfer of energy from the k.e. of the moving hammer to other forms of energy.

d) Electrical energy

This is produced by energy transfers at power stations and in batteries. It is the commonest form of energy used in homes and industry because of the ease of transmission and transfer to other forms.

e) Heat energy

This is also called **thermal** or **internal energy** and is the final fate of other forms of energy. It is transferred by conduction, convection or radiation.

f) Other forms

These include light and other forms of electromagnetic radiation, sound and nuclear energy.

Energy transfers

a) Demonstration

The apparatus in Figure 16.2 can be used to show a battery changing **chemical energy** to **electrical energy** which becomes **k.e.** in the electric motor. The motor raises a weight, giving it **p.e.** If the changeover switch is joined to the lamp and the weight allowed to fall, the motor acts as a generator in which there is an energy transfer from **k.e.** to **electrical energy**. When this is supplied to the lamp, it produces a transfer to **heat** and **light energy**.

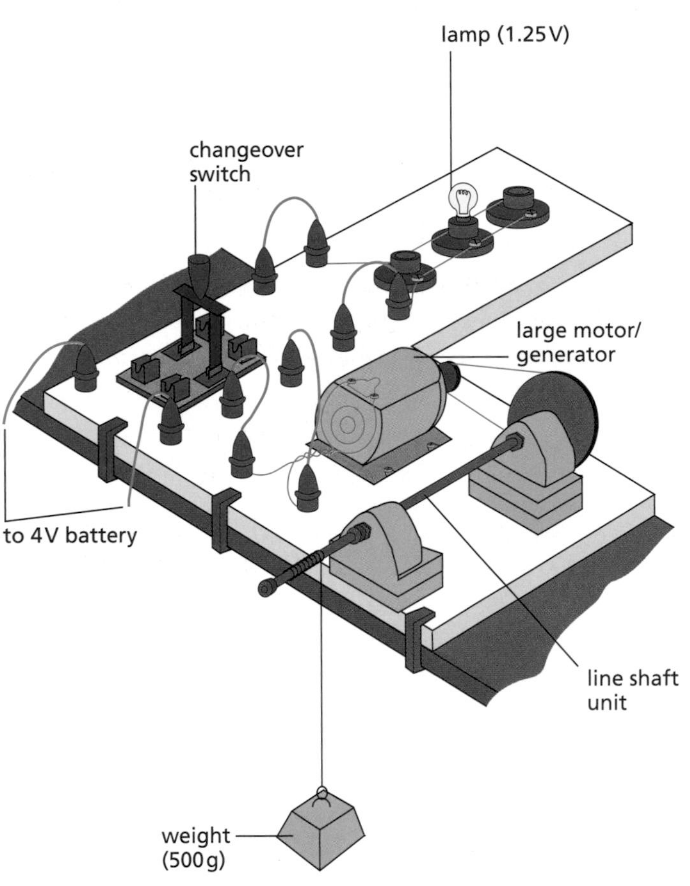

Figure 16.2 Demonstrating energy transfers

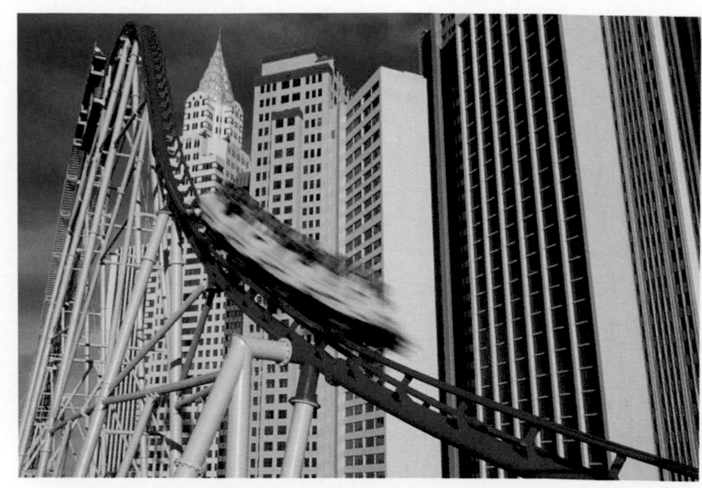

a Potential energy to kinetic energy

b Electrical energy to heat and light energy

c Chemical energy (from muscles in the arm) to p.e. (strain energy of catapult)

d Potential energy of water to kinetic energy of turbine to electrical energy from generator

Figure 16.3 Some energy transfers

b) Other examples

Study the energy transfers shown in Figures 16.3a to d. Some devices have been invented to cause particular energy transfers. For example, a **microphone** changes sound energy into electrical energy; a **loudspeaker** does the reverse. Belts, chains or gears are used to transfer energy between moving parts, e.g. in a bicycle.

Energy measurements

a) Work

In science the word **work** has a different meaning from its everyday one. **Work is done when a force moves.** No work is done in the scientific sense by someone standing still holding a heavy pile of books: an upward force is exerted, but no motion results.

If a building worker carries ten bricks up to the first floor of a building he does more work than if he carries only one brick because he has to exert a larger force. Even more work is required if he carries the ten bricks to the second floor. The amount of work done depends on the size of the force applied and the distance it moves. We therefore measure work by

work = force × distance moved in direction of force (1)

The unit of work is the **joule** (J) and is **the work done when a force of 1 newton** (N) **moves through 1 metre** (m). For example, if you have to pull with a force of 50 N to move a crate steadily 3 m in the direction of the force, Figure 16.4a, the work done is 50 N × 3 m = 150 N m = 150 J. That is

joules = newtons × metres

If you lift a mass of 3 kg vertically through 2 m, Figure 16.4b, you have to exert a vertically upward force equal to the weight of the body, i.e. 30 N (approximately) and the work done is 30 N × 2 m = 60 N m = 60 J.

Note that we must always take the distance in the direction in which the force acts.

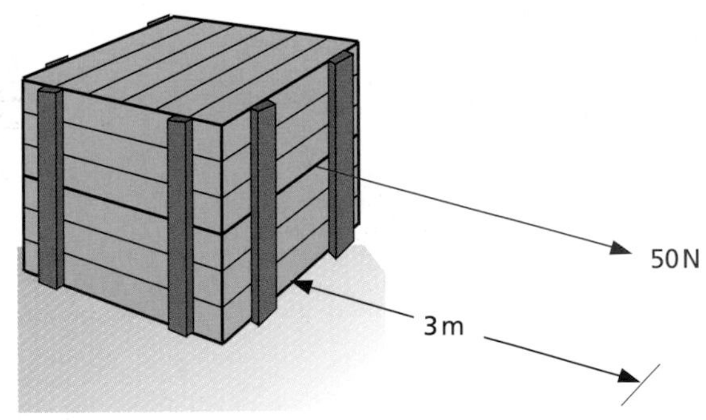

a

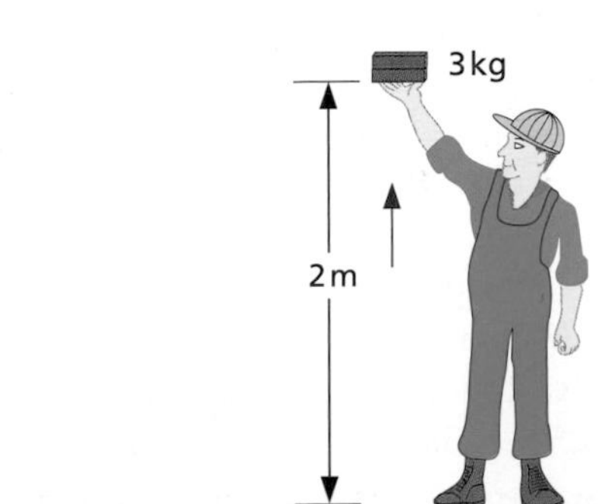

b

Figure 16.4

b) Measuring energy transfers

In an energy transfer work is done. **The work done is a measure of the amount of energy transferred.** For example, if you have to exert an upward force of 10 N to raise a stone steadily through a vertical distance of 1.5 m, the work done is 15 J. This is also the amount of chemical energy transferred from your muscles to p.e. of the stone. All forms of energy, as well as work, are measured in joules.

c) Power

The more powerful a car is, the faster it can accelerate or climb a hill, i.e. the more rapidly it does work. The power of a device is the work it does per second, i.e. the rate at which it does work. This is the same as **the rate at which it transfers energy from one form to another**.

$$\text{power} = \frac{\text{work done}}{\text{time taken}} = \frac{\text{energy transfer}}{\text{time taken}} \qquad (2)$$

The unit of power is the **watt** (W) and is **a rate of working of 1 joule per second**, i.e. 1 W = 1 J/s. Larger units are the kilowatt (kW) and the megawatt (MW):

$$1\ \text{kW} = 1000\ \text{W} = 10^3\ \text{W}$$
$$1\ \text{MW} = 1\,000\,000\ \text{W} = 10^6\ \text{W}$$

If a machine does 500 J of work in 10 s, its power is 500 J/10 s = 50 J/s = 50 W. A small car develops a maximum power of about 25 kW.

Practical work

Measuring power

a) Your own power

Get someone with a stopwatch to time you running up a flight of stairs, the more steps the better. Find your weight (in newtons). Calculate the total vertical height (in metres) you have climbed by measuring the height of one step and counting the number of steps.

The work you do (in joules) in lifting your weight to the top of the stairs is (your weight) × (vertical height of stairs). Calculate your power (in watts) from equation (2). About 0.5 kW is good.

b) Electric motor

This experiment is described in Chapter 38.

Energy conservation

a) Principle of conservation of energy

This is one of the basic laws of physics and is stated as follows.

> Energy cannot be created or destroyed; it is always conserved.

However, energy is continually being transferred from one form to another. Some forms, such as electrical and chemical energy, are more easily transferred than others, such as heat, for which it is hard to arrange a useful transfer. Ultimately all energy transfers result in the surroundings being heated (e.g. as a result of doing work against friction) and the energy is wasted, i.e. spread out and increasingly more difficult to use. For example, when a brick falls its p.e. becomes k.e.; as it hits the ground, its temperature rises and heat and sound are produced. If it seems in a transfer that some energy has disappeared, the 'lost' energy is often converted into non-useful heat. This appears to be the fate of all energy in the Universe and is one reason why new sources of useful energy have to be developed (Chapter 17).

b) Efficiency of energy transfers

The efficiency of a device is the percentage of the energy supplied to it that is usefully transferred. It is calculated from the expression:

$$\text{efficiency} = \frac{\text{useful energy transferred by device}}{\text{total energy supplied to device}} \times 100\%$$

For example, for the lever shown in Figure 13.4 (p. 61)

$$\text{efficiency} = \frac{\text{work done on load}}{\text{work done by effort}} \times 100\%$$

This will be less than 100% if there is friction in the fulcrum.

Table 16.1 lists the efficiencies of some devices and the energy transfers involved.

Table 16.1

Device	% Efficiency	Energy transfer
large electric motor	90	electrical to k.e.
large electric generator	90	k.e. to electrical
domestic gas boiler	75	chemical to heat
compact fluorescent lamp	50	electrical to light
steam turbine	45	heat to k.e.
car engine	25	chemical to k.e.
filament lamp	10	electrical to light

A device is efficient if it transfers energy mainly to useful forms and the 'lost' energy is small.

F Energy of food

When food is eaten it reacts with the oxygen we breathe into our lungs and is slowly 'burnt'. As a result chemical energy stored in food becomes thermal energy to warm the body and mechanical energy for muscular movement.

> The **energy value** of a food substance is the amount of energy released when 1 kg is completely oxidized.

Energy value is measured in J/kg. The energy values of some foods are given in Figure 16.5 in megajoules per kilogram.

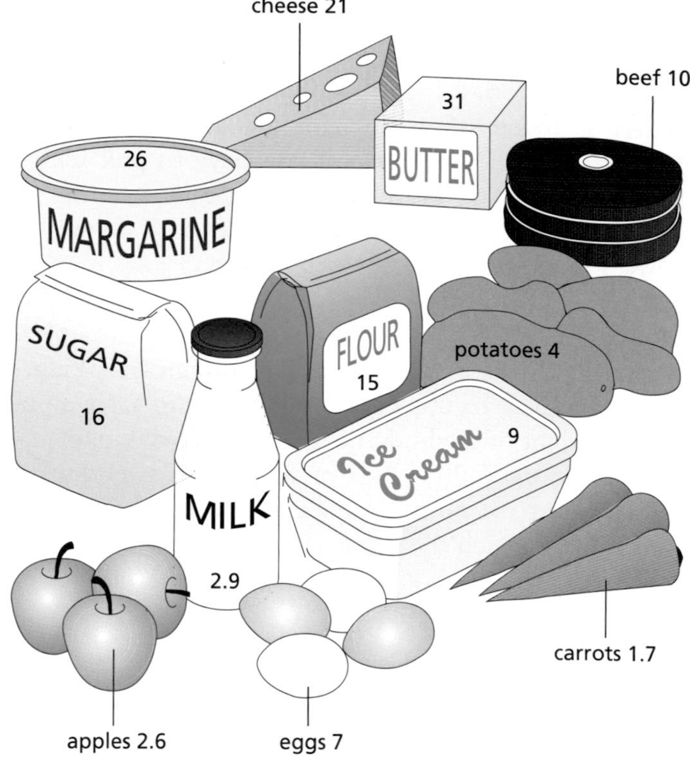

Figure 16.5 Energy values of some foods in MJ/kg

Food with high values are 'fattening' and if more food is eaten than the body really needs, the extra is stored as fat. The average adult requires about 10 MJ per day.

Our muscles change chemical energy into mechanical energy when we exert a force – to lift a weight, for example. Unfortunately, they are not very good at doing this; of every 100 J of chemical energy they use, they can convert only 25 J into mechanical energy – that is, they are only 25% efficient at changing chemical energy into mechanical energy. The other 75 J becomes thermal energy, much of which the body gets rid of by sweating.

F Combustion of fuels

Fuels can be solids such as wood and coal, liquids such as fuel oil and paraffin, or gases such as methane and butane.

Some fuels are better than others for certain jobs. For example, fuels for cooking or keeping us warm should, as well as being cheap, have a high **heating value**. This means that every gram of fuel should produce a large amount of heat energy when burnt. A fuel for a space rocket (e.g. liquid hydrogen) must also burn very quickly so that the gases created expand rapidly and leave the rocket at high speed.

The heating values of some fuels are given in Table 16.2 in kilojoules per gram.

Table 16.2 Heating values of fuels/kJ/g

Solids	Value	Liquids	Value	Gases	Value
Wood	17	Fuel oil	45	Methane	55
Coal	25–33	Paraffin	48	Butane	50

The thick dark liquid called **petroleum** or **crude oil** is the source of most liquid and gaseous fuels. It is obtained from underground deposits at oil wells in many parts of the world. Natural gas (methane) is often found with it. In an oil refinery different fuels are obtained from petroleum, including fuel oil for industry, diesel oil for lorries, paraffin (kerosene) for jet engines, petrol for cars, as well as butane (bottled gas).

Questions

1 Name the energy transfers which occur when
 a an electric bell rings,
 b someone speaks into a microphone,
 c a ball is thrown upwards,
 d there is a picture on a television screen,
 e a torch is on.

2 Name the forms of energy represented by the letters A, B, C and D in the following statement.

In a coal-fired power station, the (A) energy of coal becomes (B) energy which changes water into steam. The steam drives a turbine which drives a generator. A generator transfers (C) energy into (D) energy.

3 How much work is done when a mass of 3 kg (weighing 30 N) is lifted vertically through 6 m?

4 A hiker climbs a hill 300 m high. If she weighs 50 kg calculate the work she does in lifting her body to the top of the hill.

5 In loading a lorry a man lifts boxes each of weight 100 N through a height of 1.5 m.
 a How much work does he do in lifting one box?
 b How much energy is transferred when one box is lifted?
 c If he lifts 4 boxes per minute at what power is he working?

6 A boy whose weight is 600 N runs up a flight of stairs 10 m high in 12 s. What is his average power?

7 **a** When the energy input to a gas-fired power station is 1000 MJ, the electrical energy output is 300 MJ. What is the efficiency of the power station in changing the energy in gas into electrical energy?
 b What form does the 700 MJ of 'lost' energy take?
 c What is the fate of the 'lost' energy?

Checklist

After studying this chapter you should be able to

- recall the different forms of energy,
- describe energy transfers in given examples,
- relate work done to the magnitude of a force and the distance moved,
- use the relation **work done = force × distance moved** to calculate energy transfer,
- define the unit of work,
- relate power to work done and time taken, and give examples,
- recall that power is the rate of energy transfer, give its unit and solve problems,
- describe an experiment to measure your own power,
- state the principle of conservation of energy,
- recall the meaning of **efficiency** of a device.

17 Energy sources

Non-renewable energy sources
Renewable energy sources
Power stations
Economic, environmental and social issues

Energy is needed to heat buildings, to make cars move, to provide artificial light, to make computers work, and so on. The list is endless. This 'useful' energy needs to be produced in controllable energy transfers (Chapter 16). For example, in power stations a supply of useful energy in the form of electricity is produced. The 'raw materials' for energy production are **energy sources**. These may be **non-renewable** or **renewable**.

Non-renewable energy sources

Once used up these cannot be replaced.

a) Fossil fuels

These include coal, oil and natural gas, formed from the remains of plants and animals which lived millions of years ago and obtained energy originally from the Sun. At present they are our main energy source. Predictions vary as to how long they will last since this depends on what reserves are recoverable and on the future demands of a world population expected to increase from about 6600 million in 2007, to 7600 million by the year 2020. Some estimates say oil and gas will run low early in the present century but coal should last for 200 years or so.

Burning fossil fuels in power stations and in cars pollutes the atmosphere with harmful gases such as carbon dioxide and sulphur dioxide. Carbon dioxide emission aggravates the greenhouse effect (Chapter 32) and increases global warming. It is not immediately feasible to prevent large amounts of carbon dioxide entering the atmosphere, but less is produced by burning natural gas than by burning oil or coal; burning coal produces most carbon dioxide for each unit of energy produced. When coal and oil are burnt they also produce sulphur dioxide which causes acid rain. The sulphur dioxide can be extracted from the waste gases so it does not enter the atmosphere or the sulphur can be removed from the fuel before combustion, but these are both costly processes which increase the price of electricity produced using these measures.

b) Nuclear fuels

The energy released in a nuclear reactor (Chapter 49) from uranium, found as an ore in the ground, can be used to produce electricity. Nuclear fuels do not pollute the atmosphere with carbon dioxide or sulphur dioxide but they do generate radioactive waste materials with very long half-lives (Chapter 48); safe ways of storing this waste for perhaps thousands of years must be found. As long as a reactor is operating normally it does not pose a radiation risk, but if an accident occurs, dangerous radioactive material can leak from the reactor and spread over a large area.

Two advantages of all non-renewable fuels are

(i) their high **energy density** (i.e. they are concentrated sources) and the relatively small size of the energy transfer device (e.g. a furnace) which releases their energy, and

(ii) their ready **availability** when energy demand increases suddenly or fluctuates seasonally.

Renewable energy sources

These cannot be exhausted and are generally non-polluting.

a) Solar energy

The energy falling on the Earth from the Sun is mostly in the form of light and in an hour equals the total energy used by the world in a year. Unfortunately its low energy density requires large collecting devices and its availability varies. Its greatest potential use is as an energy source for low-temperature water heating. This uses **solar panels** as the energy transfer devices, Figure 17.1, which convert light into heat energy. They are used increasingly to produce domestic hot water at about 70 °C and to heat swimming pools.

Solar energy can also be used to produce high-temperature heating, up to 3000 °C or so, if a large curved mirror (a **solar furnace**) focuses the Sun's rays on to a small area. The energy can then be used to turn water to steam for driving the turbine of an electric generator in a power station.

Figure 17.1 Solar panels on a house provide hot water

Solar cells, made from semiconducting materials, convert sunlight into electricity directly. Panels of cells connected together supply the electronic equipment in communication and other satellites. They are also used for small-scale power generation in remote areas of developing countries where there is no electricity supply. Recent developments have made large-scale generation more cost-effective and there is now a large solar power plant in California. There are many designs for prototype light vehicles run on solar power, Figure 17.2.

Figure 17.2 Solar-powered car

b) Wind energy

Giant windmills called **wind turbines** with two or three blades each up to 30 m long drive electrical generators. 'Wind farms' of 20 to 100 turbines spaced about 400 m apart, Figure 2d, p. ix, supply about 400 MW (enough electricity for 250 000 homes) in the UK and provide a useful 'top-up' to the National Grid.

They can be noisy and may be considered unsightly so there is some environmental objection to wind farms, especially as the best sites are often in coastal or upland areas of great natural beauty.

c) Wave energy

The rise and fall of sea waves has to be transferred by some kind of wave-energy converter into the rotary motion required to drive a generator. It is a difficult problem and the large-scale production of electricity by this means is unlikely in the near future, but small systems are being developed to supply island communities with power.

d) Tidal and hydroelectric energy

The flow of water from a higher to a lower level from behind a tidal barrage (barrier) or the dam of a hydroelectric scheme is used to drive a water turbine (water wheel) connected to a generator.

One of the largest working tidal schemes is the *La Grande I* project in Canada, Figure 17.3. Feasibility studies have shown that a 10 mile long barrage across the Severn estuary could produce about 7% of today's electrical energy consumption in England and Wales. Such schemes have significant implications for the environment, as they may destroy wildlife habitats of wading birds for example, and also for shipping routes.

In the UK hydroelectric power stations generate about 2% of the electricity supply. Most are located in Scotland and Wales where the average rainfall is higher than in other areas. With good management hydroelectric energy is a reliable energy source, but there are risks connected with the construction of dams, and a variety of problems may result from the impact of a dam on the environment. Land previously used for forestry or farming may have to be flooded.

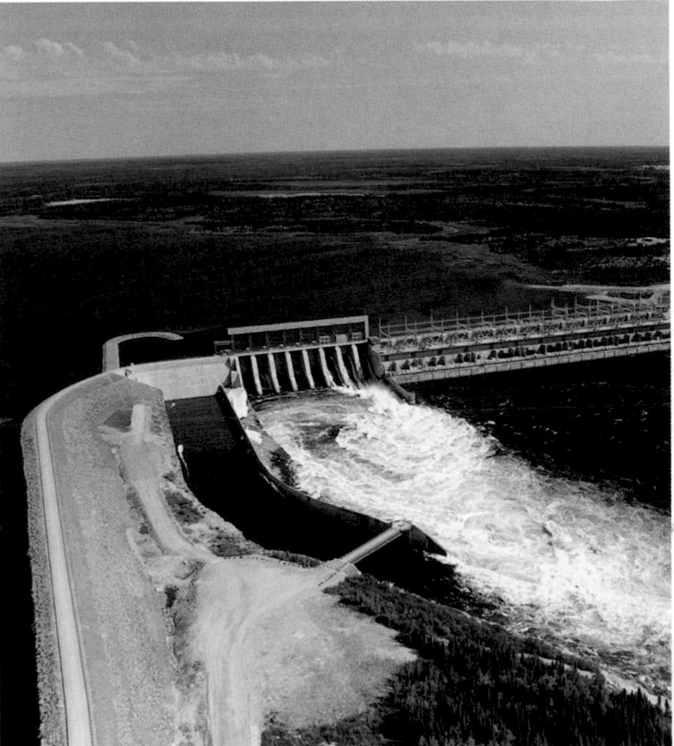

Figure 17.3 Tidal barrage in Canada

e) Geothermal energy

If cold water is pumped down a shaft into hot rocks below the Earth's surface, it may be forced up another shaft as steam. This can be used to drive a turbine and generate electricity or to heat buildings. The energy that heats the rocks is constantly being released by radioactive elements deep in the Earth as they decay (Chapter 48).

Geothermal power stations are in operation in the USA, New Zealand and Iceland.

f) Biomass (vegetable fuels)

These include cultivated crops (e.g. oil-seed rape), crop residues (e.g. cereal straw), natural vegetation (e.g. gorse), trees (e.g. spruce) grown for their wood, animal dung and sewage. **Biofuels** such as alcohol (ethanol) and methane gas are obtained from them by fermentation using enzymes or by decomposition by bacterial action in the absence of air. Liquid biofuels can replace petrol, Figure 17.4; although they have up to 50% less energy per litre, they are lead- and sulphur-free and so cleaner. **Biogas** is a mix of methane and carbon dioxide with an energy content about two-thirds that of natural gas. In developing countries it is produced from animal and human waste in 'digesters', Figure 17.5, and used for heating and cooking.

Power stations

The processes involved in the production of electricity at power stations depend on the energy source being used.

a) Non-renewable sources

These are used in **thermal** power stations to produce heat energy that turns water into steam. The steam drives turbines which in turn drive the generators that produce electrical energy as described in Chapter 45. If fossil fuels are the energy source (usually coal but natural gas is favoured in new stations), the steam is obtained from a boiler. If nuclear fuel is used, i.e. uranium or plutonium, the steam is produced in a heat exchanger as explained in Chapter 49.

The action of a **steam turbine** resembles that of a water wheel but moving steam not moving water causes the motion. Steam enters the turbine and is directed by the **stator** or **diaphragm** (sets of fixed blades) on to the **rotor** (sets of blades on a shaft that can rotate), Figure 17.6. The rotor revolves and drives the electrical generator. The steam expands as it passes through the turbine and the size of the blades increases along the turbine to allow for this.

Figure 17.4 Filling up with biofuel in Brazil

Figure 17.5 Feeding a biogas digester in rural India

Figure 17.6 The rotor of a steam turbine

The overall efficiency of thermal power stations is only about 30%. They require cooling towers to condense steam from the turbine to water and this is a waste of energy. A block diagram and an energy-transfer diagram for a thermal power station are given in Figure 17.7.

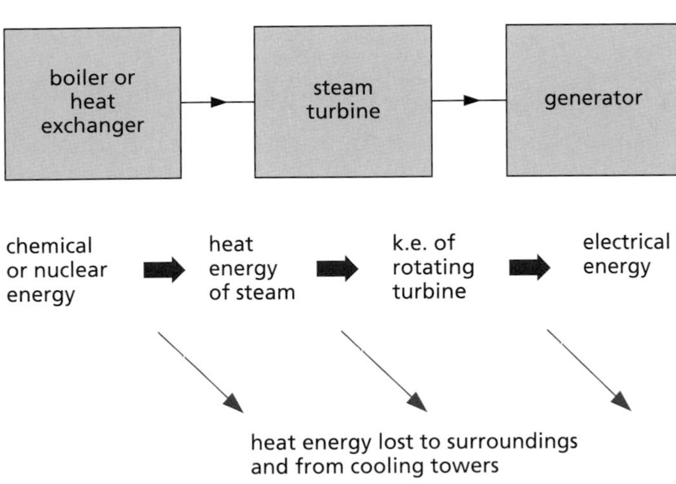

Figure 17.7 Energy transfers in a thermal power station

In gas-fired power stations, natural gas is burnt in a **gas turbine** linked directly to an electricity generator. The hot exhaust gases from the turbine are not released into the atmosphere but used to produce steam in a boiler. The steam is then used to generate more electricity from a steam turbine driving another generator. The efficiency is claimed to be over 50% without any extra fuel consumption. Furthermore, the gas turbines have a near-100% combustion efficiency so very little harmful exhaust gas (i.e. unburnt methane) is produced, and natural gas is almost sulfur-free so the environmental pollution caused is much less than for coal.

b) Renewable sources

In most cases the renewable energy source is used to drive turbines directly, as explained earlier in the cases of hydroelectric, wind, wave, tidal and geothermal schemes.

The block diagram and energy-transfer diagram for a hydroelectric scheme like that in Figure 16.3d (p. 72) are shown in Figure 17.8. The efficiency of a large installation can be as high as 85–90% since many of the causes of loss in thermal power stations (e.g. water cooling towers) are absent. In some cases the generating costs are half those of thermal stations.

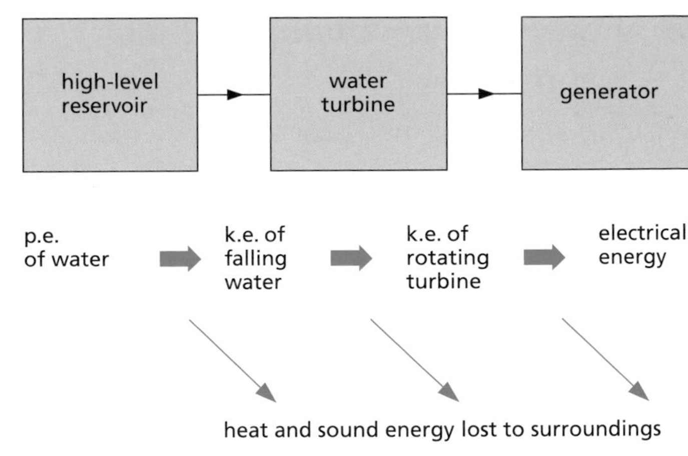

Figure 17.8 Energy transfers in a hydroelectric power station

A feature of some hydroelectric stations is **pumped storage**. Electrical energy cannot be stored on a large scale but must be used as it is generated. The demand varies with the time of day and the season, Figure 17.9, so in a pumped-storage system electricity generated at off-peak periods is used to pump water back up from a low-level reservoir to a higher-level one. It is easier to do this than to reduce the output of the generator. At peak times the p.e. of the water in the high-level reservoir is converted back into electrical energy; three-quarters of the electrical energy that was used to pump the water is generated.

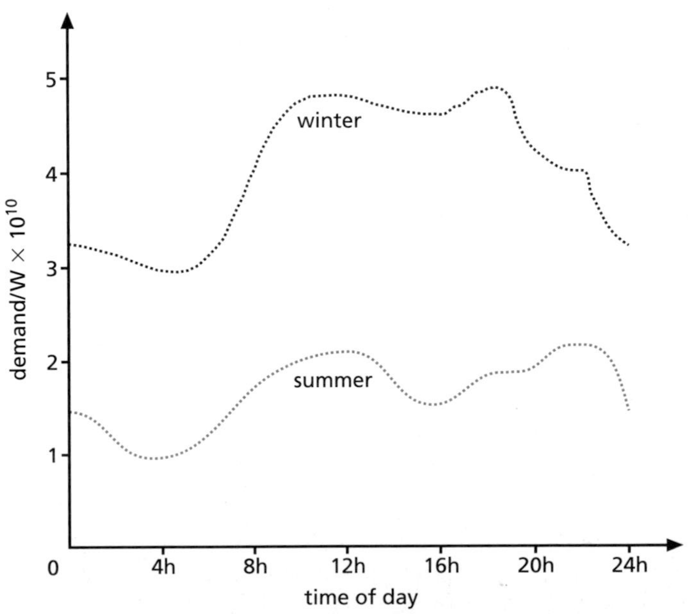

Figure 17.9 Variation in power demand

Economic, environmental and social issues

When considering the large-scale generation of electricity, the economic and environmental costs of using various energy sources have to be weighed against the benefits that electricity brings to society as a 'clean', convenient and fairly 'cheap' energy supply.

Environmental problems such as polluting emissions that arise with different energy sources were outlined when each was discussed previously. Apart from people using less energy, how far pollution can be reduced by, for example, installing desulphurization processes in coal-fired power stations, is often a matter of cost.

Although there are no fuel costs associated with electricity generation from renewable energy sources such as wind power, the energy is so 'dilute' that the capital costs of setting up the generating installation are high. Similarly, although fuel costs for nuclear power stations are relatively low, the costs of building the stations and of dismantling them at the end of their useful lives is higher than for gas- or coal-fired stations.

It has been estimated that currently (2008) it costs between 2.2p and 3.2p to produce a unit of electricity in a gas- or coal-fired power station in the UK. The cost for a nuclear power station is in excess of 2.3p per unit. Wind energy costs vary, depending upon location, but are in the range 3.7p to 5.5p per unit. In the most favourable locations wind competes with coal and gas generation.

The reliability of a source has also to be considered, as well as how easily production can be started up and shut down as demand for electricity varies. Natural gas power stations have a short start-up time, while coal and then oil power stations take successively longer to start up; nuclear power stations take longest. They are all reliable in that they can produce electricity at any time of day and in any season of the year as long as fuel is available. Hydroelectric power stations are also very reliable and have a very short start-up time which means they can be switched on when the demand for electricity peaks. The electricity output of a tidal power station, although predictable, is not as reliable because it depends on the height of the tide which varies over daily, monthly and seasonal time scales. The wind and the Sun are even less reliable sources of energy since the output of a wind turbine changes with the strength of the wind and that of a solar cell with the intensity of light falling on it; the output may not be able to match the demand for electricity at a particular time.

Renewable sources are still only being used on a small scale globally. The contribution of the main energy sources to the world's total energy consumption at present is given in Table 17.1. (The use of biofuels is not well documented.) The pattern in the UK is similar but France generates nearly three-quarters of its electricity from nuclear plants; for Japan and Taiwan the proportion is one-third, and it is in the developing economies of East Asia where interest in nuclear energy is growing most dramatically. However, the great dependence on fossil fuels worldwide is evident. It is clear the world has an energy problem, Figure 17.10.

Table 17.1 World use of energy sources

Oil	Coal	Gas	Nuclear	Hydroelectric
36%	29%	23%	6%	6%

Consumption varies from one country to another; North America and Europe are responsible for about 42% of the world's energy consumption each year. Table 17.2 shows approximate values for the annual consumption per head of population for different areas. These figures include the 'hidden' consumption in the manufacturing and transporting of goods.

The world average consumption is 69×10^9 J per head per year.

Table 17.2 Energy consumption per head per year/J $\times 10^9$

N. America	UK	Japan	S. America	China	Africa
335	156	172	60	55	20

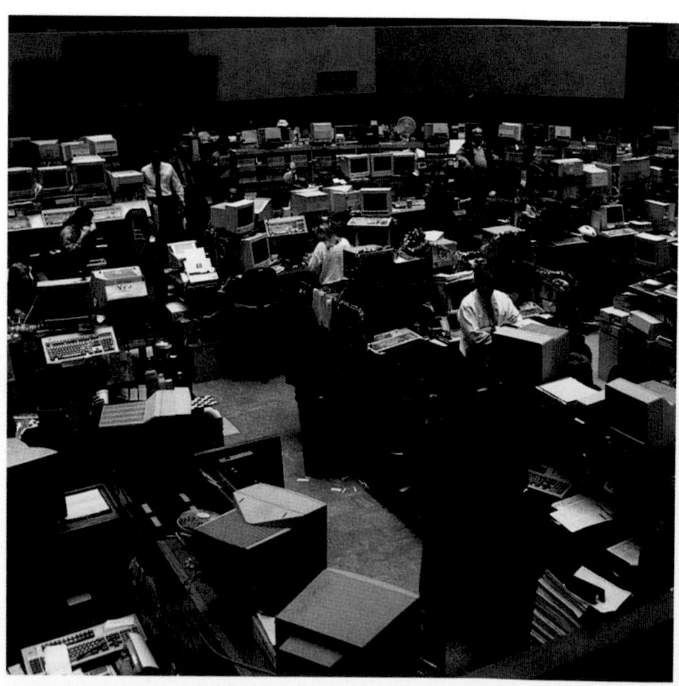

Figure 17.10 An energy supply crisis in California forces a blackout on stock exchange traders

Questions

1 The pie chart in Figure 17.11 shows the percentages of the main energy sources used by a certain country.
  a What percentage is supplied by water power?
  b Which of the sources is/are renewable?
  c What is meant by 'renewable'?
  d Name two other renewable sources.
  e Why, if energy is always conserved, is it important to develop renewable sources?

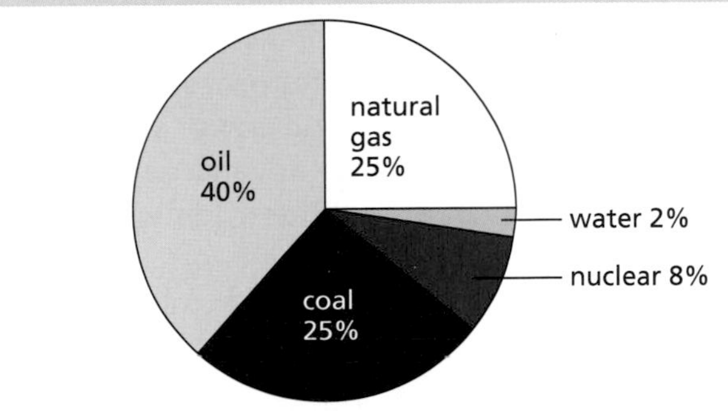

Figure 17.11

2 List *six* properties which you think the ideal energy source should have for generating electricity in a power station.

3 a List *six* social everyday benefits for which electrical energy is responsible.
  b Draw up *two* lists of suggestions for saving energy
  (i) in the home, and
  (ii) globally.

4 A group of houses uses solar panels and windmills as alternative energy sources.

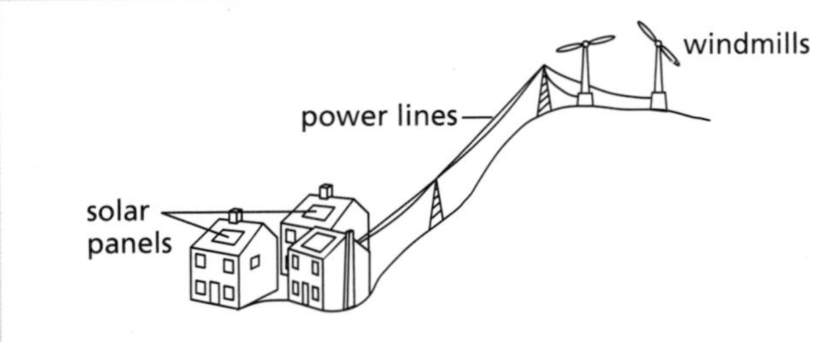

Figure 17.12

  a (i) Write down one **advantage** of using these alternative energy sources.
  (ii) Write down one **disadvantage** of using solar panels.
  (iii) Write down one **disadvantage** of using windmills.
  b Jan works out the efficiency of one of the windmills. The energy of the air hitting the blades of the windmills is 20 000 J each second. The energy transferred to the power lines is 5000 J each second.
  Calculate the efficiency of the windmill. Use the equation below. You *must* show how you work out your answer.

$$\text{energy efficiency} = \frac{\text{useful energy output}}{\text{total energy input}}$$

  c Energy can be transferred by **conduction**, **convection** and **radiation**.
  Figures 17.13a, b show details of a solar panel.

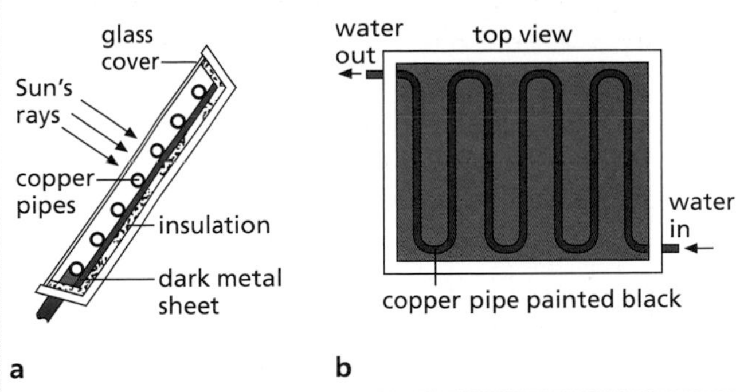

Figure 17.13

Explain why the water coming out is a lot warmer than the water going in. Use your ideas about energy transfer.

(*OCR Foundation, June 1999*)

5 This question is about generating electricity.
A dam has been built across a river where it meets the sea. The levels of the water go up and down twice a day. Figure 17.14 shows the water levels around the dam at high tide. The dam contains a tidal power station.

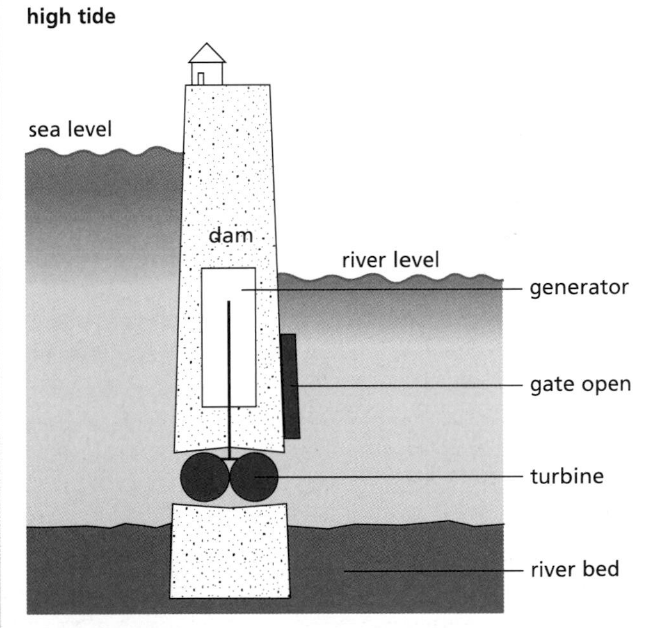

Figure 17.14

  a Describe how the energy of the water can be used to produce electricity.
  b Tidal and hydro-electric power are **renewable** sources of energy. Write down the name of *one* other renewable source of energy.
  c The gate on the dam was closed until low tide. This keeps the water level in the river high as shown in Figure 17.15.

Continued

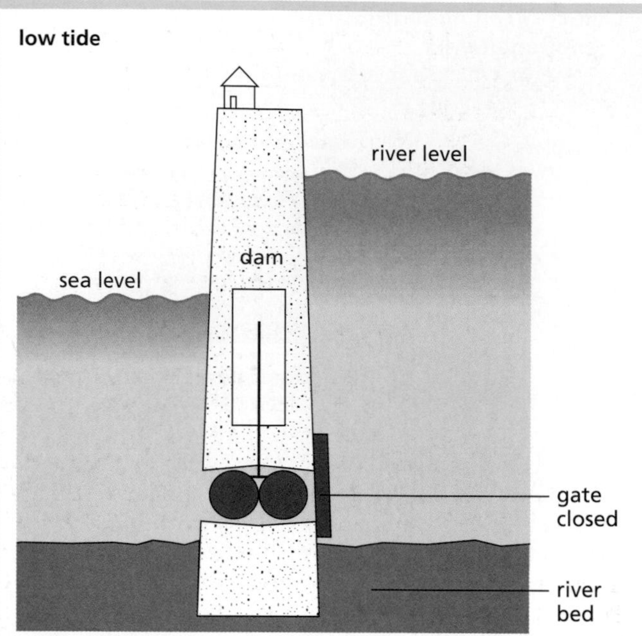

Figure 17.15

Suggest *one* problem this can cause to the environment.

(*OCR Foundation, June 1999*)

6 The map in Figure 17.16 shows the position of towns **A** and **B** on the banks of a large river estuary. **A** is an important fishing and ferry port. The wind usually blows from the west. The major roads and railways are shown.

A power station is to be built in area **X** to generate electricity for the region. The choice is between a nuclear power station and a coal-fired power station.

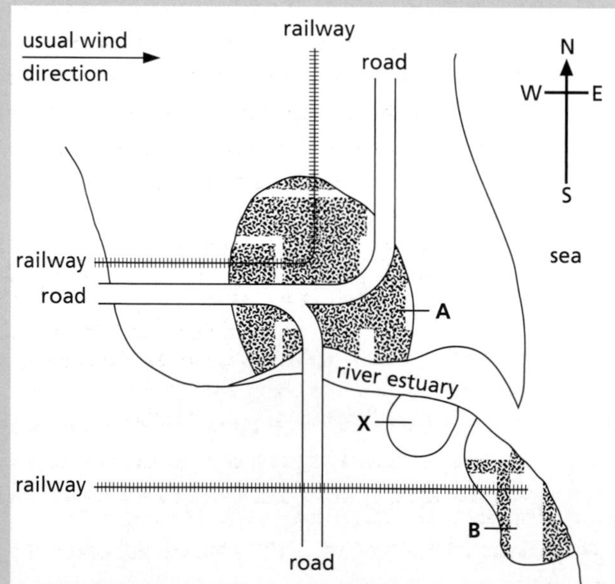

Figure 17.16

a State the advantages and disadvantages of the two methods of generating electrical energy.

b Which method would you choose for this site? Explain the reasons for your choice.

(*NEAB Higher, June 1998*)

Checklist

After studying this chapter you should be able to

- distinguish between **renewable** and **non-renewable** energy sources,
- give some advantages and some disadvantages of using non-renewable fuels,
- describe the different ways of harnessing **solar, wind, wave, tidal, hydroelectric, geothermal** and **biomass** energy,
- describe the energy transfer processes in a thermal and a hydroelectric power station,
- compare and contrast the advantages and disadvantages of using different energy sources to generate electricity,
- discuss the environmental and economic issues of electricity production and consumption.

18 Pressure and liquid pressure

Pressure

To make sense of some effects in which a force acts on a body we have to consider not only the force but also the area on which it acts. For example, wearing skis prevents you sinking into soft snow because your weight is spread over a greater area. We say the **pressure** is less.

Pressure is the force (or thrust) **acting on unit area** (i.e. 1 m²) and is calculated from

$$\text{pressure} = \frac{\text{force}}{\text{area}}$$

The unit of pressure is the **pascal** (Pa); it equals 1 newton per square metre (N/m²) and is quite a small pressure. An apple in your hand exerts about 1000 Pa.

The greater the area over which a force acts, the less the pressure. The pressure exerted on the floor by the same box, (a) standing on end, (b) lying flat, is shown in Figure 18.1. This is why a tractor with wide wheels can move over soft ground. The pressure is large when the area is small and accounts for nails being given sharp points. Walnuts can be broken in the hand by squeezing two together but not one. Why?

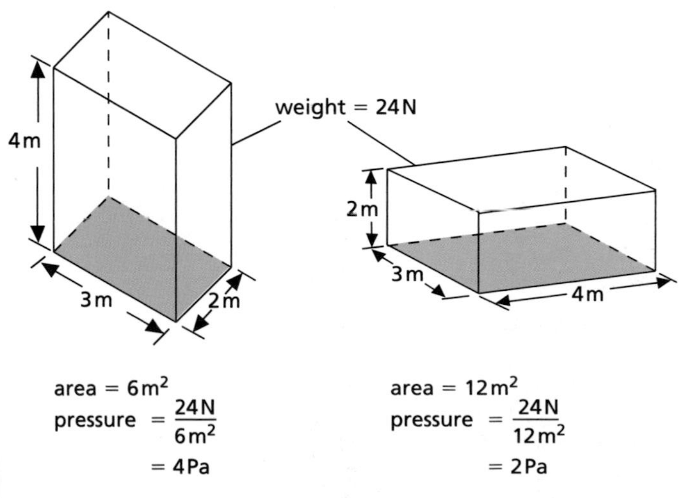

Figure 18.1

Liquid pressure

1 **Pressure in a liquid increases with depth** because the farther down you go the greater the weight of liquid above. In Figure 18.2a water spurts out fastest and furthest from the lowest hole.

2 **Pressure at one depth acts equally in all directions.** The can of water in Figure 18.2b has similar holes all round it at the same level. Water comes out equally fast and spurts equally far from each hole. Hence the pressure exerted by the water at this depth is the same in all directions.

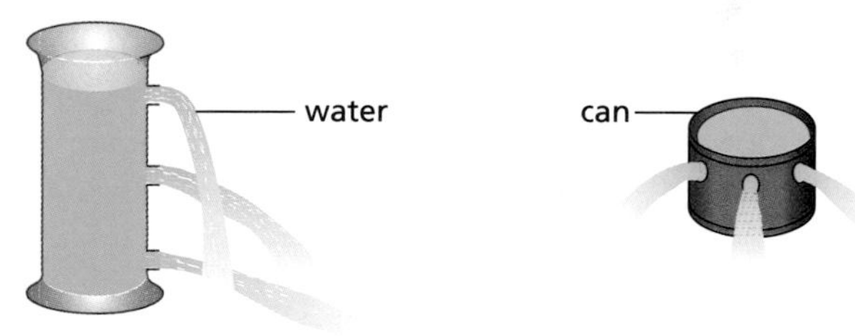

Figure 18.2

3 **A liquid finds its own level.** In the U-tube of Figure 18.3a the liquid pressure at the foot of P is greater than at the foot of Q because the left-hand column is higher than the right-hand one. When the clip is opened the liquid flows from P to Q until the pressure and the levels are the same, i.e. the liquid 'finds its own level'. Although the weight of liquid in Q is now greater than in P, it acts over a greater area because tube Q is wider.

In Figure 18.3b the liquid is at the same level in each tube and confirms that the pressure at the foot of a liquid column depends only on the *vertical* depth of the liquid and not on the tube width or shape.

4 **Pressure depends on the density of the liquid.** The denser the liquid, the greater the pressure at any given depth.

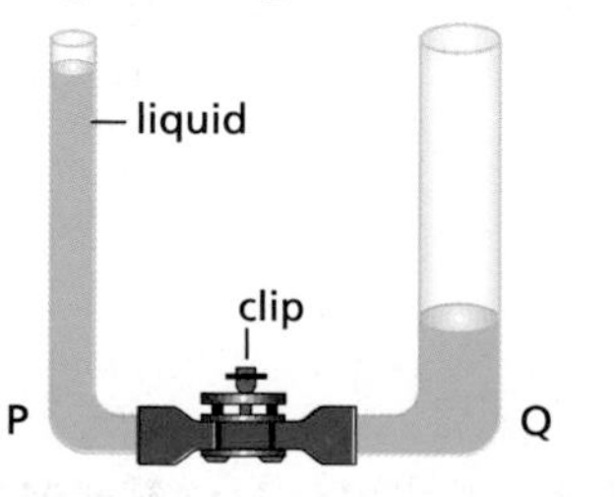

Figure 18.3

Water supply system

A town's water supply often comes from a reservoir on high ground. Water flows from it through pipes to any tap or storage tank that is below the level of water in the reservoir, Figure 18.4. The lower the place supplied, the greater the water pressure. In very tall buildings it may be necessary first to pump the water to a large tank in the roof.

Reservoirs for water supply or for hydroelectric power stations are often made in mountainous regions by building a dam at one end of a valley. The dam must be thicker at the bottom than at the top due to the large water pressure at the bottom.

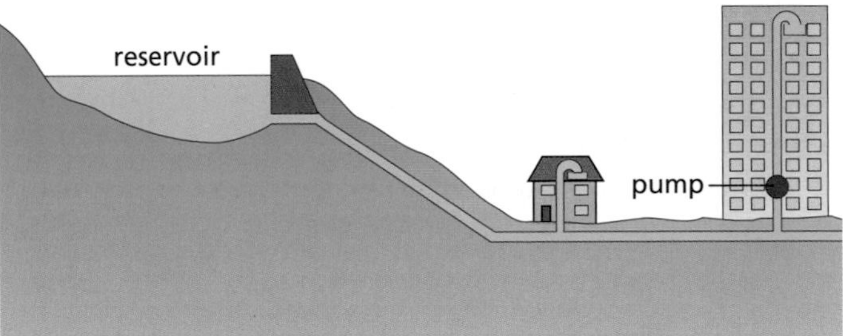

Figure 18.4 Water supply system. Why is the pump needed in the high-rise building?

Hydraulic machines

Liquids are almost incompressible (i.e. their volume cannot be reduced by squeezing) and they 'pass on' any pressure applied to them. Use is made of these facts in hydraulic machines. Figure 18.5 shows the principle on which they work.

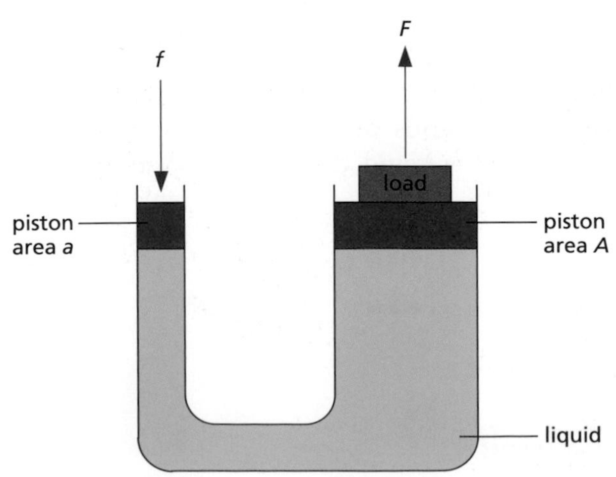

Figure 18.5 The hydraulic principle

Suppose a downward force f acts on a piston of area a. The pressure transmitted through the liquid is

$$\text{pressure} = \frac{\text{force}}{\text{area}} = \frac{f}{a}$$

This pressure acts on a second piston of larger area A, producing an upward force $F = \text{pressure} \times \text{area}$:

$$F = \frac{f}{a} \times A$$

or

$$F = f \times \frac{A}{a}$$

Since A is larger than a, F must be larger than f and the hydraulic system is a force multiplier; the **multiplying factor** is A/a.

For example if $f = 1$ N, $a = \frac{1}{100}$ m^2 and $A = \frac{1}{2}$ m^2 then

$$F = 1\,\text{N} \times \frac{\frac{1}{2}\,\text{m}^2}{\frac{1}{100}\,\text{m}^2}$$

$$= 50\,\text{N}$$

A force of 1 N could lift a load of 50 N; the hydraulic system multiplies the force 50 times.

A **hydraulic jack**, Figure 18.6, has a platform on top of piston B and is used in garages to lift cars. Both valves open only to the right and they allow B to be raised a long way when A moves up and down repeatedly. In a hydraulic press there is a fixed plate above piston B, and sheets of steel placed between B and the plate can be forged.

Hydraulic fork-lift trucks and similar machines such as loaders, Figure 18.7, work in the same way.

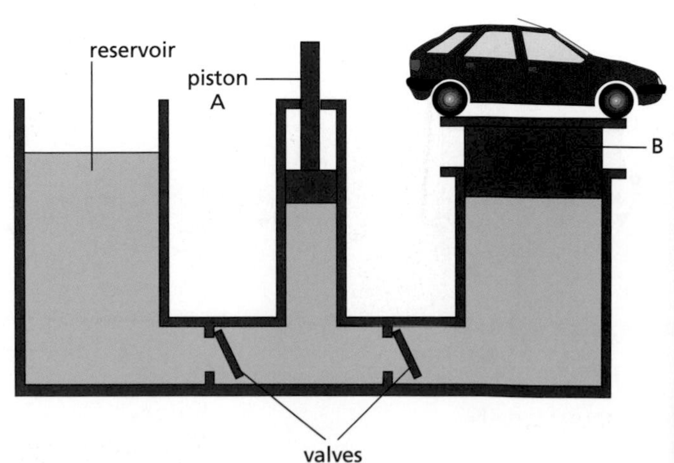

Figure 18.6 A hydraulic jack

Figure 18.7 A hydraulic machine in action

Hydraulic car brakes are shown in Figure 18.8. When the brake pedal is pushed the piston in the master cylinder exerts a force on the brake fluid and the resulting pressure is transmitted equally to eight other pistons (four are shown). These force the brake shoes or pads against the wheels and stop the car.

Expression for liquid pressure

In designing a dam an engineer has to calculate the pressure at various depths below the water surface. The pressure increases with depth and density.

An expression for the pressure at a depth h in a liquid of density ρ can be found by considering a horizontal area A, Figure 18.9. The force acting vertically downwards on A equals the weight of a liquid column of height h and cross-sectional area A above it. Then

$$\text{volume of liquid column} = hA$$

Since mass = volume × density we can say

$$\text{mass of liquid column} = hA\rho$$

Taking a mass of 1 kg to have weight 10 N,

$$\text{weight of liquid column} = 10\ hA\rho$$

$$\therefore \quad \text{force on area } A = 10\ hA\rho$$

$$\therefore \quad \text{pressure} = \text{force/area} = 10\ hA\rho/A$$

$$\therefore \quad \text{pressure} = 10\ h\rho$$

This is usually written as

$$\text{pressure} = \text{depth} \times \text{density} \times \text{g}$$

$$= \text{h}\rho\text{g}$$

where g is the acceleration of free fall (Chapter 21).

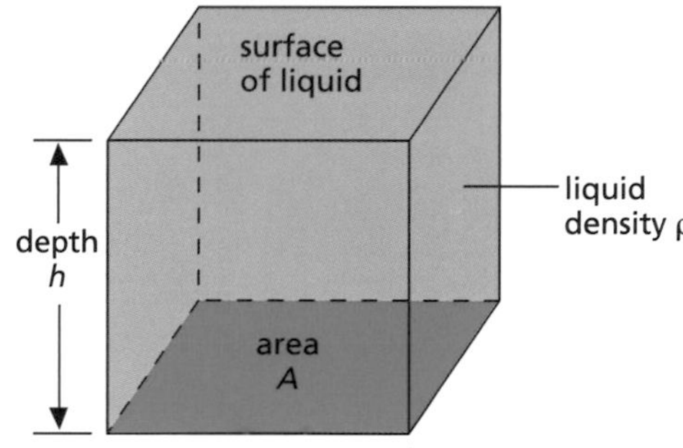

Figure 18.9

This pressure acts equally in all directions at depth h and depends only on h and ρ. Its value will be in Pa if h is in m and ρ in kg/m^3.

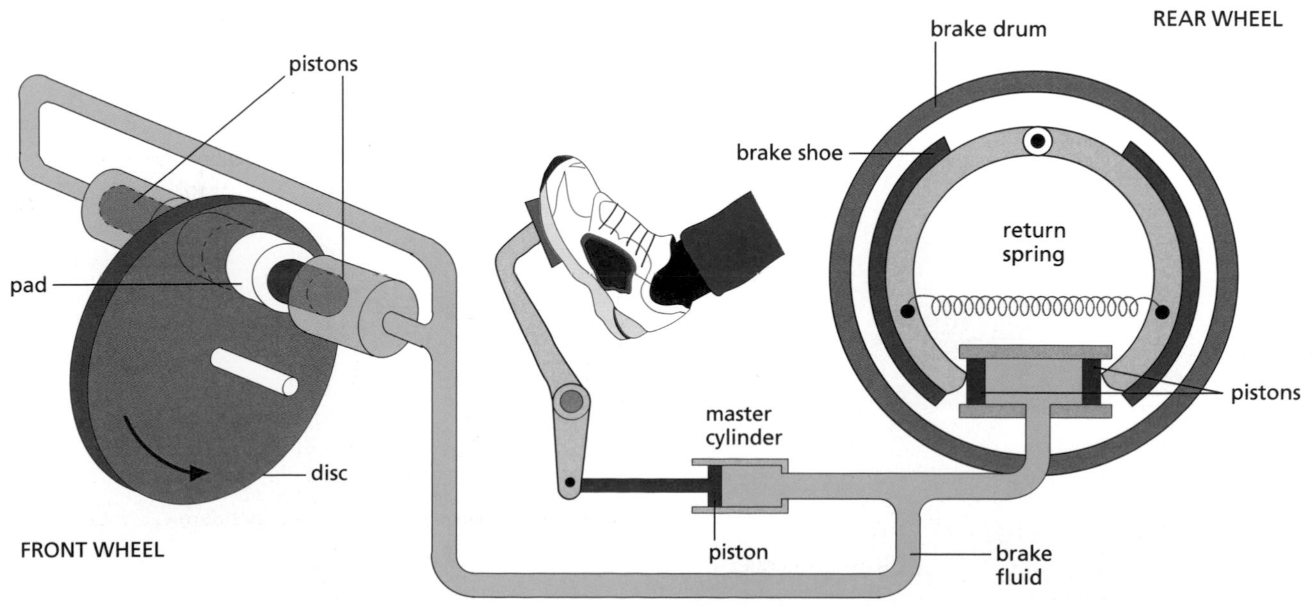

Figure 18.8 Hydraulic brakes

Pressure gauges

These measure the pressure exerted by a fluid, i.e. by a liquid or a gas.

a) Bourdon gauge

This works like the toy in Figure 18.10. The harder you blow into the paper tube, the more it uncurls. In a Bourdon gauge, Figure 18.11, when a fluid pressure is applied, the curved metal tube tries to straighten out and rotates a pointer over a scale. Car oil-pressure gauges and the gauges on gas cylinders are of this type.

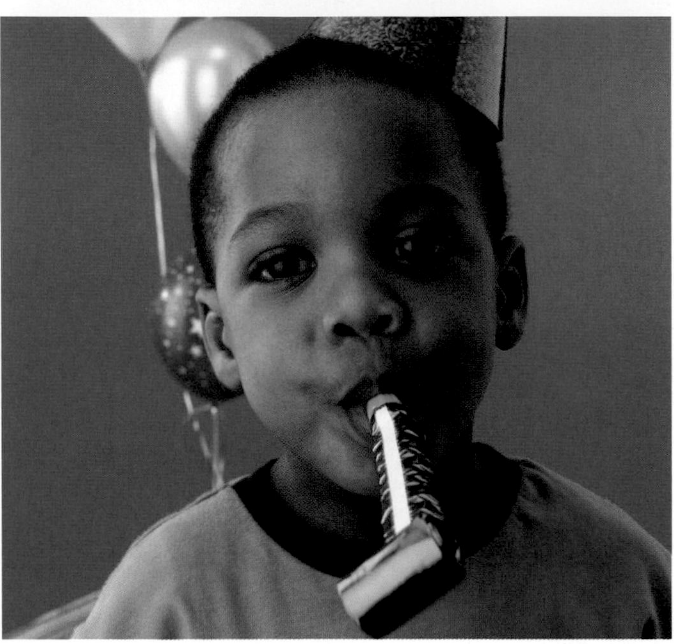

Figure 18.10 The harder you blow, the greater the pressure and the more it uncurls

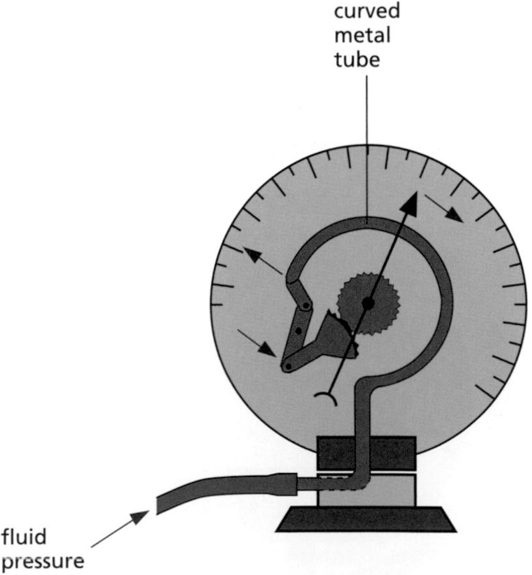

Figure 18.11 A Bourdon gauge

b) U-tube manometer

In Figure 18.12a each surface of the liquid is acted on equally by atmospheric pressure and the levels are the same. If one side is connected to, for example, the gas supply, Figure 18.12b, the gas exerts a pressure on surface A and level B rises until

pressure of gas = atmospheric pressure
+ pressure due to liquid column BC

The pressure of the liquid column BC therefore equals the amount by which the gas pressure *exceeds* atmospheric pressure. It equals $h\rho g$ (in Pa) where h is the vertical height of BC (in m) and ρ is the density of the liquid (in kg/m^3). The height h is called the **head of liquid** and sometimes, instead of stating a pressure in Pa, we say that it is so many cm of water (or mercury for higher pressures).

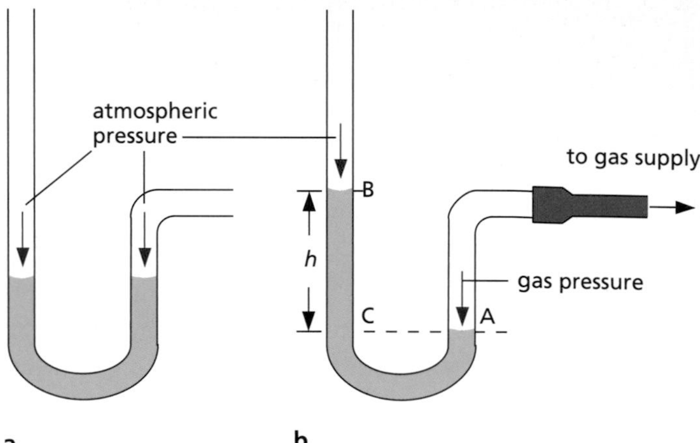

Figure 18.12 A U-tube manometer

c) Mercury barometer

A barometer is a manometer which measures atmospheric pressure. A simple barometer is shown in Figure 18.13. The pressure at X due to the weight of the column of mercury XY equals the atmospheric pressure on the surface of the mercury in the bowl. The height XY measures the atmospheric pressure in mm of mercury (mmHg).

The *vertical* height of the column is unchanged if the tube is tilted. Would it be different with a wider tube? The space above the mercury in the tube is a vacuum (except for a little mercury vapour).

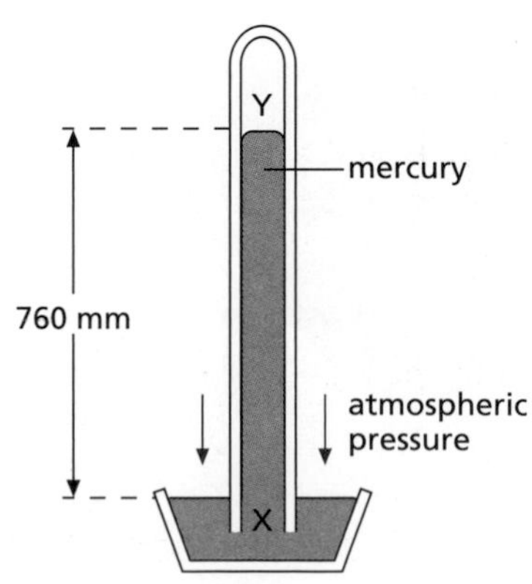

Figure 18.13 Mercury barometer

Questions

1 a What is the pressure on a surface when a force of 50 N acts on an area of
(i) 2.0 m^2,
(ii) 100 m^2,
(iii) 0.50 m^2?
b A pressure of 10 Pa acts on an area of 3.0 m^2. What is the force acting on the area?

2 In a hydraulic press a force of 20 N is applied to a piston of area 0.20 m^2. The area of the other piston is 2.0 m^2. What is
a the pressure transmitted through the liquid,
b the force on the other piston?

3 a Why must a liquid and not a gas be used as the 'fluid' in a hydraulic machine?
b On what other important property of a liquid do hydraulic machines depend?

4 What is the pressure 100 m below the surface of sea water of density 1150 kg/m^3?

5 a Heavy furniture sometimes marks the floor on which it stands. Four tables of the same weight each have four legs. Figure 18.14 shows part of a leg from each table.

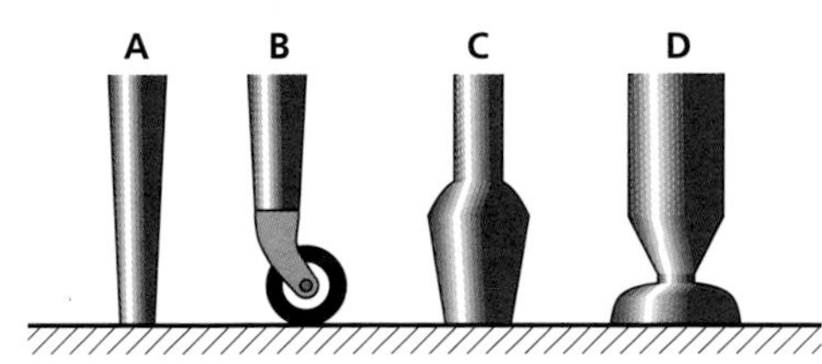

Figure 18.14

(i) Which leg is least likely to mark the floor underneath it?
(ii) Explain your answer.
b A hot flat metal sheet is placed on a horizontal surface.

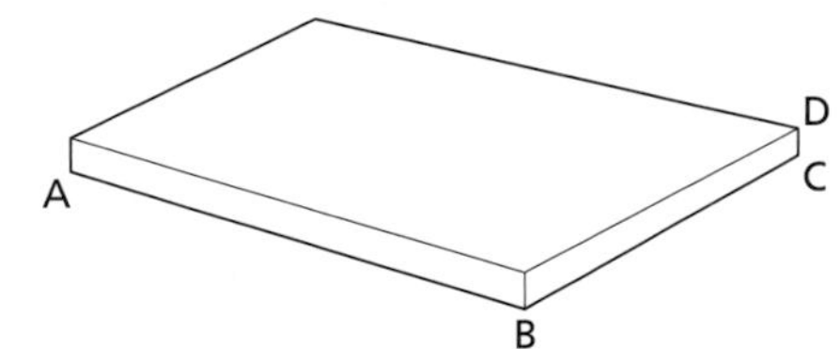

Figure 18.15

As the hot metal sheet cools, what happens to the quantities in the list below? Say for each whether the quantity *increases*, *decreases* or *stays the same*.

length AB width BC thickness CD
area touching the horizontal surface
mass of sheet weight of sheet
density of metal
pressure on horizontal surface

(UCLES IGCSE Physics Core, May 2000)

6 Figure 18.16 shows a simple barometer.
a What is the region A?
b What keeps the mercury in the tube?
c What is the value of the atmospheric pressure being shown by the barometer?
d What would happen to this reading if the barometer were taken up a high mountain? Give a reason.

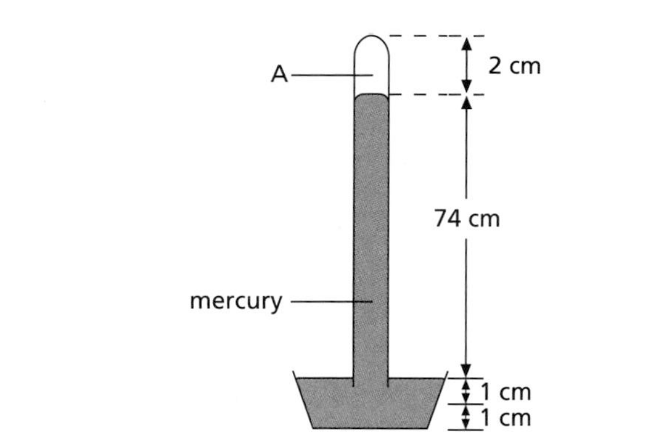

Figure 18.16

Checklist

After studying this chapter you should be able to

- relate pressure to force and area and give examples,
- define **pressure** and recall its unit,
- connect the pressure in a fluid with its depth and density,
- recall that pressure is transmitted through a liquid and use it to explain the hydraulic jack and hydraulic car brakes,
- use pressure = $h\rho g$ to solve problems,
- describe how a U-tube manometer may be used to measure fluid pressure,
- describe and use a simple mercury barometer.

Forces and resources
Additional questions

Moments and levers

1 a The diagram shows three types of water tap (faucet).

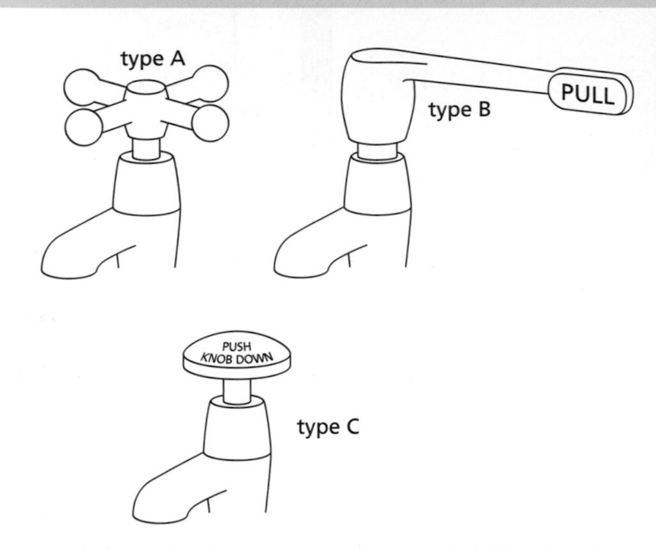

For each tap, write YES if a person would need to cause a moment on the handle in order to make the water flow, or NO if the person would not need to cause a moment.

b An old person has taps of type A in the kitchen. The person has difficulty operating the tap. What could be done to make it easier to operate the tap?

(*UCLES IGCSE Physics Core, Nov 2000*)

2 a What is meant by the *moment* of a force?

b The diagram shows a crane on a construction site. The jib of the crane is balanced about the point F.

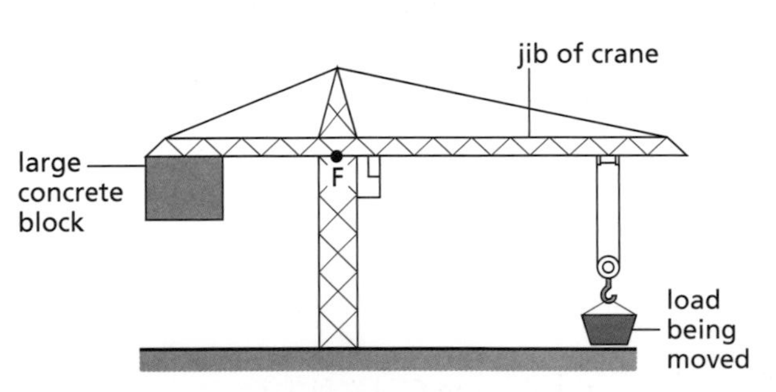

Apply your knowledge of balancing beams to
(i) suggest a reason for the large concrete block which is permanently fixed to the crane,
(ii) state whether the weight of the load, in the position shown, is likely to be greater than, smaller than, or the same as the weight of the concrete block.

(*UCLES IGCSE Physics Core, Nov 1998*)

3 a A uniform beam AB of weight W is balanced at its midpoint on a pivot. Two weights W_1 and W_2 are then hung at equal distances from the midpoint of the beam.

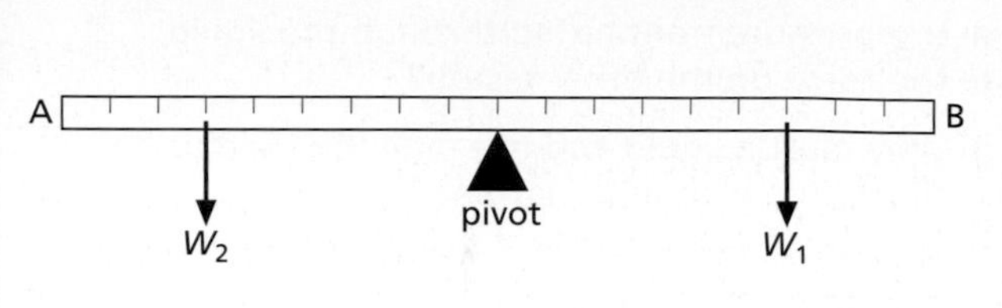

When this is done, the end B moves down.
(i) Which is the heavier weight?
(ii) Which way would W_1 have to be moved so that the beam is again balanced?

b W_2 is removed from the beam. This means that the only forces acting downwards on the beam are the weight W of the beam and W_1. W is much greater than W_1.

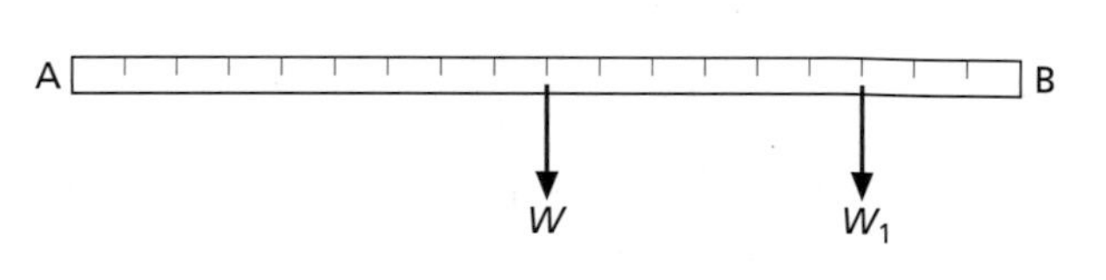

Copy the diagram and mark a possible position for the pivot to be placed so that the beam is again balanced.

(*UCLES IGCSE Physics Core, May 1999*)

Adding forces

4 Find the size of the resultant of two forces of 5 N and 12 N acting

a in opposite directions to each other,

b at 90° to each other.

Energy transfer

5 State what energy transfers occur in

a a hairdryer,

b a refrigerator,

c an audio system.

6 An escalator carries 60 people of average mass 70 kg to a height of 5 m in one minute. Find the power needed to do this.

7 A person walks from A to E, a journey which goes over the top of the hill BCD, as shown in the diagram.

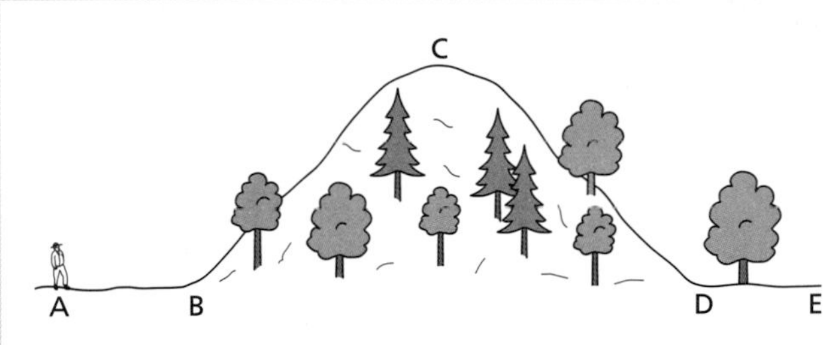

a (i) During which part of the walk does the person do most work, AB, BC, CD or DE?
(ii) Explain your answer to **a** (i).

b (i) The person now runs over the hill from A to E. How does the average power developed by the person compare with that when the person walked? Is it *greater than*, *the same as*, or *less than* when walking?
(ii) Explain your answer to **b** (i).

(*UCLES IGCSE Physics Core, Nov 1999*)

8 The diagram at the foot of the page shows one way of using water to generate electricity.

a Write out the words that are missing in the boxes.

b In other places, water is used in different ways to generate electricity. State two of these ways.

(*UCLES IGCSE Physics Core, May 2000*)

Pressure and liquid pressure

9 A pond is kept at a constant depth by a pressure-operated valve in the base.

a The pond is kept at a depth of 2.0 m. The density of water is 1000 kg/m^3. Calculate the water pressure on the valve.

b The force required to open the valve is 50 N. The valve will open when the water depth reaches 2.0 m. Calculate the area of the valve.

c The water supply is turned off and the valve is held open so that water drains out through the valve. State the energy changes of the water that occur as the depth of the water drops from 2.0 m to zero.

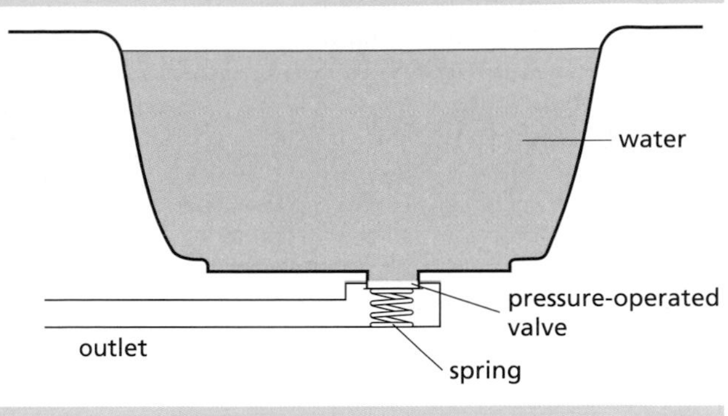

(*UCLES IGCSE Physics Extended, Nov 2005*)

10 Which of the following will damage a wood-block floor that can withstand a pressure of 2000 kPa (2000 kN/m^2)?

1 A block weighing 2000 kN standing on an area of 2 m^2.
2 An elephant weighing 200 kN standing on an area of 0.2 m^2.
3 A girl of weight 0.5 kN wearing stiletto-heeled shoes standing on an area of 0.0002 m^2.

Use the answer code:

A 1, 2, 3 **B** 1, 2 **C** 2, 3 **D** 1 **E** 3

11 The pressure at a point in a liquid

1 increases as the depth increases
2 increases if the density of the liquid increases
3 is greater vertically than horizontally.

Which statement(s) is (are) correct?

A 1, 2, 3 **B** 1, 2 **C** 2, 3 **D** 1 **E** 3

12 a (i) Draw a clear diagram of a simple mercury barometer. Now fully label your diagram.
(ii) State the physical quantity that can be determined by using a mercury barometer.
(iii) On your diagram in (i), mark clearly, using the letter *h*, the length you would measure to determine the physical quantity named in (ii).

continued

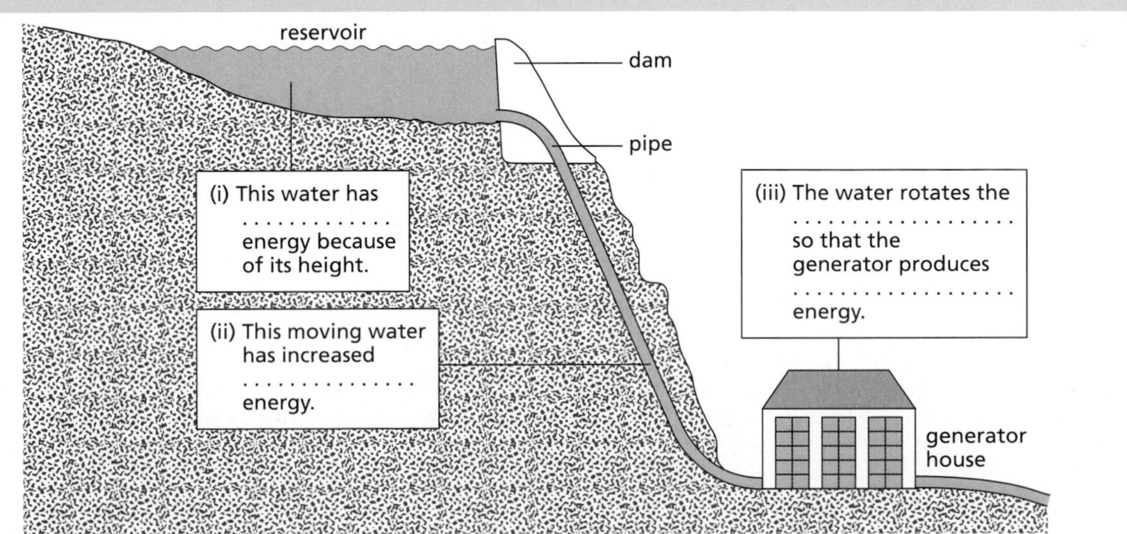

b The diagram below shows a manometer being used to measure the pressure of the gas in a container.
(i) What is the difference in height of the two mercury levels?
(ii) What does the difference in height of the mercury levels indicate?
(iii) State whether the gas pressure is greater than or less than the atmospheric pressure and how you know this.

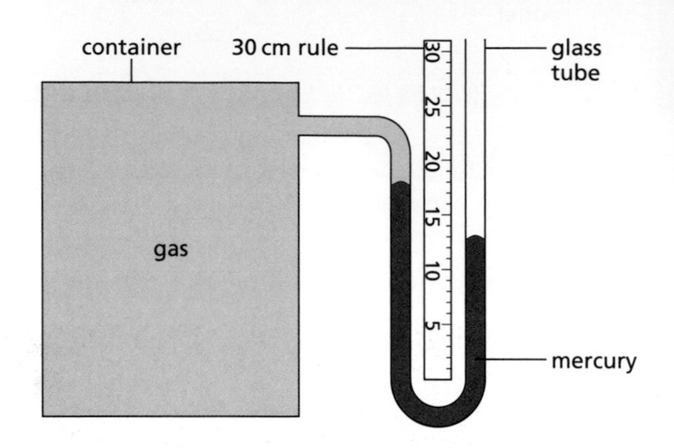

(iv) What would happen to the two mercury levels if the gas pressure increased slightly?
(v) The mercury manometer is replaced by another manometer that contains a liquid of lower density. How, if at all, does this affect the difference between the liquid levels?
(*UCLES IGCSE Physics Core, May 1999*)

Motion and energy

19 Speed, velocity and acceleration

Speed	*Timers*
Velocity	*Practical work*
Acceleration	Analysing motion.

Speed

If a car travels 300 km from Liverpool to London in five hours, its **average speed** is 300 km/5 h = 60 km/h. The speedometer would certainly not read 60 km/h for the whole journey but might vary considerably from this value. That is why we state the average speed. If a car could travel at a constant speed of 60 km/h for five hours, the distance covered would still be 300 km. It is *always* true that

$$\text{average speed} = \frac{\text{distance moved}}{\text{time taken}}$$

To find the actual speed at any instant we would need to know the distance moved in a very short interval of time. This can be done by multiflash photography. In Figure 19.1 the golfer is photographed while a flashing lamp illuminates him 100 times a second. The speed of the club-head as it hits the ball is about 200 km/h.

Velocity

Speed is the distance travelled in unit time; **velocity is the distance travelled in unit time in a stated direction**. If two trains travel due north at 20 m/s, they have the same speed of 20 m/s and the same velocity of 20 m/s *due north*. If one travels north and the other south, their speeds are the same but not their velocities since their directions of motion are different. Speed is a scalar and velocity a vector quantity (see Chapter 15).

$$\text{velocity} = \frac{\text{distance moved in a stated direction}}{\text{time taken}}$$

The velocity of a body is **uniform** or **constant** if it moves with a steady speed in a straight line. It is not uniform if it moves in a curved path. Why?

The units of speed and velocity are the same, e.g. km/h, m/s, and

$$60\ \text{km/h} = 60\,000\ \text{m}/3600\ \text{s} = 17\ \text{m/s}$$

Distance moved in a stated direction is called the **displacement**. It is a vector, unlike distance which is a scalar. Velocity may also be defined as

$$\text{velocity} = \frac{\text{displacement}}{\text{time taken}}$$

Figure 19.1 Multiflash photograph of a golf swing

Acceleration

When the velocity of a body changes we say the body **accelerates**. If a car starting from rest and moving due north has velocity 2 m/s after 1 second, its velocity has increased by 2 m/s in 1 s and its acceleration is 2 m/s per second due north. We write this as 2 m/s^2.

Acceleration is the change of velocity in unit time, or

$$\text{acceleration} = \frac{\text{change of velocity}}{\text{time taken for change}}$$

For a steady increase of velocity from 20 m/s to 50 m/s in 5 s

$$\text{acceleration} = \frac{(50 - 20)\ \text{m/s}}{5\ \text{s}} = 6\ \text{m/s}^2$$

Acceleration is also a vector and both its magnitude and direction should be stated. However, at present we will consider only motion in a straight line and so the magnitude of the velocity will equal the speed, and the magnitude of the acceleration will equal the change of speed in unit time.

The speeds of a car accelerating on a straight road are shown below.

Time/s	0	1	2	3	4	5	6
Speed/m/s	0	5	10	15	20	25	30

The speed increases by 5 m/s every second and the acceleration of 5 m/s^2 is said to be **uniform**.

An acceleration is positive if the velocity increases and negative if it decreases. A negative acceleration is also called a **deceleration** or **retardation**.

Timers

A number of different devices are useful for analysing motion in the laboratory.

a) Motion sensors

Motion sensors use the ultrasonic echo technique (see p. 38) to determine the distance of an object from the sensor. Connection of a datalogger and computer to the motion sensor then enables a distance–time graph to be plotted directly (see Figure 19.6). Further data analysis by the computer allows a velocity–time graph to be obtained, as in Figures 20.1 and 20.2, p. 96.

b) Tickertape timer: tape charts

A tickertape timer also enables us to measure speeds and hence accelerations. One type, Figure 19.2, has a marker which vibrates 50 times a second and makes dots at $\frac{1}{50}$ s intervals on the paper tape being pulled through it; $\frac{1}{50}$ s is called a 'tick'.

The distance between successive dots equals the average speed of whatever is pulling the tape in, say, cm per $\frac{1}{50}$ s, i.e. cm per tick. The 'tentick' ($\frac{1}{5}$ s) is also used as a unit of time. Since ticks and tenticks are small we drop the 'average' and just refer to the 'speed'.

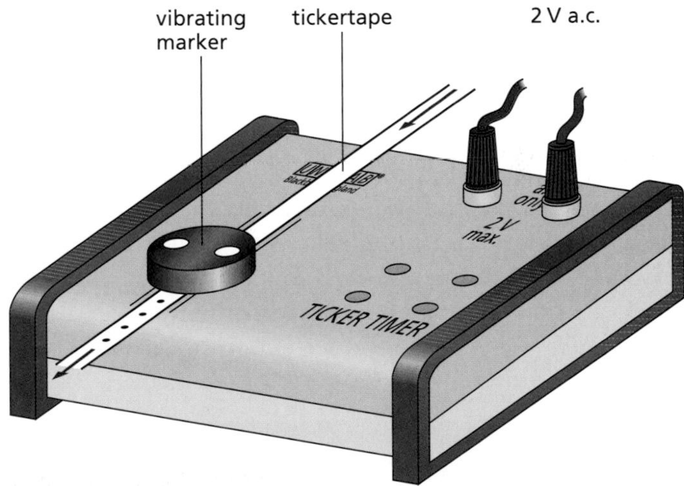

Figure 19.2 Tickertape timer

Tape charts are made by sticking successive strips of tape, usually tentick lengths, side by side. That in Figure 19.3a (overleaf) represents a body moving with **uniform speed** since equal distances have been moved in each tentick interval.

The chart in Figure 19.3b is for **uniform acceleration**: the 'steps' are of equal size showing that the speed increased by the same amount in every tentick ($\frac{1}{5}$ s). The acceleration (average) can be found from the chart as follows.

The speed during the *first* tentick is 2 cm/$\frac{1}{5}$ s or 10 cm/s. During the *sixth* tentick it is 12 cm/$\frac{1}{5}$ s or 60 cm/s. And so during this interval of 5 tenticks, i.e. 1 second, the change of speed is (60 − 10) cm/s = 50 cm/s.

$$\text{acceleration} = \frac{\text{change of speed}}{\text{time taken}}$$

$$= \frac{50 \text{ cm/s}}{1 \text{ s}}$$

$$= 50 \text{ cm/s}^2$$

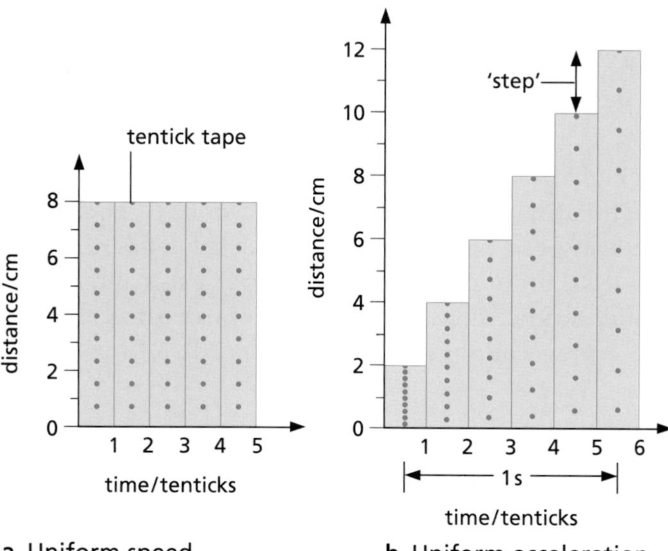

Figure 19.3 Tape charts

c) Photogate timer

Photogate timers may be used to record the time taken for a trolley to pass through the gate, Figure 19.4. If the length of the 'interrupt card' on the trolley is measured, the velocity of the trolley can then be calculated. Photogates are most useful in experiments where the velocity at only one or two positions is needed.

Figure 19.4 Use of a photogate timer

Practical work

Analysing motion

a) Your own motion

Pull a 2 m length of tape through a tickertape timer as you walk away from it quickly, then slowly, then speeding up again and finally stopping.

Cut the tape into tentick lengths and make a tape chart. Write labels on it to show where you speeded up, slowed down, etc.

b) Trolley on a sloping runway

Attach a length of tape to a trolley and release it at the top of a runway, Figure 19.5. The dots will be very crowded at the start – ignore those; but beyond them cut the tape into tentick lengths.

Make a tape chart. Is the acceleration uniform? What is its average value?

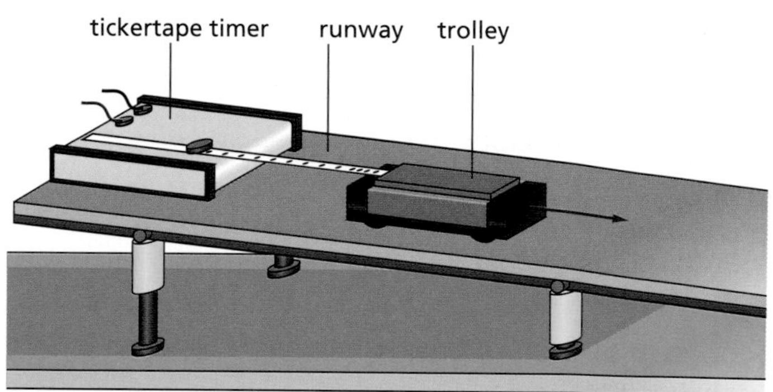

Figure 19.5

c) Datalogging

Replace the tickertape timer with a motion sensor connected to a datalogger and computer, Figure 19.6. Repeat the experiments in **a)** and **b)** and obtain distance–time and velocity–time graphs for each case; identify regions where you think the acceleration changes or remains uniform.

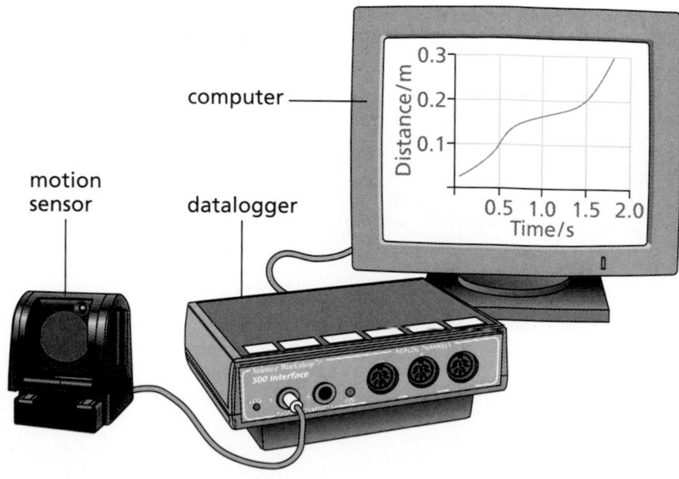

Figure 19.6 Use of a motion sensor

Questions

1 What is the average speed of
a a car which travels 400 m in 20 s,
b an athlete who runs 1500 m in 4 minutes?

2 Four runners compete in a 200 m race. The table shows the time it took for each runner to finish the race.

Runner	Time (s)
Adam	25
Ahmed	28
Neil	26
Naveed	24

a (i) Who came last in the race?
(ii) Explain how you could tell.
b (i) What is the formula for working out speed?
(ii) What is Adam's speed during the race?
(London Foundation, June 1999)

3 A train increases its speed **steadily** from 10 m/s to 20 m/s in 1 minute.
a What is its average speed during this time in m/s?
b How far does it travel while increasing its speed?

4 A motor cyclist starts from rest and reaches a speed of 6 m/s after travelling with uniform acceleration for 3 s. What is his acceleration?

5 The tape in Figure 19.7 was pulled through a timer by a trolley travelling down a runway. It was marked off in tentick lengths.
a What can you say about the trolley's motion?
b Find its acceleration in cm/s^2.

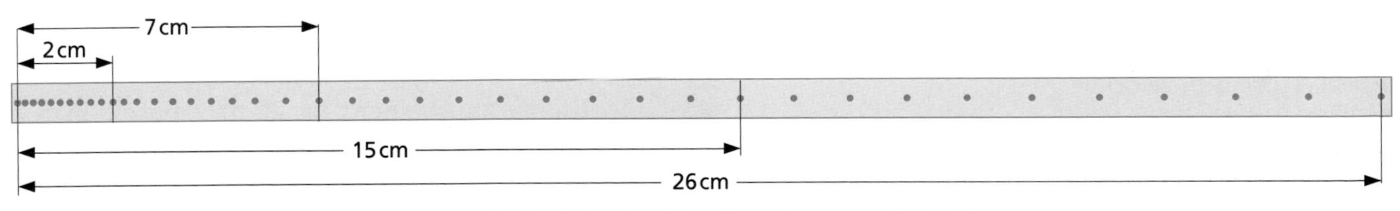

Figure 19.7

6 An aircraft travelling at 600 km/h accelerates steadily at 10 km/h per second. Taking the speed of sound as 1100 km/h at the aircraft's altitude, how long will it take to reach the 'sound barrier'?

7 A vehicle moving with a uniform acceleration of 2 m/s^2 has a velocity of 4 m/s at a certain time. What will its velocity be
a 1 s later,
b 5 s later?

8 If a bus travelling at 20 m/s is subject to a steady deceleration of 5 m/s^2, how long will it take to come to rest?

9 Each strip in the tape chart of Figure 19.8 is for a time interval of 1 tentick.
a If the timer makes 50 dots per second, what time intervals are represented by OA and AB?
b What is the acceleration between O and A in
(i) cm/tentick^2,
(ii) cm/s per tentick,
(iii) cm/s^2?
c What is the acceleration between A and B?

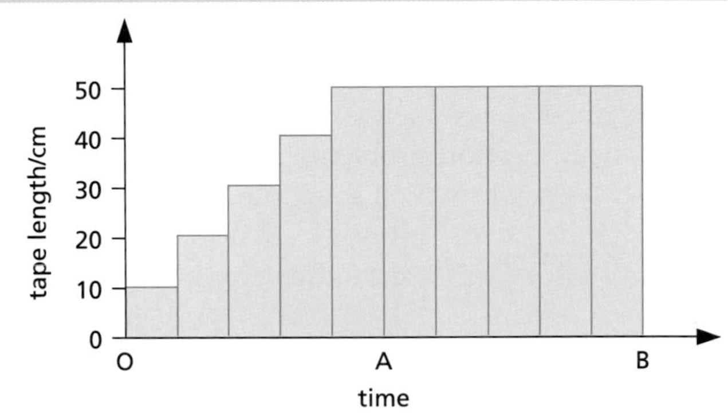

Figure 19.8

Checklist

After studying this chapter you should be able to

- explain the meaning of the terms **speed** and **acceleration**,
- distinguish between speed and velocity,
- describe how speed and acceleration may be found using tape charts and motion sensors.

20 Graphs and equations

Velocity–time graphs
Distance–time graphs
Equations for uniform acceleration

Velocity–time graphs

If the velocity of a body is plotted against the time, the graph obtained is a velocity–time graph. It provides a way of solving motion problems. Tape charts are crude velocity–time graphs which show the velocity changing in jumps rather than smoothly, as occurs in practice. A motion sensor gives a smoother plot.

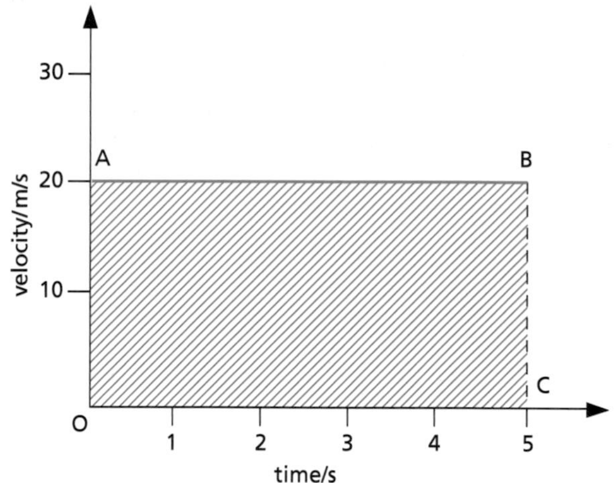

Figure 20.1 Uniform velocity

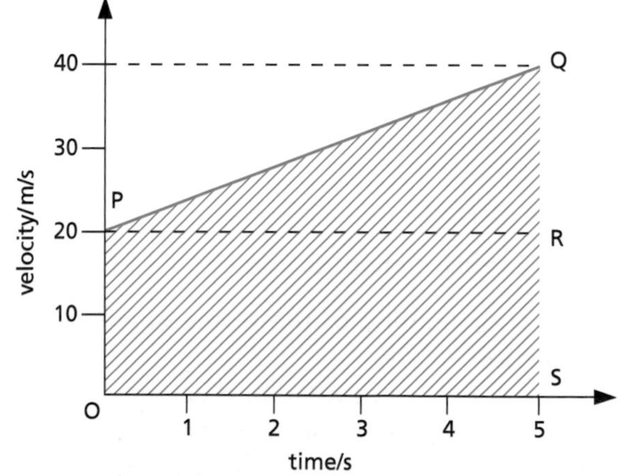

Figure 20.2a Uniform acceleration

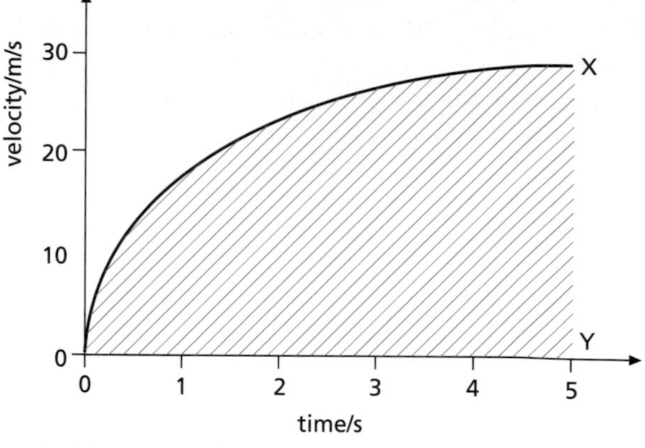

Figure 20.2b Non-uniform acceleration

The area under a velocity–time graph measures the distance travelled.

In Figure 20.1, AB is the velocity–time graph for a body moving with a **uniform velocity** of 20 m/s. Since distance = average velocity × time, after 5 s it will have moved 20 m/s × 5 s = 100 m. This is the shaded area under the graph, i.e. rectangle OABC.

In Figure 20.2a, PQ is the velocity–time graph for a body moving with **uniform acceleration**. At the start of the timing the velocity is 20 m/s but it increases steadily to 40 m/s after 5 s. If the distance covered equals the area under PQ, i.e. the shaded area OPQS, then

distance = area of rectangle OPRS + area of triangle PQR

$= OP \times OS + \frac{1}{2} \times PR \times QR$
(area of a triangle $= \frac{1}{2}$base × height)

$= 20\ \text{m/s} \times 5\ \text{s} + \frac{1}{2} \times 5\ \text{s} \times 20\ \text{m/s}$

$= 100\ \text{m} + 50\ \text{m} = 150\ \text{m}$

Notes

1 When calculating the area from the graph the unit of time must be the same on both axes.

2 This rule for finding distances travelled is true even if the acceleration is not uniform. In Figure 20.2b, the distance travelled equals the shaded area OXY.

The slope or gradient of a velocity–time graph represents the acceleration of the body.

In Figure 20.1, the slope of AB is zero, as is the acceleration. In Figure 20.2a, the slope of PQ is QR/PR = 20/5 = 4: the acceleration is 4 m/s^2. In Figure 20.2b, when the slope along OX changes, so does the acceleration.

Distance–time graphs

A body travelling with uniform velocity covers equal distances in equal times. Its distance–time graph is a straight line, like OL in Figure 20.3 for a velocity of 10 m/s. The slope of the graph is LM/OM = 40 m/4 s = 10 m/s, which is the value of the velocity. The following statement is true in general.

The slope or gradient of a distance–time graph represents the velocity of the body.

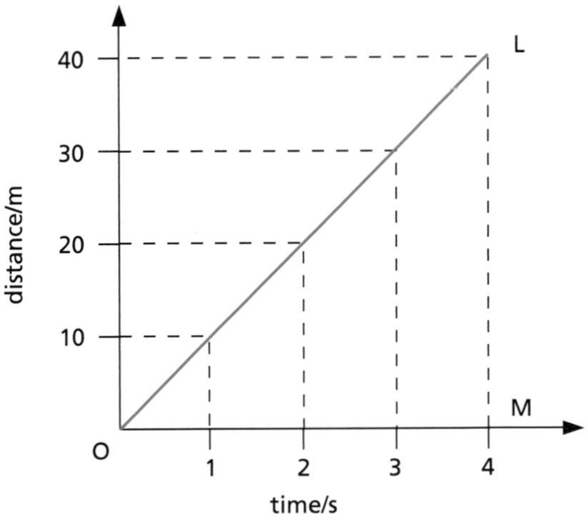

Figure 20.3 Uniform velocity

When the velocity of the body is changing, the slope of the distance–time graph varies, Figure 20.4, and at any point equals the slope of the tangent. For example, the slope of the tangent at T is AB/BC = 40 m/2 s = 20 m/s. The velocity at the instant corresponding to T is therefore 20 m/s.

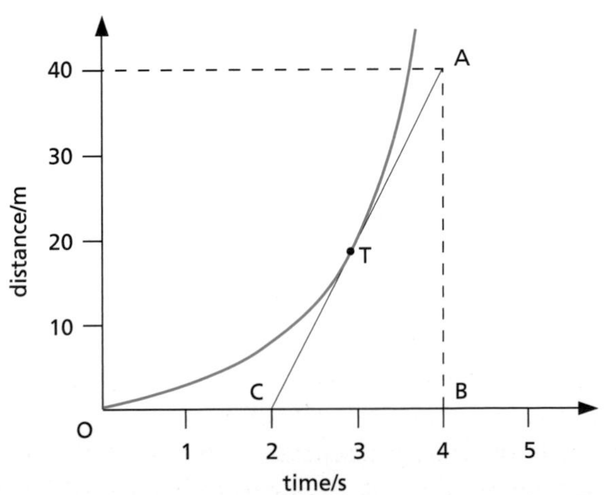

Figure 20.4 Non-uniform velocity

F Equations for uniform acceleration

Problems involving bodies moving with **uniform acceleration** can often be solved quickly using the **equations of motion**.

First equation

If a body is moving with uniform acceleration a and its velocity increases from u to v in time t, then

$$a = \frac{\text{change of velocity}}{\text{time taken}} = \frac{v - u}{t}$$

$$\therefore \quad at = v - u$$

or

$$v = u + at \qquad (1)$$

Note that the initial velocity u and the final velocity v refer to the start and the finish of the *timing* and do not necessarily mean the start and finish of the motion.

Second equation

The velocity of a body moving with uniform acceleration increases steadily. Its average velocity therefore equals half the sum of its initial and final velocities, that is,

$$\text{average velocity} = \frac{u + v}{2}$$

If s is the distance moved in time t, then since average velocity = distance/time = s/t,

$$\frac{s}{t} = \frac{u + v}{2}$$

or

$$s = \frac{(u + v)}{2}t \qquad (2)$$

Third equation

From equation (1), $v = u + at$

From equation (2),

$$\frac{s}{t} = \frac{u + v}{2}$$

$$= \frac{u + u + at}{2} = \frac{2u + at}{2}$$

$$= u + \tfrac{1}{2}at$$

and so

$$s = ut + \tfrac{1}{2}at^2 \qquad (3)$$

Fourth equation

This is obtained by eliminating t from equations (1) and (3). Squaring equation (1) we have

$$v^2 = (u + at)^2$$

$$\therefore \quad v^2 = u^2 + 2uat + a^2t^2$$

$$= u^2 + 2a(ut + \tfrac{1}{2}at^2)$$

But $\quad s = ut + \tfrac{1}{2}at^2$

$$\therefore \quad v^2 = u^2 + 2as \qquad (4)$$

If we know any *three* of u, v, a, s and t, the others can be found from the equations.

F *Worked example*

A sprint cyclist starts from rest and accelerates at 1 m/s^2 for 20 seconds. He then travels at a constant speed for 1 minute and finally decelerates at 2 m/s^2 until he stops. Find his maximum speed in km/h and the total distance covered in metres.

First stage

$$u = 0 \quad a = 1\ \text{m/s}^2 \quad t = 20\ \text{s}$$

We have $\quad v = u + at = 0 + 1\ \text{m/s}^2 \times 20\ \text{s}$

$$= 20\ \text{m/s}$$

$$= \frac{20}{1000} \times 60 \times 60 = 72\ \text{km/h}$$

The distance s moved in the first stage is given by

$$s = ut + \tfrac{1}{2}at^2 = 0 \times 20\ \text{s} + \tfrac{1}{2} \times 1\ \text{m/s}^2 \times 20^2\ \text{s}^2$$

$$= \tfrac{1}{2} \times 1\ \text{m/s}^2 \times 400\ \text{s}^2 = 200\ \text{m}$$

Second stage

$$u = 20\ \text{m/s (constant)} \quad t = 60\ \text{s}$$

$$\text{distance moved} = \text{speed} \times \text{time} = 20\ \text{m/s} \times 60\ \text{s}$$

$$= 1200\ \text{m}$$

Third stage

$$u = 20\ \text{m/s} \quad v = 0 \quad a = -2\ \text{m/s}^2\ \text{(a deceleration)}$$

We have

$$v^2 = u^2 + 2as$$

$$\therefore \quad s = \frac{v^2 - u^2}{2a} = \frac{0 - (20)^2\ \text{m}^2/\text{s}^2}{2 \times (-2)\ \text{m/s}^2} = \frac{-400\ \text{m}^2/\text{s}^2}{-4\ \text{m/s}^2}$$

$$= 100\ \text{m}$$

Answers

Maximum speed = 72 km/h

Total distance covered = 200 m + 1200 m + 100 m
= 1500 m

Questions

1 The distance–time graph for a girl on a cycle ride is shown in Figure 20.5.
  - **a** How far did she travel?
  - **b** How long did she take?
  - **c** What was her average speed in km/h?
  - **d** How many stops did she make?
  - **e** How long did she stop for altogether?
  - **f** What was her average speed *excluding* stops?
  - **g** How can you tell from the shape of the graph when she travelled fastest? Over which stage did this happen?

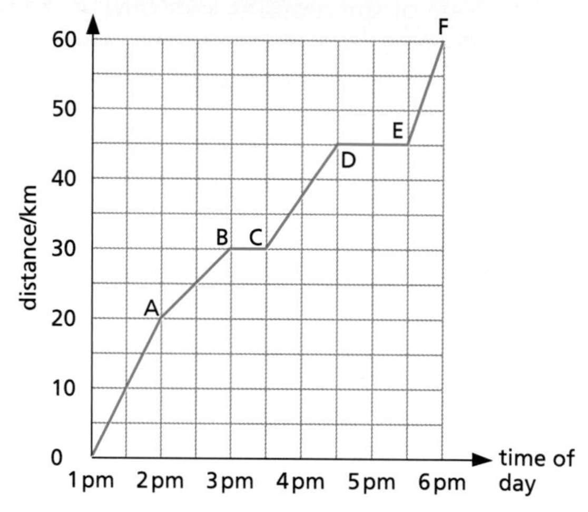

Figure 20.5

2 The graph in Figure 20.6 represents the distance travelled by a car plotted against time.
  - **a** How far has the car travelled at the end of 5 seconds?
  - **b** What is the speed of the car during the first 5 seconds?
  - **c** What has happened to the car after A?
  - **d** Draw a graph showing the speed of the car plotted against time during the first 5 seconds.

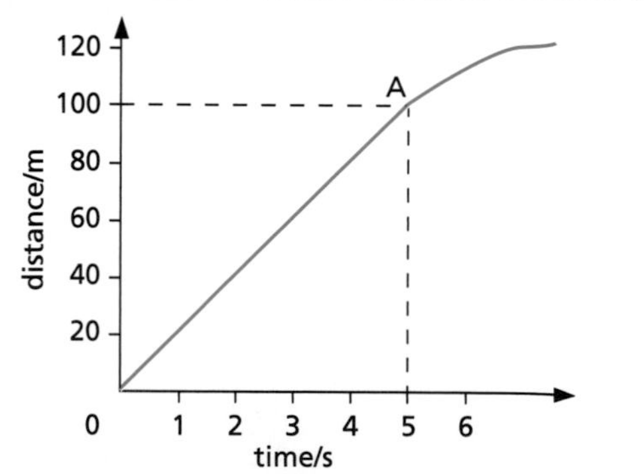

Figure 20.6

3 Figure 20.7 shows an incomplete velocity–time graph for a boy running a distance of 100 m.
  - **a** What is his acceleration during the first 4 seconds?
  - **b** How far does the boy travel during (i) the first 4 seconds, (ii) the next 9 seconds?

c Copy and complete the graph showing clearly at what time he has covered the distance of 100 m. Assume his speed remains constant at the value shown by the horizontal portion of the graph.

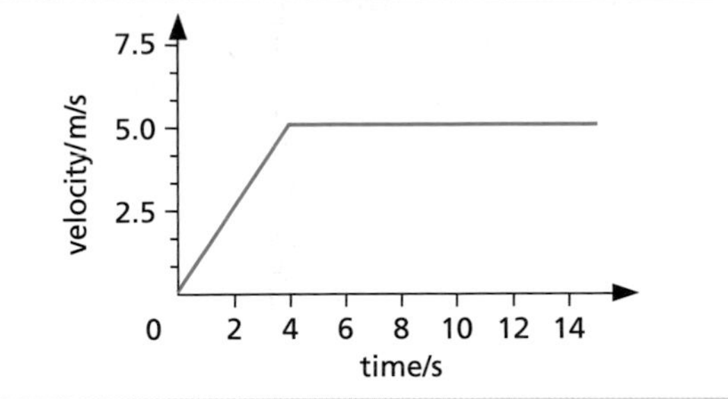

Figure 20.7

4 Figure 20.8 shows the speed–time graph for a journey travelled by a tractor.

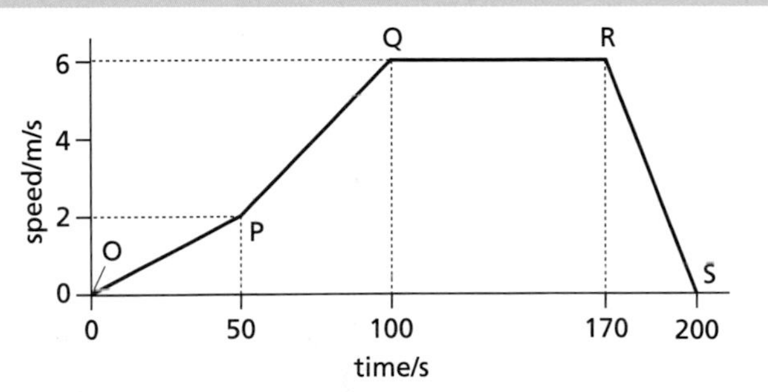

Figure 20.8

a Use the graph to describe the motion of the tractor during each of the marked sections OP, PQ, QR, RS.
b Which two points on the graph show when the tractor is stationary?
c State the greatest speed reached by the tractor.
d For how long was the tractor travelling at constant speed?
e State how the graph may be used to find the total distance travelled during the 200 s journey. Do **not** attempt a calculation.

(*UCLES IGCSE Physics Core, Nov 2006*)

Checklist

After studying this chapter you should be able to

- draw, interpret and use velocity–time and distance–time graphs to solve problems.

21 Falling bodies

Acceleration of free fall
Measuring g
Distance–time graphs
Projectiles
Practical work
Motion of a falling body.

In air, a coin falls faster than a small piece of paper. In a vacuum they fall at the same rate as may be shown with the apparatus of Figure 21.1. The difference in air is due to **air resistance** having a greater effect on light bodies than on heavy bodies. The air resistance to a light body is large when compared with the body's weight. With a dense piece of metal the resistance is negligible at low speeds.

There is a story, untrue we now think, that in the 16th century the Italian scientist Galileo dropped a small iron ball and a large cannon ball ten times heavier from the top of the Leaning Tower of Pisa, Figure 21.2. And we are told that, to the surprise of onlookers who expected the cannon ball to arrive first, they reached the ground almost simultaneously. You will learn more about air resistance in the next chapter.

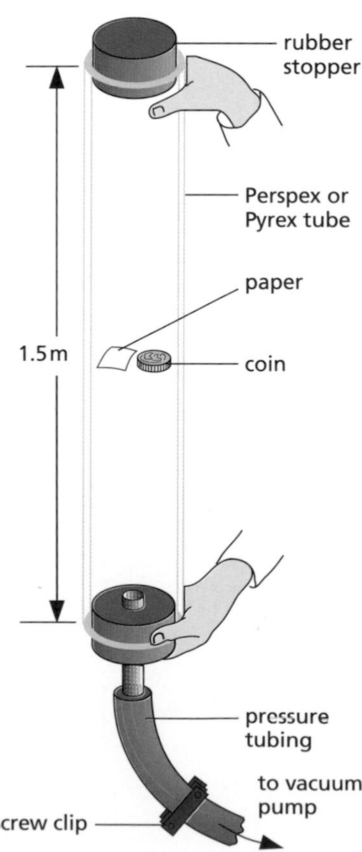

Figure 21.1 A coin and a piece of paper fall at the same rate in a vacuum

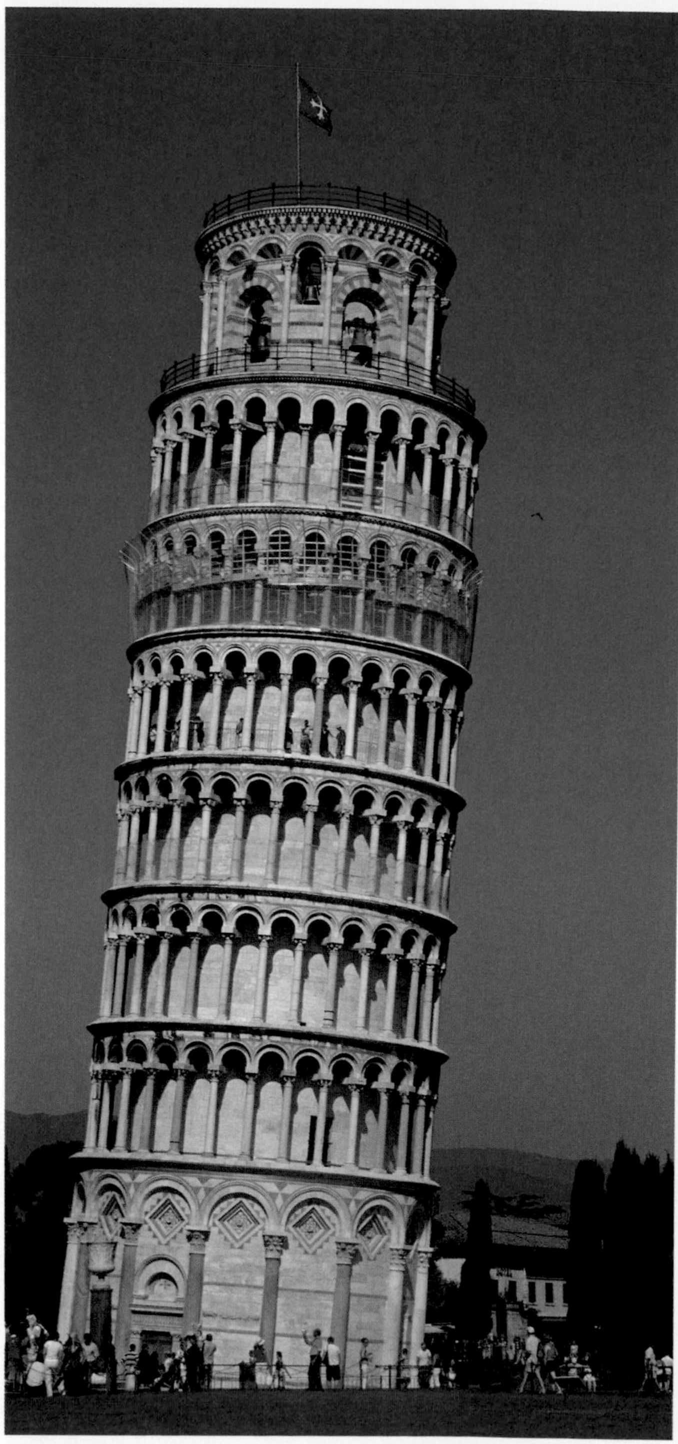

Figure 21.2 The Leaning Tower of Pisa, where Galileo is said to have experimented with falling objects

Practical work

Motion of a falling body

Arrange things as shown in Figure 21.3 and investigate the motion of a 100 g mass falling from a height of about 2 m.

Construct a tape chart using *one tick* lengths. Choose as dot '0' the first one you can distinguish clearly. What does the tape chart tell you about the motion of the falling mass? Repeat the experiment with a 200 g mass; what do you notice?

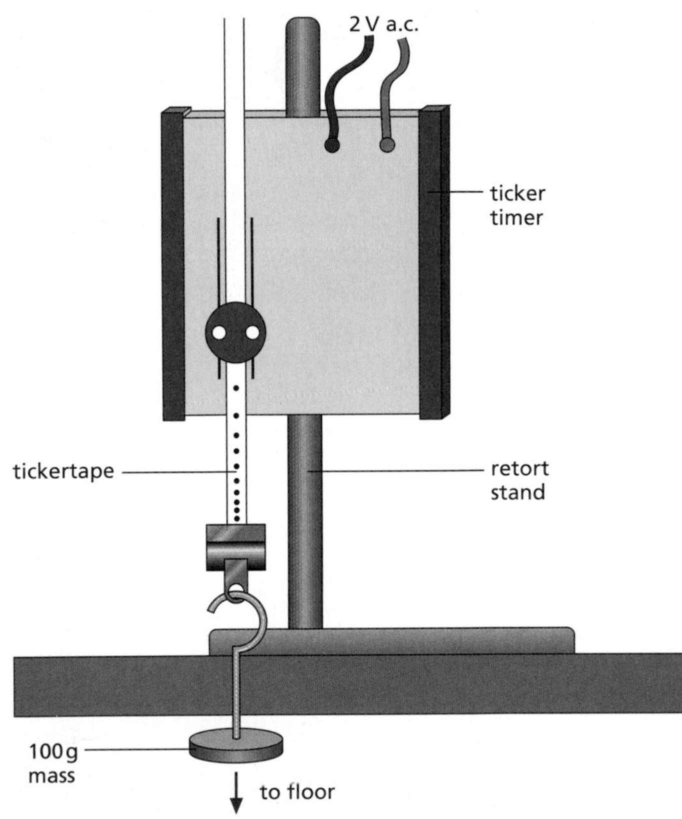

Figure 21.3

Acceleration of free fall

All bodies falling freely under the force of gravity do so with **uniform acceleration** if air resistance is negligible (i.e. the 'steps' in the tape chart from the practical work should all be equal).

This acceleration, called the **acceleration of free fall**, is denoted by the italic letter g. Its value varies slightly over the Earth but is **constant** in each place; in India it is about 9.8 m/s^2 or near enough 10 m/s^2. The velocity of a free-falling body therefore increases by 10 m/s every second. A ball shot straight upwards with a velocity of 30 m/s decelerates by 10 m/s every second and reaches its highest point after 3 s.

In calculations using the equations of motion, g replaces a. It is given a positive sign for falling bodies (i.e. $a = g = +10$ m/s^2) and a negative sign for rising bodies since they are decelerating (i.e. $a = -g = -10$ m/s^2).

F *Measuring g*

Using the arrangement in Figure 21.4 the time for a steel ball-bearing to fall a known distance is measured by an electronic timer.

When the two-way switch is changed to the 'down' position, the electromagnet releases the ball and simultaneously the clock starts. At the end of its fall the ball opens the 'trap-door' on the impact switch and the clock stops.

The result is found from the third equation of motion $s = ut + \frac{1}{2}at^2$, where s is the distance fallen (in m), t is the time taken (in s), $u = 0$ (the ball starts from rest) and $a = g$ (in m/s^2). Hence

$$s = \frac{1}{2}gt^2$$

or

$$g = 2s/t^2$$

Air resistance is negligible for a dense object such as a steel ball-bearing falling a short distance.

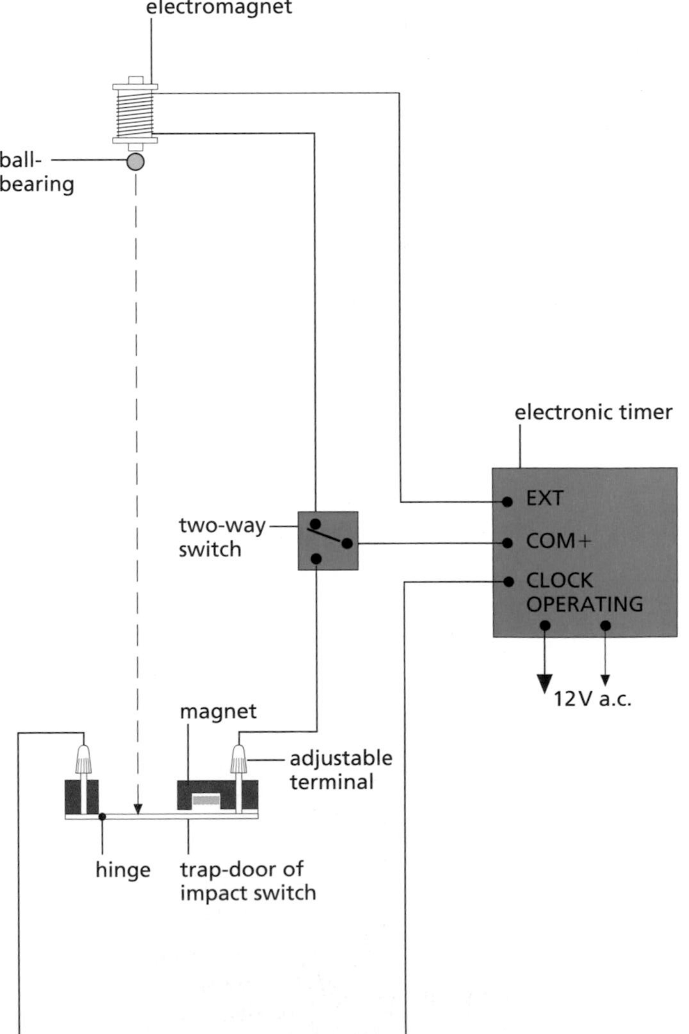

Figure 21.4

F Worked example

A ball is projected vertically upwards with an initial velocity of 30 m/s. Find **a** its maximum height and **b** the time taken to return to its starting point. Neglect air resistance and take $g = 10\ \text{m/s}^2$.

a We have $u = 30$ m/s, $a = -10\ \text{m/s}^2$ (a deceleration) and $v = 0$ since the ball is momentarily at rest at its highest point. Substituting in $v^2 = u^2 + 2as$,

$$0 = 30^2\ \text{m}^2/\text{s}^2 + 2(-10\ \text{m/s}^2) \times s$$

or

$$-900\ \text{m}^2/\text{s}^2 = -s \times 20\ \text{m/s}^2$$

$$\therefore \quad s = \frac{-900\ \text{m}^2/\text{s}^2}{-20\ \text{m/s}^2} = 45\ \text{m}$$

b If t is the time to reach the highest point, we have, from $v = u + at$,

$$0 = 30\ \text{m/s} + (-10\ \text{m/s}^2) \times t$$

or

$$-30\ \text{m/s} = -t \times 10\ \text{m/s}^2$$

$$\therefore \quad t = \frac{-30\ \text{m/s}}{-10\ \text{m/s}^2} = 3\ \text{s}$$

The downward trip takes exactly the same time as the upward one and so the answer is 6 s.

Distance–time graphs

For a body falling freely from rest we have

$$s = \tfrac{1}{2}gt^2$$

A graph of distance s against time t is shown in Figure 21.5a and for s against t^2 in Figure 21.5b. The second graph is a straight line through the origin since $s \propto t^2$ (g being constant at one place).

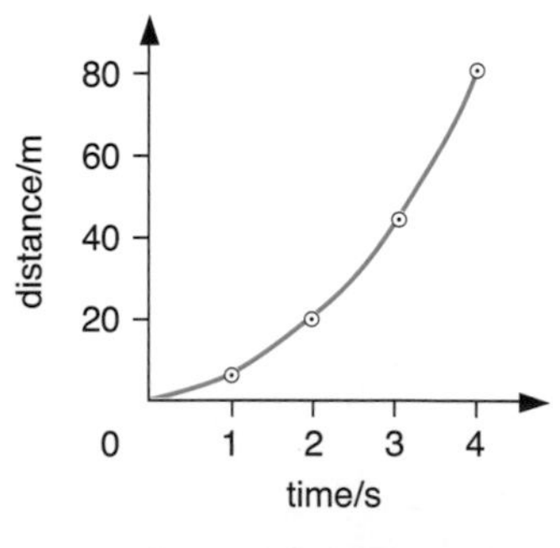

a

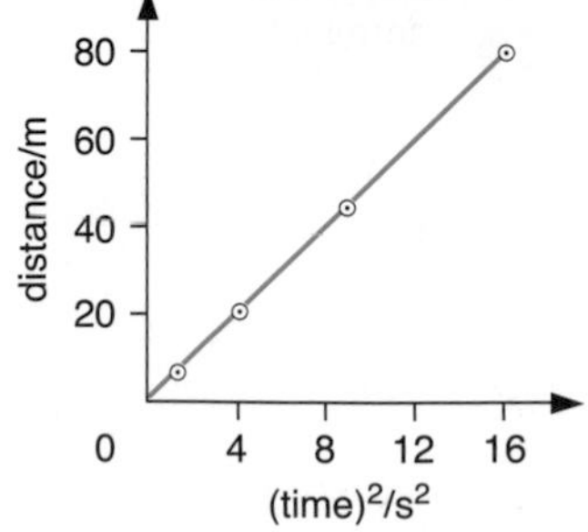

b

Figure 21.5 Graphs for a body falling freely from rest

F Projectiles

The photograph in Figure 21.6 was taken while a lamp emitted regular flashes of light. One ball was **dropped** from rest and the other, a 'projectile', was **thrown sideways** at the same time. Their vertical accelerations (due to gravity) are equal, showing that a projectile falls like a body which is dropped from rest. Its horizontal velocity does not affect its vertical motion.

> The horizontal and vertical motions of a body are independent and can be treated separately.

For example if a ball is thrown horizontally from the top of a cliff and takes 3 s to reach the beach below we can calculate the height of the cliff by considering the vertical motion only. We have $u = 0$ (since the ball has no vertical velocity initially), $a = g = +10\ \text{m/s}^2$ and $t = 3$ s. The height s of the cliff is given by

$$s = ut + \tfrac{1}{2}at^2$$

$$= 0 \times 3\ \text{s} + \tfrac{1}{2}(+10\ \text{m/s}^2)3^2\ \text{s}^2$$

$$= 45\ \text{m}$$

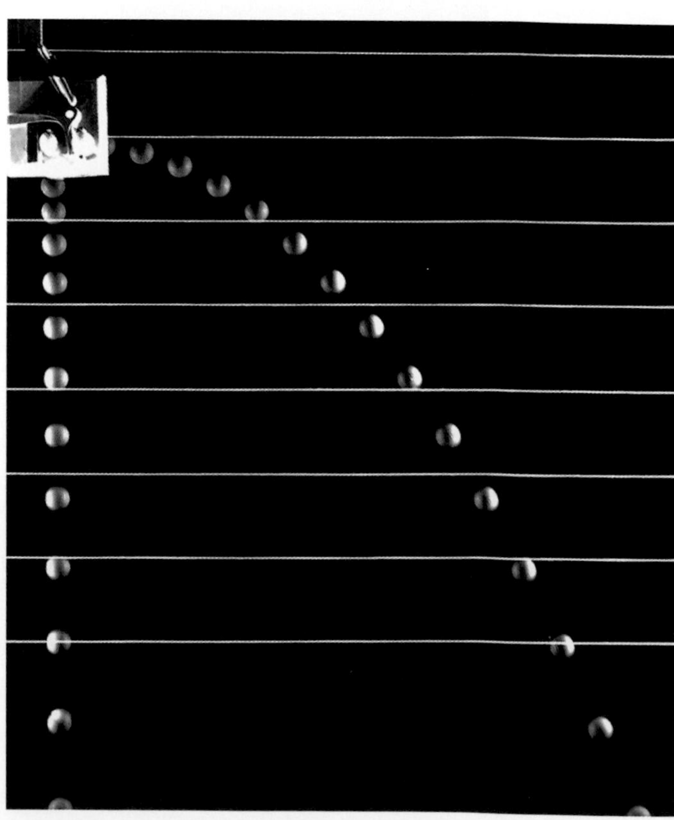

Figure 21.6 Comparing free fall and projectile motion using multiflash photography

Projectiles such as cricket balls and explosive shells are projected from near ground level and at an angle. The horizontal distance they travel, i.e. their **range**, depends on

(i) the **speed** of projection – the greater this is, the greater the range, and
(ii) the **angle** of projection – it can be shown that, neglecting air resistance, the range is a maximum when the angle is 45°, Figure 21.7.

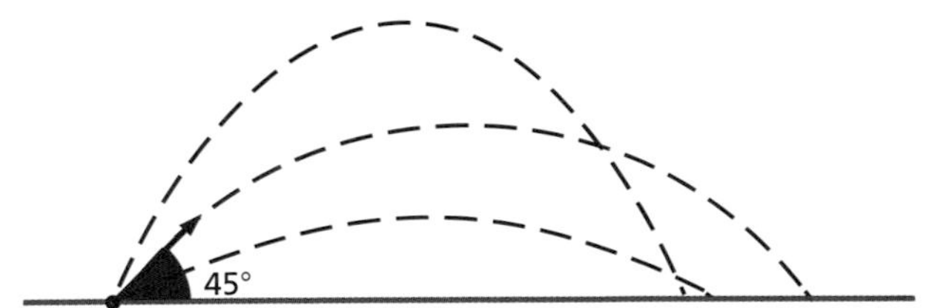

Figure 21.7 The range is greatest for an angle of projection of 45°

Questions

1 A stone falls from rest from the top of a high tower. Ignore air resistance and take $g = 10\ \text{m/s}^2$.
- **a** What is its velocity after
  - (i) 1 s,
  - (ii) 2 s,
  - (iii) 3 s,
  - (iv) 5 s?
- **b** How far has it fallen after
  - (i) 1 s,
  - (ii) 2 s,
  - (iii) 3 s,
  - (iv) 5 s?

2 An object falls from a hovering helicopter and hits the ground at a speed of 30 m/s. How long does it take the object to reach the ground and how far does it fall? Sketch a velocity–time graph for the object (ignore air resistance).

Checklist

After studying this chapter you should be able to

- □ describe the behaviour of falling objects,
- ■ state that the acceleration of free fall for a body near the Earth is constant.

22 Force and acceleration

Newton's first law	*Newton's third law*
Mass and inertia	*Air resistance: terminal velocity*
Newton's second law	*Practical work*
Weight and gravity	Effect of force and mass on acceleration.
Gravitational field	

Newton's first law

Friction and air resistance cause a car to come to rest when the engine is switched off. If these forces were absent we believe that a body, once set in motion, would go on moving forever with a constant speed in a straight line. That is, force is not needed to keep a body moving with uniform velocity so long as no opposing forces act on it.

This idea was proposed by Galileo and is summed up in Newton's first law of motion:

> A body stays at rest, or if moving it continues to move with uniform velocity, unless an external force makes it behave differently.

It seems that the question we should ask about a moving body is not 'what keeps it moving' but 'what changes or stops its motion'.

The smaller the external forces opposing a moving body, the smaller is the force needed to keep it moving with uniform velocity. An 'airboard', which is supported by a cushion of air, Figure 22.1, can skim across the ground with little frictional opposition, so that relatively little power is needed to maintain motion.

Figure 22.1 Friction is much reduced for an airboard

Mass and inertia

Newton's first law is another way of saying that all matter has a built-in opposition to being moved if it is at rest or, if it is moving, to having its motion changed. This property of matter is called **inertia** (from the Latin word for laziness).

Its effect is evident on the occupants of a car that stops suddenly; they lurch forwards in an attempt to continue moving, and this is why seat belts are needed. The reluctance of a stationary object to move can be shown by placing a large coin on a piece of card on your finger, Figure 22.2. If the card is flicked *sharply* the coin stays where it is while the card flies off.

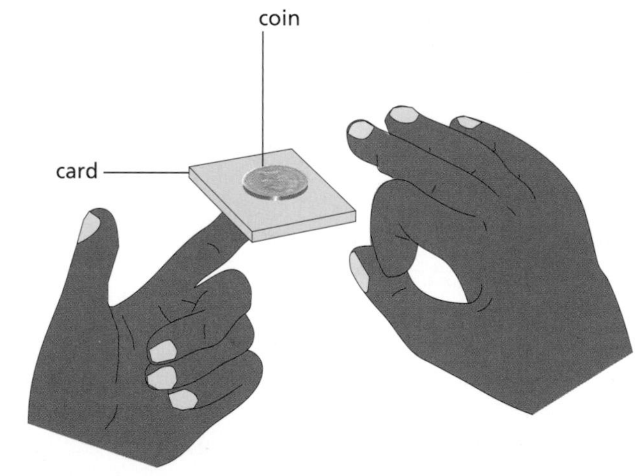

Figure 22.2 Flick the card sharply

The larger the mass of a body the greater is its inertia, i.e. the more difficult it is to move it when at rest and to stop it when in motion. Because of this we consider that **the mass of a body measures its inertia**. This is a better definition of mass than the one given earlier (Chapter 10) in which it was stated to be the 'amount of matter' in a body.

Practical work

Effect of force and mass on acceleration

The apparatus consists of a trolley to which a force is applied by a stretched length of elastic, Figure 22.3. The velocity of the trolley is found from a tickertape timer or a motion sensor, datalogger and computer (see Figure 19.6, p. 94).

First compensate the runway for friction by raising one end until the trolley runs down with uniform velocity when given a push. The dots on the tickertape should be equally spaced, or a horizontal trace obtained on a velocity–time graph. There is now no resultant force on the trolley and any acceleration produced later will be due only to the force caused by the stretched elastic.

a) Force and acceleration (mass constant)

Fix one end of a short length of elastic to the rod at the back of the trolley and stretch it until the other end is level with the front of the trolley. Practise pulling the trolley down the runway, keeping the same stretch on the elastic. After a few trials you should be able to produce a steady accelerating force.

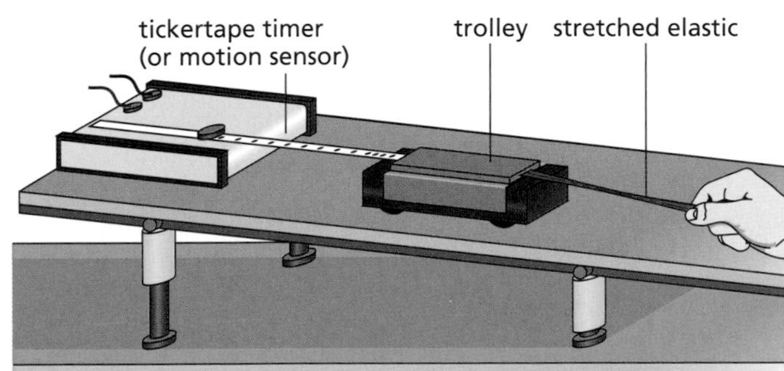

Figure 22.3

Repeat using first two and then three *identical* pieces of elastic, stretched side by side by the same amount, to give two and three units of force.

If you are using tickertape make a tape chart for each force and use it to find the acceleration produced in cm/tentick^2 (see Chapter 19). Ignore the start of the tape (where the dots are too close) and the end (where the force may not be steady). If you use a motion sensor and computer to plot a velocity–time graph, the acceleration can be obtained in m/s^2 from the slope of the graph (Chapter 20).

Does a steady force cause a steady acceleration? Put the results in a table. Do they suggest any relationship between *a* and *F*?

Force (*F*) (no. of pieces of elastic)	1	2	3
Acceleration (*a*) cm/tentick^2 or m/s^2			

b) Mass and acceleration (force constant)

Do the experiment as in **a)** using two pieces of elastic (i.e. constant *F*) to accelerate first one trolley, then two (stacked one above the other) and finally three. Check the friction compensation of the runway each time.

Find the accelerations from the tape charts or computer plots and tabulate the results. Do they suggest any relationship between *a* and *m*?

Mass (*m*) (no. of trolleys)	1	2	3
Acceleration (*a*) /cm/tentick^2 or m/s^2			

Newton's second law

The previous experiment should show roughly that the acceleration *a* is

(i) directly proportional to the applied force *F* for a fixed mass, i.e. $a \propto F$, and
(ii) inversely proportional to the mass *m* for a fixed force, i.e $a \propto 1/m$.

Combining the results into one equation, we get

$$a \propto F/m \quad \text{or} \quad F \propto ma$$

Therefore $\quad F = kma$

where *k* is the constant of proportionality.

One newton is defined as the force which gives a mass of 1 kg an acceleration of 1 m/s^2, i.e. 1 N = 1 kg m/s^2.

So if $m = 1$ kg and $a = 1$ m/s^2 then $F = 1$ N. Substituting in $F = kma$ we get $k = 1$ and so we can write

$$F = ma$$

This is Newton's second law of motion. When using it two points should be noted. First, *F* is the resultant (or unbalanced) force causing the acceleration *a*. Second, *F* must be in newtons, *m* in kilograms and *a* in metres per second squared, otherwise *k* is not 1. The law shows that *a* will be largest when *F* is large and *m* small.

You should now appreciate that when the forces acting on a body do not balance there is a net (resultant) force which causes a **change** of motion, i.e. the body accelerates or decelerates. If they balance, there is no change of motion. However, there may be a change of shape, in which case internal forces in the body (i.e. forces between neighbouring atoms) balance the external forces.

☐ Worked examples

1 A block of mass 2 kg is pushed along a table with a constant velocity by a force of 5 N. When the push is increased to 9 N what is
a the resultant force,
b the acceleration?

When the block moves with constant velocity the forces acting on it are balanced. The force of friction opposing its motion must therefore be 5 N.

a When the push is increased to 9 N the resultant (unbalanced) force F on the block is $(9-5)\text{ N} = 4\text{ N}$ (since the frictional force is still 5 N).

b The acceleration a is obtained from $F = ma$ where $F = 4\text{ N}$ and $m = 2\text{ kg}$.

$$\therefore \quad a = F/m = 4\text{ N}/2\text{ kg} = \frac{4\text{ kg m/s}^2}{2\text{ kg}} = 2\text{ m/s}^2$$

Weight and gravity

The weight W of a body is the force of gravity acting on it which gives it an acceleration g when it is falling freely near the Earth's surface. If the body has mass m, then W can be calculated from $F = ma$ if we put $F = W$ and $a = g$ to give

$$W = mg$$

Taking $g = 9.8\text{ m/s}^2$ and $m = 1\text{ kg}$, this gives $W = 9.8\text{ N}$, i.e. a body of mass 1 kg has weight 9.8 N, or near enough 10 N. Similarly a body of mass 2 kg has weight of about 20 N and so on. While the mass of a body is always the same, its weight varies depending on the value of g. On the Moon the acceleration of free fall is only about 1.6 m/s^2, and so a mass of 1 kg has a weight of just 1.6 N there.

The weight of a body is directly proportional to its mass, which explains why g is the same for all bodies. The greater the mass of a body, the greater is the force of gravity on it but it does not accelerate faster when falling because of its greater inertia (i.e. its greater resistance to acceleration).

☐ Gravitational field

The force of gravity acts through space and can cause a body, not in contact with the Earth, to fall to the ground. It is an invisible, action-at-a-distance force. We try to 'explain' its existence by saying that the Earth is surrounded by a **gravitational field** which exerts a force on any body in the field. Later, magnetic and electric fields will be considered.

The strength of a gravitational field is defined as the force acting on unit mass in the field.

Measurement shows that on the Earth's surface a mass of 1 kg experiences a force of 9.8 N, i.e. its weight is 9.8 N. The strength of the Earth's field is therefore 9.8 N/kg (near enough 10 N/kg). It is denoted by g, the letter also used to denote the acceleration of free fall. Hence

$$g = 9.8\text{ N/kg} = 9.8\text{ m/s}^2$$

We now have two ways of regarding g. When considering bodies **falling freely** we can think of it as an acceleration of 9.8 m/s^2, but when a body of known mass is **at rest** and we wish to know the force of gravity (in N) acting on it we think of g as the Earth's gravitational field strength of 9.8 N/kg.

F Newton's third law

If a body A exerts a force on body B, then body B exerts an equal but opposite force on body A.

This is Newton's third law of motion and states that forces never occur singly but always in pairs as a result of the action between two bodies. For example, when you step forwards from rest your foot pushes backwards on the Earth, and the Earth exerts an equal and opposite force forward on you. Two bodies and two forces are involved. The small force you exert on the large mass of the Earth gives no noticeable acceleration to the Earth but the equal force it exerts on your very much smaller mass causes you to accelerate.

Note that the pair of equal and opposite forces **do not act on the same body**; if they did, there could never be any resultant forces and acceleration would be impossible. For a book resting on a table, the book exerts a downward force on the table and the table exerts an equal and opposite upward force on the book; this *pair* of forces act on *different* objects and are represented by the red arrows in Figure 22.4. The weight of the book (blue arrow) *does not* form a pair with the upward force on the book (although they are equal numerically) as these two forces act on the *same* body.

An appreciation of the third law and the effect of friction is desirable when stepping from a rowing boat, Figure 22.5. You push backwards on the boat and,

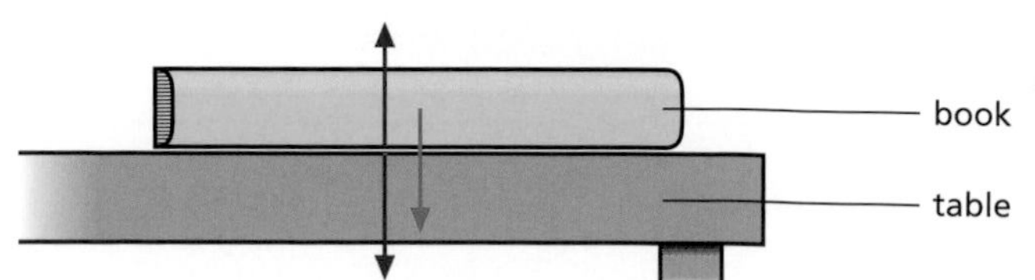

Figure 22.4 Forces between book and table

although the boat pushes you forwards with an equal force, it is itself now moving backwards (because friction with the water is slight). This reduces your forwards motion by the same amount – so you may fall in!

Figure 22.5 The boat moves backwards when you step forwards!

☐ *Air resistance: terminal velocity*

When an object falls in air, the air resistance (fluid friction) opposing its motion **increases as its speed rises**, so reducing its acceleration. Eventually, air resistance acting upwards equals the weight of the object acting downwards. The resultant force on the object is then zero since the gravitational force balances the frictional force. The object falls at a constant velocity, called its **terminal velocity**, whose value depends on the size, shape and weight of the object.

A small dense object, e.g. a steel ball-bearing, has a high terminal velocity and falls a considerable distance with a constant acceleration of $9.8\ m/s^2$ before air resistance equals its weight. A light object, e.g. a raindrop, or one with a large surface area, e.g. a parachute, has a low terminal velocity and only accelerates over a comparatively short distance before air resistance equals its weight. A sky diver, Figure 22.6, has a terminal velocity of more than 50 m/s (180 km/h).

Figure 22.6 Synchronized sky divers

Objects falling in liquids behave similarly to those falling in air.

Questions

1 Which one of the diagrams in Figure 22.7 shows the arrangement of forces that gives the block of mass M the greatest acceleration?

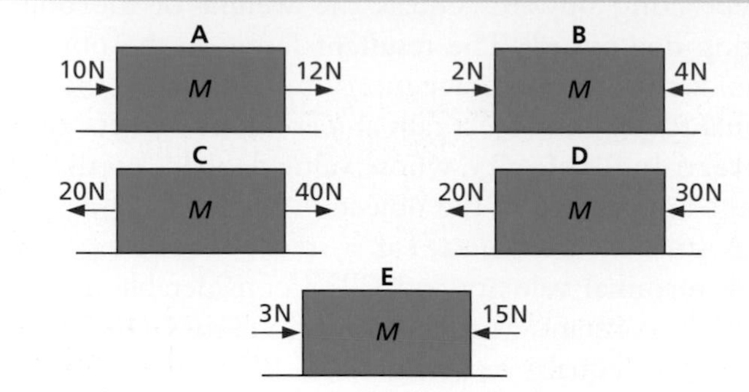

Figure 22.7

2 In Figure 22.8 if P is a force of 20 N and the object moves with **constant velocity** what is the value of the opposing force F?

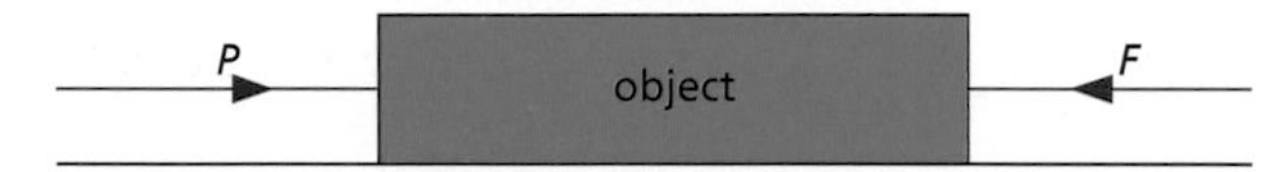

Figure 22.8

3 **a** What resultant force produces an acceleration of 5 m/s^2 in a car of mass 1000 kg?
b What acceleration is produced in a mass of 2 kg by a resultant force of 30 N?

4 A block of mass 500 g is pulled from rest on a horizontal frictionless bench by a steady force F and travels 8 m in 2 s. Find
a the acceleration,
b the value of F.

5 Starting from rest on a level road a girl can reach a speed of 5 m/s in 10 s on her bicycle. Find
a the acceleration,
b the average speed during the 10 s,
c the distance she travels in 10 s.
Eventually, even though she is still pedalling as fast as she can, she stops accelerating and her speed reaches a maximum value. Explain in terms of the forces acting why this happens.

6 What does an astronaut of mass 100 kg weigh
a on Earth where the gravitational field strength is 10 N/kg,
b on the Moon where the gravitational field strength is 1.6 N/kg?

7 A rocket has a mass of 500 kg.
a What is its weight on Earth where g = 10 N/kg?
b At lift-off the rocket engine exerts an upward force of 25 000 N. What is the resultant force on the rocket? What is its initial acceleration?

8 Figure 22.9 shows the forces acting on a raindrop which is falling to the ground.
a (i) A is the force which causes the raindrop to fall. What is this force called?
(ii) B is the total force opposing the motion of the drop. State *one* possible cause of this force.
b What happens to the drop when force A = force B?

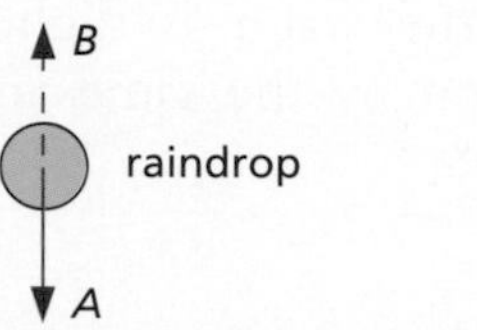

Figure 22.9

9 Explain the following using $F = ma$.
a A racing car has a powerful engine and is made of strong but lightweight material.
b A car with a small engine can still accelerate rapidly.

10 Four forces are acting on a rock, as shown in Figure 22.10.

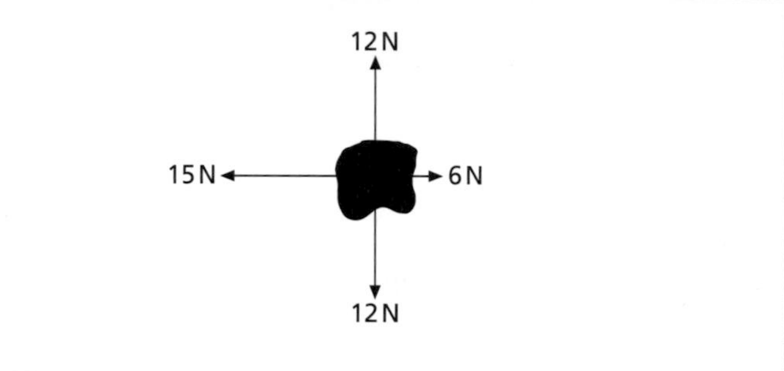

Figure 22.10

Here are some statements which students made about the rock. Write down the **two** correct statements.

The rock remains stationary.
The rock moves towards the top of the page.
The rock moves towards the bottom of the page.
The rock moves to the left.
The rock moves to the right.
The rock has a constant speed
The rock accelerates.

(*UCLES IGCSE Physics Core, May 1998*)

Checklist

After studying this chapter you should be able to

- describe an experiment to investigate the relationship between force, mass and acceleration,
- state the unit of force,
- ☐ state Newton's second law of motion and use it to solve problems,
- ☐ define the strength of the Earth's gravitational field,
- ☐ describe the motion of an object falling in air.

23 Kinetic and potential energy

Kinetic energy	*Elastic and inelastic collisions*	*Practical work*
Potential energy	*Driving and car safety*	Change of p.e. to k.e.
Conservation of energy		

Energy and its different forms were discussed earlier (Chapter 16). Here we will consider kinetic energy (k.e.) and potential energy (p.e.) in more detail.

Kinetic energy

Kinetic energy is the energy a body has because of its motion.

For a body of mass m travelling with velocity v,

$$\text{kinetic energy} = E_k = \tfrac{1}{2}mv^2$$

If m is in kg and v in m/s, then k.e. is in J. For example, a football of mass 0.4 kg (400 g) moving with velocity 20 m/s has

$$\begin{aligned}\text{k.e.} &= \tfrac{1}{2}mv^2 = \tfrac{1}{2} \times 0.4\text{ kg} \times (20)^2\text{ m}^2/\text{s}^2 \\ &= 0.2 \times 400\text{ kg m/s}^2 \times \text{m} \\ &= 80\text{ N m} = 80\text{ J}\end{aligned}$$

Since k.e. depends on v^2, a high-speed vehicle travelling at 1000 km/h, Figure 23.1, has one hundred times the k.e. it has at 100 km/h.

Potential energy

Potential energy is the energy a body has because of its position or condition.

A body above the Earth's surface is considered to have an amount of gravitational p.e. equal to the work that has been done against gravity by the force used to raise it. To lift a body of mass m through a vertical height h at a place where the Earth's gravitational field strength is g, needs a force equal and opposite to the weight mg of the body. Hence

$$\begin{aligned}\text{work done by force} &= \text{force} \times \text{vertical height} \\ &= mg \times h\end{aligned}$$

$$\therefore \quad \text{potential energy} = E_p = mgh$$

When m is in kg, g in N/kg (or m/s^2) and h in m, the p.e. is in J. For example, if g = 10 N/kg, the p.e. gained by a 0.1 kg (100 g) mass raised vertically by 1 m is 0.1 kg × 10 N/kg × 1 m = 1 N m = 1 J.

Note Strictly speaking we are concerned with *changes* in p.e. from that which a body has at the Earth's surface, rather than with actual values. The expression for p.e. is therefore more correctly written

$$\Delta E_p = mgh$$

where Δ (pronounced delta) stands for 'a change in'.

Figure 23.1 Kinetic energy depends on the square of the velocity

Practical work

Change of p.e. to k.e.

Friction-compensate a runway and arrange the apparatus as in Figure 23.2 with the bottom of the 0.1 kg (100 g) mass 0.5 m from the floor.

Start the timer and release the trolley. It will accelerate until the falling mass reaches the floor; after that it moves with *constant* velocity v.

From your results calculate v in m/s (on the tickertape 50 ticks = 1 s). Find the mass of the trolley in kg. Work out:

k.e. gained by trolley and 0.1 kg mass = . . . J
p.e. lost by 0.1 kg mass = . . . J

Compare and comment on the results.

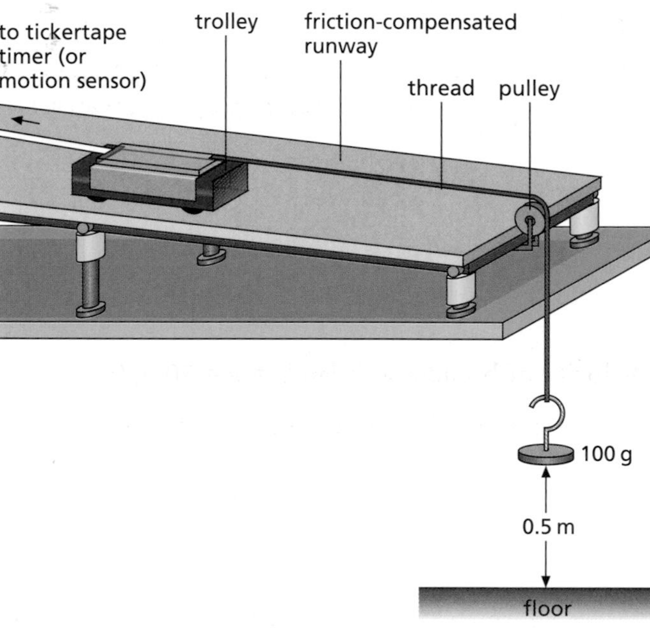

Figure 23.2

Conservation of energy

A mass m at height h above the ground has p.e. = mgh, Figure 23.3. When it falls, its velocity increases and it gains k.e. at the expense of its p.e. If it starts from rest and air resistance is negligible, its k.e. on reaching the ground equals the p.e. lost by the mass

$$\tfrac{1}{2}mv^2 = mgh$$

$\therefore$ loss of p.e. = gain of k.e.

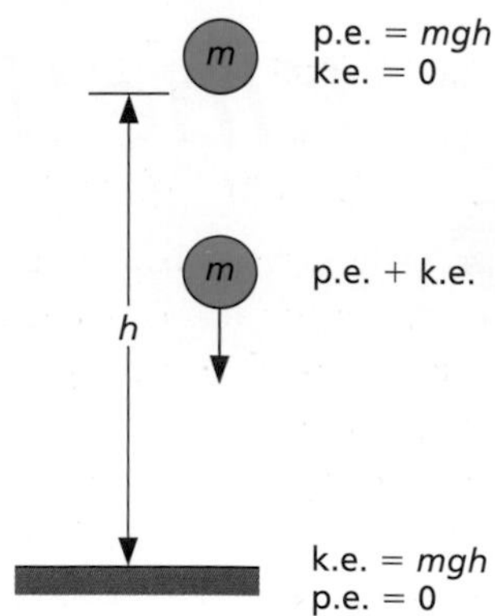

Figure 23.3 Loss of p.e. = gain of k.e.

This is an example of the **principle of conservation of energy** which was discussed in Chapter 16.

In the case of a pendulum k.e. and p.e. are interchanged continually. The energy of the bob is all p.e. at the end of the swing and all k.e. as it passes through its central position. In other positions it has both p.e. and k.e., Figure 23.4. Eventually all the energy is changed to heat as a result of overcoming air resistance.

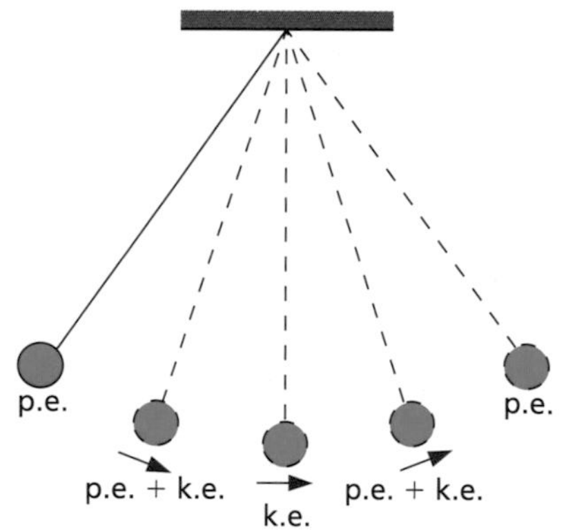

Figure 23.4 Interchange of p.e. and k.e. for a simple pendulum

F *Elastic and inelastic collisions*

In all collisions (where no external force acts) there is normally a loss of k.e., usually to heat energy and to a small extent to sound energy. The greater the proportion of k.e. lost, the less **elastic** is the collision, i.e. the more **inelastic** it is. In a perfectly elastic collision k.e. is conserved.

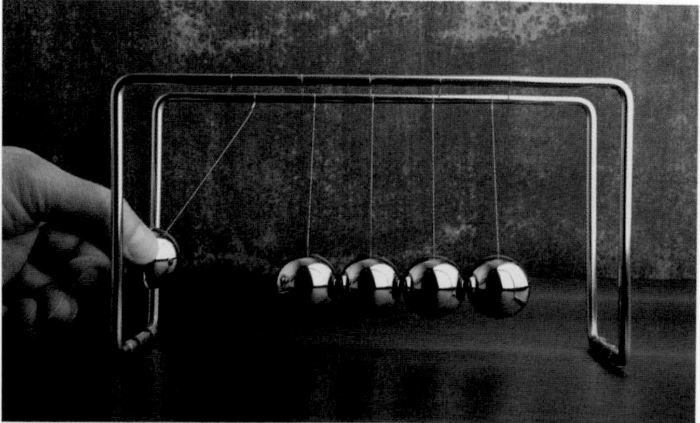

Figure 23.5 Newton's cradle is an instructive toy for studying collisions and conservation

Driving and car safety

a) Braking distance and speed

For a car moving with speed v, the brakes must be applied over a braking distance s to bring the car to rest. The **braking distance is directly proportional to the square of the speed**, i.e. if v is doubled, s is quadrupled. The **thinking distance** (i.e. the distance travelled while the driver is reacting before applying the brakes) has to be added to the **braking distance** to obtain the **overall stopping distance**, i.e.

$$\text{stopping distance} = \text{thinking distance} + \text{braking distance}$$

Typical values taken from the Highway Code are given in Table 23.1 for different speeds. The greater the speed, the greater the stopping distance for a given braking force. (To stop the car in a given distance a greater braking force is needed for higher speeds.)

Thinking distance depends on the driver's reaction time – this will vary with factors such as the driver's degree of tiredness, use of alcohol or drugs, eyesight and the visibility of the hazard. Braking distance varies with both the road conditions and the state of the car; it is longer when the road is wet or icy, when friction between the tyres and the road is low, than when conditions are dry. Efficient brakes and high tyre tread help to reduce the braking distance.

Table 23.1

Speed/mph	**20**	**40**	**60**	**80**
Thinking distance/metres	6	12	18	24
Braking distance/metres	6	24	54	96
Total stopping distance/metres	**12**	**36**	**72**	**120**

b) Car design and safety

When a car stops rapidly in a collision, large forces are produced on the car and its passengers, and their k.e. has to be dissipated.

Crumple zones at the front and rear collapse in such a way that the k.e. is absorbed gradually, Figure 23.6. This extends the collision time and reduces the decelerating force and hence the potential for injury to the passengers.

Extensible seat belts exert a backwards force (of 10 000 N or so) over about 0.5 m, which is roughly the distance between the front seat occupants and the windscreen. In a car travelling at 15 m/s (34 mph), the effect felt by anyone *not* using a seat belt is the same as that produced by jumping off a building 12 m high!

Air bags in some cars inflate and protect the driver from injury by the steering wheel.

Head restraints ensure that if the car is hit from behind, the head goes forwards with the body and not backwards over the top of the seat. This prevents damage to the top of the spine.

All these are **secondary** safety devices which aid **survival** in the event of an accident. **Primary** safety factors help to **prevent** accidents and depend on the car's roadholding, brakes, steering, handling and above all on the driver since most accidents are due to driver error.

The chance of being killed in an accident is about **five times less** if seat belts are worn and head restraints are installed.

Figure 23.6 Cars in an impact test showing the collapse of the front crumple zone

Worked example

A boulder of mass 4 kg rolls over a cliff and reaches the beach below with a velocity of 20 m/s.

a Mass of boulder = m = 4 kg
Velocity of boulder as it lands = v = 20 m/s

$$\therefore \quad \text{k.e. of boulder as it lands} = E_k = \tfrac{1}{2}mv^2$$
$$= \tfrac{1}{2} \times 4\,\text{kg} \times (20)^2\,\text{m}^2/\text{s}^2$$
$$= 800\,\text{kg m/s}^2 \times \text{m}$$
$$= 800\,\text{N m}$$
$$= 800\,\text{J}$$

b Applying the principle of conservation of energy (and neglecting energy lost in overcoming air resistance),

$$\text{p.e. of boulder on cliff} = \text{k.e. as it lands}$$
$$\therefore \quad \Delta E_p = E_k = 800\,\text{J}$$

c If h is the height of the cliff,

$$\Delta E_p = mgh$$
$$\therefore \quad h = \frac{\Delta E_p}{mg} = \frac{800\,\text{J}}{4\,\text{kg} \times 10\,\text{m/s}^2} = \frac{800\,\text{N m}}{40\,\text{kg m/s}^2}$$
$$= \frac{800\,\text{kg m/s}^2 \times \text{m}}{40\,\text{kg m/s}^2} = 20\,\text{m}$$

Questions

1 Calculate the k.e. of
 a a 1 kg trolley travelling at 2 m/s,
 b a 2 g (0.002 kg) bullet travelling at 400 m/s,
 c a 500 kg car travelling at 72 km/h.

2 a What is the velocity of an object of mass 1 kg which has 200 J of k.e.?
 b Calculate the p.e. of a 5 kg mass when it is (i) 3 m, (ii) 6 m, above the ground. (g = 10 N/kg)

3 A 100 g steel ball falls from a height of 1.8 m on to a metal plate and rebounds to a height of 1.25 m. Find
 a the p.e. of the ball before the fall (g = 10 m/s^2),
 b its k.e. as it hits the plate,
 c its velocity on hitting the plate,
 d its k.e. as it leaves the plate on the rebound,
 e its velocity of rebound.

4 It is estimated that 7×10^6 kg of water pours over the Niagara Falls every second. If the Falls are 50 m high, and if all the energy of the falling water could be harnessed, what power would be available? (g = 10 N/kg)

5 Figure 23.7 shows water falling over a dam.

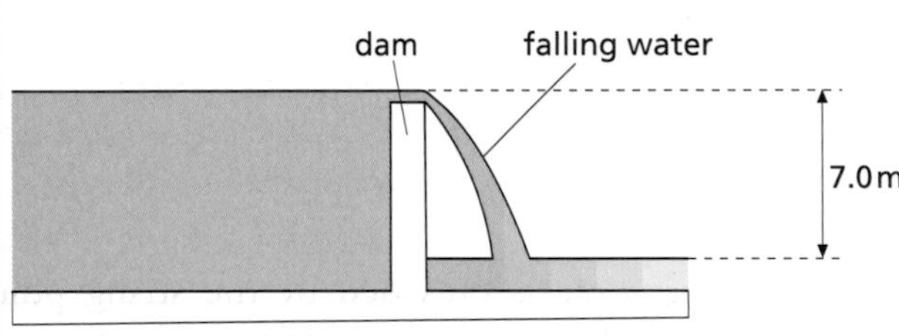

Figure 23.7

 a The vertical height that the water falls is 7.0 m. Calculate the potential energy lost by 1.0 kg of water during the fall.
 b Assuming all this potential energy loss is changed to kinetic energy of the water, calculate the speed of the water, in the vertical direction, at the end of the fall.
 c The vertical speed of the water is less than that calculated in **b**. Suggest one reason for this.

(*UCLES IGCSE Physics Extended, Nov 2006*)

6 The Highway Code gives tables of the shortest stopping distances for cars travelling at various speeds. An extract from the Highway Code is given in Figure 23.8.

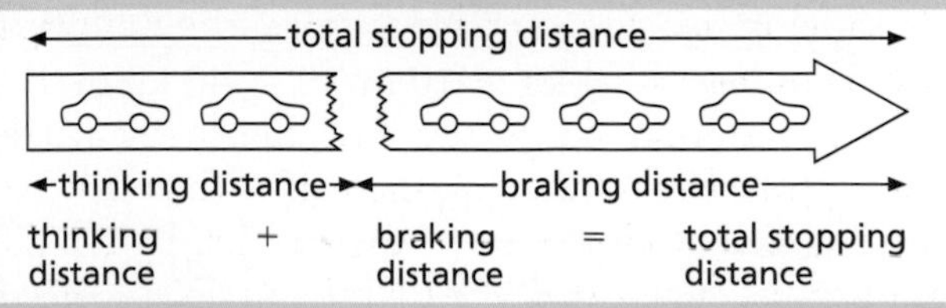

Figure 23.8

 a A driver's reaction time is 0.7 s. Calculate the 'thinking distance' when travelling at 10 m/s.
 b (i) Write down *two* factors which could increase a driver's reaction time.
 (ii) What effect does an increase in reaction time have on
 A thinking distance, B braking distance,
 C total stopping distance?
 c Explain why the braking distance would change on a wet road.
 d A car was travelling at 30 m/s. The driver braked. The graph in Figure 23.9 is a velocity–time graph showing the velocity of the car during braking.

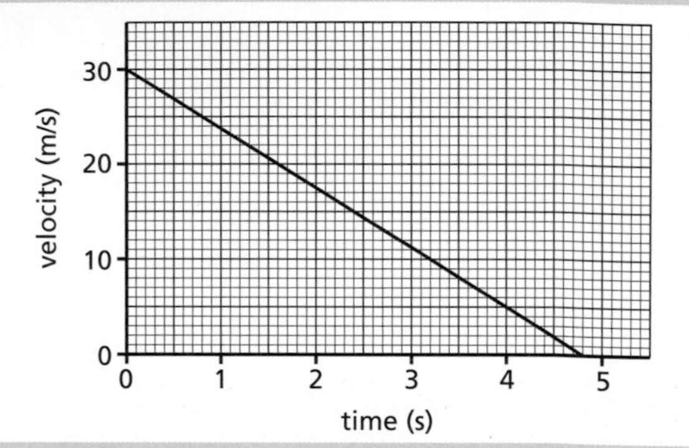

Figure 23.9

 Calculate
 (i) the rate at which the velocity decreases (deceleration),
 (ii) the braking force, if the mass of the car is 900 kg,
 (iii) the braking distance.

(*NEAB Higher, June 1998*)

Checklist

After studying this chapter you should be able to

- define **kinetic energy** (k.e.),
- perform calculations using $E_k = \frac{1}{2}mv^2$,
- define **potential energy** (p.e.),
- calculate changes in p.e. using $\Delta E_p = mgh$,
- apply the principle of conservation of energy to simple mechanical systems, e.g. a pendulum,
- recall the effect of speed on the braking distance of a vehicle,
- describe secondary safety devices in cars.

24 Circular motion

Centripetal force	*Looping the loop*	*Practical work*
Rounding a bend	*Satellites*	Investigating circular motion.

There are many examples of bodies moving in circular paths – rides at a fun fair, clothes being spun dry in a washing machine, the planets going round the Sun and the Moon circling the Earth. When a car turns a corner it may follow an arc of a circle. 'Throwing the hammer' is a sport practised at Highland Games in Scotland, Figure 24.1, in which the hammer is whirled round and round before it is released.

Figure 24.1 'Throwing the hammer'

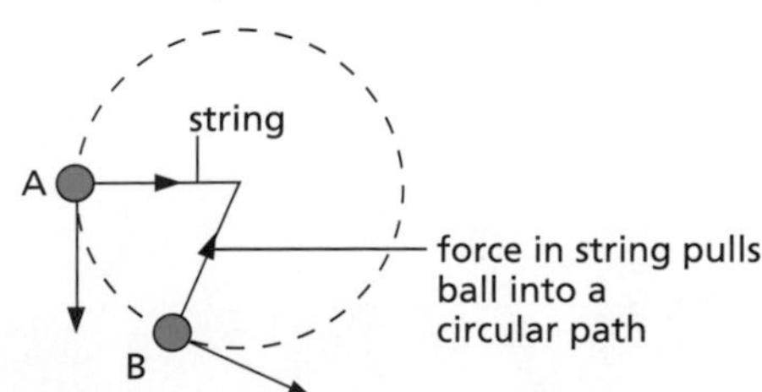

Figure 24.2

□ Centripetal force

In Figure 24.2 a ball attached to a string is being whirled round in a horizontal circle. Its direction of motion is constantly changing. At A it is along the tangent at A; shortly afterwards, at B, it is along the tangent at B; and so on.

Velocity has both size and direction; speed has only size. Velocity is speed in a stated direction and if the direction of a moving body changes, even if its speed does not, then its velocity has changed. A change of velocity is an acceleration and so during its whirling motion the ball is accelerating.

It follows from Newton's first law of motion that if we consider a body moving in a circle to be accelerating then there must be a force acting on it to cause the acceleration. In the case of the whirling ball it is reasonable to say the force is provided by the string pulling inwards on the ball. Like the acceleration, the force acts towards the centre of the circle and keeps the body at a fixed distance from the centre.

A larger force is needed if

(i) the speed v of the ball is increased,
(ii) the radius r of the circle is decreased,
(iii) the mass m of the ball is increased.

The rate of change of direction, and so the acceleration a, is increased by (i) and (ii). It can be shown that $a = v^2/r$ and so, from $F = ma$, we can write

$$F = \frac{mv^2}{r}$$

This force, which acts **towards the centre** and keeps a body moving in a circular path, is called the **centripetal force** (centre-seeking force).

Should the force be greater than the string can bear, the string breaks and the ball flies off with steady speed in a straight line **along the tangent**, i.e. in the direction of travel when the string broke (as Newton's first law of motion predicts). It is not thrown outwards.

Whenever a body moves in a circle (or circular arc) there must be a centripetal force acting on it. In throwing the hammer it is the pull of the athlete's arms acting on the hammer towards the centre of the whirling path. When a car rounds a bend a frictional force is exerted inwards by the road on the car's tyres.

Practical work

Investigating circular motion

Use the apparatus in Figure 24.3 to investigate the various factors which affect circular motion. Make sure the rubber bung is **tied securely** to the string and that **the area around you is clear of other students**. The paper clip acts as an indicator to aid keeping the radius of the circular motion constant.

Spin the rubber bung at a constant speed while adding more weights to the holder; it will be found that the radius of the orbit decreases. Show that if the rubber bung is spun faster, more weights must be added to the holder to keep the radius constant. Are these findings in agreement with the formula given above for the centripetal force $F = mv^2/r$?

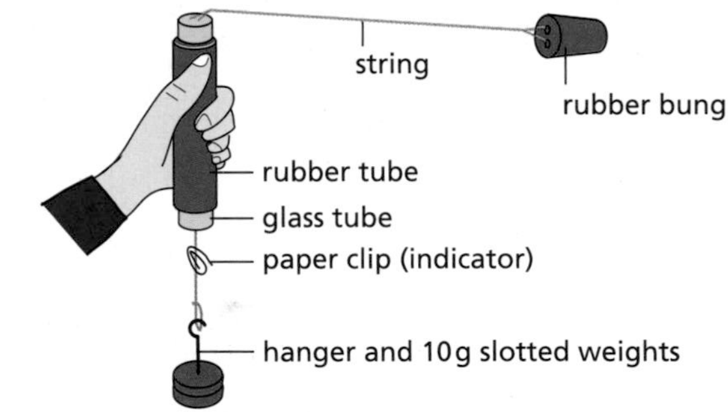

Figure 24.3

☐ *Rounding a bend*

When a car rounds a bend a frictional force is exerted **inwards** by the road on the car's tyres, so providing the centripetal force needed to keep it in its curved path, Figure 24.4a. Here friction acts as an accelerating force (towards the centre of the circle) rather than a retarding force (p. 70, p. 104). The successful negotiation of a bend on a flat road therefore depends on the tyres and the road surface being in a condition that enables them to provide a sufficiently large frictional force – otherwise skidding occurs.

Safe cornering that does not rely entirely on friction is achieved by 'banking' the road as in Figure 24.4b. Some of the centripetal force is then supplied by the part of the contact force *N* from the road surface on the car that acts horizontally. A bend in a railway track is banked so that the outer rail is not strained by having to supply the centripetal force, by pushing inwards on the wheel flanges.

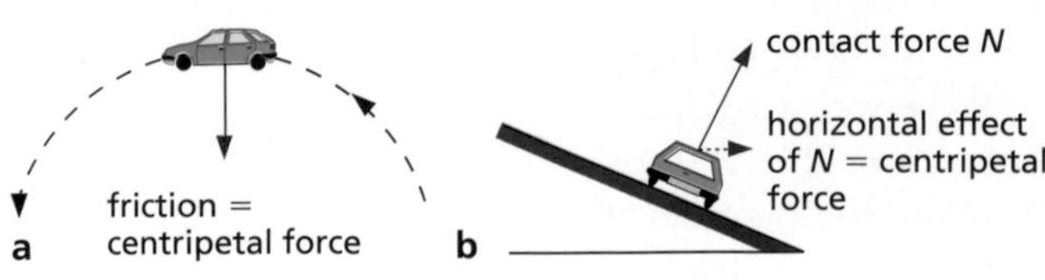

Figure 24.4

☐ *Looping the loop*

A pilot who is not strapped into his aircraft can loop the loop without falling downwards at the top of the loop. A bucket of water can be swung round in a vertical circle without spilling. Some amusement park rides, Figure 24.5, give similar effects. Can you suggest what provides the centripetal force for each of the above three cases (i) at the top of the loop and (ii) at the bottom of the loop?

Figure 24.5 Looping the loop at an amusement park

Satellites

For a satellite of mass m orbiting the Earth at radius r with orbital speed v, the centripetal force $F = mv^2/r$ is provided by gravity.

To put an artificial satellite in orbit at a certain height above the Earth it must enter the orbit at the correct speed. If it does not, the force of gravity, which decreases with height, will not be equal to the centripetal force needed for the orbit.

This can be seen by imagining a shell fired horizontally from the top of a very high mountain, Figure 24.6. If gravity did not pull it towards the centre of the Earth it would continue to travel horizontally, taking path A. In practice it might take path B. A second shell fired faster might take path C and travel farther. If a third shell is fired even faster, it might never catch up with the rate at which the Earth's surface is falling away. It would remain at the same height above the Earth (path D) and return to the mountain top, behaving like a satellite.

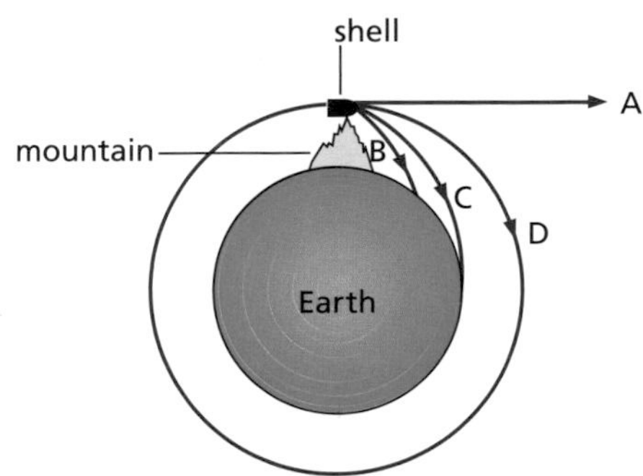

Figure 24.6

The orbital period T (the time for one orbit) of a satellite = distance/velocity. So for a circular orbit

$$T = \frac{2\pi r}{v}$$

Satellites in high orbits have longer periods than those in low orbits.

The Moon is kept in a circular orbit round the Earth by the force of gravity between it and the Earth. It has an orbital period of 27 days.

a) Communication satellites

These circle the Earth in orbits above the equator. **Geostationary** satellites have an orbit high above the equator (36 000 km); they travel with the same speed as the Earth rotates, so appear to be stationary at a particular point above the Earth's surface – their orbital period is 24 hours. They are used for transmitting television, intercontinental telephone and data signals. Geostationary satellites need to be well separated so that they do not interfere with each other; there is room for about 400.

Mobile phone networks use many satellites in much lower equatorial orbits; they are slowed by the Earth's atmosphere and their orbit has to be regularly adjusted by firing a rocket engine. Eventually they run out of fuel and burn up in the atmosphere as they fall to Earth.

b) Monitoring satellites

These circle the Earth rapidly in low **polar** orbits, i.e. passing over both poles; at a height of 850 km the orbital period is only 100 minutes. The Earth rotates below them so they scan the whole surface at short range in a 24 hour period and can be used to map or monitor regions of the Earth's surface which may be inaccessible by other means. They are widely used in weather forecasting to transmit infrared pictures of cloud patterns continuously down to Earth, Figure 24.7, which are picked up in turn by receiving stations around the world.

Figure 24.7 Satellite image of cloud over Europe

Questions

1 An apple is whirled round in a horizontal circle on the end of a string which is tied to the stalk. It is whirled faster and faster and at a certain speed the apple is torn from the stalk. Why?

2 A car rounding a bend travels in an arc of a circle.
  a What provides the centripetal force?
  b Is a larger or a smaller centripetal force required if
    (i) the car travels faster,
    (ii) the bend is less curved,
    (iii) the car has more passengers?

3 Racing cars are fitted with tyres called 'slicks', which have no tread pattern, for dry tracks and 'treads' for wet tracks. Why?

4 A satellite close to the Earth (at a height of about 200 km) has an orbital speed of 8 km/s. Taking the radius of the orbit as approximately equal to the Earth's radius of 6400 km, calculate the time it takes to make one orbit.

Checklist

After studying this chapter you should be able to

- ☐ explain circular motion in terms of an unbalanced **centripetal force,**
- ☐ describe an experiment to investigate the factors affecting circular motion,
- ☐ explain how the centripetal force arises for a car rounding a bend,
- ☐ understand satellite motion.

Motion and energy
Additional questions

Speed, velocity and acceleration; graphs and equations

1 The approximate velocity–time graph for a car on a 5-hour journey is shown below. (There is a very quick driver change midway to prevent driving fatigue!)

a State in which of the regions OA, AB, BC, CD, DE the car is (i) accelerating, (ii) decelerating, (iii) travelling with uniform velocity.

b Calculate the value of the acceleration, deceleration or constant velocity in each region.

c What is the distance travelled over each region?

d What is the total distance travelled?

e Calculate the average velocity for the whole journey.

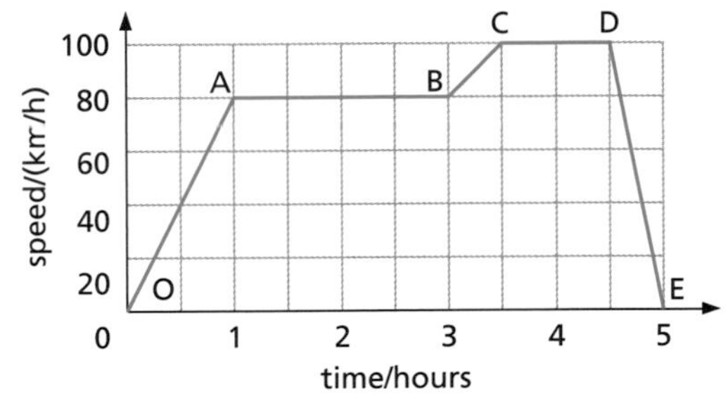

2 The distance–time graph for a motor cyclist riding off from rest is shown below.

a Describe the motion.

b How far does the motorbike move in 30 seconds?

c Calculate the speed.

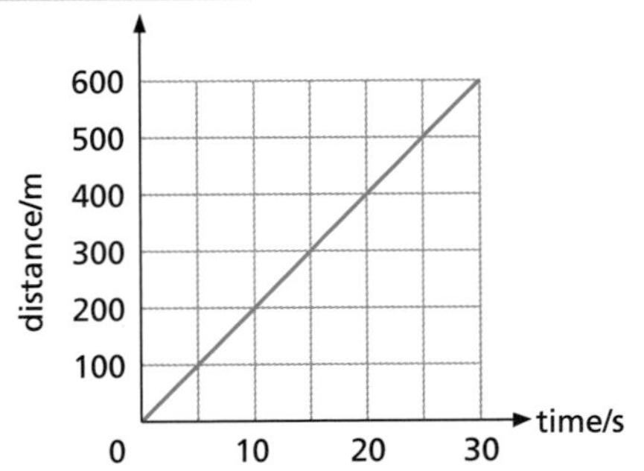

3 The speeds of a car travelling on a straight road are given below at successive intervals of 1 second.

Time/s	0	1	2	3	4
Speed/m/s	0	2	4	6	8

The car travels

1 with an average velocity of 4 m/s

2 16 m in 4 s

3 with a uniform acceleration of 2 m/s^2.

Which statement(s) is (are) correct?

A 1, 2, 3 **B** 1, 2 **C** 2, 3 **D** 1 **E** 3

4 If a train travelling at 10 m/s starts to accelerate at 1 m/s^2 for 15 s on a straight track, its final velocity in m/s is

A 5 **B** 10 **C** 15 **D** 20 **E** 25

5 Two men, Mr A and Mr B, live in the same apartment block in Hometown, and both work at the same factory in Worktown.

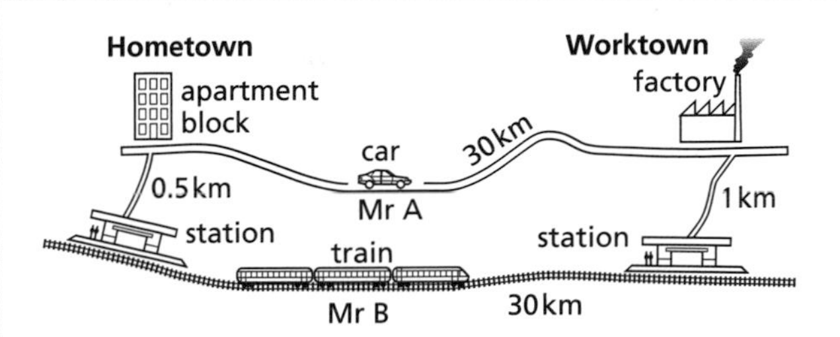

HINT: For this question, work in km and hours.

a Mr A travels to work by car, but it is a poor road and he can only average a speed of 50 km/hour. The distance by road is 30 km. How long does Mr A take to get to work?

b Mr B leaves home at the same time as Mr A, but he

1. walks the 0.5 km to Hometown station,
2. waits 6 minutes (0.1 hour) for the train,
3. travels on the train to Worktown station (journey distance 30 km),
4. walks the 1.0 km from Worktown station to the factory.

The train averages 100 km/hour and Mr B walks at 5 km/hour. Does Mr B arrive at work before or after Mr A, and by how much?

c Apart from a short time accelerating out of Hometown station and a short time slowing down whilst approaching Worktown station, the train travels at a constant speed.

On axes like those below, sketch a speed–time graph for Mr B from home to factory. On your graph, clearly label the four stages of the journey.

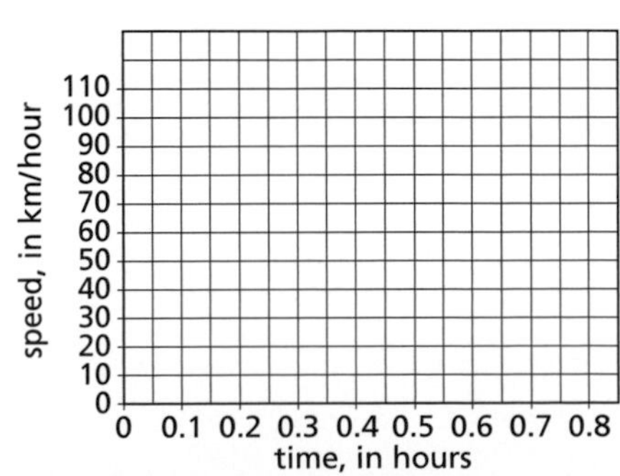

(*UCLES IGCSE Physics Core, Nov 1998*)

continued

6 A firework leaves the ground with an initial velocity of 45 m/s, travelling vertically upwards. It reaches a maximum height of 100 m.

At this point the firework fails to explode and falls back down the same vertical path to the ground.

At any point on its path, the firework has both a velocity and a speed.

a Using the terms **vector** and **scalar**, explain the difference between velocity and speed.

b The graph shows the height of the firework above the ground during the first 5 s of its journey.

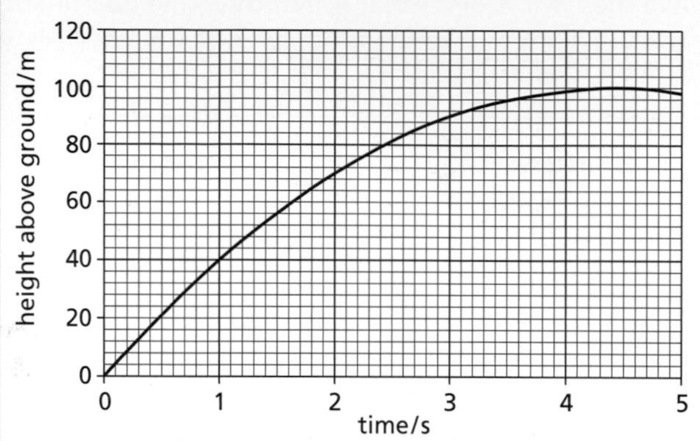

(i) Use the information on the graph to

1. find the time taken for the firework to reach its maximum height above the ground,
2. describe how the motion of the firework changes over the first 5 s of its journey.

(ii) The acceleration of free fall is 10 m/s^2 and air resistance on the firework is negligible. State

1. the deceleration of the firework as it is rising,
2. the total time taken for the firework to rise 100 m and then to fall back to the ground.

(iii) State the velocity with which the falling firework hits the ground.

(*UCLES IGCSE Physics Extended, May/June 2000*)

Force and acceleration

7 a Two skydivers jump from a plane. Each holds a different position in the air.

Complete the following sentence.

Skydiver ...... will fall faster because ......

The diagram below shows the direction of the forces acting on one of the skydivers.

b In the following sentences, *two* lines in each box are wrong. Complete each sentence using the correct line.

(i) Force **X** is caused by

air resistance friction gravity

(ii) Force **Y** is caused by

air resistance gravity weight

(iii) When force **X** is bigger than force **Y**, the speed of the skydiver will

go up stay the same go down

(iv) After the parachute opens, force **X**

goes up stays the same goes down

c How does the area of an opened parachute affect the size of force **Y**?

(*SEG Foundation, June 1998*)

8 The graph shows the speed of a small, very dense object which is falling vertically from an aeroplane, up to the point at which it hits the ground. The air resistance on the object is negligibly small for the first 5 s of its fall. The object is fitted with a parachute which springs open after a certain time of fall.

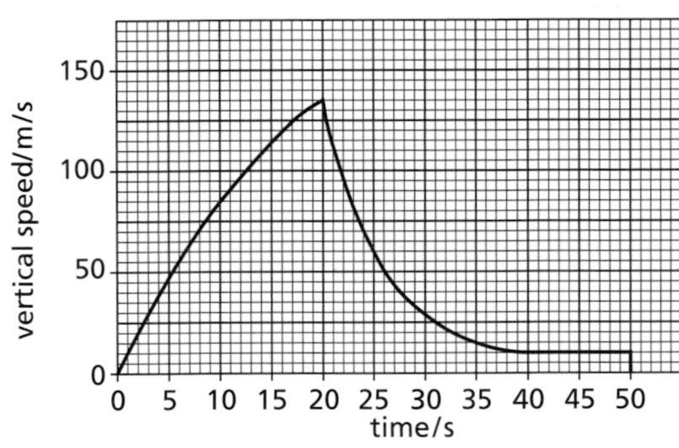

a State the type of motion
(i) between 0 and 5 s,
(ii) between 42 s and 47 s.

b Estimate the time at which the parachute opens.

c On copies of the diagrams below, indicate by labelled arrows the vertical forces acting on the falling object
(i) after 3 s of fall,

falling object

(ii) after 45 s of fall.

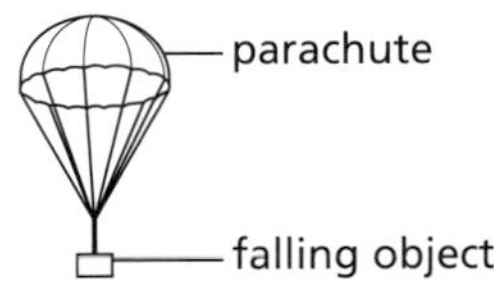

d State whether of not there is a resultant vertical force acting on the falling object
(i) after 3 s of fall,
(ii) after 45 s of fall.

e Calculate the distance fallen in the first 5 s of fall.

(*UCLES IGCSE Physics Extended, May 2001*)

Kinetic and potential energy

9 A car's speed was recorded at intervals as it travelled along a straight, level track. See the table below.

Time/s	**Speed**/(m/s)
0	0
2.0	7.0
4.0	14.0
6.0	21.0
8.0	28.0

a Plot a speed–time graph for the car's motion.

b (i) Describe the motion of the car.
(ii) Use the graph to calculate the distance the car travelled. Show your working.

c After 8.0 s, the brakes were applied. Choose words to complete the description below of the energy change that took place after the brakes were applied.

.... energy ⟶ ....energy

(*UCLES IGCSE Physics Core, Nov 1999*)

10 The diagram shows the outline of a machine for driving steel pillars (called piles) into the ground.

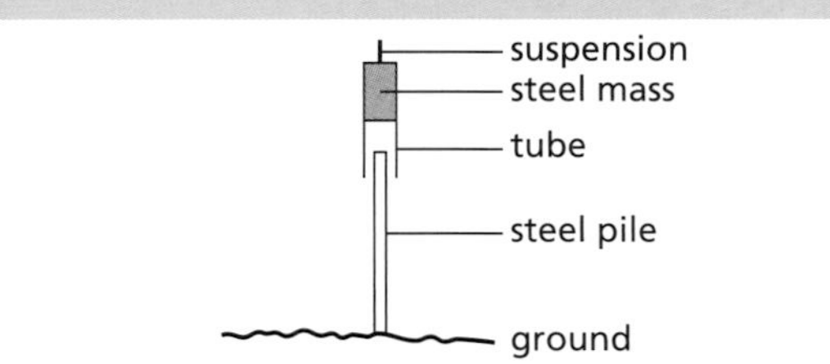

The steel mass is raised by an electric motor and then falls under gravity. The falling steel has a mass of 200 kg and falls a distance of 6.0 m.

a The acceleration of free fall is 10 m/s^2. Calculate
(i) the potential energy gained by the mass each time it is raised,
(ii) the maximum speed at which the mass hits the pile.

b When the mass hits the pile, it has kinetic energy. This energy is transformed into other forms of energy as the speed of the falling mass rapidly reduces to zero. As this happens, the pile is forced a small distance into the ground.
(i) State the energy conversions which take place, starting from the kinetic energy of the falling mass.
(ii) Explain how a large force is produced when the pile is driven a short distance into the ground.

c In raising the steel mass 6.0 m, the electric motor uses more energy than that calculated in **a** (i). Write down and explain *two* causes of this higher energy requirement.

d The equipment design is changed so that when the mass falls once, the pile is driven further into the ground than before the design was changed. Suggest *three* changes that could be made to do this.

(*UCLES IGCSE Physics Extended, Nov 1999*)

continued

11 The data in the table below shows how the stopping distance of a car depends on its speed.

Stopping distance (m)	0	4	12	22	36	52	72
Speed (m/s)	0	5	10	15	20	25	30

a Write down *two* factors, apart from speed, that affect the stopping distance of a car.
b Use the data to draw a graph of stopping distance against speed.
c The speed limit in a supermarket car park is 7.5 m/s. Use your graph to estimate the stopping distance of a car travelling at this speed.
d Describe how the stopping distance changes as the speed of a car increases.
e Explain why the stopping distance is reduced if a driver is made aware of possible hazards.
f The speed limit on roads in towns is 15 m/s. Some road safety campaigners are asking the government to change this to 10 m/s. Suggest why this may be a good idea.

(*London Higher, June 1999*)

Heat and energy

25 Molecules

Kinetic theory of matter
Crystals
Diffusion

Practical work
Brownian motion.

Matter is made up of tiny particles or **molecules** which are too small for us to see directly. But they can be 'seen' by scientific 'eyes'. One of these is the electron microscope: Figure 25.1 is a photograph taken with such an instrument showing molecules of a protein. Molecules consist of even smaller particles called **atoms** and are in continuous motion.

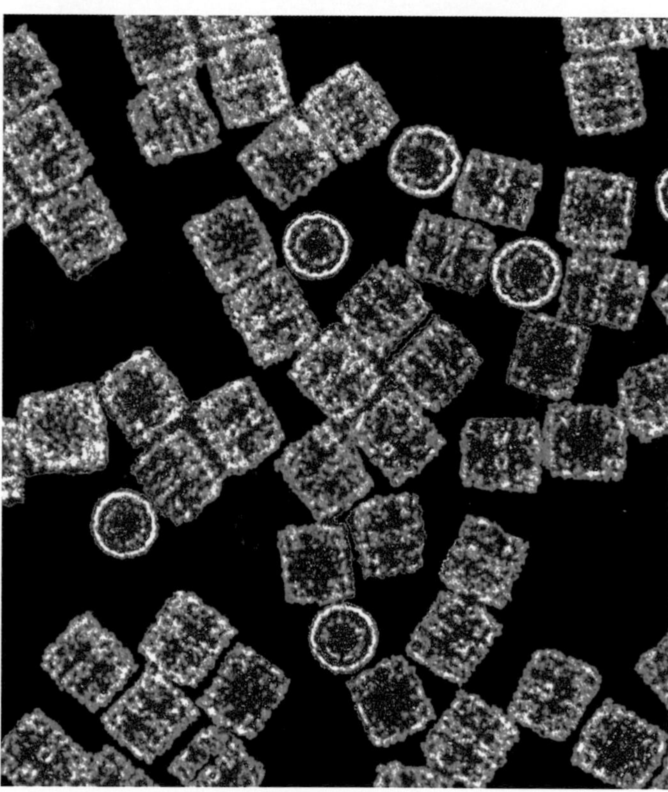

Figure 25.1 Protein molecules

Practical work

Brownian motion

The apparatus is shown in Figure 25.2a. First fill the glass cell with smoke using a match, Figure 25.2b. Replace the lid on the apparatus and set it on the microscope platform. Connect the lamp to a 12 V supply; the glass rod acts as a lens and focuses light on the smoke.

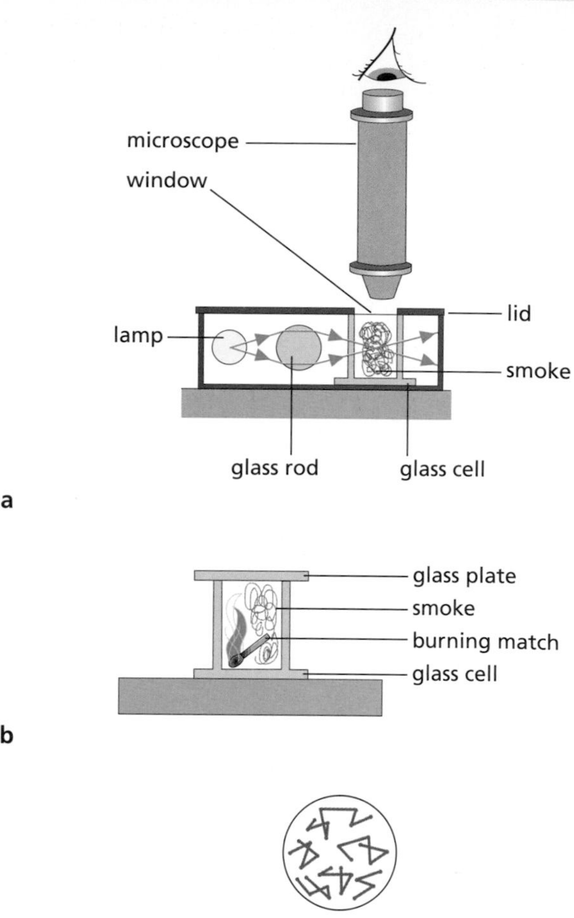

Figure 25.2

Carefully adjust the microscope until you see bright specks dancing around haphazardly, Figure 25.2c. The specks are smoke particles seen by reflected light; their random motion is called **Brownian motion**.

It is due to collisions with fast-moving air molecules in the cell. A smoke particle is massive compared with an air molecule but if there are more high-speed molecules striking one side of it than the other at a given instant, the particle will move in the direction in which there is a net force. The imbalance, and hence the direction of the net force, changes rapidly in a random manner.

Kinetic theory of matter

As well as being in continuous motion, molecules also exert strong electric forces on one another when they are close together. The forces are both attractive and repulsive. The former hold molecules together and the latter cause matter to resist compression. The **kinetic theory** can explain the existence of the solid, liquid and gaseous states.

a) Solids

The theory states that in solids the molecules are close together and the attractive and repulsive forces between neighbouring molecules balance. Also each molecule vibrates to and fro about a fixed position.

It is just as if springs, representing the electric forces between molecules, hold the molecules together, Figure 25.3. This enables the solid to keep a definite shape and volume, while still allowing the individual molecules to vibrate backwards and forwards. The theory shows that the molecules in a solid could be arranged in a regular, repeating pattern like those formed by crystalline substances.

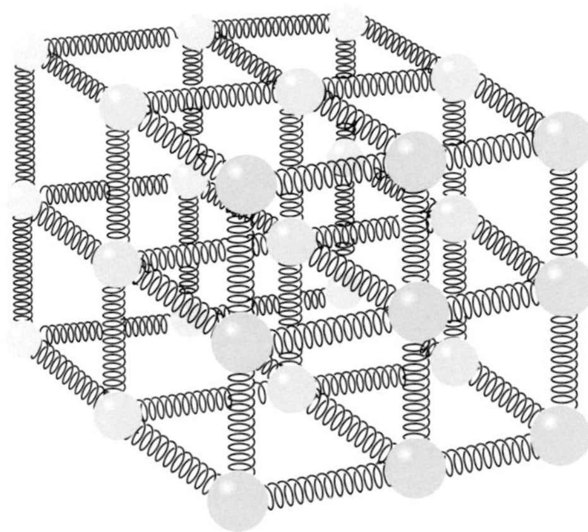

Figure 25.3 The electric forces between molecules in a solid can be represented by springs

b) Liquids

The theory considers that in liquids the molecules are slightly farther apart than in solids but still close enough together to have a definite volume. As well as vibrating, they can at the same time move rapidly over short distances, slipping past each other in all directions. They are never near another molecule long enough to get trapped in a regular pattern which would stop them flowing and taking the shape of the vessel containing them.

A model to represent the liquid state can be made by covering about a third of a tilted tray with marbles ('molecules'), Figure 25.4. It is then shaken to and fro and the motion of the marbles observed. They are able to move around but most stay in the lower half of the tray, so the liquid has a fairly definite volume. A few energetic ones 'escape' from the 'liquid' into the space above. They represent molecules that have 'evaporated' from the 'liquid' surface and become 'gas' or 'vapour' molecules. The thinning out of the marbles near the 'liquid' surface can also be seen.

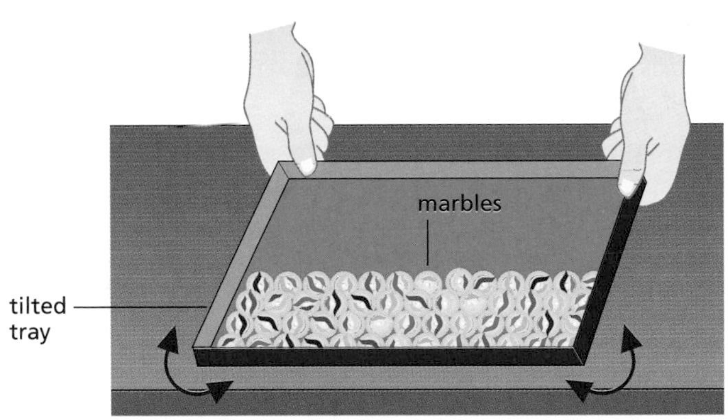

Figure 25.4 A model of molecular behaviour in a liquid

c) Gases

The molecules in gases are much farther apart than in solids or liquids (about ten times) and so gases are much less dense and can be squeezed (compressed) into a smaller space. The molecules dash around at very high speed (about 500 m/s for air molecules at 0 °C) in all the space available. It is only during the brief spells when they collide with other molecules or with the walls of the container that the molecular forces act.

A model of a gas is shown in Figure 25.5. The faster the vibrator works the more often the ball-bearings have collisions with the lid, the tube and with each other, representing a gas at a higher temperature. Adding more ball-bearings is like pumping more air into a tyre; it increases the pressure. If a polystyrene ball (1 cm diameter) is dropped into the tube its irregular motion represents Brownian motion.

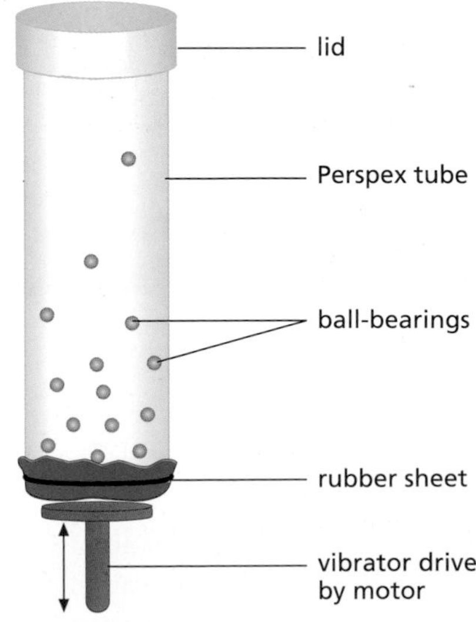

Figure 25.5 A model of molecular behaviour in a gas

F Crystals

Crystals have hard, flat sides and straight edges. Whatever their size, crystals of the same substance have the same shape. This can be seen by observing through a microscope very small cubic salt crystals growing as water evaporates from salt solution on a glass slide, Figure 25.6.

Figure 25.6 Salt crystals viewed under a microscope with polarized light

Figure 25.7 Splitting a calcite crystal

A calcite crystal will split cleanly if a trimming knife, held exactly parallel to one side of the crystal, is struck by a hammer, Figure 25.7.

These facts suggest crystals are made of small particles (e.g. atoms) arranged in an orderly way in planes. Metals have crystalline structures, but many other common solids such as glass, plastics and wood do not.

F Diffusion

Smells, pleasant or otherwise, travel quickly and are caused by rapidly moving molecules. The spreading of a substance of its own accord is called **diffusion** and is due to molecular motion.

Diffusion of gases can be shown if some brown nitrogen dioxide gas is made by pouring a mixture of equal volumes of concentrated nitric acid and water on copper turnings in a gas jar. When the reaction has stopped, a gas jar of air is inverted over the bottom jar, Figure 25.8. The brown colour spreads into the upper jar showing that nitrogen dioxide molecules diffuse upwards against gravity. Air molecules also diffuse into the lower jar.

The speed of diffusion of a gas depends on the speed of its molecules and is greater for light molecules. The apparatus of Figure 25.9 shows this. When hydrogen surrounds the porous pot, the liquid in the U-tube moves in the direction of the arrows. This is due to the lighter, faster molecules of hydrogen diffusing into the pot faster than the heavier, slower molecules of air diffuse out. The opposite happens when carbon dioxide surrounds the pot. Why?

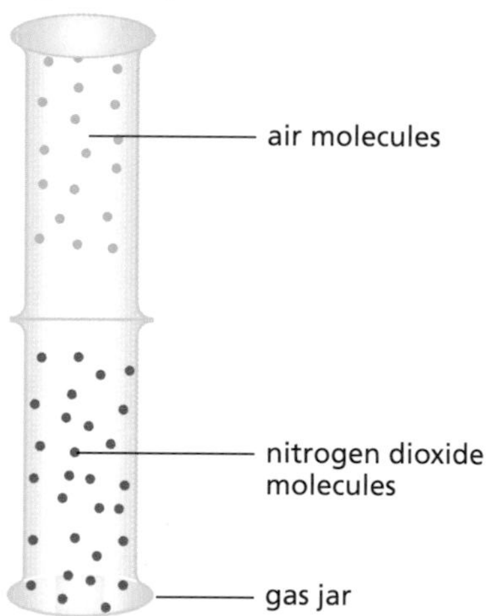

Figure 25.8 Demonstrating diffusion of a gas

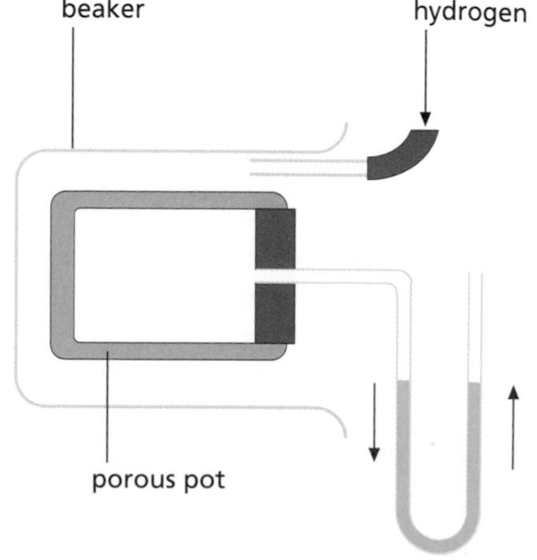

Figure 25.9 The speed of diffusion is greater for lighter molecules

Questions

1 Which one of the following statements is *not* true?

A The molecules in a solid vibrate about a fixed position.
B The molecules in a liquid are arranged in a regular pattern.
C The molecules in a gas exert negligibly small forces on each other, except during collisions.
D The densities of most liquids are about 1000 times greater than those of gases because liquid molecules are much closer together than gas molecules.
E The molecules of a gas occupy all the space available.

2 Using what you know about the compressibility (squeezability) of the different states of matter, explain why
a air is used to inflate tyres,
b steel is used to make railway lines.

Checklist

After studying this chapter you should be able to

- describe and explain an experiment to show Brownian motion,
- use the kinetic theory to explain the physical properties of solids, liquids and gases.

26 Thermometers

Liquid-in-glass thermometer	*Thermocouple thermometer*
Scale of temperature	*Other thermometers*
Clinical thermometer	*Heat and temperature*

The **temperature** of a body tells us how hot the body is. It is measured by a thermometer, usually in **degrees Celsius** (°C). The kinetic theory (Chapter 25) regards temperature as a measure of the average k.e. of the molecules of the body. The greater this is, the faster the molecules move and the higher the temperature of the body.

There are different kinds of thermometer, each type being more suitable than another for a certain job. In each type the physical property used must vary continuously over a wide range of temperature. It must be accurately measurable with simple apparatus and vary in a similar way to other physical properties. Figure 26.1 shows the temperature of a lava flow being measured.

Figure 26.1 Use of a thermocouple probe thermometer to measure a temperature of about 1160 °C in lava

Liquid-in-glass thermometer

In this type the liquid in a glass bulb expands up a capillary tube when the bulb is heated. The liquid must be easily seen and must expand (or contract) rapidly and by a large amount over a wide range of temperature. It must not stick to the inside of the tube or the reading will be too high when the temperature is falling.

Mercury and coloured alcohol are in common use. Mercury freezes at −39 °C and boils at 357 °C; alcohol freezes at −115 °C and boils at 78 °C and is therefore more suitable for low temperatures.

Scale of temperature

A scale and unit of temperature are obtained by choosing two temperatures, called the **fixed points**, and dividing the range between them into a number of equal divisions or **degrees**.

On the Celsius scale (named after the Swedish scientist who suggested it), **the lower fixed point is the temperature of pure melting ice** and is taken as 0 °C. **The upper fixed point is the temperature of the steam above water boiling at normal atmospheric pressure**, 10^5 Pa (or N/m^2), and is taken as 100 °C.

When the fixed points have been marked on the thermometer, the distance between them is divided into 100 equal degrees, Figure 26.2. The thermometer now has a **linear** scale, i.e. it has been calibrated or graduated.

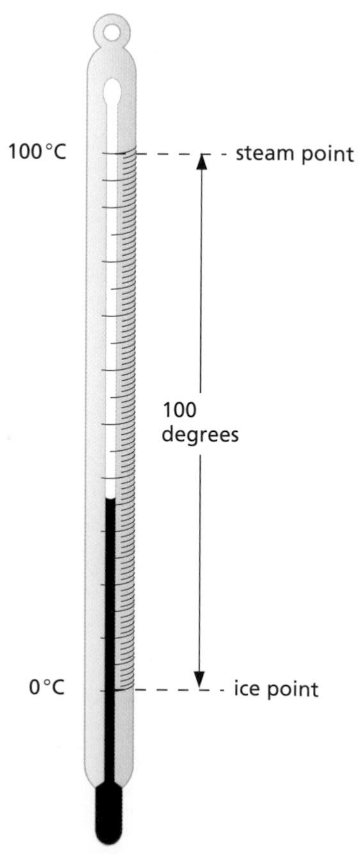

Figure 26.2 A temperature scale in degrees Celsius

Clinical thermometer

A clinical thermometer is a special type of mercury-in-glass thermometer used by doctors and nurses. Its scale only extends over a few degrees on either side of the normal body temperature of 37 °C, Figure 26.3, i.e. it has a small **range**. Because of the very narrow capillary tube, temperatures can be measured very accurately, i.e. the thermometer has a high **sensitivity**.

The tube has a constriction (i.e. a narrower part) just beyond the bulb. When the thermometer is placed under the tongue the mercury expands, forcing its way past the constriction. When the thermometer is removed (after 1 minute) from the mouth, the mercury in the bulb cools and contracts, breaking the mercury thread at the constriction. The mercury beyond the constriction stays in the tube and shows the body temperature. After use the mercury is returned to the bulb by a flick of the wrist.

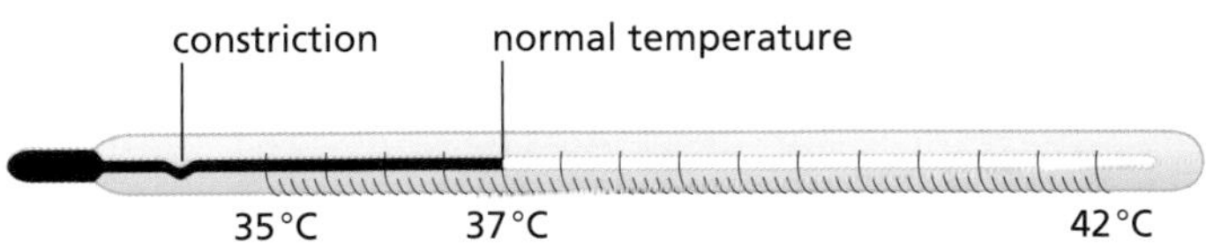

Figure 26.3 A clinical thermometer

Thermocouple thermometer

A thermocouple consists of wires of two different materials, e.g. copper and iron, joined together, Figure 26.4. When one junction is at a higher temperature than the other an electric current flows and produces a reading on a sensitive meter which depends on the temperature difference.

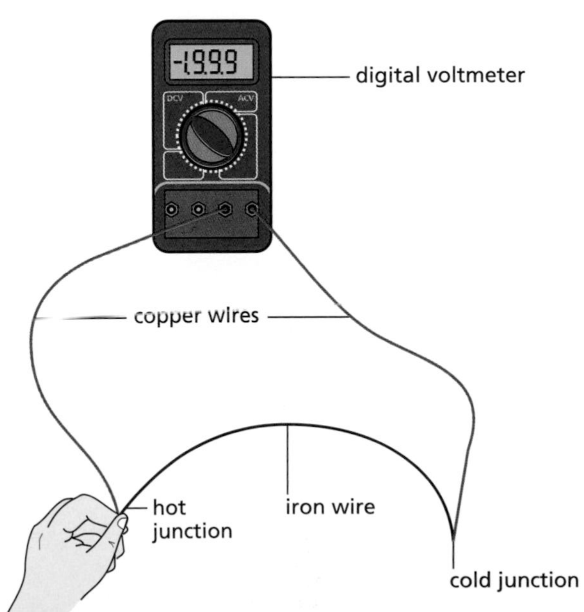

Figure 26.4 A simple thermocouple thermometer

Thermocouples are used in industry to measure a wide range of temperatures from −250 °C up to about 1500 °C, especially rapidly changing ones and those of small objects.

Other thermometers

One type of **resistance thermometer** uses the fact that the electrical resistance (Chapter 36) of a platinum wire increases with temperature.

A resistance thermometer can measure temperatures accurately in the range −200 °C to 1200 °C but it is bulky and best for steady temperatures. A **thermistor** can also be used but over a small range, e.g. −5 °C to 70 °C; its resistance decreases with temperature.

The **constant-volume gas thermometer** uses the change in pressure of a gas to measure temperatures over a wide range. It is an accurate but bulky instrument, basically similar to the apparatus of Figure 28.3 (p. 134).

Thermochromic liquids which change colour with temperature have a limited range around room temperatures.

Heat and temperature

It is important not to confuse the temperature of a body with the heat energy that can be obtained from it. For example, a red-hot spark from a fire is at a higher temperature than the boiling water in a saucepan. In the boiling water the average k.e. of the molecules is lower than in the spark; but since there are many more water molecules, their total energy is greater, and therefore more heat energy can be supplied by the water than by the spark.

Heat passes from a body at a higher temperature to one at a lower temperature. This is due to the average k.e. (and speed) of the molecules in the 'hot' body falling as a result of having collisions with molecules of the 'cold' body whose average k.e., and therefore temperature, increases. When the average k.e. of the molecules is the same in both bodies, they are at the same temperature. For example, if the red-hot spark landed in the boiling water, heat would pass from it to the water even though much more heat energy could be obtained from the water.

Heat is also called **thermal** or **internal energy**; it is the energy a body has because of the kinetic energy *and* the potential energy of its molecules. Increasing the temperature of a body increases its heat energy due to the k.e. of its molecules increasing. But as we will see later (Chapter 30), the internal energy of a body can also be increased by increasing the p.e. of its molecules.

Questions

1 1530 °C 120 °C 55 °C 37 °C 19 °C 0 °C −12 °C −50 °C

From the above list of temperatures choose the most likely value for *each* of the following:

a the melting point of iron,
b the temperature of a room that is comfortably warm,
c the melting point of pure ice at normal pressure,
d the lowest outdoor temperature recorded in London in winter,
e the normal body temperature of a healthy person.

2 In order to make a mercury thermometer that will measure small changes in temperature accurately, would you

A decrease the volume of the mercury bulb
B put the degree markings farther apart
C decrease the diameter of the capillary tube
D put the degree markings closer together
E leave the capillary tube open to the air?

3 **a** How must a property behave to measure temperature?
b Name three properties that qualify.
c Name a suitable thermometer for measuring
(i) a steady temperature of 1000 °C,
(ii) the changing temperature of a small object,
(iii) a winter temperature at the North Pole.

4 Describe the main features of a clinical thermometer.

Checklist

After studying this chapter you should be able to

- define the fixed points on the Celsius scale,
- recall the properties of mercury and alcohol as liquids suitable for use in thermometers,
- describe clinical and thermocouple thermometers,
- □ understand the meaning of range, sensitivity and linearity in relation to thermometers,
- recall some other types of thermometer and the physical properties on which they depend,
- distinguish between heat and temperature and recall that temperature decides the direction of heat flow,
- relate a rise in the temperature of a body to an increase in internal energy.

27 Expansion of solids, liquids and gases

Uses of expansion
Precautions against expansion
Bimetallic strip
Linear expansivity
Unusual expansion of water

In general, when matter is heated it expands and when cooled it contracts. If the changes are resisted large forces are created which are sometimes useful but at other times are a nuisance.

According to the kinetic theory (Chapter 25) the molecules of solids and liquids are in constant vibration. When heated they vibrate faster and force each other a little farther apart. Expansion results, and this is greater for liquids than for solids; gases expand even more. The linear (length) expansion of solids is small and for the effect to be noticed the solid must be long and/or the temperature change large.

Uses of expansion

In Figure 27.1 the axles have been shrunk by cooling in liquid nitrogen at −196 °C until the gear wheels can be slipped on to them. On regaining normal temperature the axles expand to give a very tight fit.

Figure 27.1 'Shrink-fitting' of axles into gear wheels

In the kitchen, a tight metal lid can be removed from a glass jar by immersing the lid in hot water so that it expands.

Precautions against expansion

Gaps used to be left between lengths of railway lines to allow for expansion in summer. They caused a familiar 'clickety-click' sound as the train passed over them. These days rails are welded into lengths of about 1 km and are held by concrete 'sleepers' that can withstand the large forces created without buckling. Also, at the joints the ends are tapered and overlap, Figure 27.2a. This gives a smoother journey and allows some expansion near the ends of each length of rail.

For similar reasons slight gaps are left between lengths of aluminium guttering. In central heating pipes 'expansion joints' are used to join lengths of pipe, Figure 27.2b; these allow the copper pipes to expand in length inside the joints when carrying very hot water.

Figure 27.2a Tapered overlap of rails

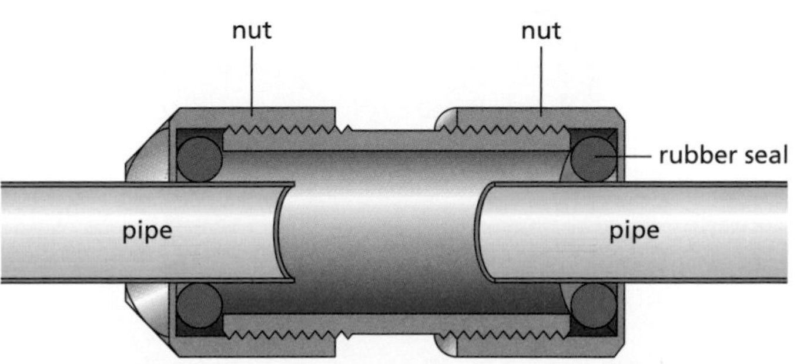

Figure 27.2b Expansion joint

Bimetallic strip

If equal lengths of two different metals, e.g. copper and iron, are riveted together so that they cannot move separately, they form a bimetallic strip, Figure 27.3a. When heated, copper expands more than iron and to allow this the strip bends with copper on the outside, Figure 27.3b. If they had expanded equally the strip would have stayed straight.

Bimetallic strips have many uses.

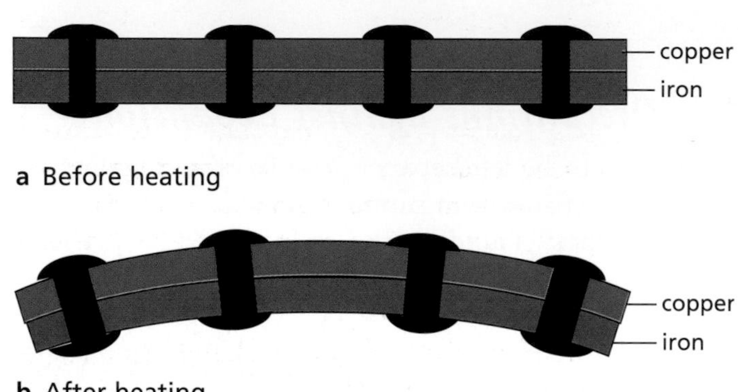

Figure 27.3 A bimetallic strip

a) Fire alarm

Heat from the fire makes the bimetallic strip bend and complete the electrical circuit, so ringing the alarm bell, Figure 27.4a.

A bimetallic strip is also used in this way to work the flashing direction indicator lamps in a car, being warmed by an electric heating coil wound round it.

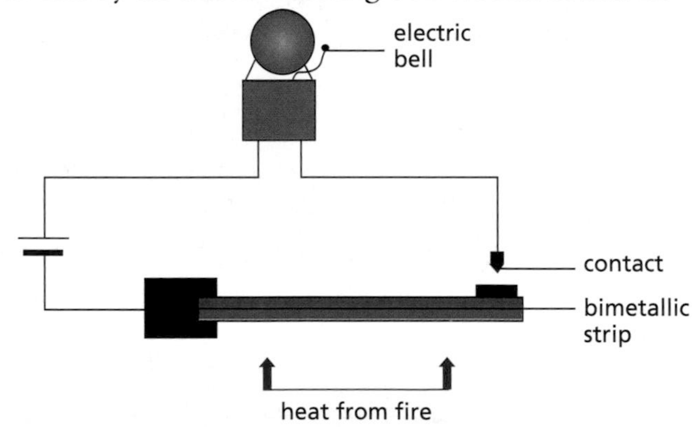

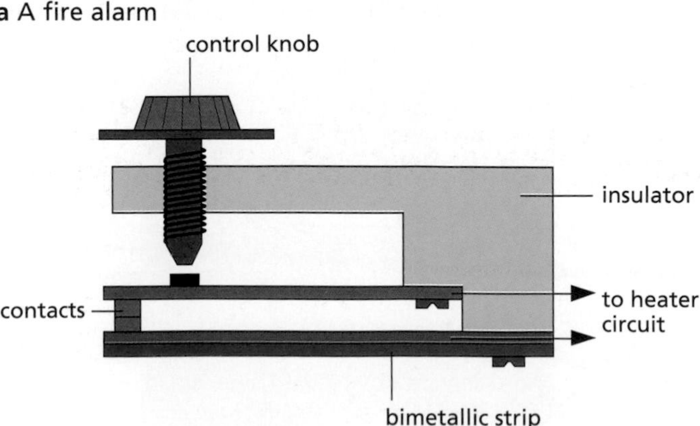

Figure 27.4 Uses of a bimetallic strip

b) Thermostat

A thermostat keeps the temperature of a room or an appliance constant. The one in Figure 27.4b uses a bimetallic strip in the electrical heating circuit of, for example, an iron.

When the iron reaches the required temperature the strip bends down, breaks the circuit at the contacts and switches off the heater. After cooling a little the strip remakes contact and turns the heater on again. A near-steady temperature results.

If the control knob is screwed down, the strip has to bend more to break the heating circuit and this needs a higher temperature.

Linear expansivity

An engineer has to allow for the linear expansion of a bridge when designing it. The expansion can be calculated if

(i) the length of the bridge,
(ii) the range of temperature it will experience, and
(iii) the **linear expansivity** of the material to be used,

are all known.

> The linear expansivity α of a substance is the increase in length of 1 m for a 1 °C rise in temperature.

The linear expansivity of a material is found by experiment. For steel it is 0.000 012 per °C. This means that 1 m will become 1.000 012 m for a temperature rise of 1 °C. A steel bridge 100 m long will expand by 0.000 012 × 100 m for each 1 °C rise in temperature. If the maximum **temperature change** Δl expected is 60 °C (e.g. from −15 °C to +45 °C), the expansion Δl will be 0.000 012 °C × 100 m × 60 °C = 0.072 m = 7.2 cm. In general,

> expansion = linear expansivity × original length × temperature rise

Values of expansivity for liquids are typically about 5 times higher than that for steel; gases have expansivity values about 100 times that of steel. These figures indicate that gases expand much more readily than liquids, and liquids expand more readily than solids. We saw in Chapter 26 how the expansion properties of liquids are used in making thermometers; in the next chapter we will look at the expansion of gases in more detail.

Unusual expansion of water

As water is cooled to 4 °C it contracts, as we would expect. However between 4 °C and 0 °C it expands, surprisingly. **Water has a maximum density at 4 °C**, Figure 27.5.

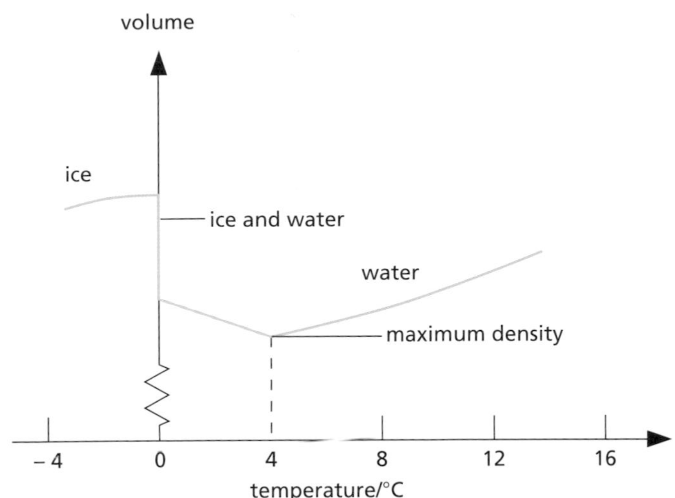

Figure 27.5 Water expands on cooling below 4 °C

At 0 °C, when it freezes, a considerable volume expansion occurs and every 100 cm³ of water becomes 109 cm³ of ice. This accounts for the bursting of unlagged water pipes in very cold weather and for the fact that ice is less dense than cold water and so floats. Figure 27.6 shows a bottle of frozen milk, the main constituent of which is water.

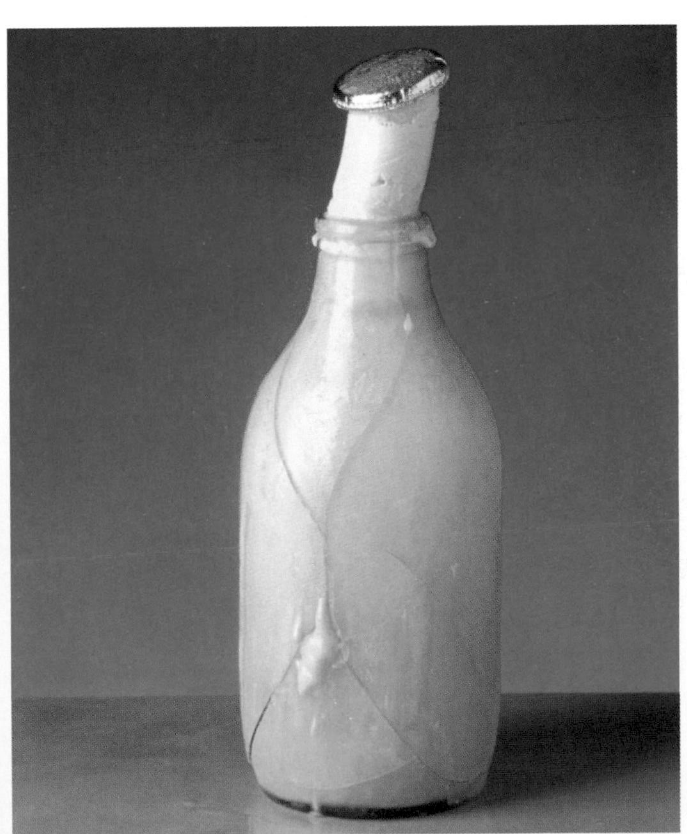

Figure 27.6 Result of the expansion of water on freezing

The unusual expansion of water between 4 °C and 0 °C explains why fish survive in a frozen pond. The water at the top of the pond cools first, contracts and being denser sinks to the bottom. Warmer less dense water rises to the surface to be cooled. When all the water is at 4 °C the circulation stops. If the temperature of the surface water falls below 4 °C, it becomes less dense and *remains at the top*, eventually forming a layer of ice at 0 °C. Temperatures in the pond are then as in Figure 27.7.

The volume expansion of water between 4 °C and 0 °C is due to the breaking up of the groups which water molecules form above 4 °C. The new arrangement requires a larger volume and more than cancels out the contraction due to the fall in temperature.

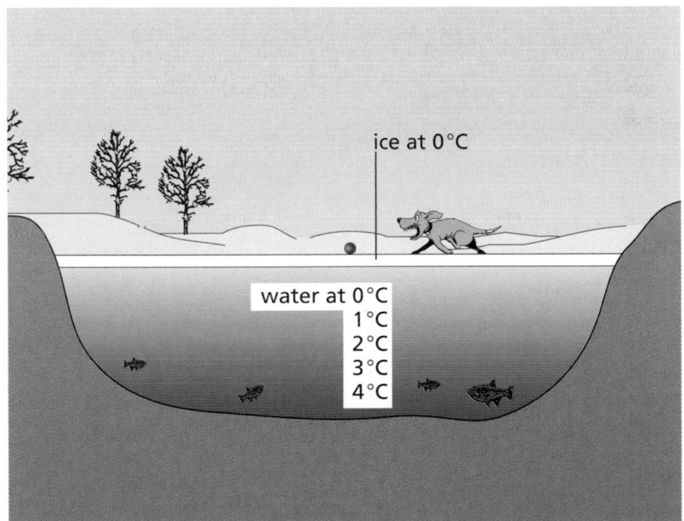

Figure 27.7 Fish can survive in a frozen pond

Questions

1 Explain why

a the metal lid on a glass jam jar can be unscrewed easily if the jar is inverted for a few seconds with the *lid* in very hot water,

b furniture may creak at night after a warm day,

c concrete roads are laid in sections with pitch between them.

2 A bimetallic strip is made from aluminium and copper. When heated it bends in the direction shown in Figure 27.8.
Which metal expands more for the same rise in temperature, aluminium or copper?

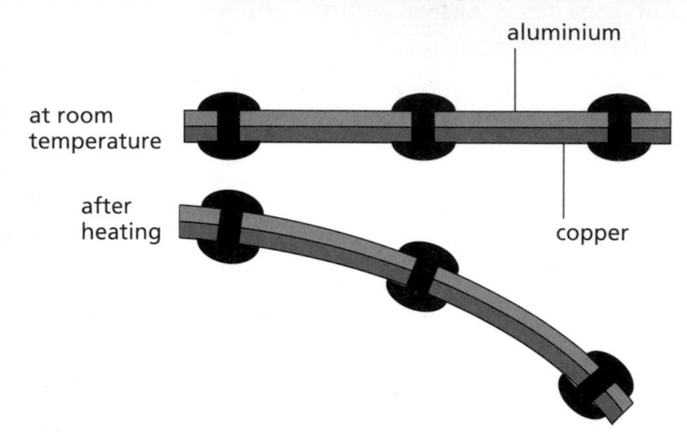

Figure 27.8

Draw a diagram to show how the bimetallic strip would appear if it were cooled to below room temperature.

Checklist

After studying this chapter you should be able to

- describe uses of expansion, including the bimetallic strip,
- describe precautions taken against expansion,
- recall that water has its maximum density at 4 °C and explain why a pond freezes at the top first,
- □ recall the relative order of magnitude of the expansion of solids, liquids and gases.

28 The gas laws

Pressure of a gas	*Practical work*
Absolute zero	Effect on volume of temperature.
The gas laws	Effect on pressure of temperature.
Gases and the kinetic theory	Effect on volume of pressure.

Pressure of a gas

The air forming the Earth's atmosphere stretches upwards a long way. Air has weight; the air in a normal room weighs about the same as you do, e.g. 500 N. Because of its weight the atmosphere exerts a large pressure at sea-level, about 100 000 N/m^2 = 10^5 Pa = 100 kPa. This pressure acts equally in all directions.

A gas in a container exerts a pressure on the walls of the container. If air is removed from a can by a vacuum pump, Figure 28.1, the can collapses because the air pressure outside is greater than that inside. A space from which all the air has been removed is a **vacuum**. Alternatively the pressure in a container can be increased, for example by pumping more gas into the can; a Bourdon gauge (p. 86) is used for measuring fluid pressures.

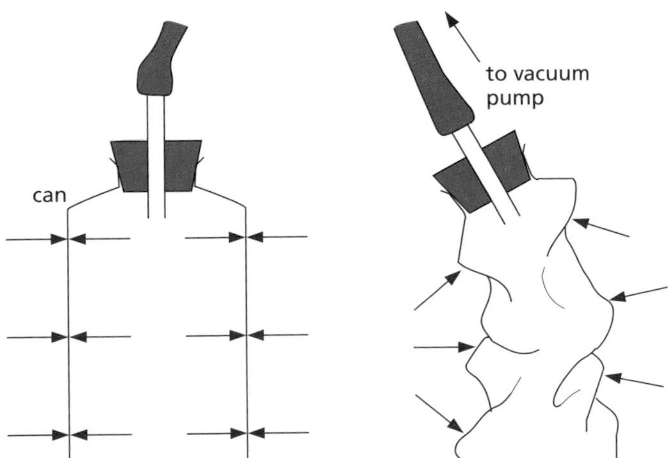

Figure 28.1 Atmospheric pressure collapses the evacuated can

When a gas is heated, as air is in a jet engine, its pressure as well as its volume may change. To study the effect of temperature on these two quantities we must keep one fixed while the other is changed.

Practical work

Effect on volume of temperature (pressure constant) – Charles' law

Arrange the apparatus as in Figure 28.2. The index of concentrated sulfuric acid traps the air column to be investigated and also dries it. Adjust the capillary tube so that the bottom of the air column is opposite a convenient mark on the ruler.

Note the length of the air column (to the *lower* end of the index) at different temperatures but, before taking a reading, stop heating and stir well to make sure that the air has reached the temperature of the water. Put the results in a table.

Plot a graph of volume (in cm, since the length of the air column is a measure of it) on the *y*-axis and temperature (in °C) on the *x*-axis.

The pressure of (and on) the air column is constant and equals atmospheric pressure plus the pressure of the acid index.

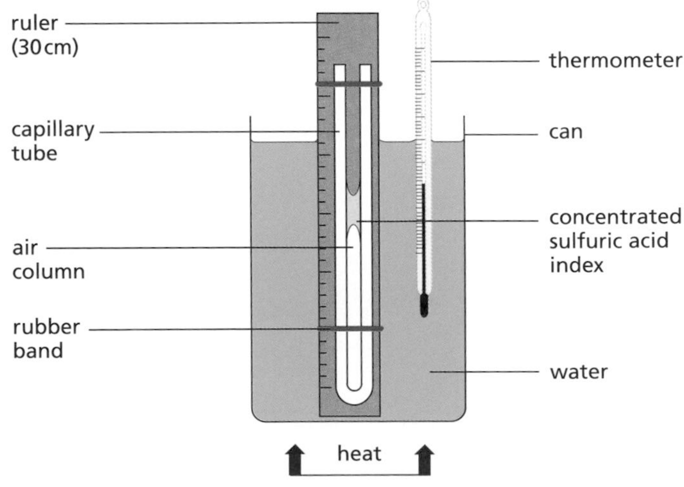

Figure 28.2

Practical work

Effect on pressure of temperature (volume constant) – the Pressure law

The apparatus is shown in Figure 28.3. The rubber tubing from the flask to the pressure gauge should be as short as possible. The flask must be in water almost to the top of its neck and be securely clamped to keep it off the bottom of the can.

Record the pressure over a wide range of temperatures but before taking a reading, stop heating, stir and allow time for the gauge reading to become steady; the air in the flask will then be at the temperature of the water. Tabulate the results.

Plot a graph of pressure on the *y*-axis and temperature on the *x*-axis.

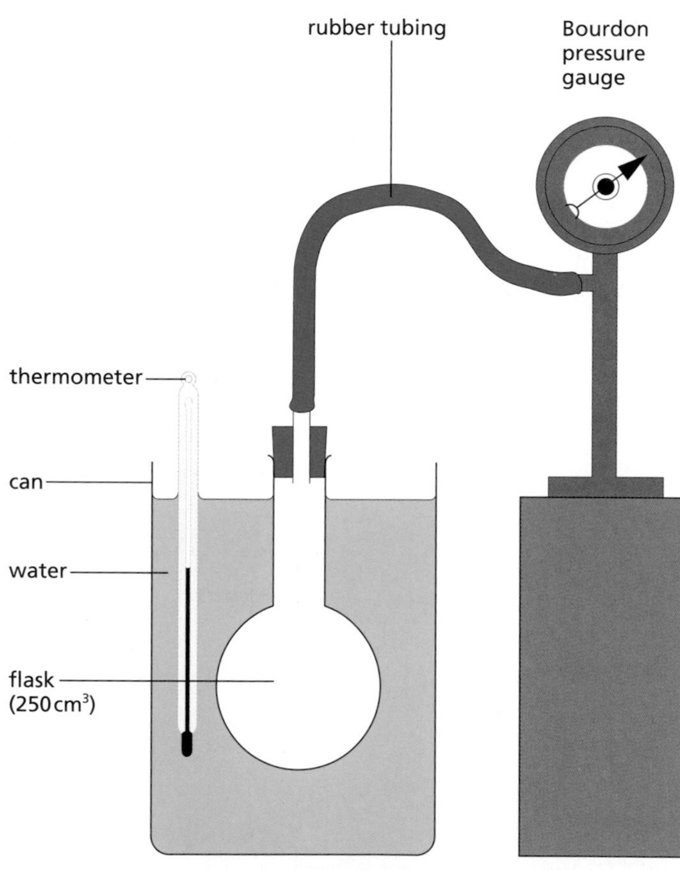

Figure 28.3

Absolute zero

The volume–temperature and pressure–temperature graphs for a gas are straight lines, Figure 28.4. They show that gases expand **linearly** with temperature as measured on a mercury thermometer, i.e. equal temperature increases cause equal volume or pressure increases.

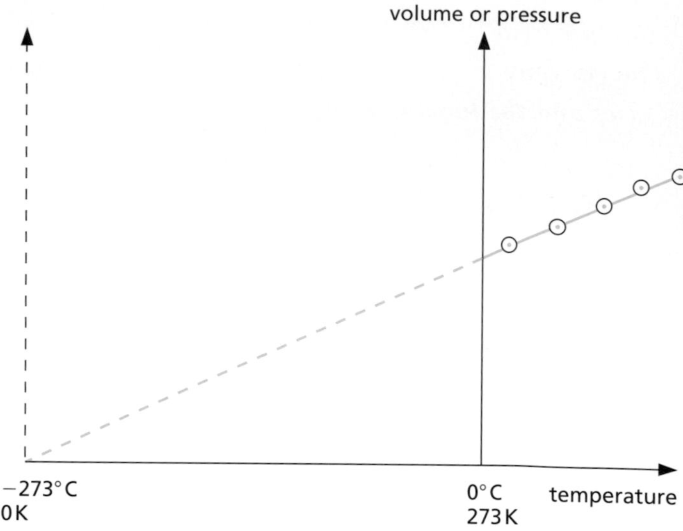

Figure 28.4

The graphs do not pass through the Celsius temperature origin (0 °C). If they are produced backwards they cut the temperature axis at about −273 °C. This temperature is called **absolute zero** because we believe it is the lowest temperature possible. It is the zero of the **absolute** or **Kelvin scale of temperature**. At absolute zero molecular motion ceases and a substance has no internal energy.

Degrees on this scale are called **kelvins** and are denoted by K. They are exactly the same size as Celsius degrees. Since −273 °C = 0 K, conversions from °C to K are made by adding 273. For example

$$0\ °C = 273\ K$$
$$15\ °C = 273 + 15 = 288\ K$$
$$100\ °C = 273 + 100 = 373\ K$$

Kelvin or absolute temperatures are represented by the letter T and if θ stands for a Celsius scale temperature then, in general,

$$T = 273 + \theta$$

Near absolute zero strange things occur. Liquid helium becomes a 'superfluid'. It cannot be kept in an open vessel because it flows up the inside of the vessel, over the edge and down the outside. Some metals and compounds become 'superconductors' of electricity and a current once started in them flows for ever without a battery. Figure 28.5 shows research equipment that is being used to create materials which are superconductors at very much higher temperatures, e.g. −23 °C.

Figure 28.5 Equipment that is being used to make films of complex composite materials which are superconducting at temperatures far above absolute zero

Practical work

Effect on volume of pressure (temperature constant) – Boyle's law

Changes in the volume of a gas due to pressure changes can be studied using the apparatus in Figure 28.6. The volume V of air trapped in the glass tube is read off on the scale behind. The pressure is altered by pumping air from a foot pump into the space above the oil reservoir. This forces more oil into the glass tube and increases the pressure p on the air in it; p is measured by the Bourdon gauge.

If a graph of pressure against volume is plotted using the results, a curve like that in Figure 28.7a is obtained. Close examination of it shows that if p is doubled, V is halved. That is, **p is inversely proportional to V.** In symbols

$$p \propto \frac{1}{V} \quad \text{or} \quad p = \text{constant} \times \frac{1}{V}$$

$$\therefore \qquad pV = \text{constant}$$

If several pairs of readings p_1V_1, p_2V_2, etc. are taken, then it can be confirmed that $p_1V_1 = p_2V_2 = $ constant. This is **Boyle's law** which is stated as follows.

The pressure of a fixed mass of gas is inversely proportional to its volume if its temperature is kept constant.

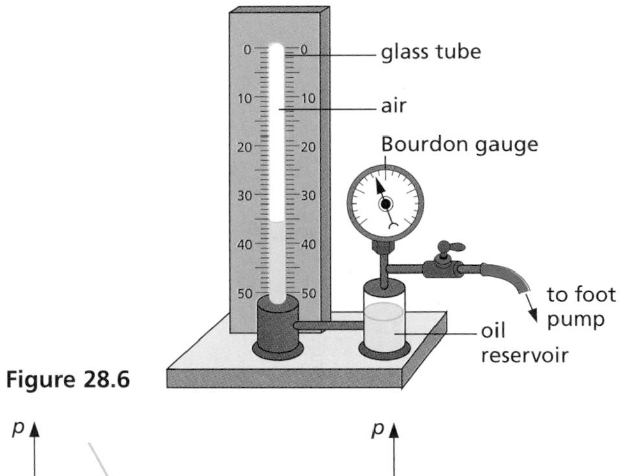

Figure 28.6

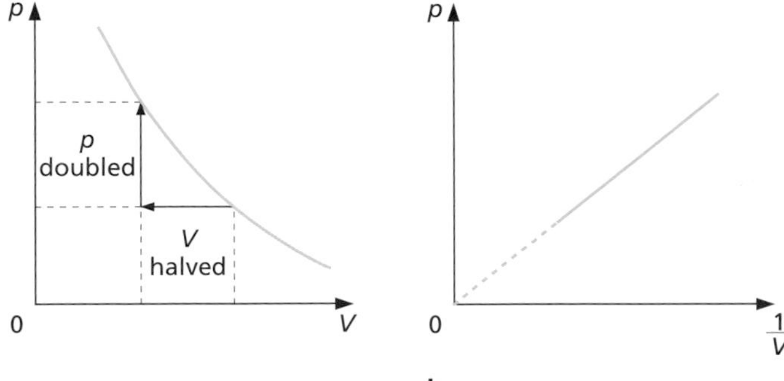

Figure 28.7

Since p is inversely proportional to V, then p is directly proportional to $1/V$. A graph of p against $1/V$ is therefore a straight line through the origin, Figure 28.7b.

F The gas laws

Using absolute temperatures the gas laws can be stated in a convenient form for calculations.

a) Charles' law

In Figure 28.4 (p. 134) the volume–temperature graph passes through the origin if temperatures are measured on the Kelvin scale, that is, if we take 0 K as the origin. We can then say that the volume V is directly proportional to the absolute temperature T, i.e. doubling T doubles V, etc. Therefore

$$V \propto T \quad \text{or} \quad V = \text{constant} \times T$$

or

$$V/T = \text{constant} \qquad (1)$$

Charles' law may be stated as follows.

> The volume of a fixed mass of gas is directly proportional to its absolute temperature if the pressure is kept constant.

b) Pressure law

From Figure 28.4 we can say similarly for the pressure p that

$$p \propto T \quad \text{or} \quad p = \text{constant} \times T$$

or

$$p/T = \text{constant} \qquad (2)$$

The Pressure law may be stated as follows.

> The pressure of a fixed mass of gas is directly proportional to its absolute temperature if the volume is kept constant.

c) Boyle's law

For a fixed mass of gas at constant temperature

$$pV = \text{constant} \qquad (3)$$

d) Combining the laws

The three equations can be combined giving

$$\frac{pV}{T} = \text{constant}$$

For cases in which p, V and T all change from, say, p_1, V_1 and T_1 to p_2, V_2 and T_2, then

$$\frac{p_1V_1}{T_1} = \frac{p_2V_2}{T_2} \qquad (4)$$

F Worked example

A bicycle pump contains 50 cm³ of air at 17 °C and at 1.0 atmosphere pressure. Find the pressure when the air is compressed to 10 cm³ and its temperature rises to 27 °C.

We have

$p_1 = 1.0$ atm $\quad p_2 = ?$
$V_1 = 50$ cm³ $\quad V_2 = 10$ cm³
$T_1 = 273 + 17 = 290$ K $\quad T_2 = 273 + 27 = 300$ K

From equation (4) we get

$$p_2 = p_1 \times \frac{V_1}{V_2} \times \frac{T_2}{T_1}$$

$$= 1 \times \frac{50}{10} \times \frac{300}{290} = 5.2 \text{ atm}$$

Notes

1. All temperatures must be in K.
2. Any units can be used for p and V so long as they are the same on both sides of the equation.
3. In some calculations the volume of the gas has to be found at s.t.p. (standard temperature and pressure). This is 0 °C and 1 atm = 10^5 Pa pressure.

Gases and the kinetic theory

The kinetic theory can explain the behaviour of gases.

a) Cause of gas pressure

All the molecules in a gas are in rapid random motion, with a wide range of speeds, and repeatedly hit and rebound from the walls of the container in huge numbers per second. The average force and hence the pressure they exert on the walls is constant since pressure is force on unit area.

b) Boyle's law

If the volume of a fixed mass of gas is halved by halving the volume of the container, Figure 28.8, the number of molecules per cm³ will be doubled. There will be twice as many collisions per second with the walls, i.e. the pressure is doubled. This is Boyle's law.

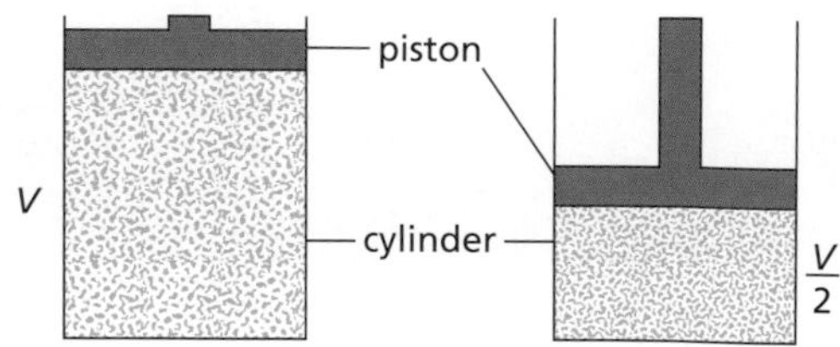

Figure 28.8 Halving the volume doubles the pressure

c) Temperature

When a gas is heated and its temperature rises, the average speed of its molecules increases. If the volume of the gas is to remain constant, its pressure increases due to more frequent and more violent collisions of the molecules with the walls. If the pressure of the gas is to remain constant, the volume must increase so that the frequency of collisions does not.

Questions

1 The apparatus shown in Figure 28.9 is set up in a laboratory during a morning science lesson.

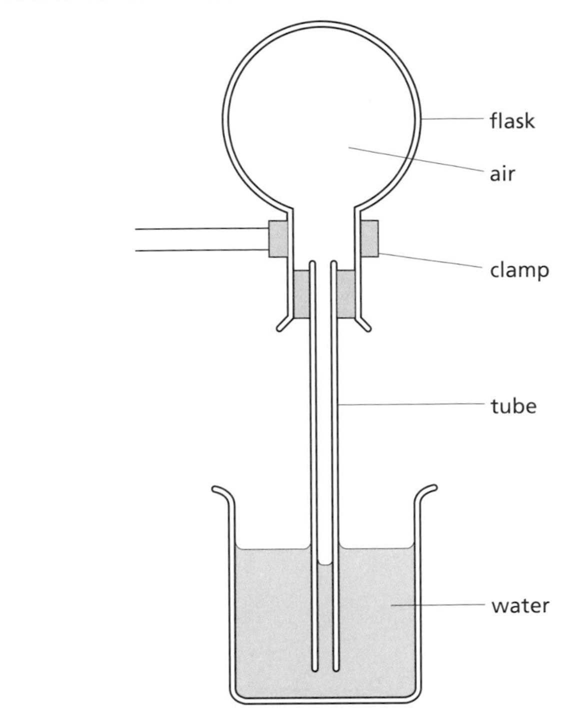

Figure 28.9

Later in the day, the room temperature is higher than in the morning.

a What change is observed in the apparatus?

b Explain why this change happens.

c Suggest one disadvantage of using this apparatus to measure temperature.

(UCLES IGCSE Physics Core, Nov 2005)

2 Figure 28.10 shows a way of indicating the positions and direction of movement of some molecules in a gas at one instant.

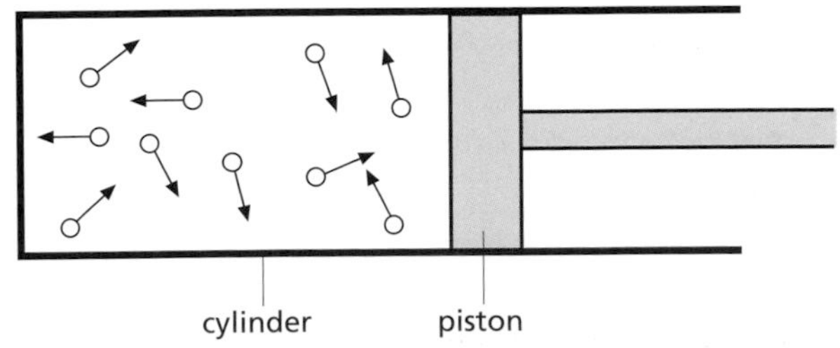

Figure 28.10

a (i) Describe the movement of the molecules.
(ii) Explain how the molecules exert a pressure on the container walls.

b When the gas in the cylinder is heated, it pushes the piston further out of the cylinder. State what happens to
(i) the average spacing of the molecules,
(ii) the average speed of the molecules.

c The gas shown in Figure 28.10 is changed into a liquid and then into a solid by cooling. Compare the gaseous and solid states in terms of
(i) the movement of the molecules,
(ii) the average separation of the molecules.

(UCLES IGCSE Physics Extended, Nov 2005)

3 If a certain quantity of gas has a volume of 30 cm^3 at a pressure of 1×10^5 Pa, what is its volume when the pressure is

a 2×10^5 Pa,

b 5×10^5 Pa?

Assume the temperature remains constant.

4 **a** Figure 28.11 shows a sealed gas syringe in a heated water bath.

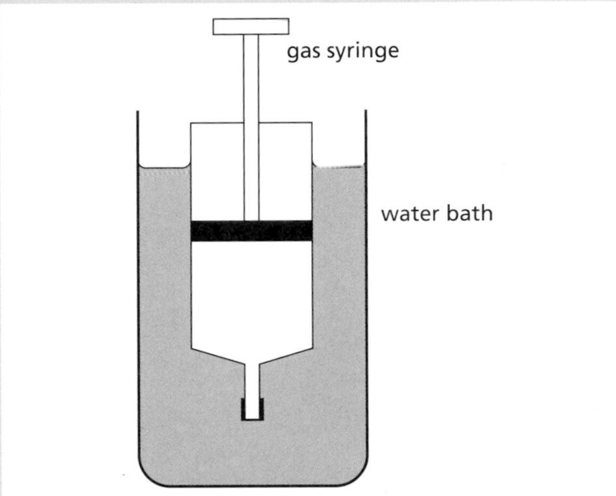

Figure 28.11

Explain in terms of the gas molecules, why, if the pressure inside the syringe is to remain constant the gas must expand as its temperature rises.

b A student was asked to investigate the relationship between the volume and pressure of a gas. The student used a fixed mass of oxygen kept at a constant temperature of 300 kelvins. The results of the investigation were used to plot a graph, Figure 28.12.

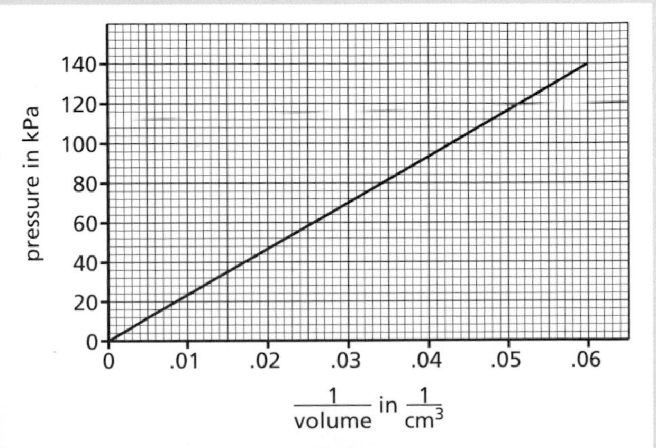

Figure 28.12

continued

(i) Write down the relationship between the volume and pressure of a gas as found by this investigation.

(ii) Copy the graph and draw the line that would be obtained if it were possible to repeat the investigation at a temperature of 600 kelvins.

c A gas cylinder contains 0.05 m^3 of helium at a pressure of 16 000 kPa. The gas is used to blow up balloons. Each balloon when fully inflated contains 0.04 m^3 of helium at a pressure of 102 kPa. The temperature of the helium inside both the cylinder and the balloons is the same.

Use the following equation to calculate the maximum number of balloons that can be fully inflated using this gas cylinder. Show clearly how you work out your final answer.

$$\frac{pV}{T} = \text{constant}$$

(AQA (SEG) Higher, June 1999)

Checklist

After studying this chapter you should be able to

- describe experiments to study the relationships between the pressure, volume and temperature of a gas,
- explain the establishment of the Kelvin (absolute) temperature scale from graphs of pressure or volume against temperature and recall the equation connecting the Kelvin and Celsius scales, i.e. $T = 273 + \theta$,
- recall that pV = constant and use it to solve problems,
- explain the behaviour of gases using the kinetic theory.

29 Specific heat capacity

The heat equation
Thermal capacity
Importance of the high specific heat capacity of water
Practical work
Finding specific heat capacities: water, aluminium.

If 1 kg of water and 1 kg of paraffin are heated in turn for the same time by the same heater, the temperature rise of the paraffin is about *twice* that of the water. Since the heater gives equal amounts of heat energy to each liquid, it seems that different substances require different amounts of heat to cause the same temperature rise in the same mass, say 1 °C in 1 kg.

The 'thirst' of a substance for heat is measured by its **specific heat capacity** (symbol c).

The specific heat capacity of a substance is the heat required to produce a 1 °C rise in 1 kg.

Heat, like other forms of energy, is measured in joules (J) and the unit of specific heat capacity is the joule per kilogram per °C, i.e. J/(kg °C).

In physics the word 'specific' means 'unit mass' is being considered.

The heat equation

If a substance has a specific heat capacity of 1000 J/(kg °C) then

1000 J raises the temperature of 1 kg by 1 °C

∴ 2 × 1000 J raises the temperature of 2 kg by 1 °C

∴ 3 × 2 × 1000 J raises the temperature of 2 kg by 3 °C

That is 6000 J will raise the temperature of 2 kg of this substance by 3 °C. We have obtained this answer by multiplying together:

(i) the **mass** in kg,
(ii) the **temperature rise** in °C, and
(iii) the **specific heat capacity** in J/(kg °C).

If the temperature of the substance fell by 3 °C, the heat given out would also be 6000 J. In general, we can write the 'heat equation' as

heat received or given out
= mass × temperature change
× specific heat capacity

In symbols

$$Q = m \times \Delta\theta \times c$$

For example, if the temperature of a 5 kg mass of material of specific heat capacity 400 J/(kg °C) rises from 15 °C to 25 °C, the heat received Q is

$$\begin{aligned} Q &= 5\text{ kg} \times (25 - 15)\text{ °C} \times 400\text{ J/(kg °C)} \\ &= 5\text{ kg} \times 10\text{ °C} \times 400\text{ J/(kg °C)} \\ &= 20\,000\text{ J} \end{aligned}$$

Thermal capacity

The **thermal capacity** of a body is the quantity of heat needed to raise the temperature of the whole body by 1 °C.

For a temperature rise of 1 °C the heat equation becomes:

heat received = mass × 1 × specific heat capacity

so that

thermal capacity = mass × specific heat capacity
= $m \times c$

Thermal capacity is measured in joules per °C, i.e. J/°C.

For a copper block of mass 0.1 kg and specific heat capacity 390 J/(kg °C),

$$\begin{aligned} \text{thermal capacity} &= m \times c \\ &= 0.1\text{ kg} \times 390\text{ J/(kg °C)} \\ &= 39\text{ J/°C} \end{aligned}$$

Practical work

Finding specific heat capacities

You need to know the power of the 12 V electric immersion heater to be used. (**Precaution.** Do not use one with a cracked seal.) A 40 W heater converts 40 joules of electrical energy into heat energy per second. If the power is not marked on the heater ask about it.[1]

a) Water

Weigh out 1 kg of water into a container, e.g. an aluminium saucepan. Note the temperature of the water, insert the heater, Figure 29.1, switch on the 12 V supply and start timing. Stir the water and after 5 minutes switch off, but continue stirring and note the *highest* temperature reached.

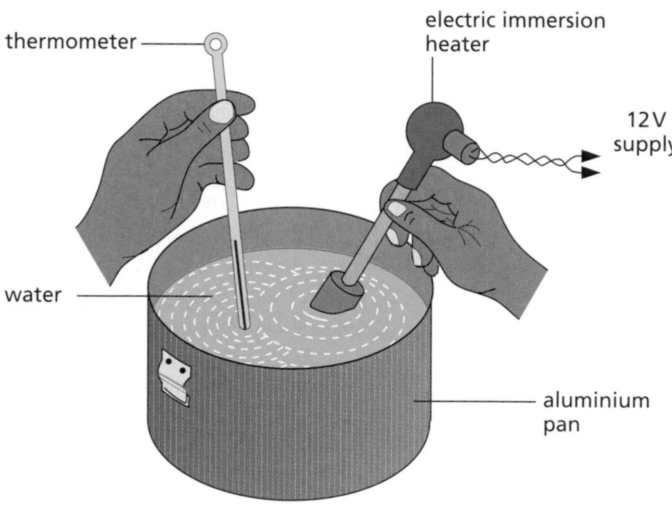

Figure 29.1

Assuming that the heat supplied by the heater equals the heat received by the water, work out the specific heat capacity of water in J/(kg °C), as shown below:

heat received by water (J)
= power of heater (J/s) × time heater on (s)

Rearranging the 'heat equation' we get

$$\text{specific heat capacity of water} = \frac{\text{heat received by water (J)}}{\text{mass (kg)} \times \text{temp. rise (°C)}}$$

Suggest causes of error in this experiment.

b) Aluminium

An aluminium cylinder weighing 1 kg and having two holes drilled in it is used. Place the immersion heater in the central hole and a thermometer in the other hole, Figure 29.2.

Note the temperature, connect the heater to a 12 V supply and switch it on for 5 minutes. When the temperature stops rising record its highest value.

Calculate the specific heat capacity as before.

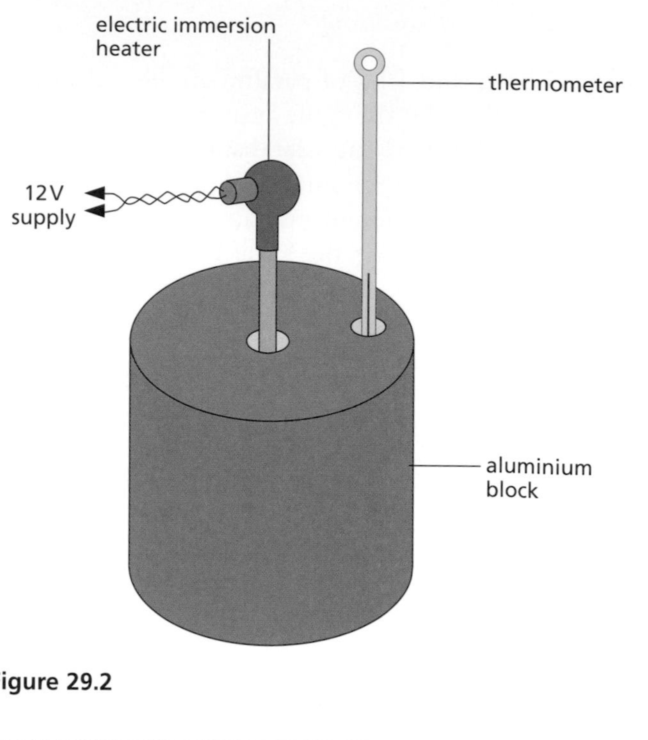

Figure 29.2

Importance of the high specific heat capacity of water

The specific heat capacity of water is 4200 J/(kg °C) and that of soil is about 800 J/(kg °C). As a result, the temperature of the sea rises and falls more slowly than that of the land. A certain mass of water needs five times more heat than the same mass of soil for its temperature to rise by 1 °C. Water also has to give out more heat to fall 1 °C. Since islands are surrounded by water they experience much smaller changes of temperature from summer to winter than large land masses such as Central Asia.

The high specific heat capacity of water (as well as its cheapness and availability) accounts for its use in cooling engines and in the radiators of central heating systems.

[1] The power is found by immersing the heater in water, connecting it to a 12 V d.c. supply and measuring the current taken (usually 3–4 amperes). Then power in watts = volts × amperes.

Worked examples

1 A tank holding 60 kg of water is heated by a 3 kW electric immersion heater. If the specific heat capacity of water is 4200 J/(kg °C), estimate the time for the temperature to rise from 10 °C to 60 °C.

A 3 kW (3000 W) heater supplies 3000 J of heat energy per second.

Let t = time taken in seconds to raise the temperature of the water by (60 − 10) = 50 °C,

$\therefore$ heat supplied to water in time $t = 3000 \times t$ J

From the 'heat equation', we can say

heat received by water
$$= 60\text{ kg} \times 4200\text{ J/(kg °C)} \times 50\text{ °C}$$

Assuming heat supplied = heat received

$$3000\text{ J/s} \times t = (60 \times 4200 \times 50)\text{ J}$$

$$\therefore \quad t = \frac{(60 \times 4200 \times 50)\text{ J}}{3000\text{ J/s}} = 4200\text{ s (70 min)}$$

2 A piece of aluminium of mass 0.5 kg is heated to 100 °C and then placed in 0.4 kg of water at 10 °C. If the resulting temperature of the mixture is 30 °C, what is the specific heat capacity of aluminium if that of water is 4200 J/(kg °C)?

When two substances at different temperatures are mixed, heat flows from the one at the higher temperature to the one at the lower temperature until both are at the same temperature – the temperature of the mixture. If there is no loss of heat, then in this case

heat given out by aluminium
= heat taken in by water

Using the 'heat equation' and letting c be the specific heat capacity of aluminium in J/(kg °C), we have

$$\text{heat given out} = 0.5\text{ kg} \times c \times (100 - 30)\text{ °C}$$

heat taken in
$$= 0.4\text{ kg} \times 4200\text{ J/(kg °C)} \times (30 - 10)\text{ °C}$$

$$\therefore \quad 0.5\text{ kg} \times c \times 70\text{ °C} = 0.4\text{ kg} \times 4200\text{ J/(kg °C)} \times 20\text{ °C}$$

$$\therefore \quad c = \frac{(4200 \times 8)\text{ J}}{35\text{ kg °C}} = 960\text{ J/(kg °C)}$$

Questions

1 How much heat is needed to raise the temperature by 10 °C of 5 kg of a substance of specific heat capacity 300 J/(kg °C)? What is the thermal capacity of the substance?

2 The same quantity of heat was given to different masses of three substances A, B and C. The temperature rise in each case is shown in the table. Calculate the specific heat capacities of A, B and C.

Material	Mass /kg	Heat given /J	Temp. rise /°C
A	1.0	2000	1.0
B	2.0	2000	5.0
C	0.5	2000	4.0

3 a The table shows how much energy is needed to make the temperature of 1 kg of different substances go up by 1°C.

Substance	Energy in joules
copper	390
mercury	140
silver	240
steel	450

Which *one* of these four substances has the highest specific heat capacity? Give a reason for your answer.

b A microwave oven is used to heat a cold mug of coffee. In a few seconds the temperature of the coffee goes up by 70°C. Use the following equation

energy exchanged = mass × specific heat capacity × temperature change

to calculate the energy in joules gained by the coffee. Show clearly how you work out your final answer.
(Take the mass of the coffee to be 0.2 kg and the specific heat capacity of the coffee to be 4000 J/kg °C.)

(SEG Foundation, June 1998)

4 The jam in a hot pop tart always seems hotter than the pastry. Why?

Checklist

After studying this chapter you should be able to

- define **specific heat capacity**, c,
- define **thermal capacity**,
- solve problems on specific heat capacity using the heat equation $Q = m \times \Delta\theta \times c$,
- describe experiments to measure the specific heat capacity of metals and liquids by electrical heating,
- explain the importance of the high specific heat capacity of water.

30 Specific latent heat

Specific latent heat of fusion
Specific latent heat of vaporization
Latent heat and the kinetic theory
Evaporation and boiling
Condensation and solidification
Cooling by evaporation
Liquefaction of gases and vapours
Practical work
Cooling curve of ethanamide.
Specific latent heat of fusion for ice.
Specific latent heat of vaporization for steam.

When a solid is heated, it may melt and **change its state** from solid to liquid. If ice is heated it becomes water. The opposite process, freezing, occurs when a liquid solidifies.

A pure substance melts at a definite temperature, called the **melting point**; it solidifies at the same temperature – sometimes then called the **freezing point**.

Practical work

Cooling curve of ethanamide

Half-fill a test-tube with ethanamide (acetamide) and place it in a beaker of water, Figure 30.1a. Heat the water until all the ethanamide has melted and its temperature reaches about 90 °C.

Remove the test-tube and arrange it as in Figure 30.1b with a thermometer in the liquid ethanamide. Record the temperature every minute until it has fallen to 70 °C.

Plot a cooling curve of temperature against time. What is the freezing (melting) point of ethanamide?

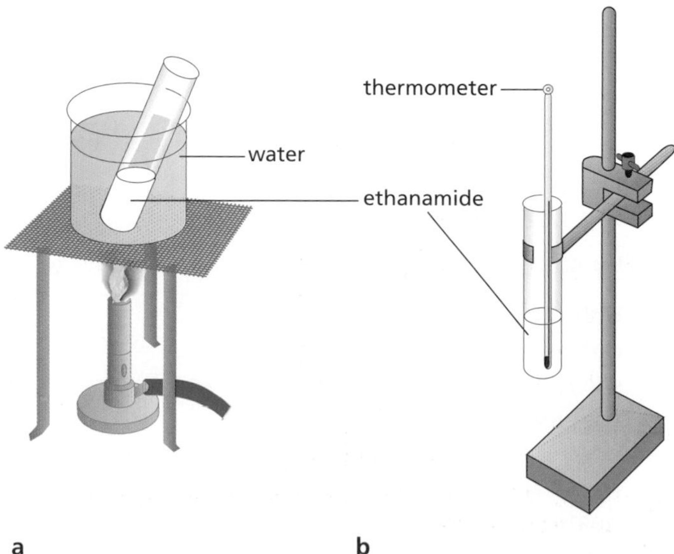

Figure 30.1

Specific latent heat of fusion

The previous experiment shows that the temperature of liquid ethanamide falls until it starts to solidify (at 82 °C) and remains constant until it has all solidified. The cooling curve in Figure 30.2 is for a pure substance; the flat part AB occurs at the melting point when the substance is solidifying.

During solidification a substance loses heat to its surroundings but its temperature does not fall. Conversely when a solid is melting, the heat supplied does not cause a temperature rise; heat is added but the substance does not get hotter. For example, the temperature of a well-stirred ice–water mixture remains at 0 °C until all the ice is melted.

Heat that is **absorbed** by a solid during melting or **given out** by a liquid during solidification is called **latent heat of fusion**. 'Latent' means hidden and 'fusion' means melting. Latent heat does not cause a temperature change; it seems to disappear.

The **specific latent heat of fusion** (l_f) of a substance is the quantity of heat needed to change *unit mass* from solid to liquid without temperature change.

Specific latent heat is measured in J/kg or J/g. In general, the quantity of heat Q to change a mass *m* from solid to liquid is given by

$$Q = m \times l_f$$

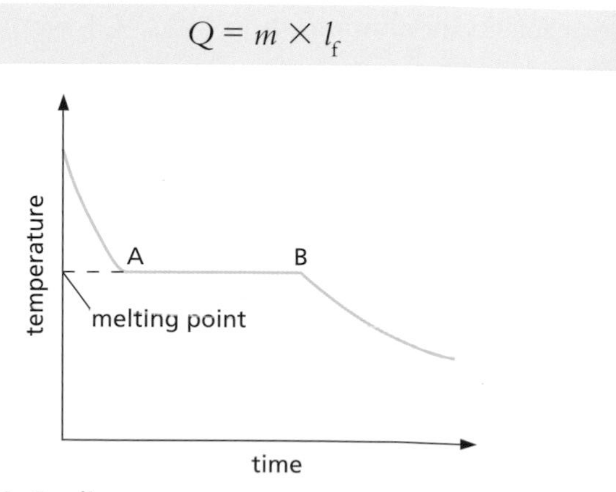

Figure 30.2 Cooling curve

Practical work

Specific latent heat of fusion for ice

Through measurement of the mass of water m produced when energy Q is transferred to melting ice, the specific latent heat of fusion for ice can be calculated.

Insert a 12 V electric immersion heater of known power P into a funnel, and pack crushed ice around it as shown in Figure 30.3. To correct for heat transferred from the surroundings, collect the melted ice in a beaker for time t (4 minutes); weigh the beaker plus the melted ice, m_1. Empty the beaker, switch on the heater, and collect the melted ice for the same time t; reweigh the beaker plus the melted ice, m_2. The mass of ice melted by the heater is then

$$m = m_2 - m_1$$

The electrical energy supplied by the heater is given by $Q = P \times t$ where P is in J/s and t is in seconds; Q will be in joules. Alternatively, a joulemeter can be used to record Q directly.

Calculate the specific latent heat of fusion l_f for ice using

$$Q = m \times l_f$$

How does it compare with the accepted value of 340 J/g? How could the experiment be improved?

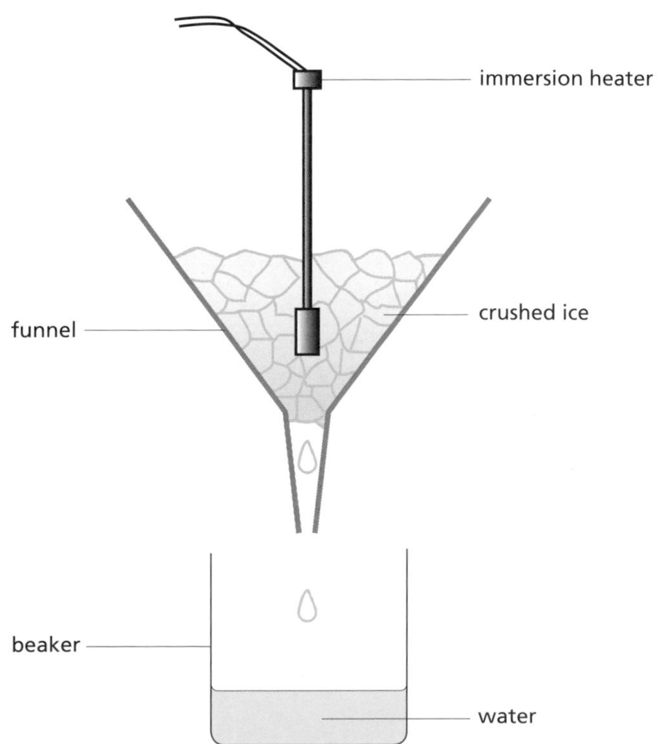

Figure 30.3

Specific latent heat of vaporization

Latent heat is also needed to change a liquid into a vapour. The reading of a thermometer placed in water which is boiling remains constant at 100 °C even though heat, called **latent heat of vaporization**, is still being absorbed by the water from whatever is heating it. When steam condenses to form water, latent heat is given out.

The **specific latent heat of vaporization** (l_v) of a substance is the quantity of heat needed to change unit mass from liquid to vapour without change of temperature.

Again, the specific latent heat is measured in J/kg or J/g. In general, the quantity of heat Q to change a mass m from liquid to vapour is given by

$$Q = m \times l_v$$

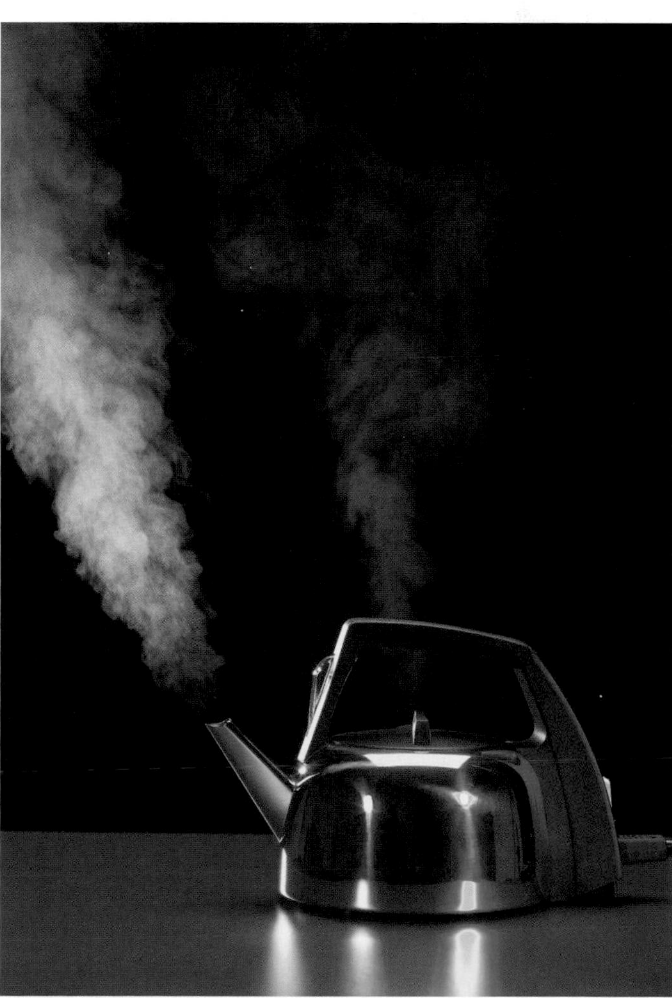

Figure 30.4 Why is a scald from steam often more serious than one from boiling water?

Practical work

Specific latent heat of vaporization for steam

Through measurement of the mass of vapour m produced when energy Q is transferred to boiling water, the specific latent heat of vaporization for steam can be calculated.

Water in the flask, Figure 30.5, is heated to boiling point by an immersion heater of power P. Steam passes out through the holes in the top of the flask, down the outside of the flask and into the inner tube of a condenser, where it changes back to liquid (because cold water is flowing through the outer tube), and is collected in a beaker.

After the water has been boiling for some time, it becomes enclosed by a 'jacket' of vapour at the boiling point, which helps to reduce loss of heat to the surroundings. The rate of vaporization becomes equal to the rate of condensation, and the electrical energy is only being used to transfer latent heat to the water (not to raise its temperature).

The electrical energy Q supplied by the heater is given by

$$Q = P \times t = ItV$$

where I is the steady current through the heater and V is the p.d across it. Q is in joules if P is in J/s and t is in seconds. Alternatively, a joulemeter can be used to record Q directly.

If a mass of water m is collected in time t, then the latent heat of vaporization l_v can be calculated using

$$Q = m \times l_v$$

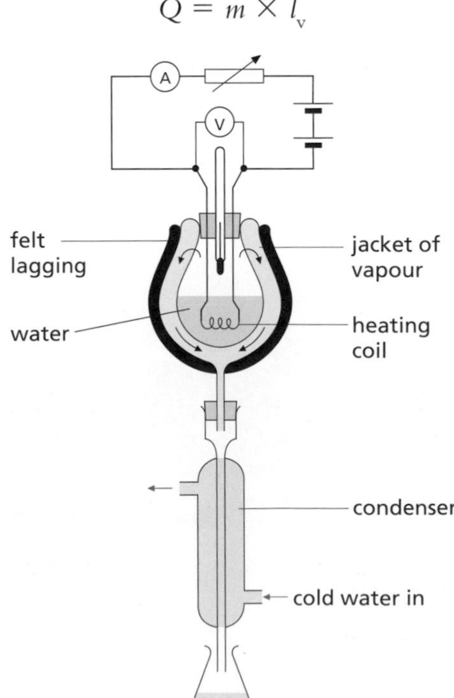

Figure 30.5

Latent heat and the kinetic theory

a) Fusion

The kinetic theory explains latent heat of fusion as being the energy that enables the molecules of a solid to overcome the intermolecular forces that hold them in place and when it exceeds a certain value they break free. Their vibratory motion about fixed positions changes to the slightly greater range of movement they have as liquid molecules and the solid melts. The energy input is used to increase the p.e. of the molecules but not their average k.e. as happens when the heat causes a temperature rise.

b) Vaporization

If liquid molecules are to overcome the forces holding them together and gain the freedom to move around independently as gas molecules, they need a large amount of energy. They receive this as latent heat of vaporization which, like latent heat of fusion, increases the p.e. of the molecules but not their k.e. It also gives the molecules the energy required to push back the surrounding atmosphere in the large expansion that occurs when a liquid vaporizes.

To change 1 kg of water at 100 °C to steam at 100 °C needs over *five* times as much heat as is needed to raise the temperature of 1 kg of water at 0 °C to water at 100 °C (see *Worked example* 1, p. 145).

Evaporation and boiling

a) Evaporation

A few energetic molecules close to the surface of a liquid may escape and become gas molecules. This process occurs at all temperatures and is called evaporation. It happens more rapidly when

(i) the **temperature is higher**, since then more molecules in the liquid are moving fast enough to escape from the surface,

(ii) the **surface area** of the liquid is large so giving more molecules a chance to escape because more are near the surface, and

(iii) a **wind** or **draught** is blowing over the surface carrying vapour molecules away from the surface, thus stopping them from returning to the liquid and making it easier for more liquid molecules to break free. (Evaporation into a vacuum occurs much more rapidly than into a region where there are gas molecules.)

b) Boiling

This occurs for a pure liquid at a definite temperature called its **boiling point** and is accompanied by bubbles that form within the liquid, containing the gaseous or vapour form of the particular substance.

Latent heat is needed in both evaporation and boiling and is stored in the vapour, from which it is released when the vapour is cooled or compressed and changes to liquid again.

Condensation and solidification

In condensation, a gas changes to a liquid state and latent heat of vaporization is released. In solidification, a liquid changes to a solid and latent heat of fusion is given out. In each case the p.e. of the molecules decreases. Condensation of steam is easily achieved by contact with a cold surface, for example a cold window-pane. In Figure 30.5, the latent heat released when the steam condenses to water is transferred to the cold water flowing through the condenser.

Cooling by evaporation

In evaporation latent heat is obtained by the liquid from its surroundings, as may be shown by the following demonstration, **done in a fume cupboard**.

a) Demonstration

Dichloromethane is a **volatile** liquid, i.e. it has a low boiling point and evaporates readily at room temperature, especially when air is blown through it, Figure 30.6. Latent heat is taken first from the liquid itself and then from the water below the can. The water soon freezes causing the block and can to stick together.

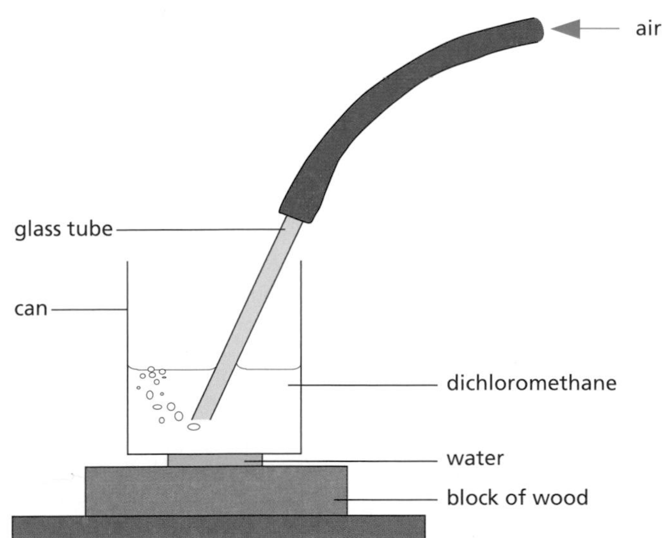

Figure 30.6 Demonstrating cooling by evaporation

b) Explanation

Evaporation occurs when faster-moving molecules escape from the surface of the liquid. The average speed and therefore the average k.e. of the molecules left behind decreases, i.e. the temperature of the liquid falls.

c) Uses

Water evaporates from the skin when we sweat. This is the body's way of losing unwanted heat and keeping a constant temperature. After vigorous exercise there is a risk of the body being overcooled, especially in a draught; it is then less able to resist infection.

Ether acts as a local anaesthetic by chilling (as well as cleaning) your arm when you are having an injection. Refrigerators, freezers and air-conditioning systems use cooling by evaporation on a large scale.

Volatile liquids are used in perfumes.

F *Liquefaction of gases and vapours*

A vapour can be liquefied if it is compressed enough. However, a gas must be cooled below a certain **critical temperature** T_c before liquefaction by pressure can occur. The critical temperatures for some gases are given in Table 30.1.

Table 30.1 Critical temperatures of some gases

Gas	carbon dioxide	oxygen	air	nitrogen	hydrogen	helium
T_c/K	304	154	132	126	33.3	5.3
T_c/°C	+31	−119	−141	−147	−239.7	−267.7

Low-temperature liquids have many uses. Liquid hydrogen and oxygen are used as the fuel and oxidant respectively in space rockets. Liquid nitrogen is used in industry as a coolant in, for example, shrink-fitting (see Figure 27.1, p. 129). Materials that behave as superconductors (see Chapter 28) when cooled by liquid nitrogen are increasingly being used in electrical power engineering and electronics.

Worked examples

The values in Table 30.2 are required.

Table 30.2

	water	ice	aluminium
specific heat capacity/ J/(g °C)	4.2	2.0	0.90
specific latent heat/ J/g	2300	340	

1 How much heat is needed to change 20 g of ice at 0 °C to steam at 100 °C?

There are three stages in the change.

Heat to change 20 g **ice at 0 °C** to **water at 0 °C**
= mass of ice × specific latent heat of ice
= 20 g × 340 J/g = 6800 J

Heat to change 20 g **water at 0 °C** to **water at 100 °C**

= mass of water × specific heat capacity of water × temperature rise

= 20 g × 4.2 J/(g °C) × 100 °C = 8400 J

Heat to change 20 g **water at 100 °C** to **steam at 100 °C**

= mass of water × specific latent heat of steam

= 20 g × 2300 J/g = 46 000 J

∴ Total heat supplied

= 6800 + 8400 + 46 000 = 61 200 J

2 An aluminium can of mass 100 g contains 200 g of water. Both, initially at 15 °C, are placed in a freezer at −5.0 °C. Calculate the quantity of heat that has to be removed from the water and the can for their temperatures to fall to −5.0 °C.

Heat lost by **can** in falling from 15 °C to −5.0 °C

= mass of can × specific heat capacity of aluminium × temperature fall

= 100 g × 0.90 J/(g °C) × [15 − (−5)] °C

= 100 g × 0.90 J/(g °C) × 20 °C

= 1800 J

Heat lost by **water** in falling from 15 °C to 0 °C

= mass of water × specific heat capacity of water × temperature fall

= 200 g × 4.2 J/(g °C) × 15 °C

= 12 600 J

Heat lost by **water** at 0 °C freezing to ice at 0 °C

= mass of water × specific latent heat of ice

= 200 g × 340 J/g

= 68 000 J

Heat lost by **ice** in falling from 0 °C to −5.0 °C

= mass of ice × specific heat capacity of ice × temperature fall

= 200 g × 2.0 J/(g °C) × 5.0 °C

= 2000 J

∴ Total heat removed

= 1800 + 12 600 + 68 000 + 2000 = 84 400 J

Questions

Use values given in Table 30.2.

1 **a** How much heat will change 10 g of ice at 0 °C to water at 0 °C?
b What quantity of heat must be removed from 20 g of water at 0 °C to change it to ice at 0 °C?

2 **a** How much heat is needed to change 5 g of ice at 0 °C to water at 50 °C?
b If a freezer cools 200 g of water from 20 °C to its freezing point in 10 minutes, how much heat is removed per minute from the water?

3 How long will it take a 50 W heater to melt 100 g of ice at 0 °C?

4 Some small aluminium rivets of total mass 170 g and at 100 °C are emptied into a hole in a large block of ice at 0 °C.
a What will be the final temperature of the rivets?
b How much ice will melt?

5 **a** How much heat is needed to change 4 g of water at 100 °C to steam at 100 °C?
b Find the heat given out when 10 g of steam at 100 °C condenses and cools to water at 50 °C.

6 A 3 kW electric kettle is left on for 2 minutes after the water starts to boil. What mass of water is boiled off in this time?

7 **a** Why is ice good for cooling drinks?
b Why do engineers often use superheated steam (steam above 100 °C) to transfer heat?

8 Some water is stored in a bag of porous material, e.g. canvas, which is hung where it is exposed to a draught of air. Explain why the temperature of the water is lower than that of the air.

9 Explain why a bottle of milk keeps better when it stands in water in a porous pot in a draught.

10 **a** Figure 30.7 shows one method of measuring the specific latent heat of fusion of ice. Two funnels, A and B, contain crushed ice at 0 °C. The mass of melted ice from each funnel is measured after 12 minutes. The joulemeter measures the energy supplied to the immersion heater.

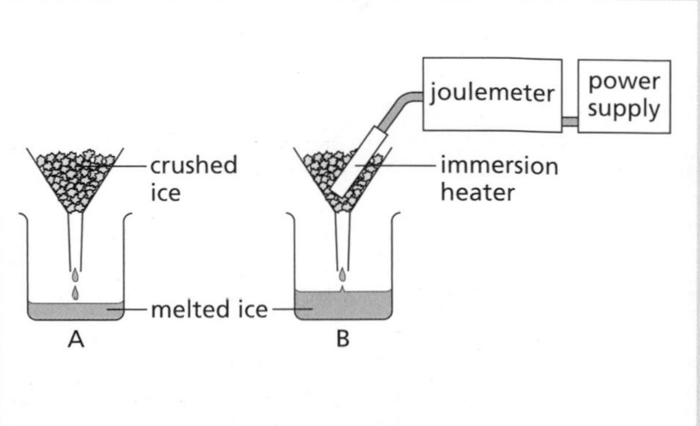

Figure 30.7

(i) What is the reason for setting up funnel A?
(ii) The measurements taken are listed below.

Mass of melted ice collected from funnel A	= 24 g
Mass of melted ice collected from funnel B	= 63 g
Joulemeter reading	= 17 160 J

Use these measurements and the following equation to calculate the specific latent heat of fusion of ice. Show clearly how you work out your answer and give the unit.

$$\text{energy exchanged} = \text{mass} \times \text{specific latent heat}$$

(iii) The measurements obtained in this experiment give a value for the specific latent heat which is higher than the accepted value. Give a reason why.

b (i) What happens to the average kinetic energy of the molecules as ice melts at 0 °C?
(ii) Energy must be given to ice if it is to change into water. Why?

(AQA (SEG) Higher, June 1999)

11 a The table gives the melting and boiling points for lead and oxygen.

	Melting point in °C	**Boiling point** in °C
lead	327	1744
oxygen	−219	−183

(i) At 450 °C will the lead be a solid, a liquid or a gas?
(ii) At −200 °C will the oxygen be a solid, a liquid or a gas?

b The graph in Figure 30.8 shows how the temperature of a pure substance changes as it is heated.

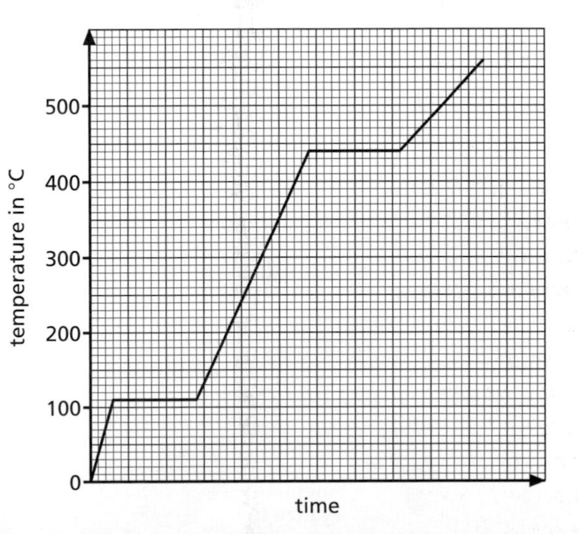

Figure 30.8

(i) At what temperature does the substance boil?
(ii) Copy the graph and mark with an **X** any point where the substance exists as both a liquid and gas at the same time.

c (i) All substances consist of particles. What happens to the average kinetic energy of these particles as the substance changes from a liquid to a gas?
(ii) Explain, in terms of particles, why energy must be given to a liquid if it is to change to a gas.

(SEG Higher, June 1998)

Checklist

After studying this chapter you should be able to

- describe an experiment to show that during a change of state the temperature stays constant,
- state the meaning of **melting point** and **boiling point**,
- describe condensation and solidification,
- define **specific latent heat of fusion**, l_f,
- define **specific latent heat of vaporization**, l_v,
- explain latent heat using the kinetic theory,
- solve problems on latent heat using $Q = ml$,
- distinguish between evaporation and boiling,
- describe an experiment to measure specific latent heats for ice and steam,
- explain cooling by evaporation using the kinetic theory.

31 Conduction and convection

To keep a building or a house at a comfortable temperature in winter and in summer requires a knowledge of how heat travels, if it is to be done economically and efficiently.

Conduction

The handle of a metal spoon held in a hot drink soon gets warm. Heat passes along the spoon by **conduction**.

Conduction is the flow of heat through matter from places of higher temperature to places of lower temperature without movement of the matter as a whole.

A simple demonstration of the different conducting powers of various metals is shown in Figure 31.1. A match is fixed to one end of each rod using a little melted wax. The other ends of the rods are heated by a burner. When the temperatures of the far ends reach the melting point of wax, the matches drop off. The match on copper falls first showing it is the best conductor, followed by aluminium, brass and then iron.

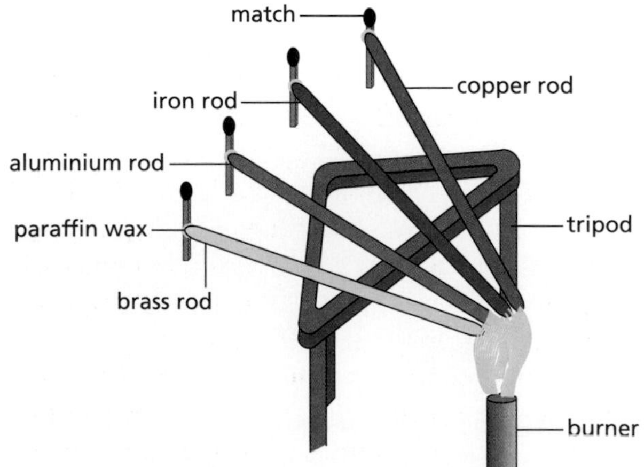

Figure 31.1 Comparing conducting powers

Heat is conducted faster through a rod if it has a large cross-section area, is short and has a large temperature difference between its ends.

Most metals are good conductors of heat; materials such as wood, glass, cork, plastics and fabrics are bad conductors. The arrangement in Figure 31.2 can be used to show the difference between brass and wood. If the rod is passed through a flame several times, the paper over the wood scorches but not over the brass. The brass conducts the heat away from the paper quickly and prevents it reaching the temperature at which it burns. The wood conducts it away only very slowly.

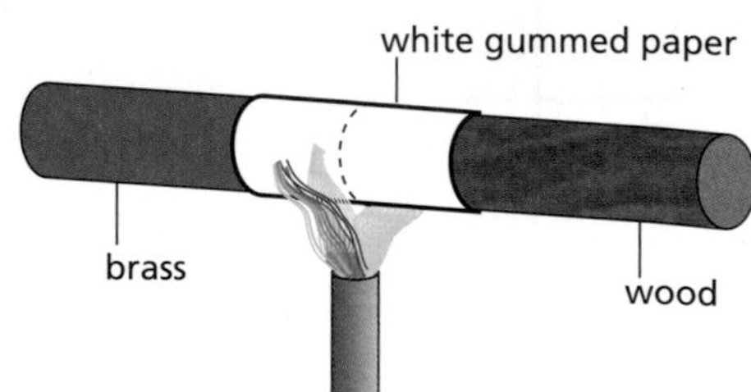

Figure 31.2 The paper over the brass does not burn

Metal objects below body temperature *feel* colder than those made of bad conductors – even if all the objects are at exactly the same temperature – because they carry heat away faster from the hand.

Liquids and gases also conduct heat but only very slowly. Water is a very poor conductor as shown in Figure 31.3. The water at the top of the tube can be boiled before the ice at the bottom melts.

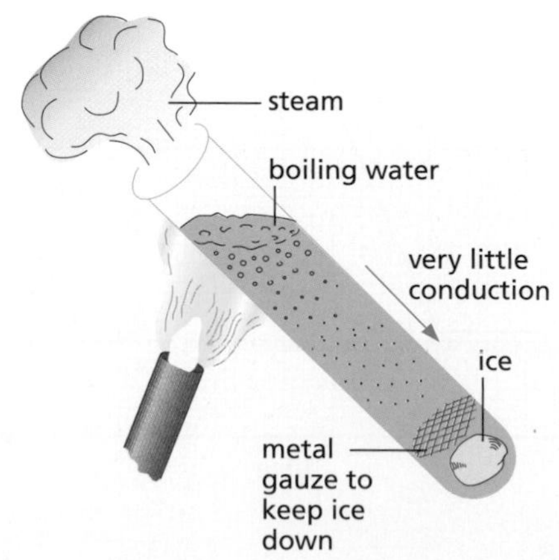

Figure 31.3 Water is a poor conductor of heat

Uses of conductors

a) Good conductors

These are used whenever heat is required to travel quickly through something. Saucepans, boilers and radiators are made of metals such as aluminium, iron and copper.

b) Bad conductors (insulators)

The handles of some saucepans are made of wood or plastic. Cork is used for table mats.

Air is one of the worst conductors, i.e. best insulators. This is why houses with cavity walls (i.e. two layers of bricks separated by an air space) and double-glazed windows keep warmer in winter and cooler in summer.

Materials that trap air, e.g. wool, felt, fur, feathers, polystyrene foam, fibreglass, are also very bad conductors. Some of these materials are used as 'lagging' to insulate water pipes, hot water cylinders, ovens, refrigerators and the walls and roofs of houses, Figures 31.4a, b. Others are used to make warm winter clothes like 'fleece' jackets, Figure 31.4c.

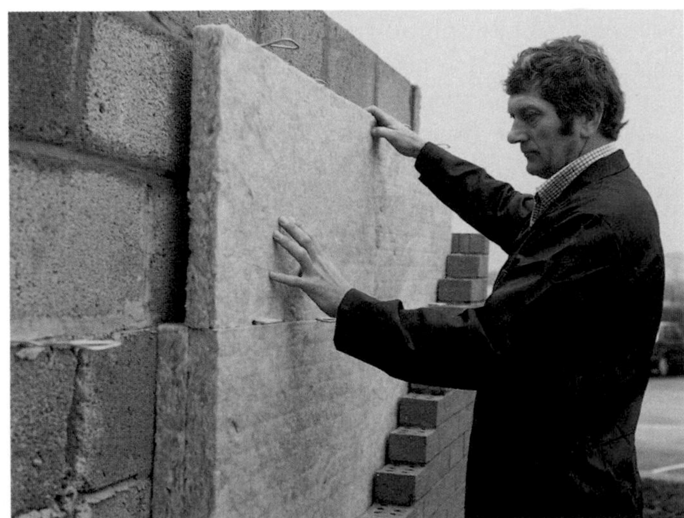

Figure 31.4a Lagging in a cavity wall provides extra insulation

Figure 31.4b Laying lagging in a house loft

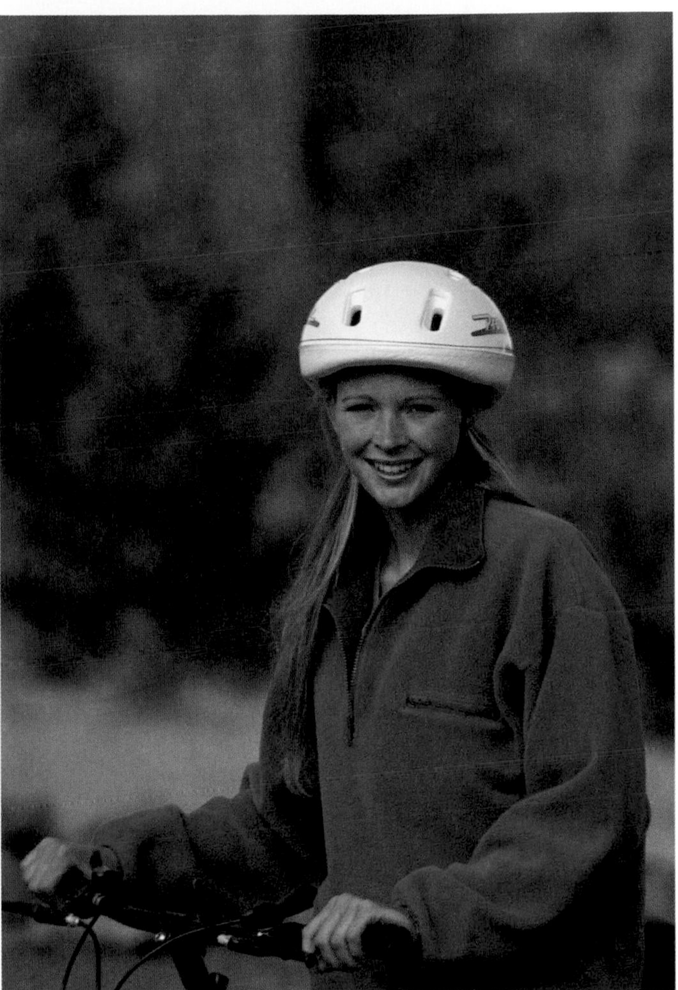

Figure 31.4c Fleece jackets help you to retain your body warmth

'Wet suits' are worn by divers and water skiers to keep them warm. The suit gets wet and a layer of water gathers between the person's body and the suit. The water is warmed by body heat and stays warm because the suit is made of an insulating fabric, e.g. neoprene, a synthetic rubber.

Conduction and the kinetic theory

Two processes occur in metals. Metals have a large number of 'free' electrons (Chapter 34) which wander about inside them. When one part of a metal is heated, the electrons there move faster (i.e. their k.e. increases) and farther. As a result they 'jostle' atoms in cooler parts, so passing on their energy and raising the temperature of these parts. This process occurs quickly.

The second process is much slower. The atoms themselves at the hot part make 'colder' neighbouring atoms vibrate more vigorously. This is less important in metals but is the only way conduction occurs in non-metals since these do not have 'free' electrons; hence non-metals are poor conductors of heat.

Convection in liquids

Convection is the usual method by which heat travels through fluids, i.e. liquids and gases. It can be shown in water by dropping a few crystals of potassium permanganate down a tube to the bottom of a beaker or flask of water. When the tube is removed and the beaker heated just below the crystals by a *small* flame, Figure 31.5a, purple streaks of water rise upwards and fan outwards.

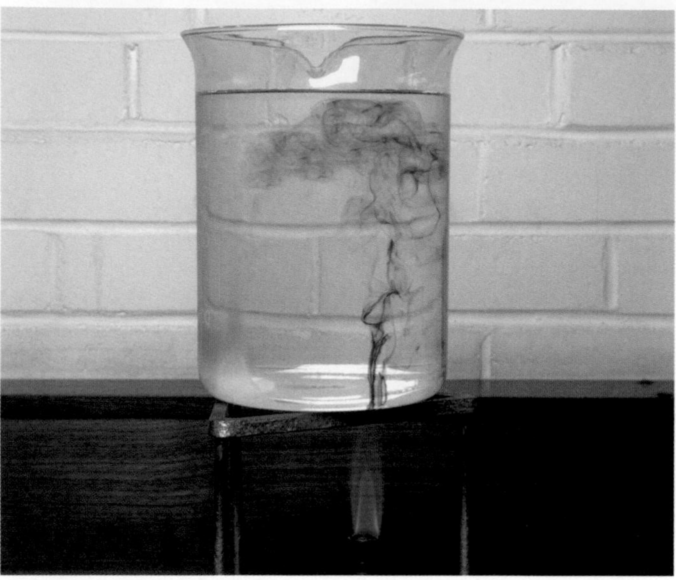

Figure 31.5a Convection currents shown by potassium permanganate in water

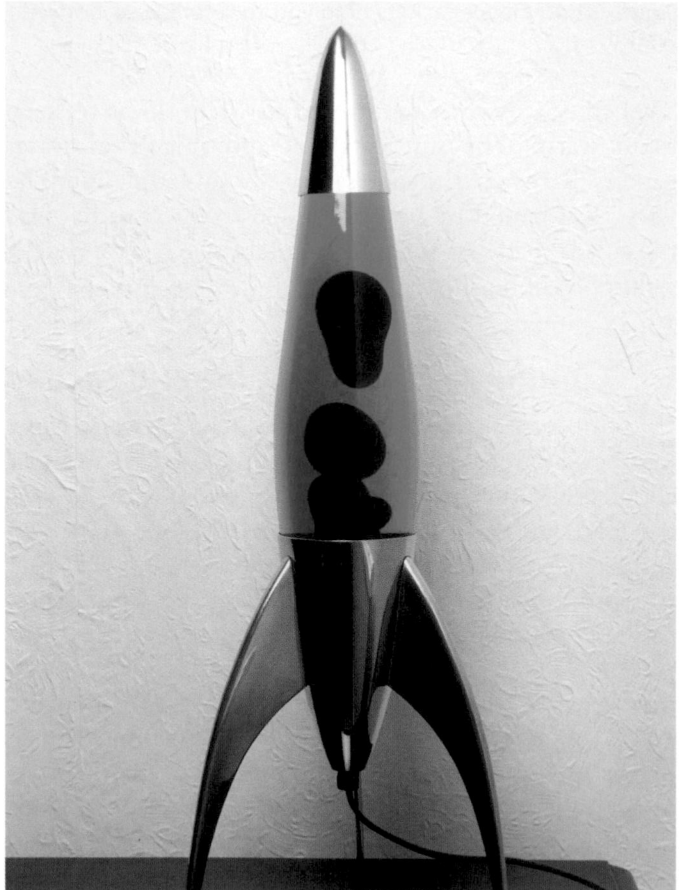

Figure 31.5b Lava lamps make use of convection

Streams of warm moving fluids are called **convection currents**. They arise when a fluid is heated because it expands, becomes less dense and is forced upwards by surrounding cooler, denser fluid which moves under it. We say 'hot water (or hot air) rises'. Warm fluid behaves like a cork released under water: being less dense it bobs up. Lava lamps, Figure 31.5b, use this principle.

Convection is the flow of heat through a fluid from places of higher temperature to places of lower temperature by movement of the fluid itself.

Convection in air

Black marks often appear on the wall or ceiling above a lamp or a radiator. They are caused by dust being carried upwards in air convection currents produced by the hot lamp or radiator.

A laboratory demonstration of convection currents in air can be given using the apparatus of Figure 31.6. The direction of the convection current created by the candle is made visible by the smoke from the touch paper (made by soaking brown paper in strong potassium nitrate solution and drying it).

Convection currents set up by electric, gas and oil heaters help to warm our homes. Many so-called 'radiators' are really convector heaters.

Where should the input and extraction ducts for cold/hot air be located in a room?

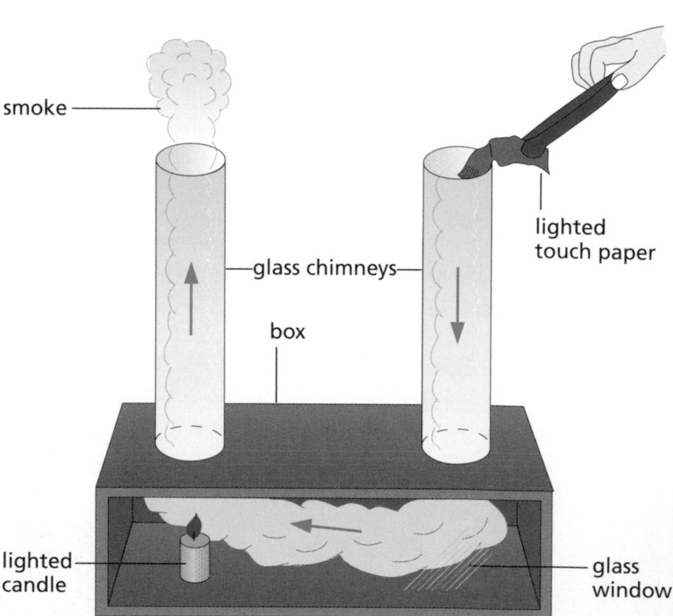

Figure 31.6 Demonstrating convection in air

Natural convection currents

a) Coastal breezes

During the day the temperature of the land increases more quickly than that of the sea (because the specific heat capacity of the land is much smaller, Chapter 29). The hot air above the land rises and is replaced by colder air from the sea. A breeze from the sea results, Figure 31.7a.

At night the opposite happens. The sea has more heat to lose and cools more slowly. The air above the sea is warmer than that over the land and a breeze blows from the land, Figure 31.7b.

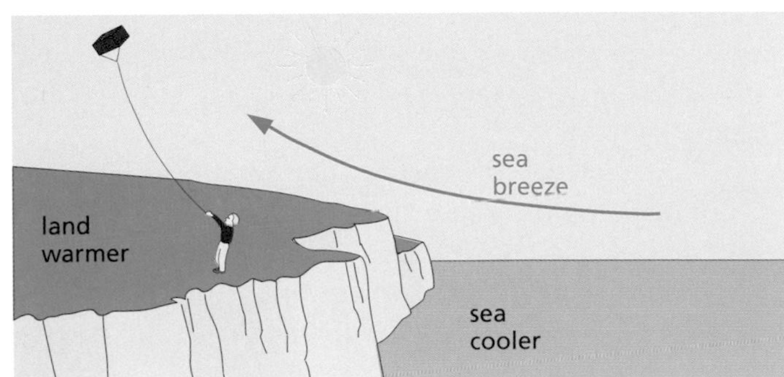

a Day

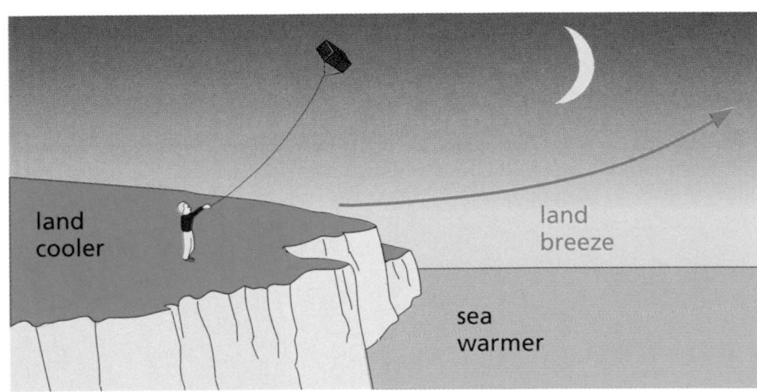

b Night

Figure 31.7 Coastal breezes are due to convection

b) Gliding

Gliders, including 'hang-gliders', Figure 31.8, depend on hot air currents, called **thermals**.

Figure 31.8 Once airborne, a hang-glider pilot can stay aloft for several hours by flying from one thermal to another

Energy losses from buildings

The inside of a building can only be kept at a steady temperature above that outside by heating it at a rate which equals the rate at which it is losing energy. The loss occurs mainly by conduction through the walls, roof, floors and windows. For a typical house in the UK where no special precautions have been taken, the contribution each of these makes to the total loss is shown in Table 31.1a.

As fuels (and electricity) become more expensive and the burning of fuels becomes of greater environmental concern (Chapter 17), more people are considering it worthwhile to reduce heat losses from their homes. The substantial reduction of this loss which can be achieved, especially by wall and roof insulation, is shown in Table 31.1b.

Table 31.1 Energy losses from a typical house

a

Percentage of total energy loss due to				
walls	roof	floors	windows	draughts
35	25	15	10	15

b

Percentage of each loss saved by				
insulating				
walls	roof	carpets on floors	double glazing	draught excluders
65	80	≈30	50	≈60
Percentage of total loss saved = 60				

F Ventilation

In addition to supplying heat to compensate for the energy losses from a building, a heating system has also to warm the ventilated cold air, needed for comfort, which comes in to replace stale air.

If the rate of heat loss is, say, 6000 J/s, i.e. 6 kW, and the warming of ventilated air requires 2 kW, then the total power needed to maintain a certain temperature (e.g. 20 °C) in the building is 8 kW. Some of this is supplied by each person's 'body heat', estimated to be roughly equal to a 100 W heater.

Questions

1 Explain why
   a newspaper wrapping keeps hot things hot, e.g. fish and chips, and cold things cold, e.g. ice cream,
   b fur coats would keep their owners warmer if they were worn inside out,
   c a string vest helps to keep a person warm even though it is a collection of holes bounded by string.

2 Figure 31.9 illustrates three ways of reducing heat losses from a house.
   a As far as you can, explain how each of the three methods reduces heat losses. Draw diagrams where they will help your explanations.
   b Why are fibreglass and plastic foam good substances to use?
   c Air is one of the worst conductors of heat. What is the point of replacing it by the plastic foam in (ii)?
   d A vacuum is an even better heat insulator than air. Suggest one (scientific) reason why the double glazing should not have a vacuum between the sheets of glass.
   e The manufacturers of roof lagging suggest that two layers of fibreglass are more effective than one. Describe how you might set up an experiment in the laboratory to test whether this is true.

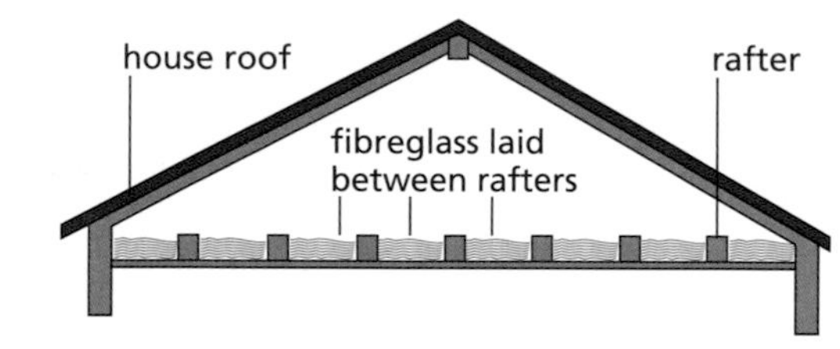

(i) Roof insulation

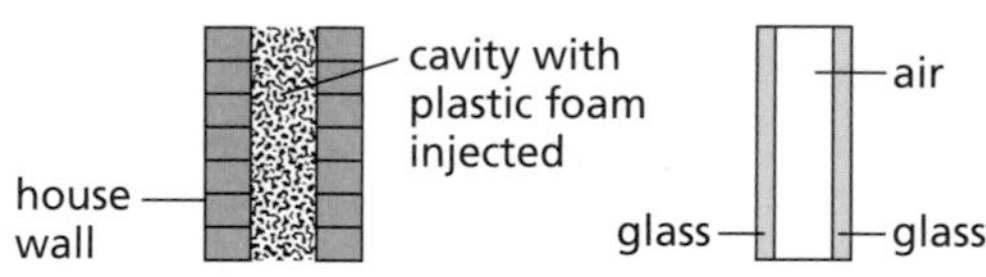

(ii) Cavity wall insulation (iii) Double glazing

Figure 31.9

3 What is the advantage of placing an electric immersion heater
   a near the top,
   b near the bottom,
   of a tank of water?

Checklist

After studying this chapter you should be able to

- describe experiments to show the different conducting powers of various substances,
- name good and bad conductors and state uses for each,
- explain conduction using the kinetic theory,
- describe experiments to show convection in fluids (liquids and gases),
- relate convection to phenomena such as land and sea breezes,
- explain the importance of insulating a building.

32 Radiation

Good and bad absorbers	*The greenhouse*
Good and bad emitters	*Rate of cooling of an object*
Vacuum flask	

Radiation is a third way in which heat can travel, but whereas conduction and convection both need matter to be present, radiation can occur in a vacuum; particles of matter are not involved. Radiation is the way heat reaches us from the Sun.

Radiation has all the properties of electromagnetic waves (Chapter 8), e.g. it travels at the speed of radio waves and gives interference effects. When it falls on an object, it is partly reflected, partly transmitted and partly absorbed; the absorbed part raises the temperature of the object.

Radiation is the flow of heat from one place to another by means of electromagnetic waves.

Radiation is emitted by all bodies above absolute zero and consists mostly of infrared radiation (Chapter 8) but light and ultraviolet are also present if the body is very hot (e.g. the Sun).

Figure 32.1 Why are buildings in hot countries often painted white?

Good and bad absorbers

Some surfaces absorb radiation better than others as may be shown using the apparatus in Figure 32.2. The inside surface of one lid is shiny and of the other dull black. The coins are stuck on the outside of each lid with candle wax. If the heater is midway between the lids they each receive the same amount of radiation. After a few minutes the wax on the black lid melts and the coin falls off. The shiny lids stay cool and the wax unmelted.

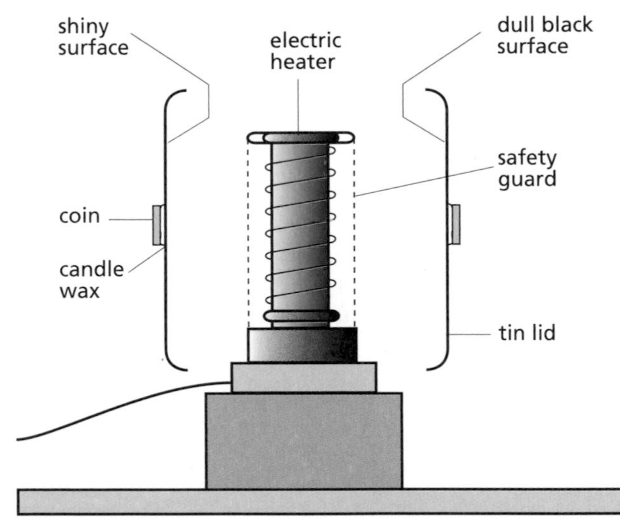

Figure 32.2 Comparing absorbers of radiation

Dull black surfaces are better absorbers of radiation than white shiny surfaces – the latter are good **reflectors** of radiation. Reflectors on electric fires are made of polished metal because of its good reflecting properties.

Good and bad emitters

Some surfaces also emit radiation better than others when they are hot. If you hold the backs of your hands on either side of a hot copper sheet that has one side polished and the other side blackened, Figure 32.3, it will be found that the **dull black surface is a better emitter of radiation than the shiny one**.

The cooling fins on the heat exchangers at the back of a refrigerator are painted black so that they lose heat more quickly. By contrast, saucepans etc. that are polished are poor emitters and keep their heat longer.

In general, surfaces that are good absorbers of radiation are good emitters when hot.

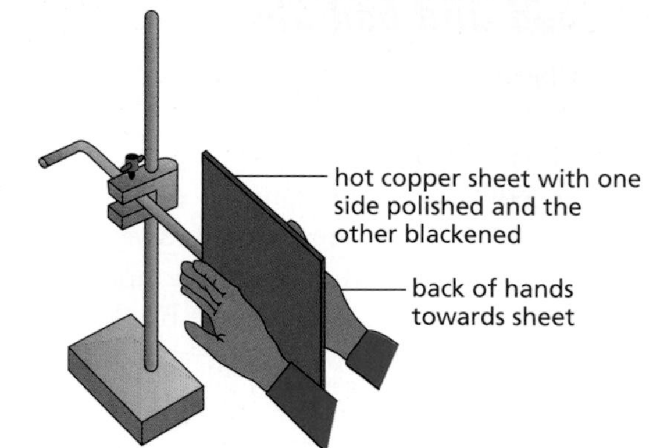

Figure 32.3 Comparing emitters of radiation

Vacuum flask

A vacuum or Thermos flask keeps hot liquids hot or cold liquids cold. It is very difficult for heat to travel into or out of the flask.

Transfer by conduction and convection is minimized by making the flask a double-walled glass vessel with a vacuum between the walls, Figure 32.4. Radiation is reduced by silvering both walls on the vacuum side. Then if, for example, a hot liquid is stored, the small amount of radiation from the hot inside wall is reflected back across the vacuum by the silvering on the outer wall. The slight heat loss that does occur is by conduction up the thin glass walls and through the stopper.

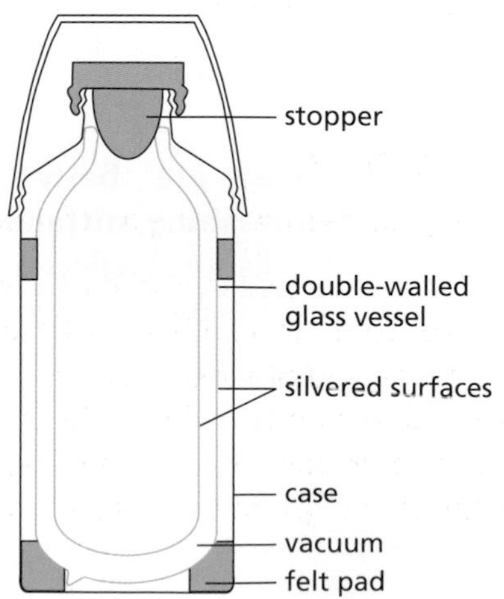

Figure 32.4 A vacuum flask

The greenhouse

The warmth from the Sun is not cut off by a sheet of glass but the warmth from a red-hot fire is. The radiation from very hot bodies like the Sun is mostly in the form of light and short-wavelength infrared. The radiation from less hot objects, e.g. a fire, is largely long-wavelength infrared which, unlike light and short-wavelength infrared, cannot pass through glass.

Light and short-wavelength infrared from the Sun penetrate the glass of a greenhouse and are absorbed by the soil, plants, etc., raising their temperature. These in turn emit infrared but, because of their relatively low temperature, this has a long wavelength and is not transmitted by the glass. The greenhouse thus acts as a 'heat-trap' and its temperature rises.

Carbon dioxide and other gases such as methane in the Earth's atmosphere act in a similar way to the glass of a greenhouse in trapping heat; this has serious implications for the global climate.

F Rate of cooling of an object

The rate at which an object cools, i.e. at which its temperature falls, can be shown to be proportional to the ratio of its surface area A to its volume V.

For a cube of side l

$$A_1/V_1 = 6 \times l^2/l^3 = 6/l$$

For a cube of side $2\,l$

$$A_2/V_2 = 6 \times 4\,l^2/8\,l^3 = 3/l$$

$$= \tfrac{1}{2} \times 6/l = \tfrac{1}{2} A_1/V_1$$

The larger cube has the smaller A/V ratio and so cools more slowly.

You could investigate this using two aluminium cubes, one having twice the length of side of the other. Each needs holes for a thermometer and an electric heater to raise them to the same starting temperature. Temperature against time graphs for both blocks can then be obtained.

Questions

1 a Choose words from the list below to complete the following sentences.

conduction convection diffusion evaporation radiation

(i) Thermal (heat) energy is transferred through a vacuum by ......
(ii) Transfer of energy by hot liquids moving is called ......
(iii) Transfer of energy by a substance, without the substance itself moving is called ......

b Figure 32.5a shows a science toy. Figure 32.5b shows a close-up of one of its vanes.

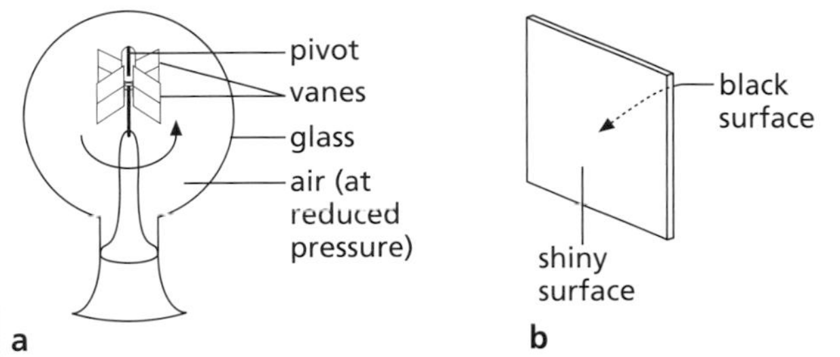

Figure 32.5

When the sun shines on the toy the vanes spin round. This happens because one side of each vane becomes warmer than the other.
(i) When the sun shines, which side becomes warmer, the shiny one or the black one?
(ii) Give *two* reasons for your answer.

(*AQA* (*NEAB*) *Foundation, June 1999*)

2 The door canopy in Figure 32.6 shows in a striking way the difference between white and black surfaces when radiation falls on them. Explain why.

Figure 32.6

3 a The Earth has been warmed by the radiation from the Sun for millions of years yet we think its average temperature has remained fairly steady. Why is this?

b Why is frost less likely on a cloudy night than a clear one?

4 The diagram in Figure 32.7 comes from a leaflet about a 'fuel effect' gas fire. It shows how air circulates through the fire.

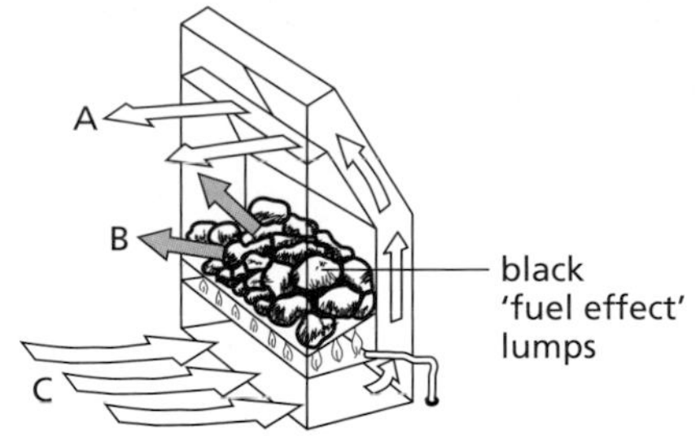

Figure 32.7

a Explain in detail why the air travels from C to A.

b The black 'fuel effect' lumps become very hot.
(i) Name the process by which the lumps transfer thermal energy to the room, as shown at B.
(ii) Suggest a feature of the black 'fuel effect' lumps which make them efficient at transferring energy.

(*AQA* (*NEAB*) *Higher, June 1999*)

5 The vacuum flask shown in Figure 32.8 has five features labelled, each one designed to reduce heat transfer.

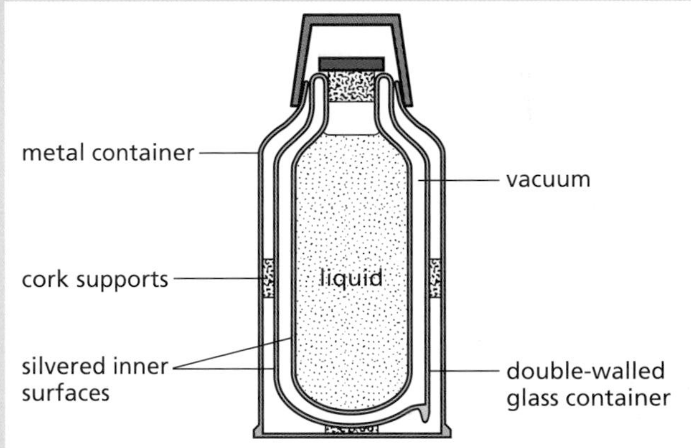

Figure 32.8

a (i) Which labelled feature of the vacuum flask reduces heat transfer by both conduction and convection?
(ii) Explain how this feature reduces heat transfer by **both** conduction and convection.

b (i) Which labelled feature of the vacuum flask reduces heat transfer by radiation?
(ii) Explain how this feature reduces heat transfer by radiation.

(*SEG Higher, June 1998*)

Checklist

After studying this chapter you should be able to

- describe experiments to study factors affecting the absorption and emission of radiation,
- recall that good absorbers are also good emitters,
- explain how a knowledge of heat transfer affects the design of a vacuum flask,
- explain how a greenhouse acts as a 'heat-trap'.

Heat and energy
Additional questions

Molecules

1 This question is about particles.

a The diagrams show how particles are arranged in solids, liquids and gases.

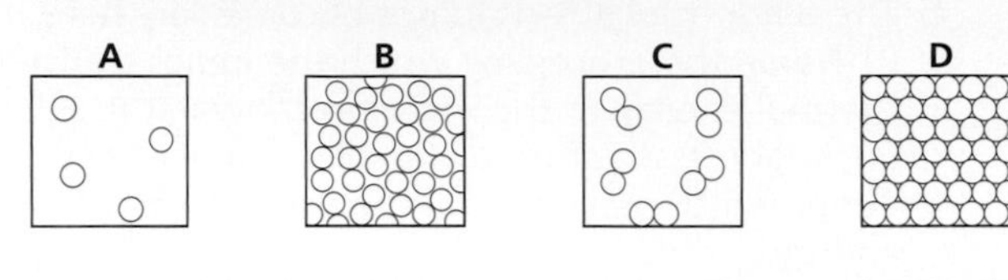

(i) Which diagram shows how particles are arranged in solids?
(ii) Which diagram shows how particles are arranged in liquids?

b Harry uses a microscope to look at smoke particles. He sees the smoke particles moving randomly. What makes the smoke particles move randomly?

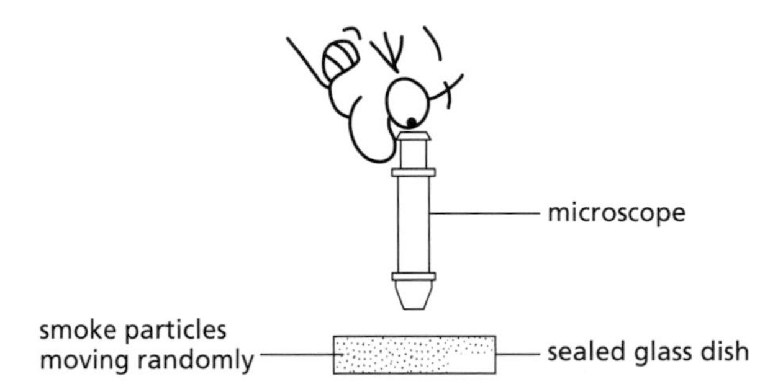

c The temperature inside the dish rises.
(i) What effect does this have on how the smoke particles move?
(ii) What effect does the rise in temperature have on the gas pressure inside the dish?
(iii) Explain how a gas exerts pressure. Use your ideas of particles.

(*OCR Foundation, June 1999*)

Thermometers

2 The diagram shows a mercury-in-glass thermometer.

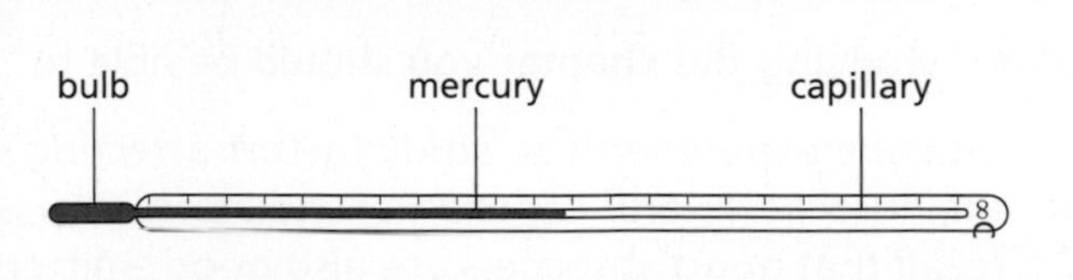

a Explain how the thermometer works.

b The thermometer has a scale which shows *linearity*. What do you understand by this statement?

c What modification would you make to the design of a mercury-in-glass thermometer in order to increase its sensitivity?

d Name two possible advantages that a thermocouple would have over a mercury-in-glass thermometer.

(*UCLES IGCSE Physical Science Extended, May 2000*)

Expansion of solids, liquids and gases

3 When a metal bar is heated the increase in length is greater if

1 the bar is long
2 the temperature rise is large
3 the bar has a large diameter.

Which statement(s) is (are) correct?

A 1, 2, 3 **B** 1, 2 **C** 2, 3 **D** 1 **E** 3

4 A bimetallic thermostat for use in an iron is shown in the diagram below.

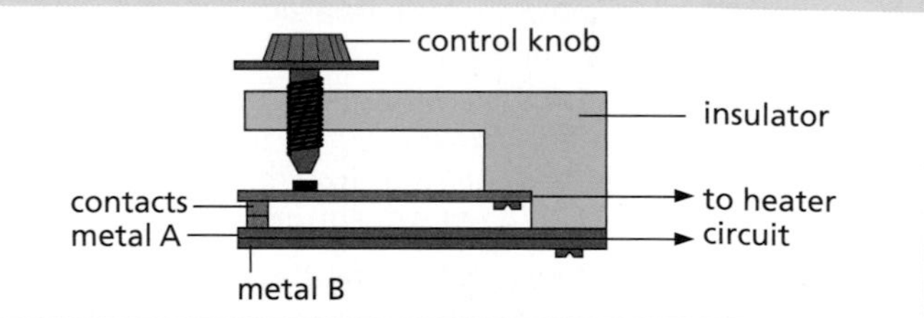

1 It operates by the bimetallic strip bending away from the contact.
2 Metal A has a greater expansivity than metal B.
3 Screwing in the control knob raises the temperature at which the contacts open.

Which statement(s) is (are) correct?

A 1, 2, 3 **B** 1, 2 **C** 2, 3 **D** 1 **E** 3

The gas laws

5 Carbon dioxide is used to make fizzy drinks. It is stored at high pressure in a cast iron cylinder. The diagram represents the particles in a cylinder of carbon dioxide.

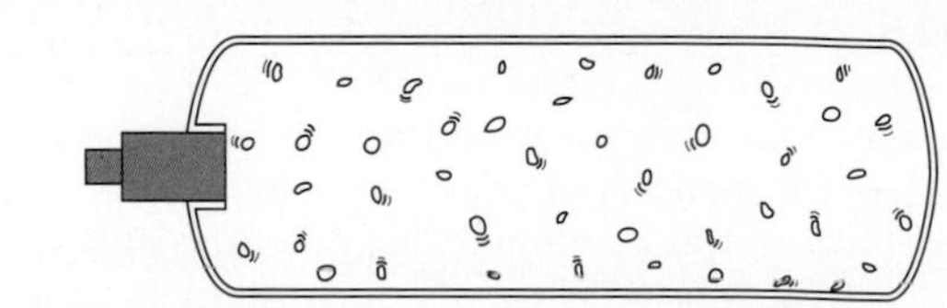

a Describe how the particles of carbon dioxide exert pressure.

b The temperature of the gas in the cylinder is increased.

(i) Describe the effect this has on the movement of the carbon dioxide particles.
(ii) Explain how this affects the pressure exerted by the gas.
(iii) The cylinders are painted black. Explain why the cylinders should *not* be stored outside in direct sunlight.

(*London Foundation, June 1999*)

Specific heat capacity; specific latent heat

6 A certain liquid has a specific heat capacity of 4.0 J/(g °C). How much heat must be supplied to raise the temperature of 10 g of the liquid from 20 °C to 50 °C?

7 A student attempted to find the specific heat capacity of water using the following data obtained from the heating system of a small swimming pool:

mass of water in the pool, heating system and circulation pipes, 54 000 kg;
power of the heating system, 30 kW;
rise in temperature, 2 °C in 5 hours (18 000 s).

a Assuming no energy loss, use these data to calculate a value for the specific heat capacity of water. Show your working.

b The student found that the value for the specific heat capacity of water, worked out by this method, was higher than the accepted value.

The average temperature of the water in the pool during the test period was 24 °C, whilst the average temperature of the air was 19 °C.
(i) Describe, in molecular terms, ways in which the water loses heat from its surface.
(ii) Explain why the loss of heat from the water led to the student's higher value.

(*UCLES IGCSE Physics Extended, May 1999*)

8 Some water is heated electrically in a glass beaker in an experiment to find the specific heat capacity of water. The temperature of the water is taken at regular intervals.

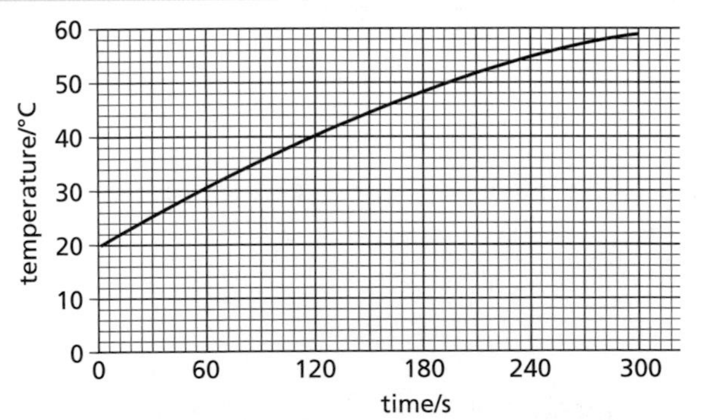

The temperature–time graph for this heating is shown below.

a (i) Use the graph to find
 1 the temperature rise in the first 120 s,
 2 the temperature rise in the second 120 s interval.
(ii) Explain why these values are different.

b The experiment is repeated in an insulated beaker. This time, the temperature of the water increases for 20 °C to 60 °C in 210 s. The beaker contains 75 g of water. The power of the heater is 60 W. Calculate the specific heat capacity of water.

c In order to measure the temperature during the heating, a thermocouple is used. Draw a labelled diagram of a thermocouple connected to measure temperature.

(*UCLES IGCSE Physics Extended, November 2006*)

Conduction and convection; radiation

9 Explain why on a cold day the metal handlebars of a bicycle feel colder than the rubber grips.

10 This question is about keeping a house warm. A house has been insulated in these two ways.

1 The windows are double glazed.
2 There is shiny aluminium foil on the wall behind the radiators.

a Describe how each of these ways helps to keep the house warm. Use your ideas about conduction, convection and radiation.
(i) Double glazing.

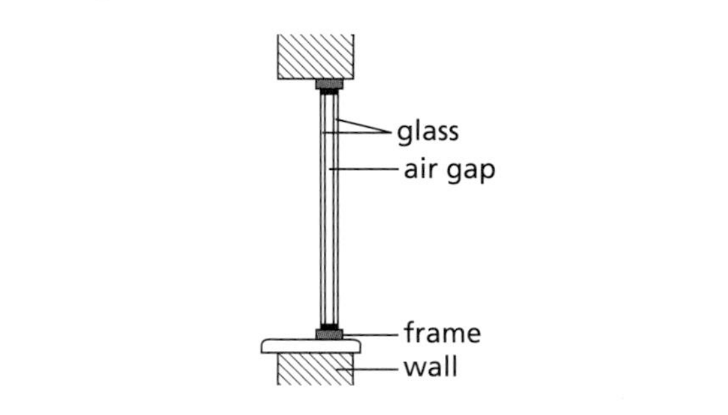

(ii) Putting shiny aluminium foil on the wall behind a radiator.

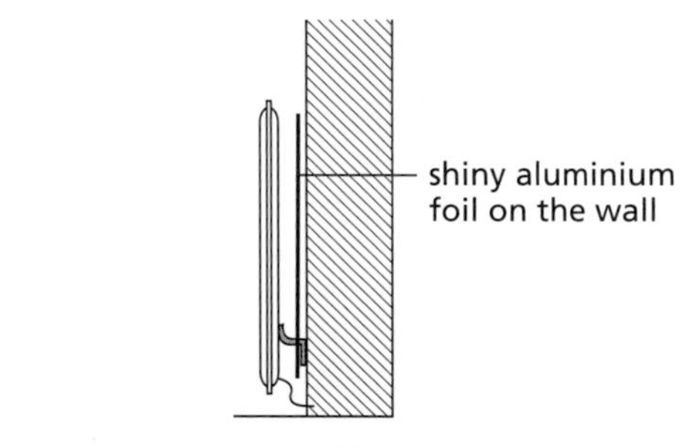

b Write down one *other* type of insulation that can be used in houses.

(*OCR Foundation, June 1999*)

11 Explain the following situations in terms of the various methods of heat transfer.

a If you put your hand above a burning match, your hand feels hot. However, your hand does not feel particularly hot when it is underneath the burning match (see diagram overleaf).

continued

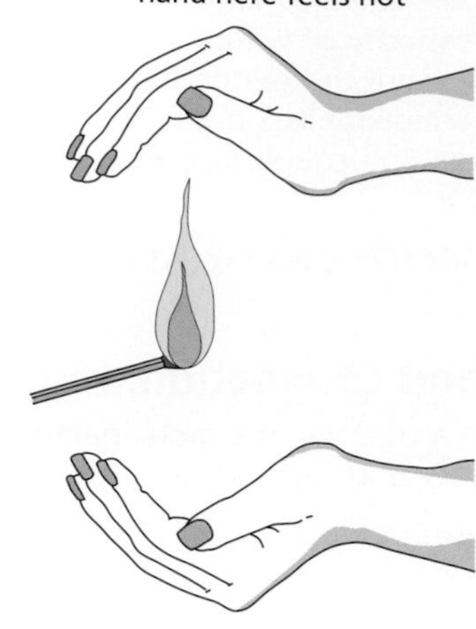

b Putting a thick carpet on a concrete floor, as shown in the diagram below, helps to keep the room warm.

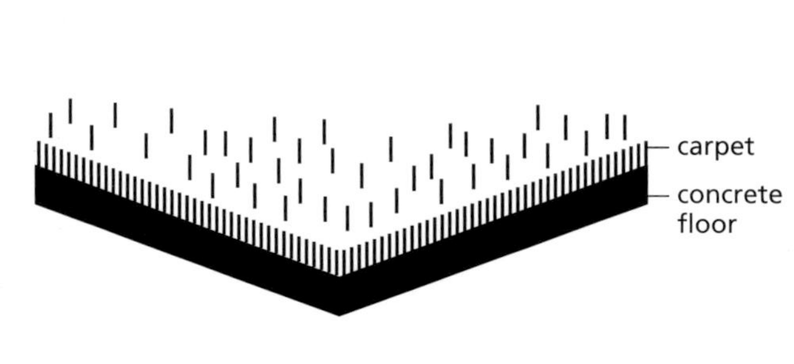

c Heat from the Sun can reach us, even though there is a vacuum between the Earth and the Sun.

d If you get lost in a snow storm, you can keep warm by digging yourself into a hole in the snow as illustrated below.

(*UCLES IGCSE Physics Core, Nov 1997*)

12

a A copper rod AB is being heated at one end.

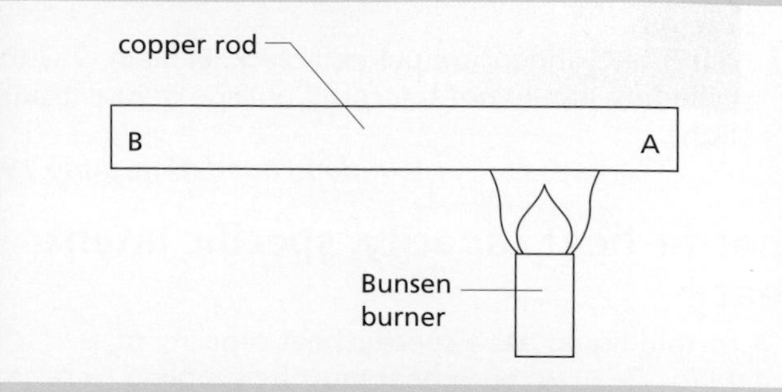

(i) Name the process by which heat moves from A to B.

(ii) By reference to the behaviour of the particles of copper along AB, state how this process happens.

b Give an account of an experiment that is designed to show which of four surfaces will absorb most heat radiation. The four surfaces are all the same metal, but one is a polished black surface, one is a polished silver surface, one is a dull black surface and the fourth one is painted white. Give your answer under the three headings below.

1 labelled diagram of the apparatus,
2 readings to be taken,
3 one precaution to try to achieve a fair comparison between the various surfaces.

(*UCLES IGCSE Physics Extended, Nov 2006*)

Electricity

33 Static electricity

Figure 33.1 A flash of lightning is nature's most spectacular static electricity effect

Clothes containing nylon often crackle when they are taken off. We say they are 'charged with static electricity'; the crackles are caused by tiny electric sparks which can be seen in the dark. Pens and combs made of certain plastics become charged when rubbed on your sleeve and can then attract scraps of paper.

Positive and negative charges

When a strip of polythene (white) is rubbed with a cloth it becomes charged. If it is hung up and another rubbed polythene strip is brought near, repulsion occurs, Figure 33.2. Attraction occurs when a rubbed strip of cellulose acetate (clear) approaches.

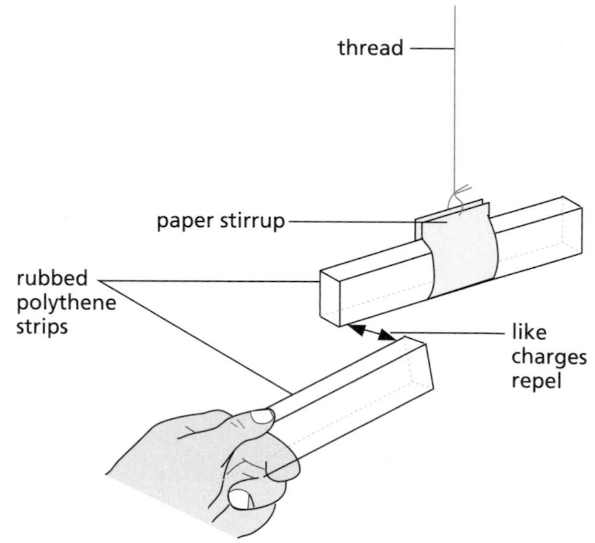

Figure 33.2 Investigating charges

This shows there are two kinds of electric charge. That on cellulose acetate is taken as **positive** (+) and that on polythene is **negative** (−). It also shows that:

> Like charges (+ and + or − and −) repel, while unlike charges (+ and −) attract.

The force between electric charges decreases as their separation increases.

Charges, atoms and electrons

There is evidence (Chapter 49) that we can picture an atom as being made up of a small central nucleus containing positively charged particles called **protons**, surrounded by an equal number of negatively charged **electrons**. The charges on a proton and an electron are equal and so an atom as a whole is normally electrically neutral, i.e. has no net charge.

Hydrogen is the simplest atom with one proton and one electron, Figure 33.3. A copper atom has 29 protons in the nucleus and 29 surrounding electrons. Every nucleus except hydrogen also contains uncharged particles called **neutrons**.

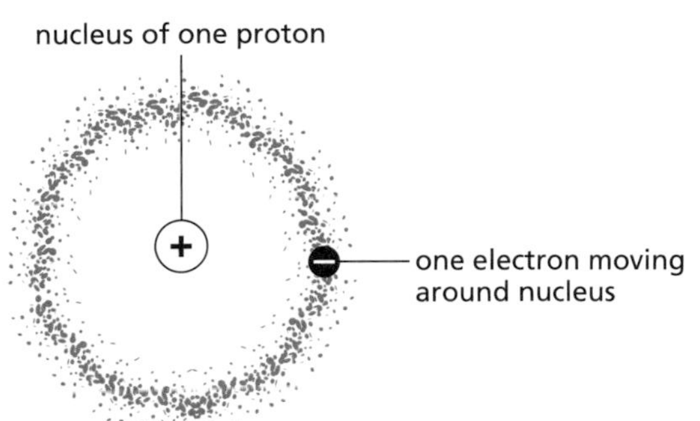

Figure 33.3 Hydrogen atom

The production of charges by rubbing can be explained by supposing that electrons are transferred from one material to the other. For example, when cellulose acetate is rubbed, electrons go from the acetate to the cloth, leaving the acetate short of electrons, i.e. positively charged. The cloth now has more electrons than protons and becomes negatively charged. Note that it is only electrons which move; the protons remain fixed in the nucleus.

How does polythene become charged when rubbed?

Practical work

Gold-leaf electroscope

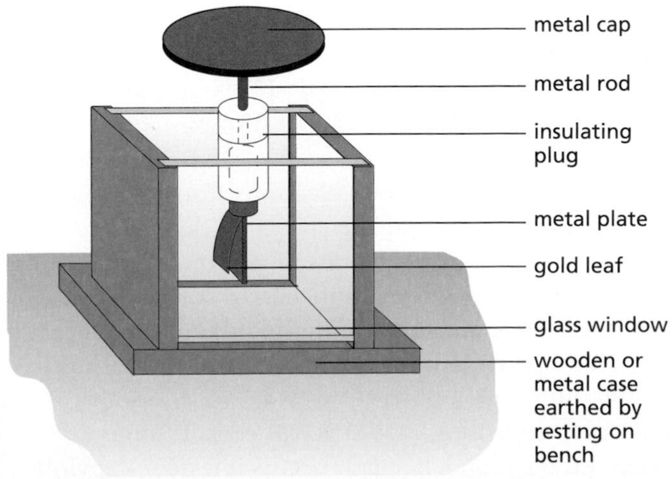

Figure 33.4 Gold-leaf electroscope

A gold-leaf electroscope consists of a metal cap on a metal rod at the foot of which is a metal plate having a leaf of gold foil attached, Figure 33.4. The rod is held by an insulating plastic plug in a case with glass sides to protect the leaf from draughts.

a) Detecting a charge

Bring a charged polythene strip towards the cap: the leaf rises away from the plate. On removing the charged strip the leaf falls again. Repeat with a charged acetate strip.

b) Charging by contact

Draw a charged polythene strip *firmly across the edge* of the cap. The leaf should rise and stay up when the strip is removed. If it does not, repeat the process but press harder. The electroscope has now become negatively charged by contact with the polythene strip, from which electrons have been transferred.

c) Insulators and conductors

Touch the cap of the charged electroscope with different things, e.g. a piece of paper, a wire, your finger, a comb, a cotton handkerchief, a piece of wood, a glass rod, a plastic pen, rubber tubing.

When the leaf falls, charge is passing to or from the ground through you and the material touching the cap. If the fall is rapid the material is a **good conductor**, if slow, it is a poor conductor and if the leaf does not alter the material is a **good insulator**. Record your results.

Electrons, insulators and conductors

In an insulator all electrons are bound firmly to their atoms; in a conductor some electrons can move freely from atom to atom. An insulator can be charged by rubbing because the charge produced cannot move from where the rubbing occurs, i.e. the electric charge is static. A conductor will become charged only if it is held with an insulating handle; otherwise electrons are transferred between the conductor and the ground via the person's body.

Good insulators are plastics such as polythene, cellulose acetate, Perspex and nylon. All metals and carbon are good conductors. In between are materials that are both poor conductors and (because they conduct to some extent) poor insulators. Examples are wood, paper, cotton, the human body and the Earth. Water conducts and if it were not present in materials like wood and on the surface of, for example, glass, these would be good insulators. Dry air insulates well.

□ *Electrostatic induction*

This effect may be shown by bringing a negatively charged polythene strip near to an insulated metal sphere X which is touching a similar sphere Y, Figure 33.5a. Electrons in the spheres are repelled to the far side of Y.

If X and Y are separated, with the charged strip still in position, X is left with a positive charge (deficient of electrons) and Y with a negative charge (excess of electrons), Figure 33.5b. The signs of the charges can be tested by removing the charged strip, Figure 33.5c, and taking X up to the cap of a positively charged electroscope. Electrons will be drawn towards X, making the leaf more positive so that it rises. If Y is taken towards the cap of a *negatively* charged electroscope the leaf again rises; can you explain why, in terms of electron motion?

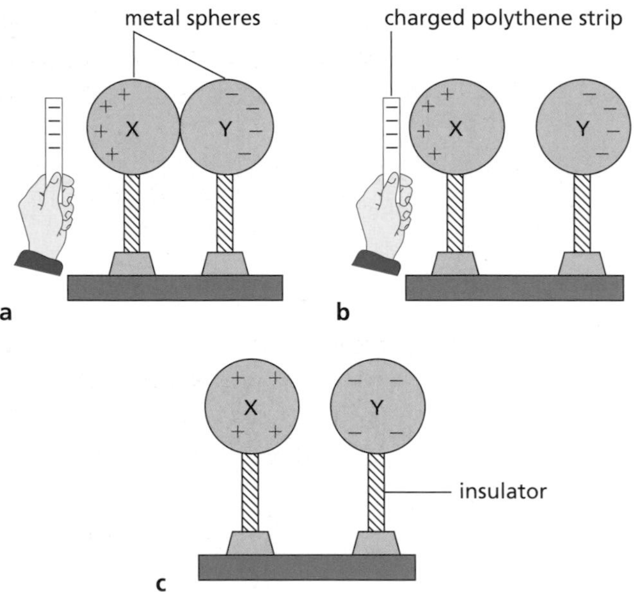

Figure 33.5 Electrostatic induction

□ *Attraction between uncharged and charged objects*

The attraction of an uncharged object by a charged object near it is due to electrostatic induction.

In Figure 33.6a a small piece of aluminium foil is attracted to a negatively charged polythene rod held just above it. The charge on the rod pushes free electrons to the bottom of the foil (aluminium is a conductor), leaving the top of the foil short of electrons, i.e. with a net positive charge, and the bottom negatively charged. The top of the foil is nearer the rod than the bottom. Hence the force of attraction between the negative charge on the rod and the positive charge on the top of the foil is greater than the force of repulsion between the negative charge on the rod and the negative charge on the bottom of the foil. The foil is pulled to the rod.

A small scrap of paper, although an insulator, is also attracted by a charged rod. There are no free electrons in the paper but the charged rod pulls the electrons of the atoms in the paper slightly closer (by electrostatic induction) and so distorts the atoms. In the case of a negatively charged polythene rod, the paper behaves as if it had a positively charged top and a negative charge at the bottom.

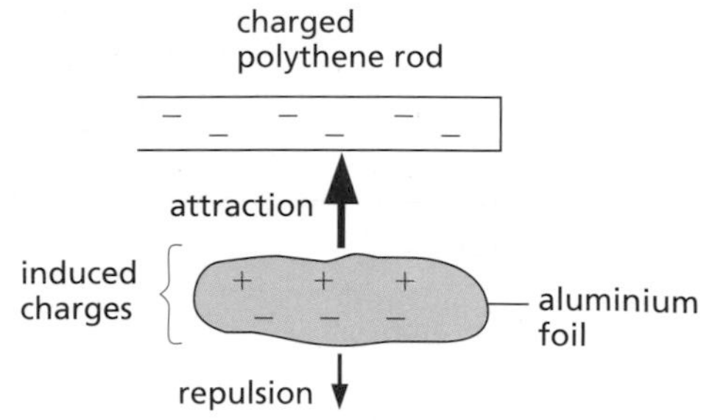

Figure 33.6a An uncharged object is attracted to a charged one

Figure 33.6b A slow stream of water being bent by electrostatic attraction

In Figure 33.6b a slow, uncharged stream of water is attracted by a charged polythene rod, due to distortion of the water molecules.

□ *Dangers of static electricity*

a) Lightning

A tall building is protected by a lightning conductor consisting of a thick copper strip fixed on the outside of the building connecting metal spikes at the top to a metal plate in the ground, Figure 33.7.

Thunderclouds carry charges; a negatively charged one passing overhead repels electrons from the spikes to the Earth. The points of the spikes are left with a large positive charge (charge concentrates on sharp points) which removes electrons from nearby air molecules, so charging them positively and causing them to be repelled from the spike. This effect, called **action at points**, results in an 'electric wind' of positive air molecules streaming upwards to cancel some of the charge on the cloud. If a flash occurs it is now less violent and the conductor gives it an easy path to ground.

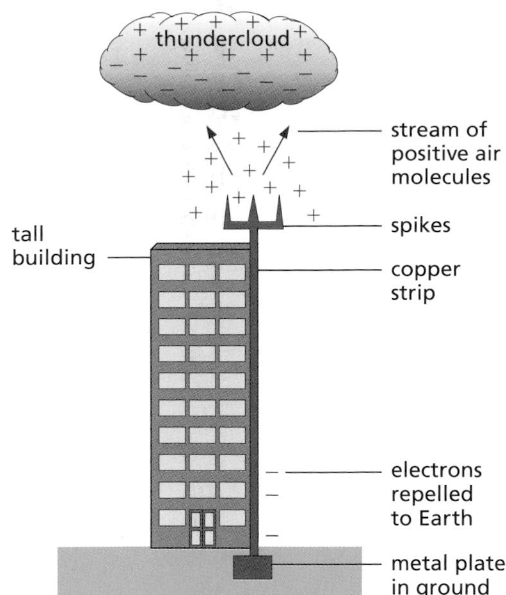

Figure 33.7 Lightning conductor

b) Refuelling

Sparks from static electricity can be dangerous when flammable vapour is present. For this reason, the tanks in an oil-tanker may be cleaned in an atmosphere of nitrogen – otherwise oxygen in the air could promote a fire.

An aircraft in flight may become charged by 'rubbing' the air. Its tyres are made of conducting rubber which lets the charge pass harmlessly to ground on landing, otherwise an explosion could be 'sparked off' when the aircraft refuels. What precautions are taken at petrol pumps when a car is refuelled?

c) Operating theatres

Dust and germs are attracted by charged objects and so it is essential to ensure that equipment and medical personnel are well 'earthed' allowing electrons to flow to and from the ground, e.g. by conducting rubber.

d) Computers

Computers require similar 'anti-static' conditions as they are vulnerable to electrostatic damage.

☐ *Uses of static electricity*

a) Flue-ash precipitation

An electrostatic precipitator removes the dust and ash that goes up the chimneys of coal-burning power stations. It consists of a charged fine wire mesh which gives a similar charge to the rising particles of ash. They are then attracted to plates with an opposite charge. These are tapped from time to time to remove the ash which falls to the bottom of the chimney, from where it is removed.

b) Photocopiers

These contain a charged drum and when the paper to be copied is laid on the glass plate, the light reflected from the white parts of the paper causes the charge to disappear from the corresponding parts of the drum opposite. The charge pattern remaining on the drum corresponds to the dark-coloured printing on the original. Special **toner** powder is then dusted over the drum and sticks to those parts which are still charged. When a sheet of paper passes over the drum, the particles of toner are attracted to it and fused into place by a short burst of heat.

c) Inkjet printers

In an inkjet printer tiny drops of ink are forced out of a fine nozzle, charged electrostatically and then passed between two oppositely charged plates; a negatively charged drop will be attracted towards the positive plate causing it to be deflected as shown in Figure 33.8. The amount of deflection and hence the position at which the ink strikes the page is determined by the charge on the drop and the p.d. between the plates; both of these are controlled by a computer. About 100 precisely located drops are needed to make up an individual letter but very fast printing speeds can be achieved.

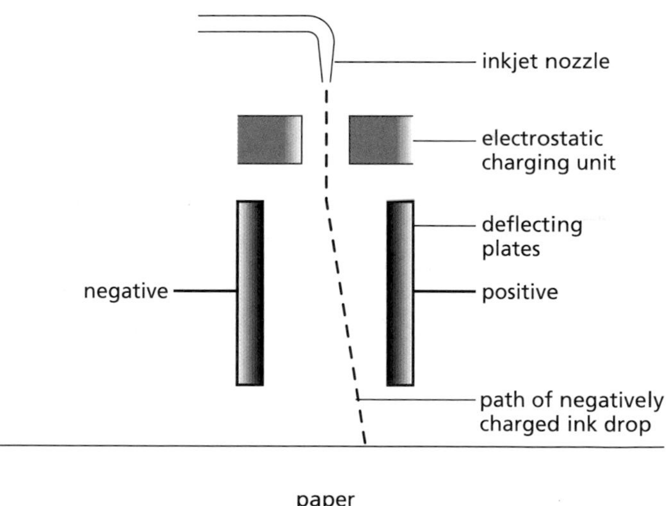

Figure 33.8 Inkjet printer

F *van de Graaff generator*

This produces a continuous supply of charge on a large metal dome when a rubber belt is driven by an electric motor or by hand, Figure 33.9a (overleaf).

a) Demonstrations

In Figure 33.9a sparks jump between the dome and the discharging sphere. Electrons flow round a complete path (circuit) from the dome. Can you trace it? In part b why does the 'hair' stand on end? In part c the 'windmill' revolves due to the reaction that arises from the 'electric wind' caused by the **action at points** effect, explained above for the lightning conductor.

In Figure 33.9d the 'body' on the insulating stool first gets charged by touching the dome and then lights a neon lamp.

The dome can be discharged harmlessly by bringing your elbow close to it.

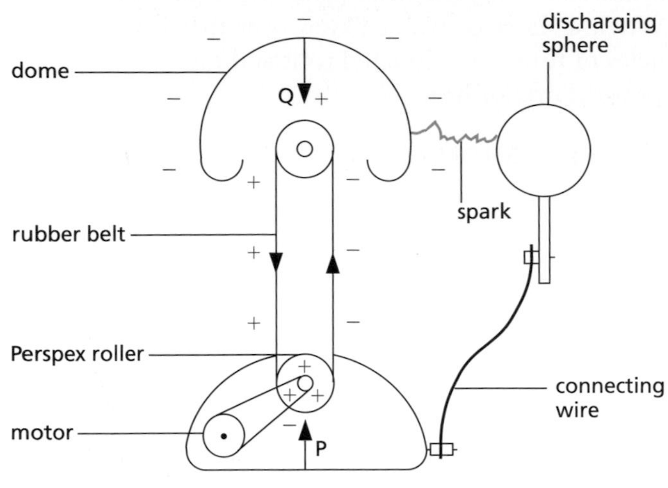

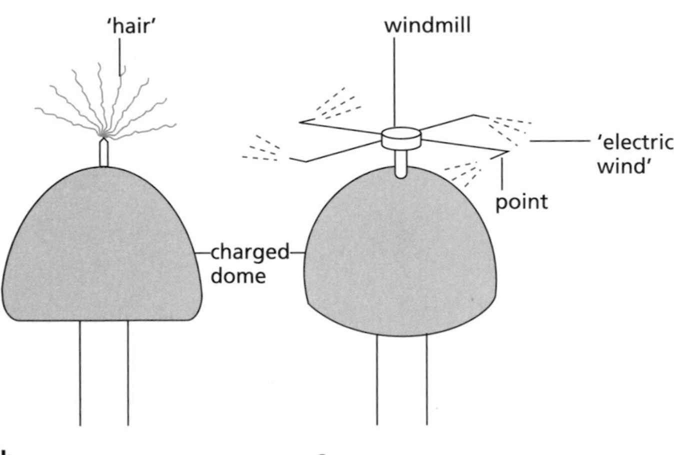

Figure 33.9

b) Action

Initially a positive charge is produced on the motor-driven Perspex roller due to it rubbing the belt. This induces a negative charge on the 'comb' of metal points P, Figure 33.9a, which are sprayed off by 'action at points' on to the outside of the belt and carried upwards. A positive charge is then induced in the comb of metal points Q and negative charge is repelled to the dome.

Electric fields

When an electric charge is placed near to another electric charge it experiences a force. The electric force does not require contact between the two charges so we call it an 'action-at-a-distance force' – it acts through space. The region of space where an electric charge experiences a force due to other charges is called an **electric field**. If the electric force felt by a charge is the same everywhere in a region, the field is uniform; a uniform electric field is produced between two oppositely charged parallel metal plates, Figure 33.10. It can be represented by evenly spaced parallel lines drawn perpendicular to the metal surfaces. The direction of the field, denoted by arrows, is the direction in which a small positive charge placed in the field would move (negative charges move in the opposite direction to the field).

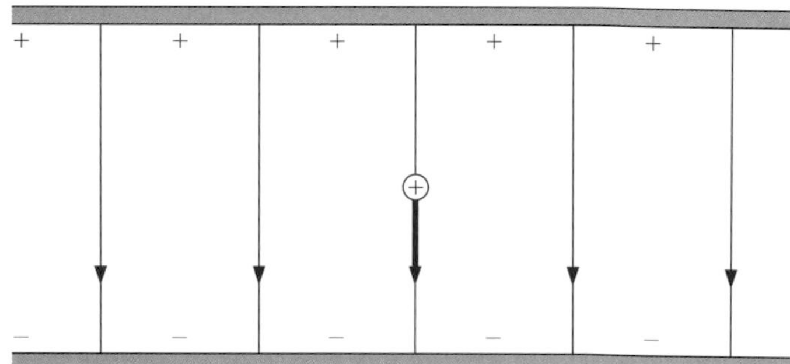

Figure 33.10 Uniform electric field

Moving charges are deflected by an electric field due to the electric force exerted on them; this occurs in the inkjet printer (Figure 33.8).

The electric field lines radiating from an isolated positively charged conducting sphere and a point charge are shown in Figure 33.11a, b; again the field lines emerge at right angles to the conducting surface.

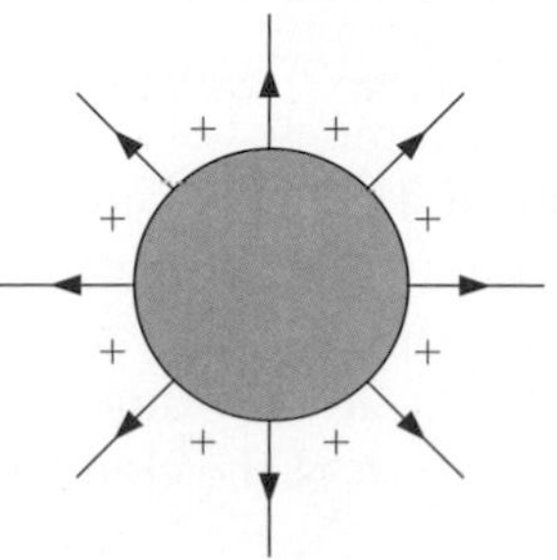

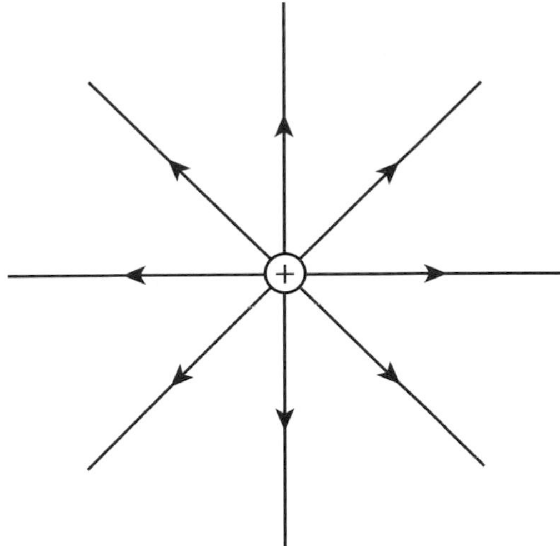

b
Figure 33.11 Radial electric field

Questions

1 Two identical conducting balls, suspended on nylon threads, come to rest with the threads making equal angles with the vertical, Figure 33.12. This shows that:

A the balls are equally and oppositely charged
B the balls are oppositely charged but not necessarily equally charged
C one ball is charged and the other is uncharged
D the balls both carry the same type of charge
E one is charged and the other may or may not be charged.

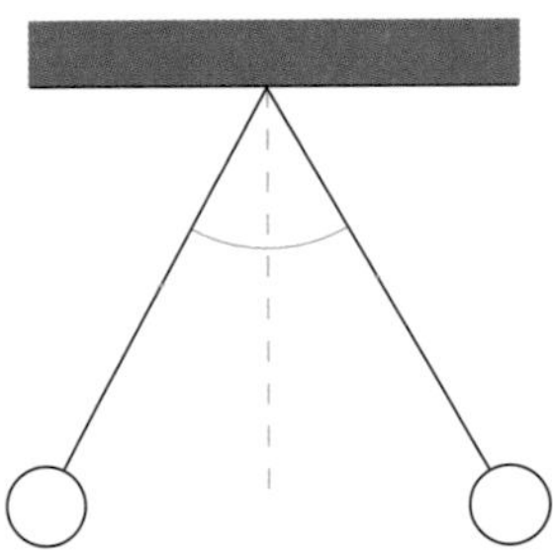

Figure 33.12

2 Explain in terms of electron movement what happens when a polythene rod becomes charged negatively by being rubbed with a cloth.

3 a One method of painting a car uses electrostatics. A paint spray produces paint droplets, all of which are given a positive charge. The car body is given a negative charge, Figure 33.13.
(i) Explain why it is important to give all of the paint droplets a positive charge.
(ii) Explain why it is important to give the car body a negative charge.

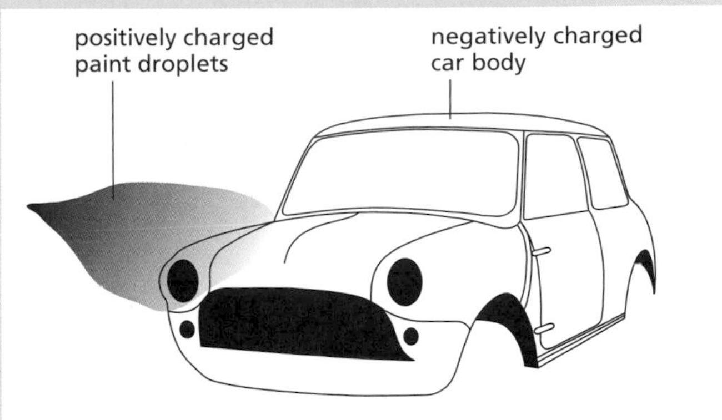

Figure 33.13

b Figure 33.14 shows a light aircraft being refuelled after a flight.

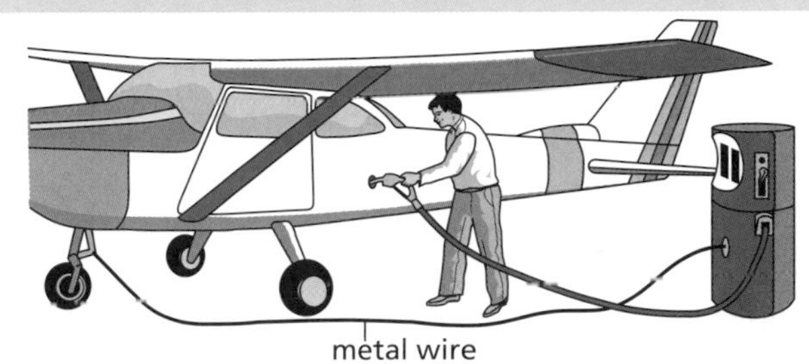

Figure 33.14

Explain why it is important that, before refuelling starts, the aircraft is first connected to the fuel pump by a metal wire.

(*SEG Foundation, June 1998*)

Checklist

After studying this chapter you should be able to

- ■ describe how positive and negative charges are produced by rubbing,
- ■ recall that like charges repel and unlike charges attract,
- ■ explain the charging of objects in terms of the motion of negatively charged electrons,
- ■ describe the gold-leaf electroscope, and explain how it can be used to compare electrical conductivities of different materials,
- ■ explain the differences between insulators and conductors,
- □ describe how a conductor can be charged by induction,
- □ explain how a charged object can attract uncharged objects,
- □ give examples of the dangers and the uses of static electricity,
- ■ explain what is meant by an electric field.

34 Electric current

Effects of a current
The ampere and the coulomb
Circuit diagrams
Series and parallel circuits
Direct and alternating current
Practical work
Measuring current.

An electric current consists of moving electric charges. In Figure 34.1, when the van de Graaff is working, the table-tennis ball shuttles rapidly to and fro between the plates and the meter records a small current. As the ball touches each plate it becomes charged and is repelled to the other plate. In this way charge is carried across the gap. This also shows that 'static' charges cause a deflection on a meter just as current electricity produced by a battery does.

In a metal, each atom has one or more loosely held electrons that are free to move. When a van de Graaff or a battery is connected across the ends of such a conductor, the free electrons drift slowly along it in the direction from the negative to the positive terminal of a battery. There is then a current of negative charge.

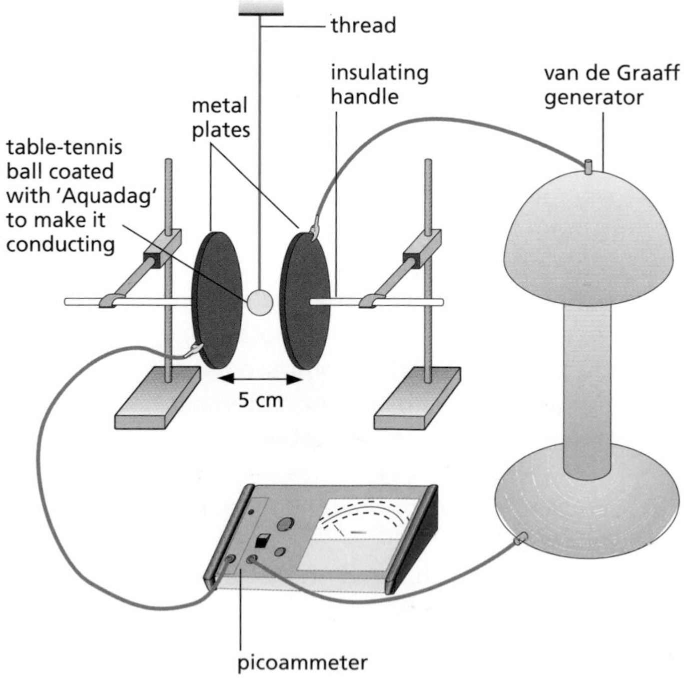

Figure 34.1 Demonstrating that an electric current consists of moving charges

Effects of a current

An electric current has three effects that reveal its existence and which can be shown with the circuit of Figure 34.2.

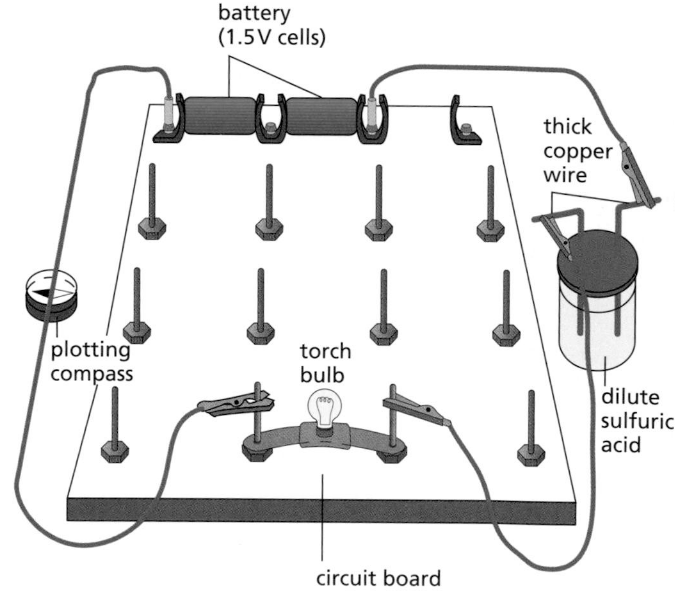

Figure 34.2 Investigating the effects of a current

a) Heating and lighting

The bulb lights due to a small wire in it (the filament) being made white hot by the current.

b) Magnetic

The plotting compass is deflected when it is placed in the magnetic field produced round any wire carrying a current.

c) Chemical

Bubbles of gas are given off at the wires in the acid because of the chemical action of the current.

The ampere and the coulomb

The unit of current is the **ampere** (A) which is defined using the magnetic effect. One milliampere (mA) is one-thousandth of an ampere. Current is measured by an **ammeter**.

The unit of charge, the **coulomb** (C), is defined in terms of the ampere.

One coulomb is the charge passing any point in a circuit when a steady current of 1 ampere flows for 1 second. That is, 1 C = 1 A s.

A charge of 3 C would pass each point in 1 s if the current were 3 A. In 2 s, 3 A × 2 s = 6 A s = 6 C would pass. In general, if a steady current I (amperes) flows for time t (seconds) the charge Q (coulombs) passing any point is given by

$$Q = I \times t$$

This is a useful expression connecting charge and current.

Circuit diagrams

Current must have a complete path (a circuit) of conductors if it is to flow. Wires of copper are used to connect batteries, lamps, etc. in a circuit since copper is a good electrical conductor. If the wires are covered with insulation, e.g. plastic, the ends are bared for connecting up.

The signs or symbols used for various parts of an electric circuit are shown in Figure 34.3.

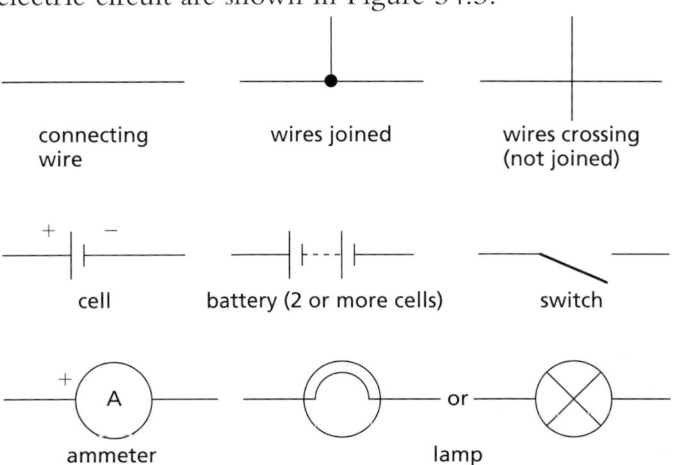

Figure 34.3 Circuit symbols

Before the electron was discovered scientists agreed to think of current as positive charges moving round a circuit in the direction from positive to negative of a battery. This agreement still stands. Arrows on circuit diagrams show the direction of what we call the **conventional current**, i.e. the direction in which **positive** charges would flow. Electrons flow in the opposite direction to the conventional current.

Practical work

Measuring current

(a) Connect the circuit of Figure 34.4a (on a circuit board if possible) ensuring that the + of the cell (the metal stud) goes to the + of the ammeter (marked red). Note the current.

(b) Connect the circuit of Figure 34.4b. The cells are **in series** (+ of one to − of the other), as are the lamps. Record the current. Measure the current at B, C and D by disconnecting the circuit at each point in turn and inserting the ammeter. What do you find?

(c) Connect the circuit of Figure 34.4c. The lamps are **in parallel**. Read the ammeter. Also measure the currents at P, Q and R. What is your conclusion?

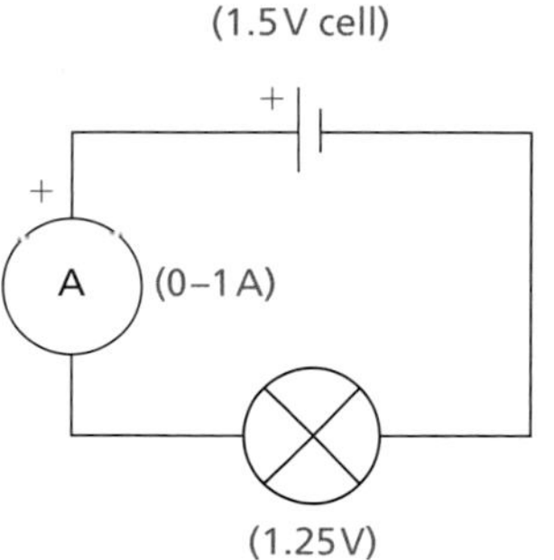

a

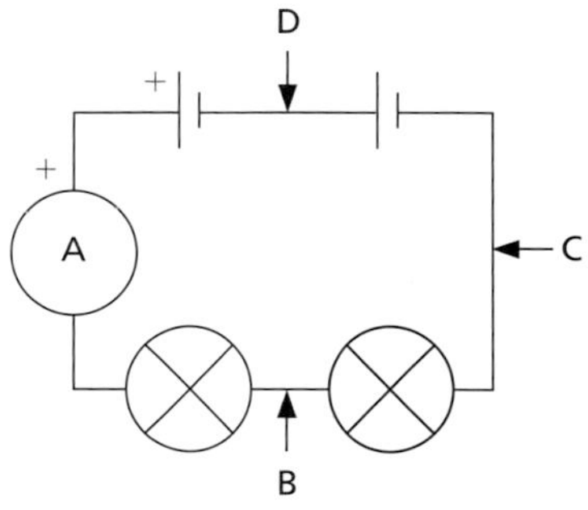

b

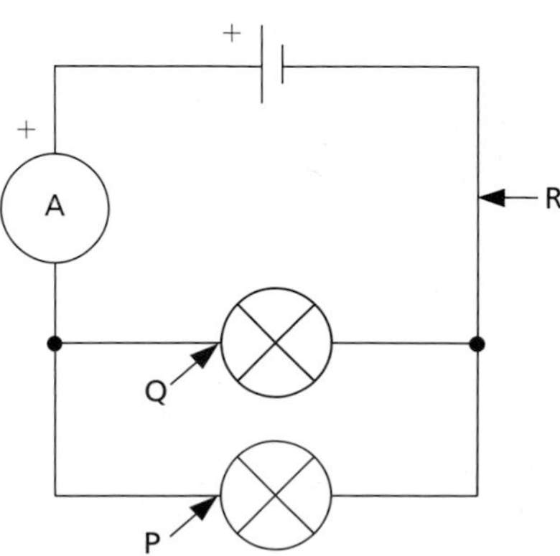

c

Figure 34.4

Series and parallel circuits

a) Series

In a series circuit, Figure 34.4b, the different parts follow one after the other and there is just one path for the current to follow. You should have found in the previous experiment that the reading on the ammeter when in the position shown (e.g. 0.2 A) is also obtained at B, C and D. That is, current is not used up.

The current is the same at all points in a series circuit.

b) Parallel

In a parallel circuit, Figure 34.4c, the lamps are side by side and alternative paths are provided for the current which splits. Some goes through one lamp and the rest through the other. The current from the source is larger than the current in each branch. For example, if the ammeter reading was 0.4 A in the position shown, then if the lamps are identical the reading at P would be 0.2 A, as it would be at Q, giving a total of 0.4 A. Whether the current splits equally or not depends on the lamps (as we will see later); it might divide so that 0.3 A goes one way and 0.1 A by the other branch.

The sum of the currents in the branches of a parallel circuit equals the current entering or leaving the parallel section.

Direct and alternating current

a) Difference

In a **direct current** (**d.c.**) the electrons flow in one direction only. Graphs for steady and varying d.c. are shown in Figure 34.5.

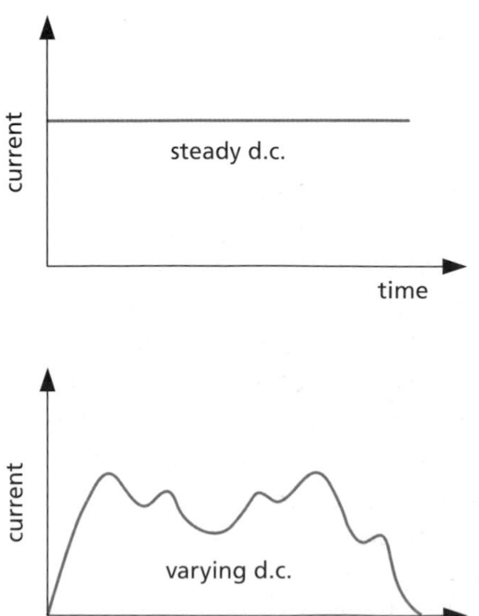

Figure 34.5 Direct current

In an **alternating current** (**a.c.**) the direction of flow reverses regularly, Figure 34.6. The circuit sign for a.c. is —∘∿∘—

The pointer of an ammeter for measuring d.c. is deflected one way by d.c.; a.c. makes it move to and fro about the zero if the changes are slow enough, otherwise there is no deflection.

Batteries give d.c.; generators can produce either d.c. or a.c.

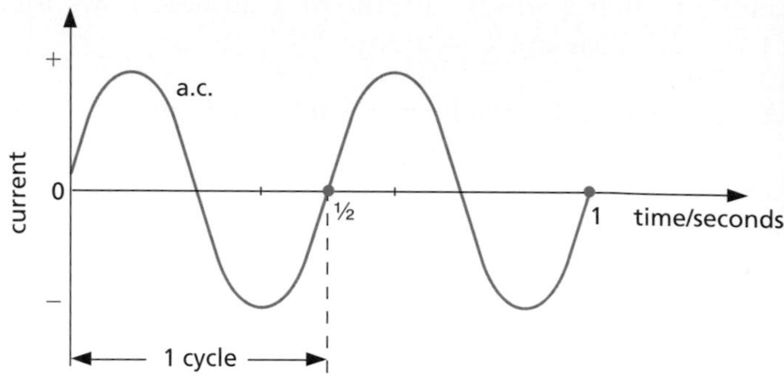

Figure 34.6 Alternating current

b) Frequency of a.c.

The number of complete alternations or cycles in 1 second is the **frequency** of the a.c. The unit of frequency is the **hertz** (Hz). The frequency of the a.c. in Figure 34.6 is 2 Hz, which means there are two cycles per second, or one cycle lasts 1/2 = 0.5 s. The mains supply in the UK is a.c. of frequency 50 Hz; each cycle lasts 1/50th of a second. This regularity was used in the tickertape timer (Chapter 19) and is relied upon in mains-operated clocks.

Questions

1 If the current through a floodlamp is 5 A, what charge passes in
   a 1 s,
   b 10 s,
   c 5 minutes?

2 What is the current in a circuit if the charge passing each point is
   a 10 C in 2 s,
   b 20 C in 40 s,
   c 240 C in 2 minutes?

3 Here are some circuit symbols.

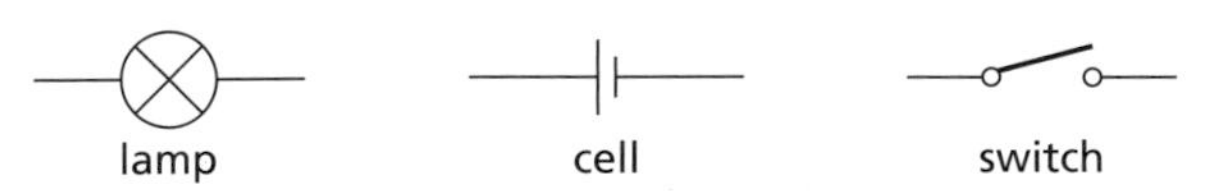

Figure 34.7

   a (i) Draw a circuit diagram showing two lamps, one cell and a switch connected in series.
   (ii) How can you change the brightness of the lamps?
   b (i) Draw a circuit diagram showing two lamps in parallel, together with a cell and a switch that works both lamps.
   (ii) An ammeter is used to measure the current in the cell. Mark an (A) on your diagram to show where the ammeter should be placed.
   (*London Foundation, June 1999*)

4 Study the circuits in Figure 34.8. The switch S is open (there is a break in the circuit at this point). In which circuit would lamps Q and R light but not lamp P?

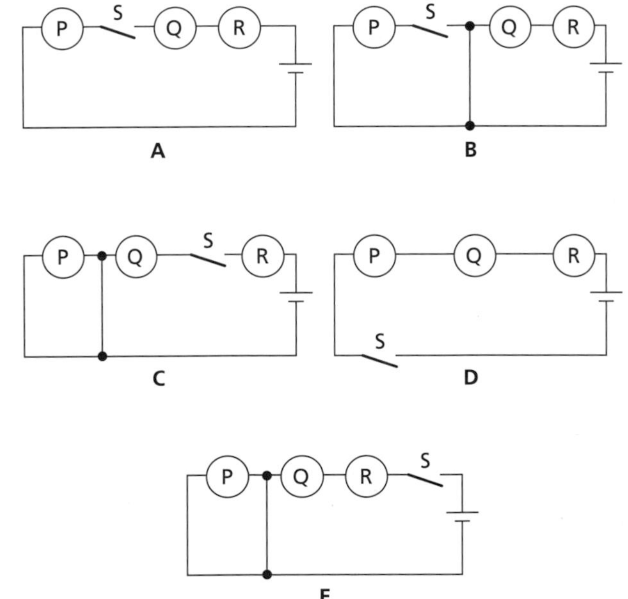

Figure 34.8

5 Using the circuit in Figure 34.9, which of the following statements is correct?

   A When S_1 and S_2 are closed, lamps A and B are lit.
   B With S_1 open and S_2 closed, A is lit and B is not lit.
   C With S_2 open and S_1 closed, A and B are lit.

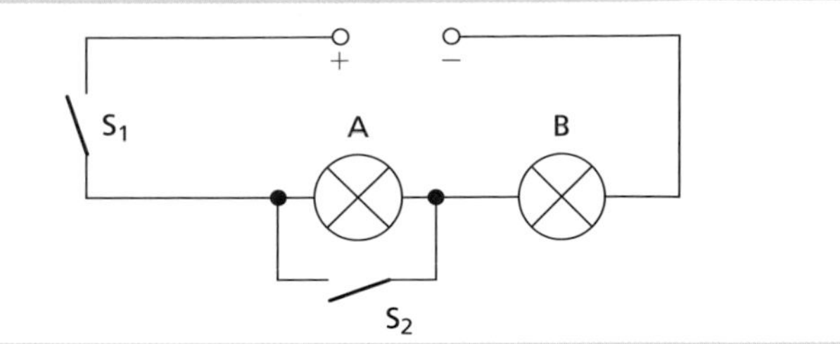

Figure 34.9

6 If the lamps are both the same in Figure 34.10 and if (A_1) reads 0.50 A, what do (A_2), (A_3), (A_4) and (A_5) read?

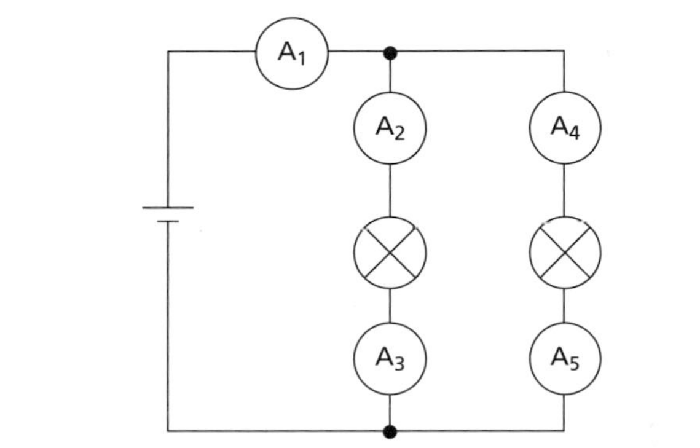

Figure 34.10

Checklist

After studying this chapter you should be able to

- ■ describe a demonstration which shows that an electric current is a flow of charge,
- ■ recall that an electric current in a metal is a flow of electrons from the negative to the positive terminal of the battery round a circuit,
- ■ state the three effects of an electric current,
- ■ state the unit of electric current and recall that current is measured by an ammeter,
- □ define the unit of charge in terms of the unit of current,
- □ recall the relation $Q = It$ and use it to solve problems,
- ■ use circuit symbols for wires, cells, switches, ammeters and lamps,
- ■ draw and connect simple series and parallel circuits, observing correct polarities for meters,
- ■ recall that the current in a series circuit is the same everywhere in the circuit,
- □ recall that the sum of the currents in the branches of a parallel circuit equals the current entering or leaving the parallel section,
- □ distinguish between electron flow and conventional current,
- ■ distinguish between direct current and alternating current,
- ■ recall that frequency of a.c. is the number of cycles per second.

35 Potential difference

Energy transfers and p.d.
Model of a circuit
The volt
Cells, batteries and e.m.f.
Voltages round a circuit
Practical work
Measuring voltage.

A battery transforms chemical energy to electrical energy. Because of the chemical action going on inside it, it builds up a surplus of electrons at one of its terminals (the negative) and creates a shortage at the other (the positive). It is then able to maintain a **flow of electrons**, i.e. an **electric current**, in any circuit connected across its terminals so long as the chemical action lasts.

The battery is said to have a **potential difference** (**p.d.** for short) at its terminals. Potential difference is measured in **volts** (V) and the term **voltage** is sometimes used instead of p.d. The p.d. of a car battery is 12 V and the domestic mains supply in the UK is 230 V.

Energy transfers and p.d.

In an electric circuit electrical energy is supplied from a source such as a battery and is transferred to other forms of energy by devices in the circuit. A lamp produces heat and light.

If the circuits of Figure 35.1 are connected up, it will be found from the ammeter readings that the current is about the same (0.4 A) in each lamp. However, the mains lamp with a p.d. of 230 V applied to it gives much more light and heat than the car lamp with 12 V across it. In terms of energy, the mains lamp transfers a great deal more electrical energy in a second than the car lamp.

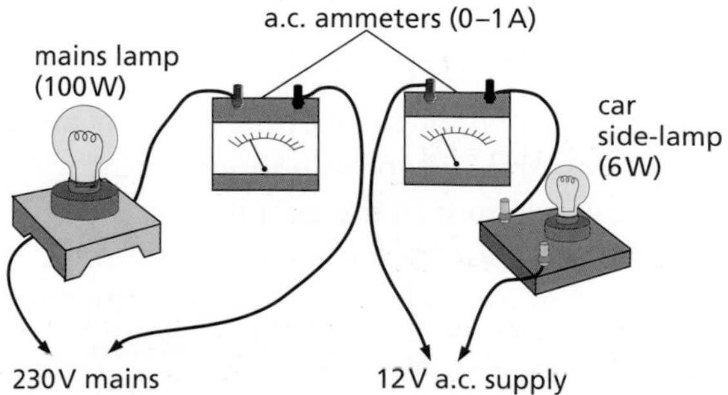

Figure 35.1 Investigating the effect of p.d. on energy transfer

Evidently the p.d. across a device affects the rate at which it transfers electrical energy. This gives us a way of defining the unit of p.d. – the volt.

Model of a circuit

It may help you to understand the definition of the volt, i.e. what a volt is, if you *imagine* that the current in a circuit is formed by 'drops' of electricity, each having a charge of 1 coulomb and carrying equal-sized 'bundles' of electrical energy. In Figure 35.2, Mr Coulomb represents one such 'drop'. As a 'drop' moves around the circuit it gives up all its energy which is changed to other forms of energy. **Note that electrical energy, not charge or current, is 'used up'.**

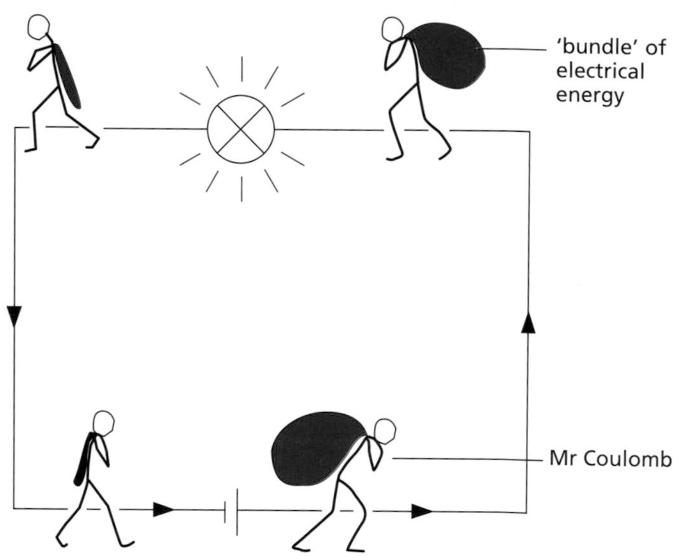

Figure 35.2 Model of a circuit

In our imaginary representation, Mr Coulomb travels round the circuit and unloads energy as he goes, most of it in the lamp. We think of him receiving a fresh 'bundle' every time he passes through the battery, which suggests he must be travelling very fast. In fact, as we found earlier (Chapter 34), the electrons drift along quite slowly. As soon as the circuit is complete, energy is delivered at once to the lamp, not by electrons directly from the battery but from electrons that were in the connecting wires. The model is helpful but not an exact representation.

The volt

The demonstrations of Figure 35.1 show that the greater the voltage at the terminals of a supply, the larger is the 'bundle' of electrical energy given to each coulomb and the greater is the rate at which light and heat are produced in a lamp.

The p.d. between two points in a circuit is 1 volt if 1 joule of electrical energy is transferred to other forms of energy when 1 coulomb passes from one point to the other.

That is, 1 volt = 1 joule per coulomb (1 V = 1 J/C). If 2 J are given up by each coulomb, the p.d. is 2 V. If 6 J are transferred when 2 C pass, the p.d. is 6 J/2 C = 3 V.

In general if E (joules) is the energy transferred (i.e. the work done) when charge Q (coulombs) passes between two points, the p.d. V (volts) between the points is given by

$$V = E/Q \qquad \text{or} \qquad E = Q \times V$$

If Q is in the form of a steady current I (amperes) flowing for time t (seconds) then $Q = I \times t$ (Chapter 34) and

$$E = I \times t \times V$$

Cells, batteries and e.m.f.

A 'battery' (Figure 35.3) consists of two or more **electric cells**. Greater voltages are obtained when cells are joined in series, i.e. + of one to − of next. In Figure 35.4a the two 1.5 V cells give a voltage of 3 V at the terminals A, B. Every coulomb in a circuit connected to this battery will have 3 J of electrical energy.

Figure 35.3 Compact batteries

The cells in Figure 35.4b are in opposition and the voltage at X, Y is zero.

If two 1.5 V cells are connected in parallel, Figure 35.4c, the voltage at terminals P, Q is still 1.5 V but the arrangement behaves like a larger cell and will last longer.

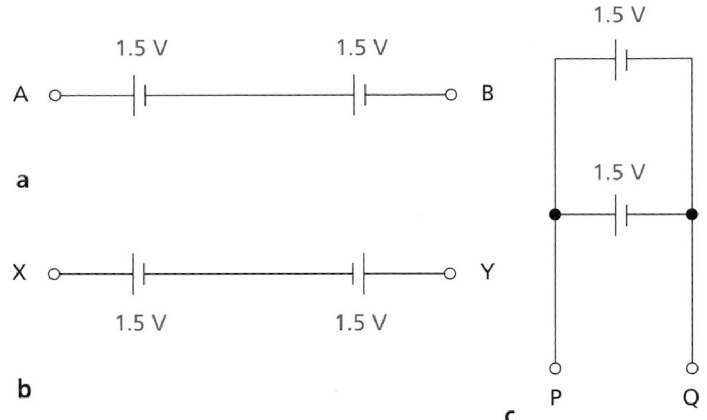

Figure 35.4

The p.d. at the terminals of a battery decreases slightly when current is drawn from it. This effect is due to the internal resistance of the battery which transfers electrical energy to heat as current flows through it. The greater the current drawn, the larger the 'lost' voltage. When no current is drawn from a battery it is said to be on 'open circuit' and its terminal p.d. is a maximum. This maximum voltage is termed the **electromotive force** (e.m.f.) of the battery; e.m.f., like p.d., is measured in volts and can be written as

$$\text{e.m.f.} = \text{'lost' volts} + \text{terminal p.d.}$$

In energy terms, the e.m.f. is defined as the number of joules of chemical energy transferred to electrical energy and heat when one coulomb of charge passes through the battery (or cell).

In Figure 35.2 the size of the energy bundle Mr Coulomb is carrying when he leaves the cell would be smaller if the internal resistance were larger.

Practical work

Measuring voltage

A **voltmeter** is an instrument for measuring voltage or p.d. It looks like an ammeter but has a scale marked in volts. Whereas an ammeter is inserted **in series** in a circuit to measure the current, a voltmeter is connected across that part of the circuit where the voltage is required, i.e. **in parallel**. (We will see later that a voltmeter should have a high resistance and an ammeter a low resistance.)

To prevent damage the + terminal (marked red) must be connected to the point nearest the + of the battery.

(a) Connect the circuit of Figure 35.5a. The voltmeter gives the voltage across the lamp. Read it.

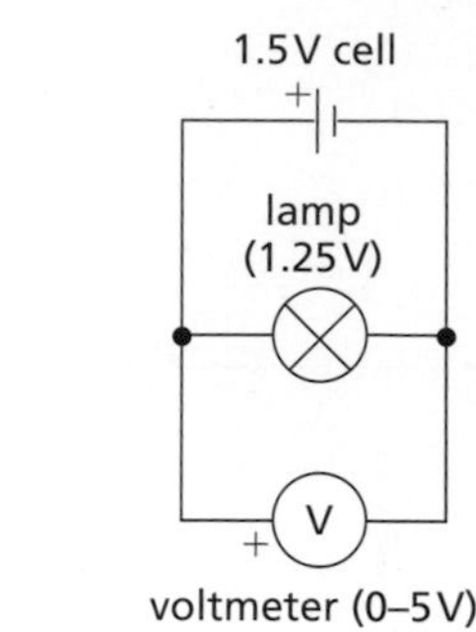

a

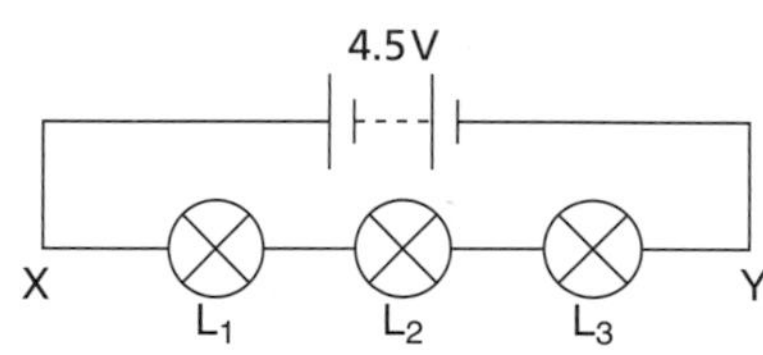

b

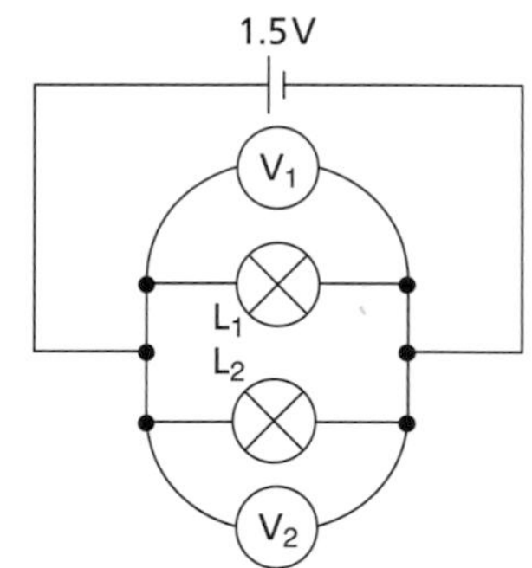

c

Figure 35.5

(b) Connect the circuit of Figure 35.5b. Measure:

(i) the voltage V between X and Y,
(ii) the voltage V_1 across lamp L_1,
(iii) the voltage V_2 across lamp L_2,
(iv) the voltage V_3 across lamp L_3.

How does the value of V compare with $V_1 + V_2 + V_3$?

(c) Connect the circuit of Figure 35.5c, so that two lamps L_1 and L_2 are in parallel across one 1.5 V cell. Measure the voltages, V_1 and V_2, across each lamp in turn. How do V_1 and V_2 compare?

□ *Voltages round a circuit*

a) Series

In the previous experiment you should have found in the circuit of Figure 35.5b that

$$V = V_1 + V_2 + V_3$$

For example, if $V_1 = 1.4$ V, $V_2 = 1.5$ V and $V_3 = 1.6$ V, then V will be $(1.4 + 1.5 + 1.6) = 4.5$ V.

The voltage at the terminals of a battery equals the sum of the voltages across the devices in the external circuit from one battery terminal to the other.

b) Parallel

In the circuit of Figure 35.5c

$$V_1 = V_2$$

The voltages across devices in parallel in a circuit are equal.

Questions

1 The p.d. across the lamp in Figure 35.6 is 12 V. How many joules of electrical energy are changed into light and heat when
a a charge of 1 C passes through it,
b a charge of 5 C passes through it,
c a current of 2 A flows through it for 10 s?

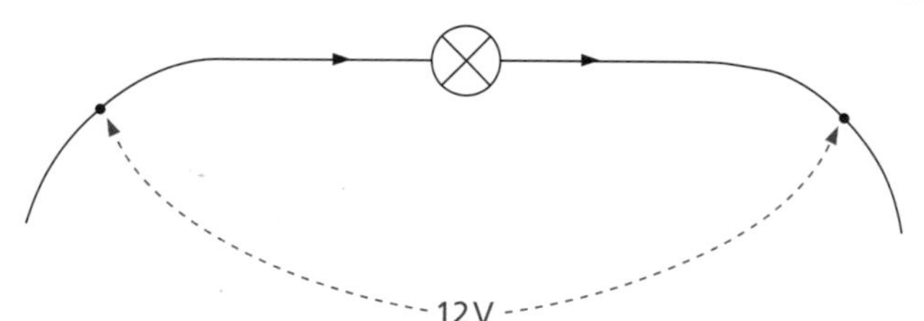

Figure 35.6

2 Three 2 V cells are connected in series and used as the supply for a circuit.
a What is the p.d. at the terminals of the supply?
b How many joules of electrical energy does 1 C gain on passing through (i) one cell, (ii) all three cells?

3 The symbol for a 1.5 V cell is ─┤├─. Which of the arrangements in Figure 35.7 would produce a battery with a p.d. of 6 V?

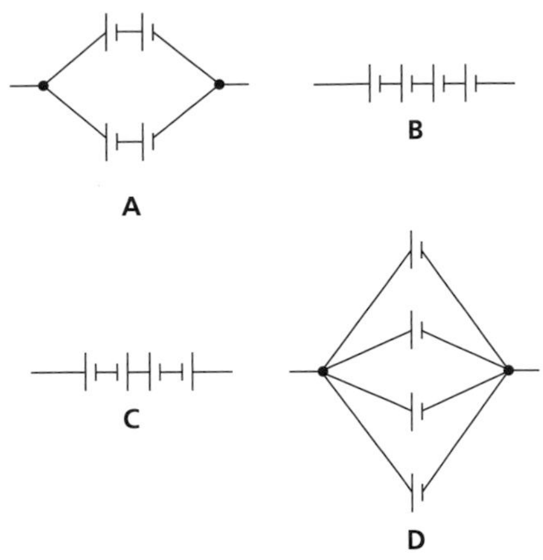

Figure 35.7

4 The lamps and the cells in all the circuits of Figure 35.8 are the same. If the lamp in **a** has its full, normal brightness, what can you say about the brightness of the lamps in **b, c, d, e** and **f**?

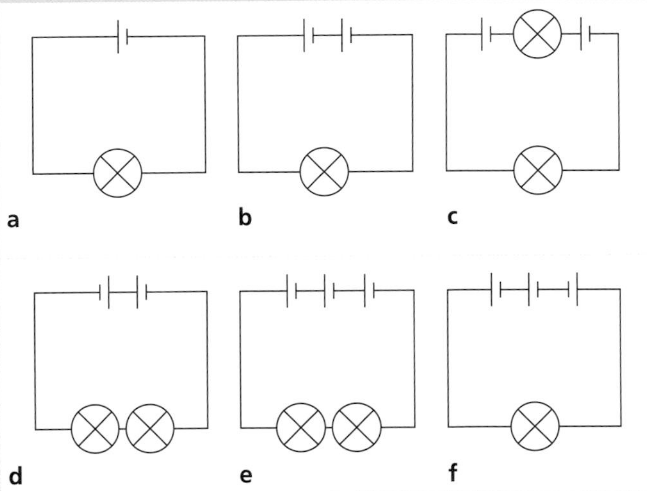

Figure 35.8

5 Three voltmeters (V), (V_1), (V_2) are connected as in Figure 35.9.

a If (V) reads 18 V and (V_1) reads 12 V, what does (V_2) read?

b If the ammeter (A) reads 0.5 A, how much electrical energy is changed to heat and light in lamp L_1 in one minute?

c Copy Figure 35.9 and mark with a + the positive terminals of the ammeter and voltmeters for correct connection.

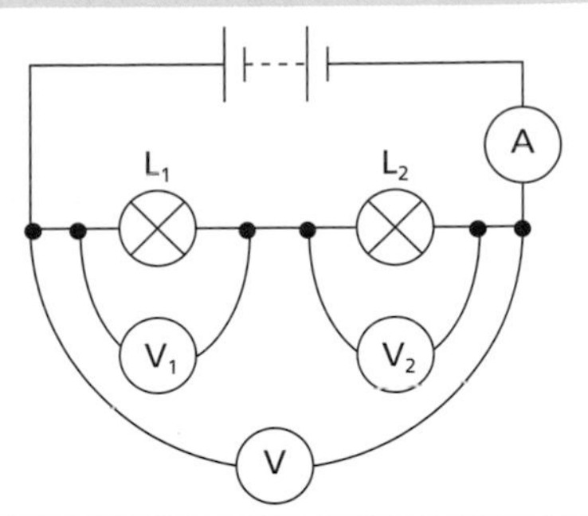

Figure 35.9

6 Three voltmeters are connected as in Figure 35.10. What are the voltmeter readings *x*, *y* and *z* in the table below (which were obtained with three different batteries)?

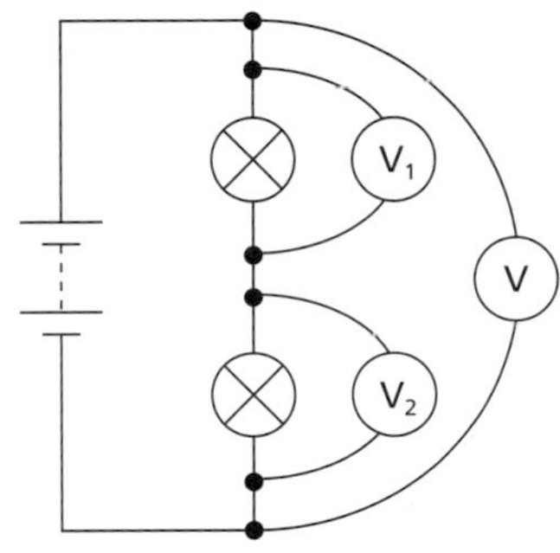

Figure 35.10

V/V	V_1/V	V_2/V
x	12	6
6	4	y
12	z	4

Checklist

After studying this chapter you should be able to

- describe simple experiments to show the transfer of electrical energy to other forms (e.g. in a lamp),
- recall the definition of the unit of p.d. and that p.d. (also called 'voltage') is measured by a voltmeter,
- demonstrate that the sum of the voltages across any number of components in series equals the voltage across all of those components,
- demonstrate that the voltages across any number of components in parallel are the same,
- work out the voltages of cells connected in series and parallel,
- explain the meaning of the term **electromotive force** (e.m.f.).

36 Resistance

The ohm	***Resistor colour code***
Resistors	***Resistivity***
I–V graphs: Ohm's law	***Potential divider***
Resistors in series	***Practical work*** Measuring resistance.
Resistors in parallel	

Electrons move more easily through some conductors than others when a p.d. is applied. The opposition of a conductor to current is called its **resistance**. A good conductor has a low resistance and a poor conductor has a high resistance. The resistance of a wire of a certain material

(i) increases as its length increases,
(ii) increases as its cross-section area decreases,
(iii) depends on the material.

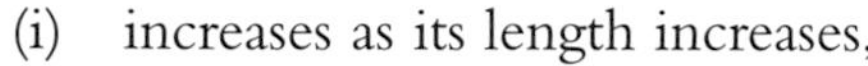

A long thin wire has more resistance than a short thick one of the same material. Silver is the best conductor, but copper, the next best, is cheaper and is used for connecting wire and domestic electric cables.

The ohm

If the current through a conductor is I when the voltage across it is V, Figure 36.1a, its resistance R is defined by

$$R = \frac{V}{I}$$

This is a reasonable way to measure resistance since the smaller I is for a given V, the greater is R. If V is in volts and I in amperes, R is in **ohms** (Ω: the Greek letter omega). For example, if $I = 2$ A when $V = 12$ V, then $R = 12\text{ V}/2\text{ A} = 6\ \Omega$.

The ohm is the resistance of a conductor in which the current is 1 ampere when a voltage of 1 volt is applied across it.

a

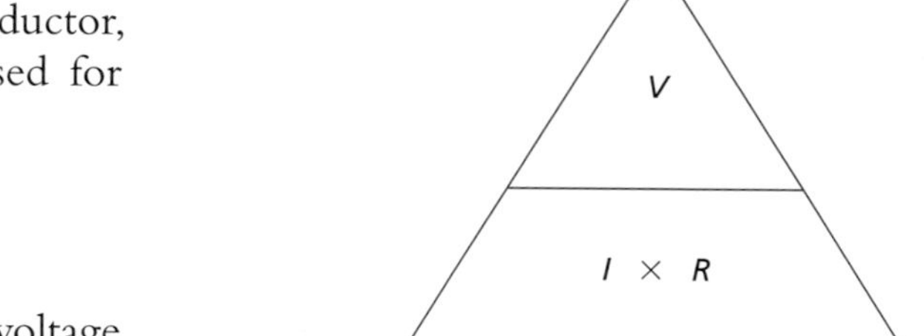

b

Figure 36.1

Alternatively, if R and I are known, V can be found from

$$V = IR$$

Also, knowing V and R, I can be calculated from

$$I = \frac{V}{R}$$

The triangle in Figure 36.1b is an aid to remembering the three equations. It is used like the 'density triangle' in Chapter 11.

Resistors

Conductors intended to have resistance are called **resistors** (symbol —▭—) and are made either from wires of special alloys or from carbon. Those in radio and television sets have values from a few ohms up to millions of ohms, Figure 36.2a.

Variable resistors are used in electronics (and are then called **potentiometers**) as volume and other controls, Figure 36.2b. Larger current versions are useful in laboratory experiments and consist of a coil of constantan wire (an alloy of 60% copper, 40% nickel) wound on a tube with a sliding contact on a metal bar above the tube, Figure 36.3.

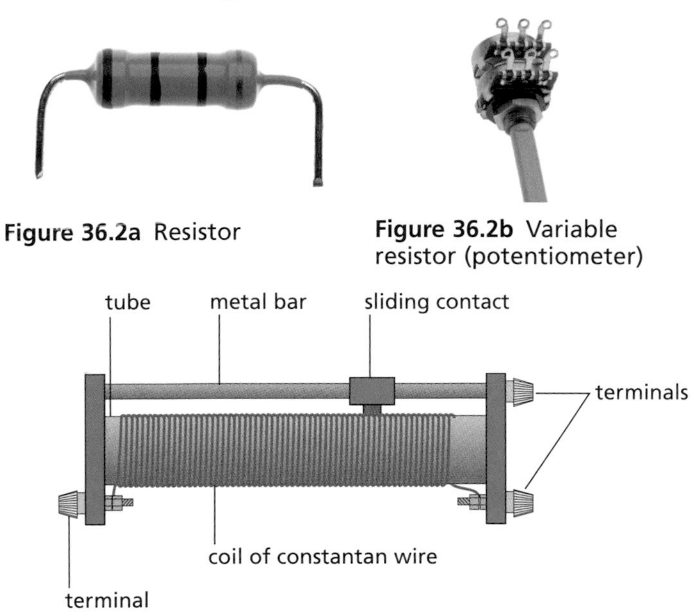

Figure 36.2a Resistor

Figure 36.2b Variable resistor (potentiometer)

Figure 36.3 Large variable resistor

There are two ways of using such a variable resistor. It may be used as a **rheostat** for changing the current in a circuit; only one end connection and the sliding contact are then required. In Figure 36.4a moving the sliding contact to the left reduces the resistance and increases the current. It can also act as a **potential divider** for changing the p.d. applied to a device, all three connections being used. In Figure 36.4b any fraction from the total p.d. of the battery to zero can be 'tapped off' by moving the sliding contact down. For a variable resistor being used in rheostat mode, the symbol —⧄— is used in circuit diagrams.

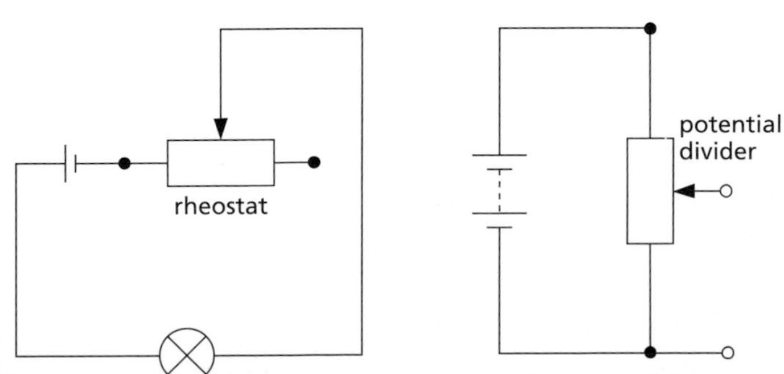

Figure 36.4 A variable resistor can be used as a rheostat or as a potential divider

Practical work

Measuring resistance

The resistance R of a conductor can be found by measuring the current I through it when a p.d. V is applied across it and then using $R = V/I$. This is called the **ammeter–voltmeter method**.

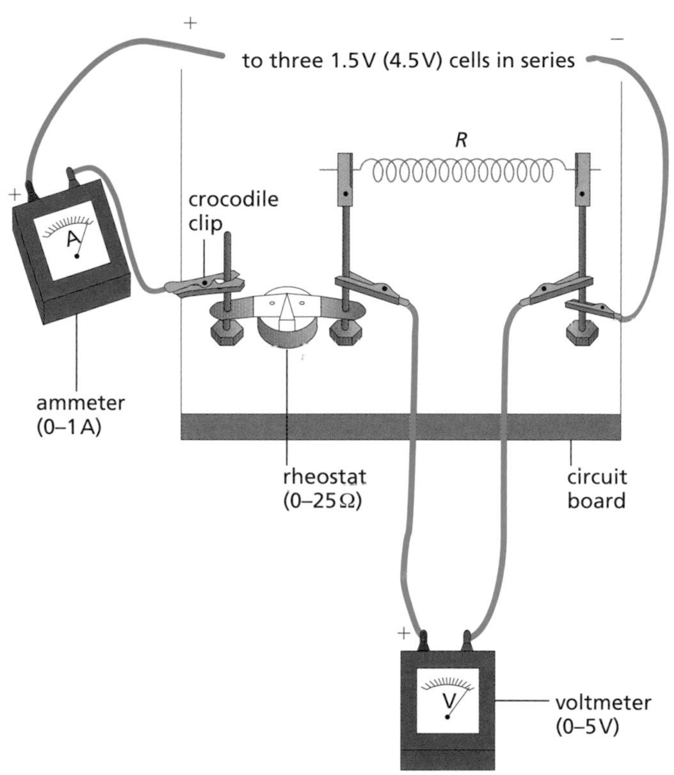

Figure 36.5

Set up the circuit of Figure 36.5 in which the unknown resistance R is 1 metre of SWG 34 constantan wire. Altering the rheostat changes both the p.d. V and the current I. Record in a table, with three columns, five values of I (e.g. 0.10, 0.15, 0.20, 0.25 and 0.30 A) and the corresponding values of V. Work out R for each pair of readings.

Repeat the experiment, but instead of the wire use (i) a torch bulb (e.g. 2.5 V, 0.3 A), (ii) a semiconductor diode (e.g. 1 N4001) connected first one way then the other way round, (iii) a thermistor (e.g TH 7). (Semiconductor diodes and thermistors are considered in Chapter 39 in more detail.)

I–V graphs: Ohm's law

The results of the previous experiment allow graphs of I against V to be plotted for different conductors.

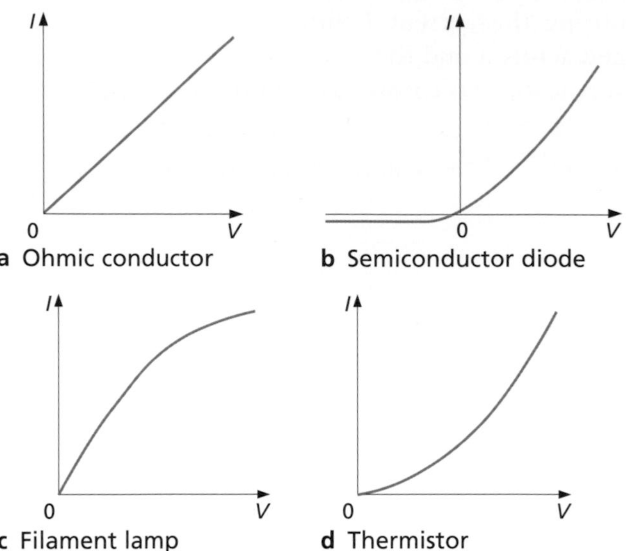

Figure 36.6 *I–V* graphs

a) Metallic conductors

Metals and some alloys give *I–V* graphs that are a straight line through the origin, Figure 36.6a, so long as their temperature is constant. I is directly proportional to V, i.e. $I \propto V$. Doubling V doubles I, etc. Such conductors obey **Ohm's law**, stated as follows.

> The current through a metallic conductor is directly proportional to the p.d. across its ends if the temperature and other conditions are constant.

They are called **ohmic** or **linear** conductors and since $I \propto V$, it follows that V/I = a constant (obtained from the slope of the *I–V* graph). The resistance of an ohmic conductor therefore does not change when the p.d. does.

b) Semiconductor diode

The typical *I–V* graph in Figure 36.6b shows that current passes when the p.d. is applied in one direction but is almost zero when it acts in the opposite direction. A diode has a small resistance when connected one way round but a very large resistance when the p.d. is reversed. It conducts in one direction only and is a **non-ohmic** conductor.

c) Filament lamp

For a filament lamp, e.g. a torch bulb, the *I–V* graph bends over as V and I increase, Figure 36.6c. That is, the resistance (V/I) increases as I increases and makes the filament hotter.

d) Variation of resistance with temperature

In general, an increase of temperature increases the resistance of metals, as for the filament lamp in Figure 36.6c, but it decreases the resistance of semiconductors. The resistance of semiconductor **thermistors** (see Chapter 39) decreases if their temperature rises, i.e. their *I–V* graph bends up, Figure 36.6d.

If a resistor and a thermistor are connected as a potential divider, Figure 36.7, the p.d. across the resistor increases as the temperature of the thermistor increases; the circuit can be used to monitor temperature, for example in a car radiator.

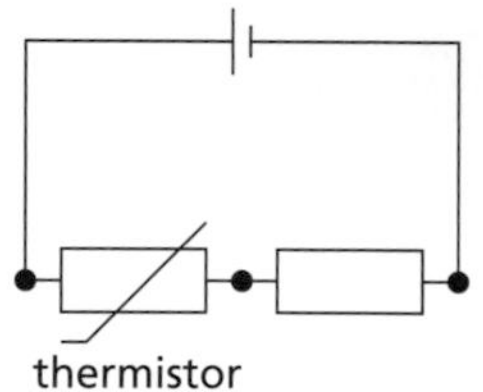

Figure 36.7 Potential divider circuit for monitoring temperature

e) Variation of resistance with light intensity

The resistance of some semiconducting materials decreases when the intensity of light falling on them increases. This property is made use of in light dependent resistors, LDRs (see Chapter 39). The *I–V* graph for an LDR is similar to that shown in Figure 36.6d for a thermistor.

Resistors in series

The resistors in Figure 36.8 are in series. The **same current *I* flows through each** and the total voltage V across all three equals the separate voltages across them, i.e.

$$V = V_1 + V_2 + V_3$$

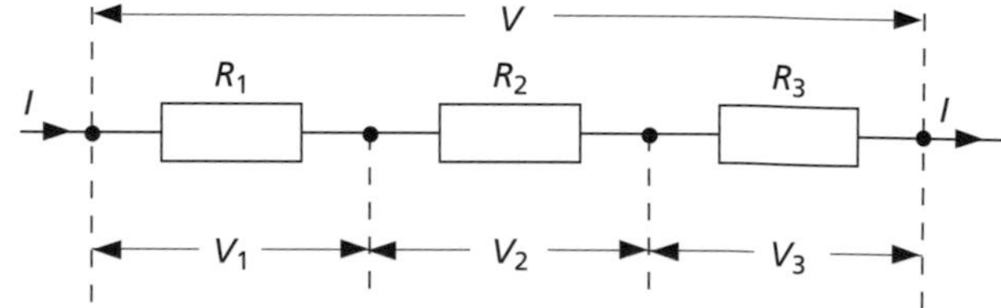

Figure 36.8 Resistors in series

But $V_1 = IR_1$, $V_2 = IR_2$ and $V_3 = IR_3$. Also, if R is the combined resistance, $V = IR$, and so

$$IR = IR_1 + IR_2 + IR_3$$

Dividing both sides by I,

$$R = R_1 + R_2 + R_3$$

Resistors in parallel

The resistors in Figure 36.9 are in parallel. The **voltage V between the ends of each is the same** and the total current I equals the sum of the currents in the separate branches, i.e.

$$I = I_1 + I_2 + I_3$$

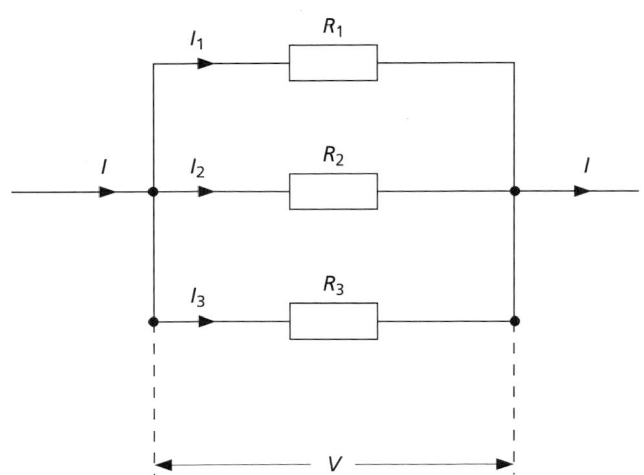

Figure 36.9 Resistors in parallel

But $I_1 = V/R_1$, $I_2 = V/R_2$ and $I_3 = V/R_3$. Also, if R is the combined resistance, $V = IR$, and so

$$\frac{V}{R} = \frac{V}{R_1} + \frac{V}{R_2} + \frac{V}{R_3}$$

Dividing both sides by V,

$$\frac{1}{R} = \frac{1}{R_1} + \frac{1}{R_2} + \frac{1}{R_3}$$

For the simpler case of *two* resistors in parallel

$$\frac{1}{R} = \frac{1}{R_1} + \frac{1}{R_2} = \frac{R_2}{R_1R_2} + \frac{R_1}{R_1R_2}$$

$$\therefore \quad \frac{1}{R} = \frac{R_2 + R_1}{R_1R_2}$$

Inverting both sides,

$$R = \frac{R_1R_2}{R_1 + R_2} = \frac{\text{product of resistances}}{\text{sum of resistances}}$$

The combined resistance of two resistors in parallel is less than the value of either resistor alone. Check this is true in the following *Worked example*. Lamps are connected in parallel rather than in series in a lighting circuit. Can you suggest why? (See p. 186 for the advantages.)

Worked example

A p.d. of 24 V from a battery is applied to the network of resistors in Figure 36.10a.

a What is the combined resistance of the 6 Ω and 12 Ω resistors in parallel?
b What is the current in the 8 Ω resistor?
c What is the voltage across the parallel network?
d What is the current in the 6 Ω resistor?

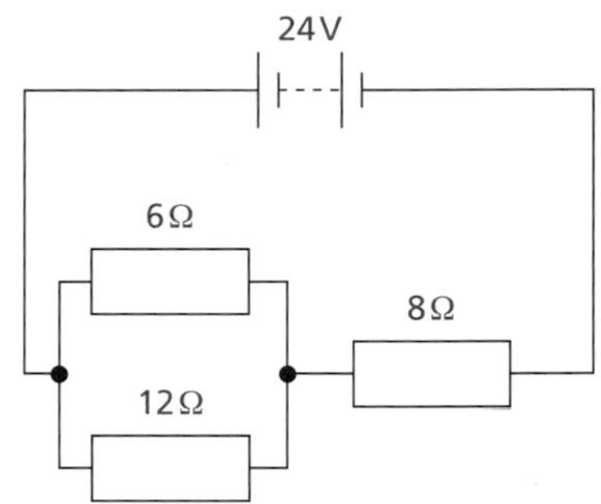

Figure 36.10a

a Let R_1 = resistance of 6 Ω and 12 Ω in parallel

$$\therefore \quad \frac{1}{R_1} = \frac{1}{6} + \frac{1}{12} = \frac{2}{12} + \frac{1}{12} = \frac{3}{12}$$

$$\therefore \quad R_1 = \frac{12}{3} = 4\ \Omega$$

b Let R = *total* resistance of circuit = 4 + 8 = 12 Ω. The equivalent circuit is shown in Figure 36.10b and if I is the current in it then since V = 24 V

$$I = \frac{V}{R} = \frac{24\ \text{V}}{12\ \Omega} = 2\ \text{A}$$

$\therefore$ current in 8 Ω resistor = 2 A

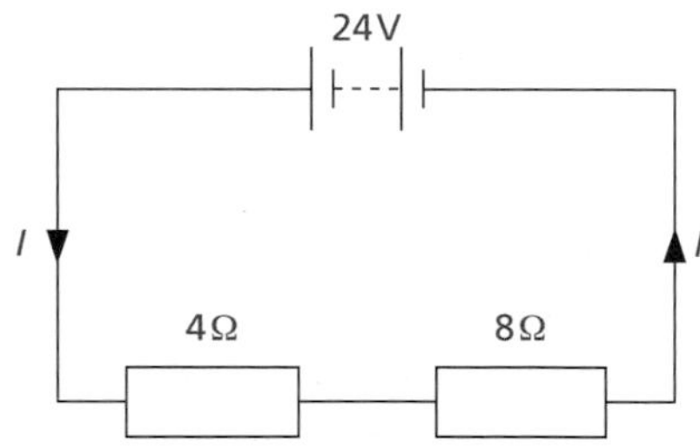

Figure 36.10b

c Let V_1 = voltage across parallel network in Figure 40.10a.

$$\therefore \quad V_1 = I \times R_1 = 2\ \text{A} \times 4\ \Omega = 8\ \text{V}$$

d Let I_1 = current in 6 Ω resistor, then since V_1 = 8 V

$$I_1 = \frac{V_1}{6\ \Omega} = \frac{8\ \text{V}}{6\ \Omega} = \frac{4}{3}\ \text{A}$$

Resistor colour code

Resistors have colour coded bands as shown in Figure 36.11. In the orientation shown the first two bands on the left give digits 2 and 7; the third band gives the number of noughts (3) and the fourth band gives the resistor's 'tolerance' (or accuracy, here ±10%). So the resistor has a value of 27 000 Ω (±10%).

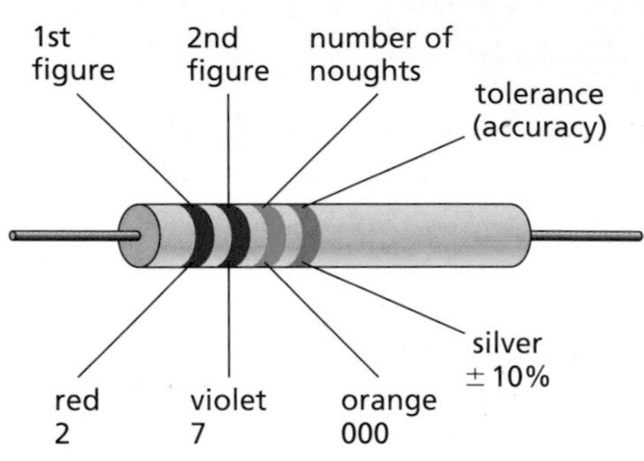

Figure	Colour
0	black
1	brown
2	red
3	orange
4	yellow
5	green
6	blue
7	violet
8	grey
9	white

Tolerance	
±5%	gold
±10%	silver
±20%	no band

Figure 36.11 Colour code for resistors

Resistivity

Experiments show that the resistance R of a wire of a given material is

(i) directly proportional to its length l, i.e. $R \propto l$,
(ii) inversely proportional to its cross-section area A, i.e. $R \propto 1/A$ (doubling A halves R).

Combining these two statements, we get

$$R \propto \frac{l}{A} \quad \text{or} \quad R = \frac{\rho l}{A}$$

where ρ is a constant, called the **resistivity** of the material. If we put $l = 1$ m and $A = 1$ m^2, then $\rho = R$.

> The resistivity of a material is numerically equal to the resistance of a 1 m length of it of cross-section area 1 m^2.

The unit of ρ is the **ohm-metre** (Ω m) as can be seen by rearranging the equation to give $\rho = AR/l$ and inserting units for A, R and l. Knowing ρ for a material, the resistance of any sample of it can be calculated. The resistivities of metals increase at higher temperatures; for most other materials they decrease.

Worked example

Calculate the resistance of a copper wire 1.0 km long and 0.50 mm diameter if the resistivity of copper is 1.7×10^{-8} Ω m.

Converting all units to metres, we get

$$\text{length } l = 1.0 \text{ km} = 1000 \text{ m} = 10^3 \text{ m}$$
$$\text{diameter } d = 0.50 \text{ mm} = 0.50 \times 10^{-3} \text{ m}$$

If r is the radius of the wire, the cross-section area $A = \pi r^2 = \pi(d/2)^2 = (\pi/4)d^2$

$$\therefore \quad A = \frac{\pi}{4}\,(0.50 \times 10^{-3})^2 \text{ m}^2 \approx 0.20 \times 10^{-6} \text{ m}^2$$

$$R = \frac{\rho l}{A} = \frac{(1.7 \times 10^{-8}\ \Omega\text{ m}) \times (10^3 \text{ m})}{0.20 \times 10^{-6} \text{ m}^2} = 85\ \Omega$$

Potential divider

In the circuit shown in Figure 36.12, two resistors R_1 and R_2 are in series with a supply of voltage V. The current in the circuit is

$$I = \frac{\text{supply voltage}}{\text{total resistance}} = \frac{V}{(R_1 + R_2)}$$

So the voltage across R_1 is

$$V_1 = I \times R_1 = \frac{V \times R_1}{(R_1 + R_2)}$$

and the voltage across R_2 is

$$V_2 = I \times R_2 = \frac{V \times R_2}{(R_1 + R_2)}$$

Also the ratio of the voltages across each resistor is

$$\frac{V_1}{V_2} = \frac{R_1}{R_2}$$

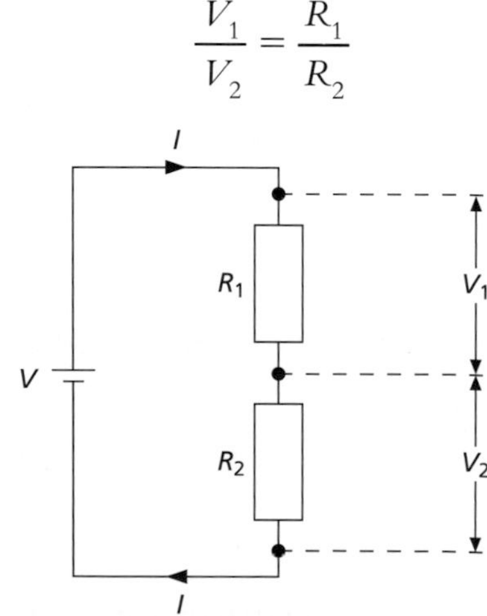

Figure 36.12 Potential divider circuit

Returning to Figure 36.7 (p. 176), can you now explain why the voltage across the resistor increases when the resistance of the thermistor decreases?

Questions

1 What is the resistance of a lamp when a voltage of 12 V across it causes a current of 4 A?

2 Calculate the p.d. across a 10 Ω resistor carrying a current of 2 A.

3 The p.d. across a 3 Ω resistor is 6 V. What is the current flowing (in ampere)?

A $\frac{1}{2}$ **B** 1 **C** 2 **D** 6 **E** 8

4 The resistors R_1, R_2, R_3 and R_4 in Figure 36.13 are all equal in value. What would you expect the voltmeters A, B and C to read, assuming that the connecting wires in the circuit have negligible resistance?

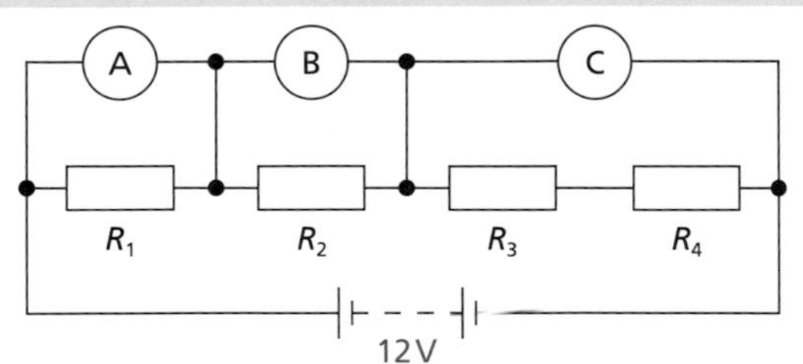

Figure 36.13

5 Calculate the effective resistance between A and B in Figure 36.14.

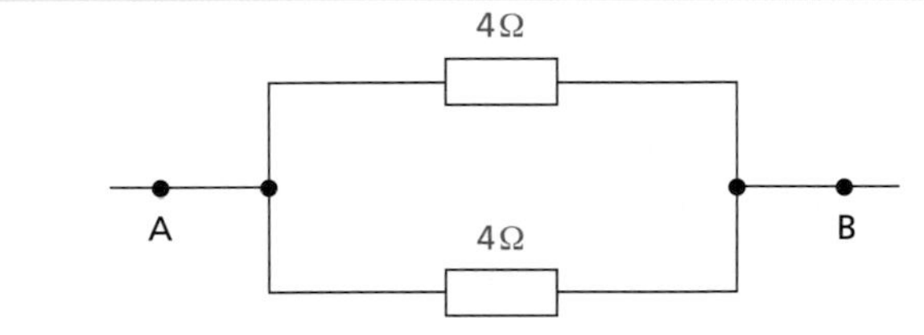

Figure 36.14

6 What is the effective resistance in Figure 36.15 between
a A and B,
b C and D?

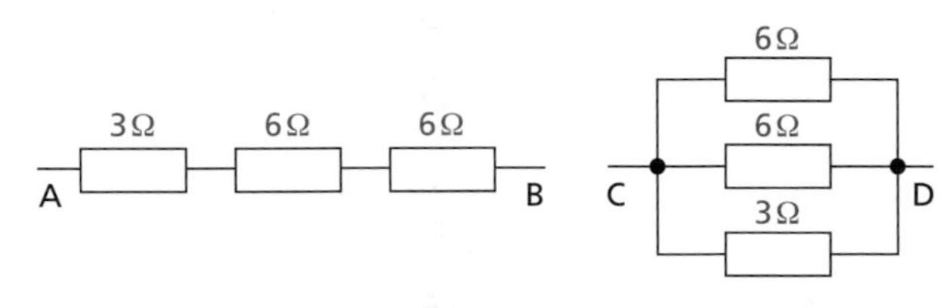

Figure 36.15

7 Figure 36.16 shows three resistors. Their combined resistance in ohms is

A $1\frac{5}{7}$ **B** 14 **C** $1\frac{1}{5}$ **D** $7\frac{1}{2}$ **E** $6\frac{2}{3}$

Figure 36.16

8 A student investigates how the current flowing through a filament lamp changes with the voltage across it. She is given a filament lamp and connecting wire. She decides to use a 15 V power supply, a variable resistor, an ammeter, a voltmeter and a switch.

a Complete the circuit diagram, started in Figure 36.17, to show how she should set up the circuit.

Figure 36.17

b The student obtains the following results.

Voltage (V)	0	3.0	5.0	7.0	9.0	11.0
Current (A)	0	1.0	1.4	1.7	1.9	2.1

(i) Plot a graph of current against voltage.
(ii) Use your graph to find the current when the voltage is 10 V.
(iii) Use your answer to (ii) to calculate the resistance of the lamp when the voltage is 10 V.

c (i) What happens to the resistance of the lamp as the current through it increases?
(ii) Explain your answer.

(AQA (NEAB) Higher, June 1999)

Checklist

After studying this chapter you should be able to

- define **resistance** and state the factors on which it depends,
- recall the unit of resistance,
- solve simple problems using $R = V/I$,
- describe experiments using the ammeter–voltmeter method to measure resistance and study the relationship between current and p.d. for (a) metallic conductors, (b) semiconductor diodes, (c) filament lamps, (d) thermistors, (e) LDRs,
- plot I–V graphs from the results of such experiments and draw appropriate conclusions from them,
- use the formulae for resistors in series,
- recall that the combined resistance of two resistors in parallel is less than that of either resistor alone,
- □ calculate the effective resistance of two resistors in parallel,
- relate the resistance of a wire to its length and diameter,
- calculate voltages in a potential divider circuit.

37 Capacitors

Capacitance
Types of capacitor
Charging and discharging a capacitor
Effect of capacitors in d.c. and a.c. circuits

A capacitor stores electric charge and is useful in many electronic circuits. In its simplest form it consists of two parallel metal plates separated by an insulator, called the **dielectric**, Figure 37.1.

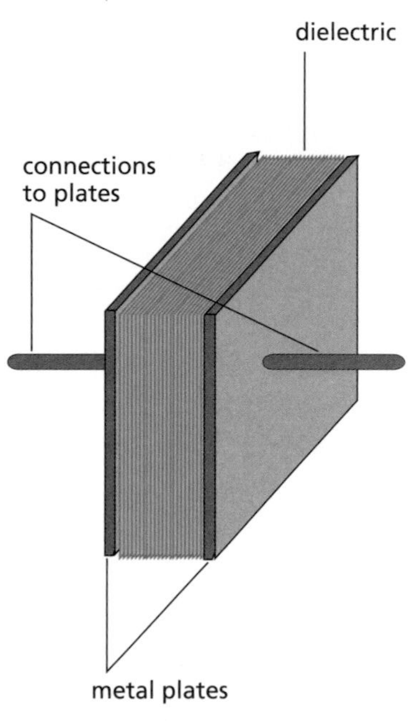

a Parallel-plate capacitor

b Symbol for a capacitor

Figure 37.1

Capacitance

The more charge a capacitor can store, the greater is its **capacitance** (C). The capacitance is large when the plates have a large area and are close together. It is measured in **farads** (F) but smaller units such as the **microfarad** (μF) are more convenient.

$$1\ \mu F = 1 \text{ millionth of a farad} = 10^{-6}\ F$$

F Types of capacitor

Practical capacitors, with values ranging from about 0.01 μF to 100 000 μF, often consist of two long strips of metal foil separated by long strips of dielectric, rolled up like a 'Swiss roll' as in Figure 37.2. The arrangement allows plates of large area to be close together in a small volume. Plastics (e.g. polyesters) are commonly used as the dielectric, with films of metal being deposited on the plastic to act as the plates, Figure 37.3.

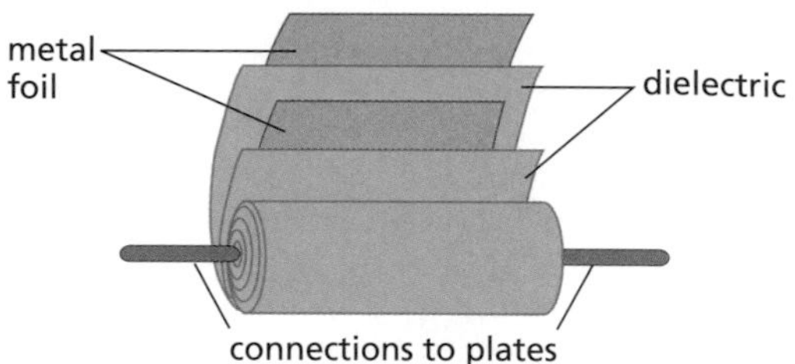

Figure 37.2 Construction of a practical capacitor

Figure 37.3 Polyester capacitor

The **electrolytic** type, Figure 37.4a, has a very thin layer of aluminium oxide as the dielectric between two strips of aluminium foil, giving large capacitances. It is polarized, i.e. it has positive and negative terminals, Figure 37.4b, and these *must* be connected to the + and − respectively of the voltage supply.

a Electrolytic capacitor

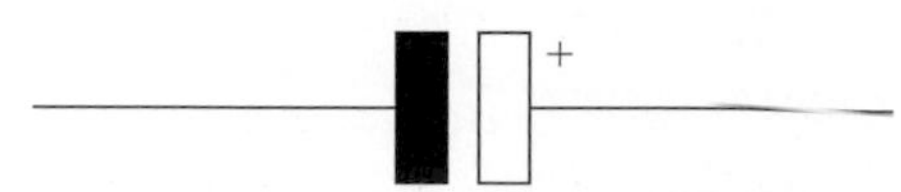

b Symbol for an electrolytic capacitor showing polarity

Figure 37.4

Charging and discharging a capacitor

a) Charging

A capacitor can be charged by connecting a battery across it. In Figure 37.5a, the + of the battery attracts electrons (since they have a negative charge) from plate X and the − of the battery repels electrons to plate Y. A positive charge builds up on plate X (since it loses electrons) and an equal negative charge builds up on Y (since it gains electrons).

During the charging, there is a *brief* flow of electrons round the circuit from X to Y (but not through the dielectric). A momentary current would be detected by a sensitive ammeter. The voltage builds up between X and Y and opposes the battery voltage. Charging stops when these two voltages are equal; the electron flow, i.e. the charging current, is then zero. The variation of *current* with time (for both charging or discharging a capacitor) has a similar shape to the curve shown in Figure 37.7b.

During the charging process, electrical energy is transferred from the battery to the capacitor, which then stores the energy.

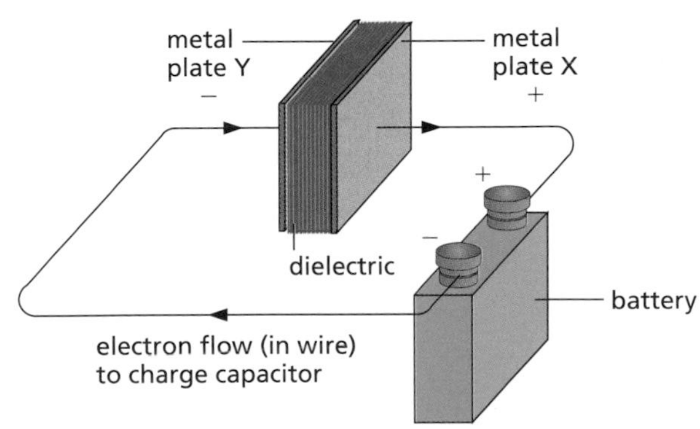

a

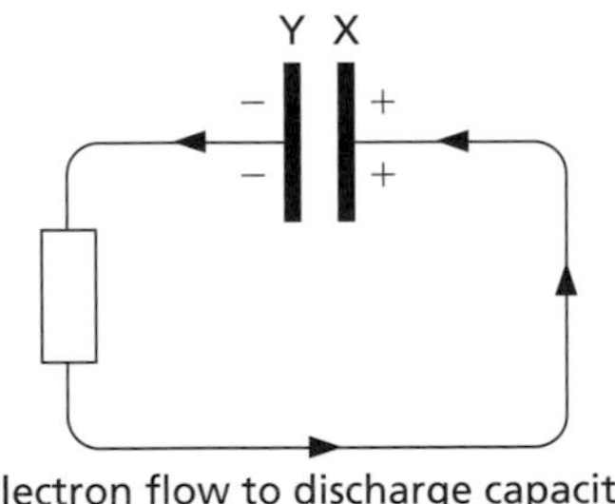

b

Figure 37.5 Charging and discharging a capacitor

b) Discharging

When a conductor is connected across a charged capacitor, there is a brief flow of electrons from the negatively charged plate to the positively charged one, i.e. from Y to X in Figure 37.5b. The charge stored by the capacitor falls to zero, as does the voltage across it. The capacitor has transferred its stored energy to the conductor. The 'delay' time taken for a capacitor to fully charge or discharge through a resistor is made use of in many electronic circuits.

c) Demonstration

The circuit in Figure 37.6 uses a two-way switch S which charges C in position 1 and discharges it in position 2. The larger the values of R and C the longer it takes for the capacitor to charge or discharge; with the values shown the capacitor will take 2 to 3 minutes to fully charge or discharge. The direction of the deflection of the centre-zero milliammeter reverses for each process. The corresponding changes of capacitor charge (measured by the voltage across it) with time are shown by the graphs in Figures 37.7a and b. These can be plotted directly if the voltmeter is replaced by a data-logger and computer.

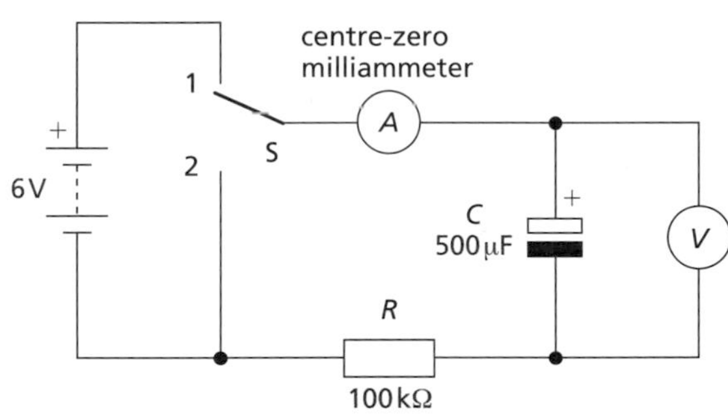

Figure 37.6 Demonstration circuit

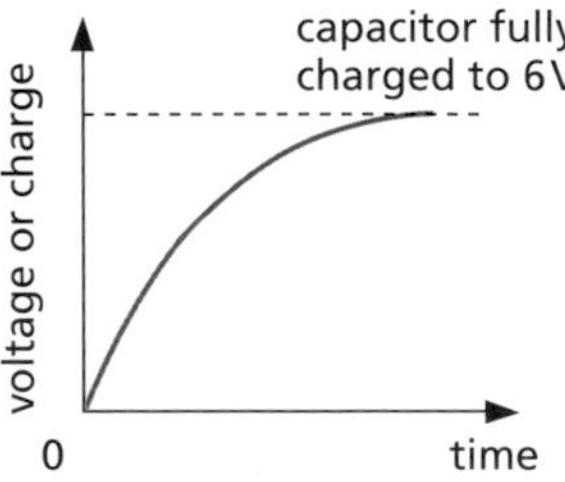

a Charging

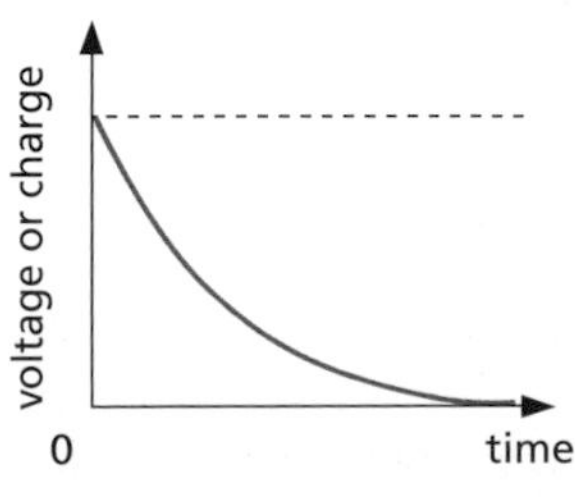

b Discharging

Figure 37.7 Graphs

Effect of capacitors in d.c. and a.c. circuits

a) d.c.

In Figure 37.8a the supply is d.c. but the lamp does not light, i.e. a capacitor blocks d.c.

b) a.c.

In Figure 37.8b the supply is a.c. and the lamp lights, suggesting that a capacitor passes a.c. In fact, no current actually passes *through* the capacitor since its plates are separated by an insulator. But as the a.c. reverses direction, the capacitor charges and discharges causing electrons to flow to and fro rapidly in the wires joining the plates. Thus effectively a.c. flows round the circuit, lighting the lamp.

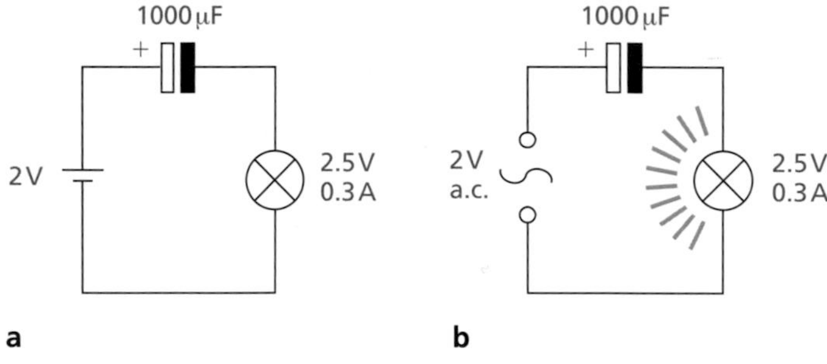

Figure 37.8 A capacitor blocks d.c. and passes a.c.

Questions

1 a Describe the basic construction of a capacitor.
b What does a capacitor do?
c State two ways of increasing the capacitance of a capacitor.
d Name a unit of capacitance.

2 a When a capacitor is being charged, is the value of the charging current maximum or zero
(i) at the start and
(ii) at the end of charging?
b Repeat **a** for a capacitor discharging.

3 How does a capacitor behave in a circuit with
a a d.c. supply,
b an a.c. supply?

Checklist

After studying this chapter you should be able to

- state what a capacitor does,
- state the unit of capacitance,
- describe in terms of electron motion how a capacitor can be charged and discharged, and sketch graphs of the capacitor voltage with time for charging and discharging through a resistor,
- recall that a capacitor blocks d.c. but passes a.c. and explain why.

38 Electric power

Power in electric circuits
Electric lighting
Electric heating
Joulemeter
House circuits
Paying for electricity
Dangers of electricity
Practical work
Measuring electric power.

Power in electric circuits

In many circuits it is important to know the rate at which electrical energy is transferred into other forms of energy. Earlier (Chapter 16) we said that **energy transfers were measured by the work done** and power was defined by the equation

$$\text{power} = \frac{\text{work done}}{\text{time taken}} = \frac{\text{energy transfer}}{\text{time taken}}$$

In symbols $$P = \frac{E}{t} \qquad (1)$$

where if E is in joules (J) and t in seconds (s) then P is in J/s or watts (W).

From the definition of p.d. (Chapter 35) we saw that if E is the electrical energy transferred when a steady current I (in amperes) passes for time t (in seconds) through a device (e.g. a lamp) with a p.d. V (in volts) across it, Figure 38.1, then

$$E = ItV \qquad (2)$$

Substituting for E in (1) we get

$$P = \frac{E}{t} = \frac{ItV}{t}$$

or

$$P = IV$$

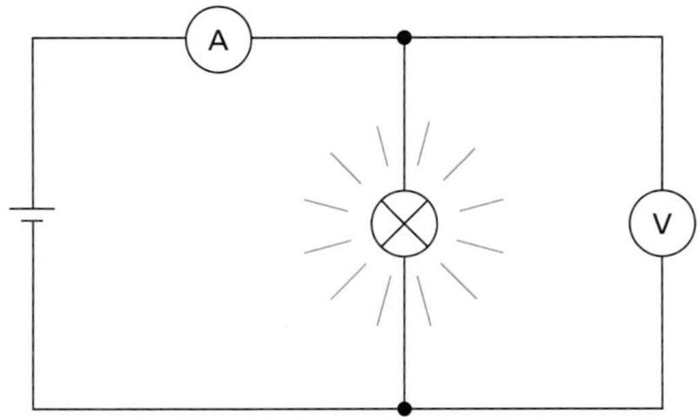

Figure 38.1

Therefore to calculate the power P of an electrical appliance we multiply the current I through it by the p.d. V across it. For example if a lamp on a 240 V supply has a current of 0.25 A through it, its power is 240 V × 0.25 A = 60 W. The lamp is transferring 60 J of electrical energy into heat and light each second. Larger units of power are the **kilowatt** (kW) and the **megawatt** (MW) where

$$1\text{ kW} = 1000\text{ W} \quad \text{and} \quad 1\text{ MW} = 1\,000\,000\text{ W}$$

In units

$$\text{watts} = \text{amperes} \times \text{volts} \qquad (3)$$

It follows from (3) that since

$$\text{volts} = \frac{\text{watts}}{\text{amperes}} \qquad (4)$$

the volt can be defined as a **watt per ampere** and p.d. calculated from (4).

If all the energy is transferred to heat in a resistor of resistance R, then $V = IR$ and the rate of production of heat is given by

$$P = V \times I = IR \times I = I^2R$$

That is, if the current is doubled, four times as much heat is produced per second. Also, $P = V^2/R$.

Practical work

Measuring electric power

a) Lamp

Connect the circuit of Figure 38.2. Note the ammeter and voltmeter readings and work out the electric power supplied to the bulb in watts.

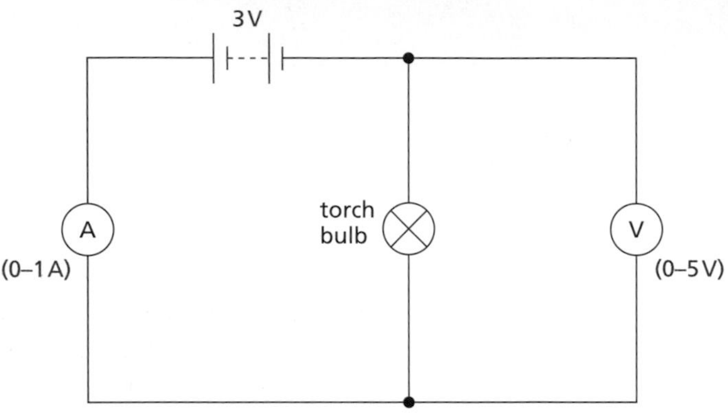

Figure 38.2

b) Motor

Replace the bulb in Figure 38.2 by a small electric motor. Attach a known mass m (in kg) to the axle of the motor with a length of thin string and find the time t (in s) required to raise the mass through a known height h (in m) at a steady speed. Then the power output P_o (in W) of the motor is given by

$$P_o = \frac{\text{work done in raising mass}}{\text{time taken}} = \frac{mgh}{t}$$

If the ammeter and voltmeter readings I and V are noted while the mass is being raised, the power input P_i (in W) can be found from

$$P_i = IV$$

The efficiency of the motor is given by

$$\text{efficiency} = \frac{P_o}{P_i} \times 100\%$$

Also investigate the effect of a greater mass on (i) the speed, (ii) the power output and (iii) the efficiency of the motor at its rated p.d.

Electric lighting

a) Filament lamps

The filament is a small coiled coil of tungsten wire, Figure 38.3, which becomes white hot when current flows through it. The higher the temperature of the filament the greater is the proportion of electrical energy transferred to light and for this reason it is made of tungsten, a metal with a high melting point (3400 °C).

Most lamps are gas-filled and contain nitrogen and argon, not air. This reduces evaporation of the tungsten which would otherwise condense on the bulb and blacken it. The coil is coiled compactly so that it is cooled less by convection currents in the gas.

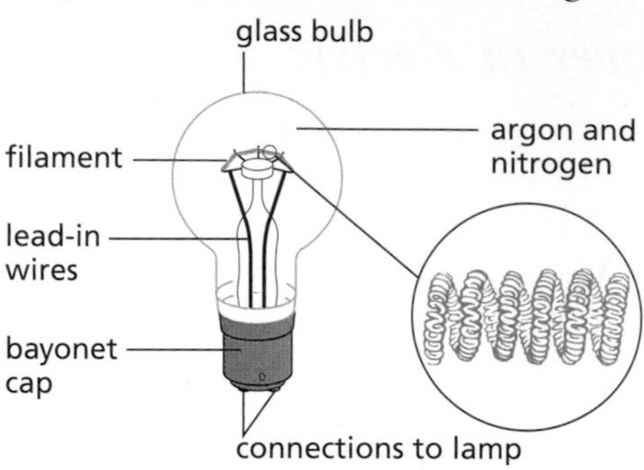

Figure 38.3 A filament lamp

b) Fluorescent strips

A filament lamp transfers only 10% of the electrical energy supplied to light; the other 90% becomes heat. Fluorescent strip lamps, Figure 38.4a, are five times as efficient and may last 3000 hours compared with the 1000-hour life of filament lamps. They cost more to install but running costs are less.

When a fluorescent strip lamp is switched on, the mercury vapour emits invisible ultraviolet radiation which makes the powder on the inside of the tube fluoresce (glow), i.e. visible light is emitted. Different powders give different colours.

c) Compact fluorescent lamps

These energy-saving fluorescent lamps, Figure 38.4b, are available to fit straight into normal light sockets, either bayonet or screw-in. They last up to eight times longer (typically 8000 hours) and use about five times less energy than filament lamps for the *same* light output. For example, a 20 W compact fluorescent is equivalent to a 100 W filament lamp.

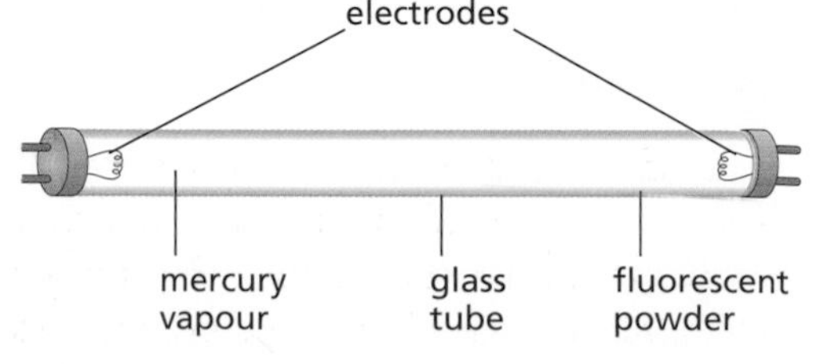

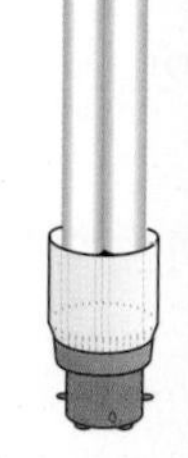

Figure 38.4 Fluorescent lamps

Electric heating

a) Heating elements

In domestic appliances such as electric fires, cookers, kettles and irons the 'elements', Figure 38.5, are made from Nichrome wire. This is an alloy of nickel and chromium which does not oxidize (and so become brittle) when the current makes it red hot.

The elements in **radiant** electric fires are at red heat (about 900 °C) and the radiation they emit is directed into the room by polished reflectors. In **convector** types the element is below red heat (about 450 °C) and is designed to warm air which is drawn through the heater by natural or forced convection. In **storage** heaters the elements heat fire-clay bricks during the night using 'off-peak' electricity. On the following day these cool down, giving off the stored heat to warm the room.

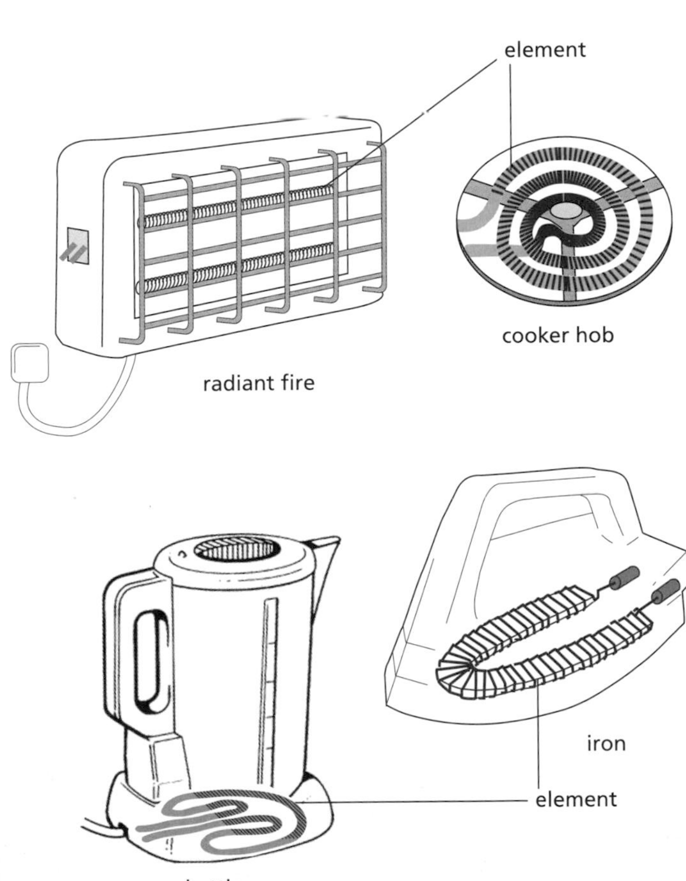

Figure 38.5 Heating elements

b) Three-heat switch

This is sometimes used to control heating appliances. It has three settings and uses two identical elements. On 'high', the elements are in parallel across the supply voltage, Figure 38.6a; on 'medium', current only passes through one, Figure 38.6b; on 'low', they are in series, Figure 38.6c.

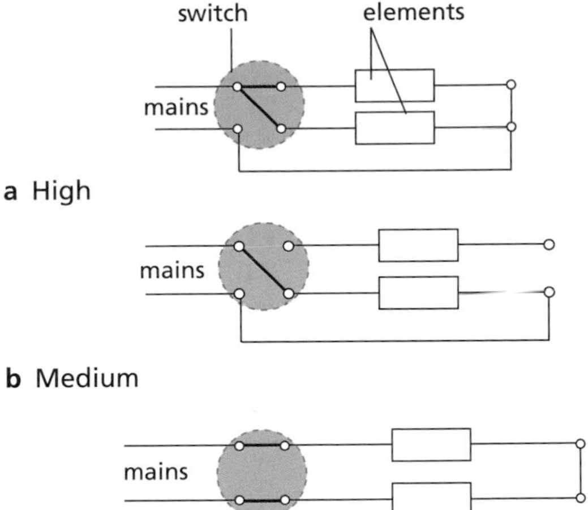

Figure 38.6 Three-heat switch

c) Fuses

A fuse is a short length of wire of material with a low melting point, often tinned copper, which melts and breaks the circuit when the current through it exceeds a certain value. Two reasons for excessive currents are 'short circuits' due to worn insulation on connecting wires, and overloaded circuits. Without a fuse the wiring would become hot in these cases and could cause a fire. **A fuse should ensure that the current-carrying capacity of the wiring is not exceeded.** In general the thicker a cable is, the more current it can carry, but each size has a limit.

Two types of fuse are shown in Figure 38.7. **Always switch off before replacing a fuse**, and always replace with one of the same value as recommended by the manufacturer of the appliance.

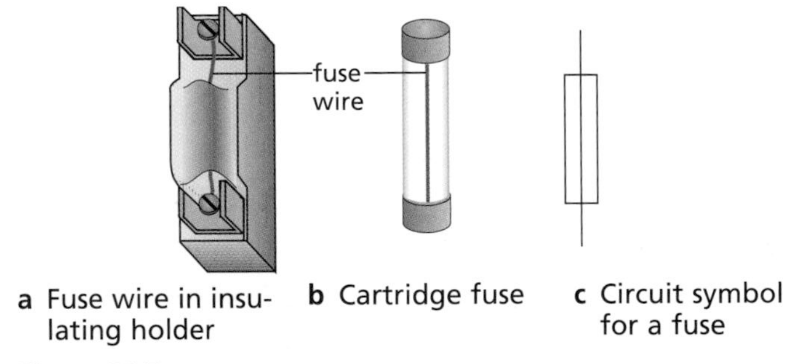

Figure 38.7

Joulemeter

Instead of using an ammeter and a voltmeter to measure the electrical energy transferred by an appliance, a **joulemeter** can be used to obtain it directly in joules. The circuit connections are shown in Figure 38.8. A household electricity meter, Figure 38.12, is a joulemeter.

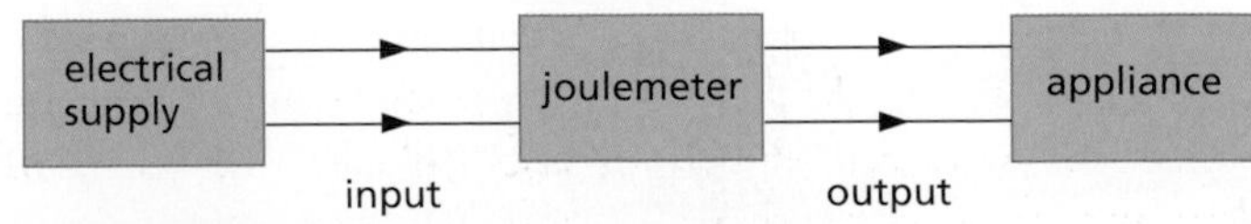

Figure 38.8 Connections to a joulemeter

House circuits

Electricity usually comes to our homes by an underground cable containing two wires, the **live** (L) and the **neutral** (N). The neutral is earthed at the local sub-station and so there is no p.d. between it and earth. The supply is a.c. (Chapter 34) and the live wire is alternately positive and negative. Study the typical house circuits shown in Figure 38.9.

a) Circuits in parallel

Every circuit is connected in parallel with the supply, i.e. across the live and neutral, and receives the full mains p.d. (in the UK) of 230 V. The advantages of having appliances connected in parallel, rather than in series, can be seen by studying the lighting circuit in Figure 38.9.

(i) The p.d. across each lamp is fixed (at the mains p.d.), so the lamp shines with the same brightness irrespective of how many other lamps are switched on.

(ii) Each lamp can be turned on and off independently; if one lamp fails, the others can still be operated.

b) Switches and fuses

These are always in the live wire. If they were in the neutral, light switches and power sockets would be 'live' when switches were 'off' or fuses 'blown'. A fatal shock could then be obtained by, for example, touching the element of an electric fire when it was switched off.

c) Staircase circuit

The light is controlled from two places by the two two-way switches.

d) Ring main circuit

The live and neutral wires each run in two complete rings round the house, and the power sockets, each rated at 13 A, are tapped off from them. Thinner wires can be used since the current to each socket flows by two paths, i.e. in the whole ring. The ring has a 30 A fuse and if it has, say, ten sockets all can be used so long as the total current does not exceed 30 A, otherwise the wires overheat. A house may have several ring circuits, each serving a different area.

e) Fused plug

Only one type of plug is used in a UK ring main circuit. It is wired as in Figure 38.10a. Note the colours of the wire coverings: L – brown, N – blue, E – green and yellow. It has its own cartridge fuse, 3 A (red) for appliances with powers up to 720 W, or 13 A (brown) for

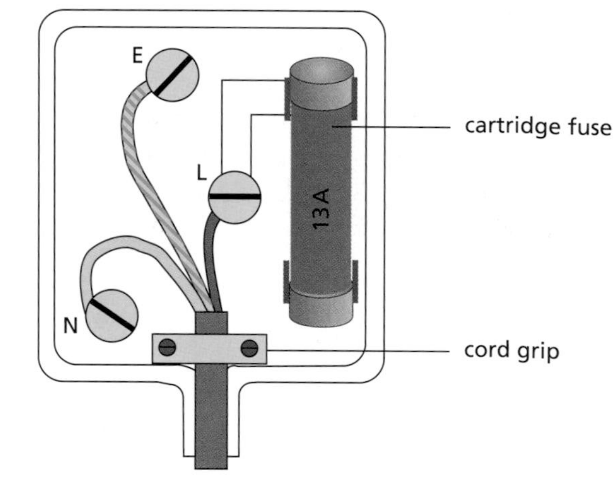

a Wiring of a plug

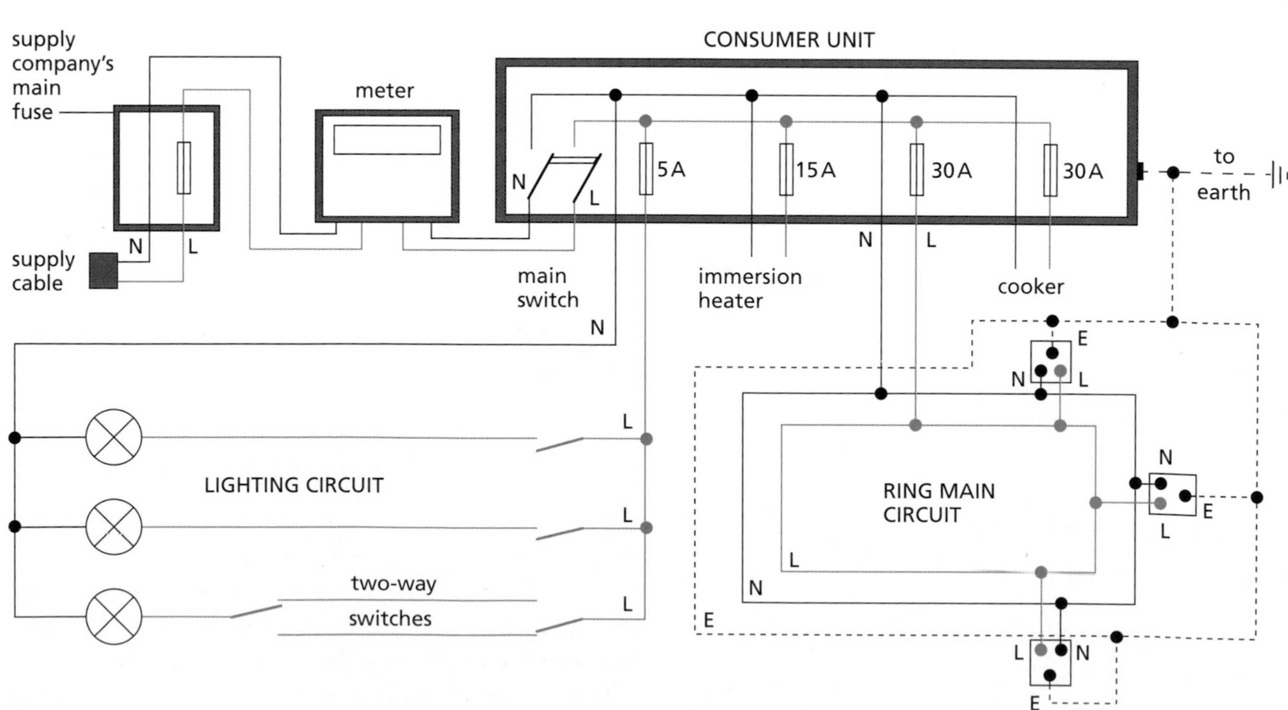

Figure 38.9 Electric circuits in a house

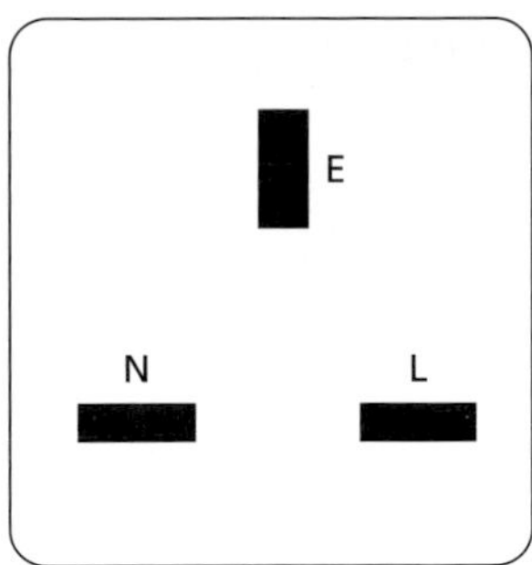

b Socket

Figure 38.10

those between 720 W and 3 kW. Typical power ratings for various appliances are shown in Table 38.1, p. 188.

In some countries the fuse is placed in the appliance rather than in the plug.

f) Earthing and safety

A ring main has a third wire which goes to the top sockets on all power points, Figure 38.9, and is earthed by being connected either to a **metal** water pipe entering the house or to an earth connection on the supply cable. This third wire is a safety precaution to prevent electric shock should an appliance develop a fault.

The earth pin on a three-pin plug is connected to the metal case of the appliance which is thus joined to earth by a path of almost zero resistance. If then, for example, the element of an electric fire breaks or sags and touches the case, a large current flows to earth and 'blows' the fuse. Otherwise the case would become 'live' and anyone touching it would receive a shock which might be fatal, especially if they were 'earthed' by, say, standing in a damp environment, e.g. on a wet concrete floor.

g) Circuit breakers

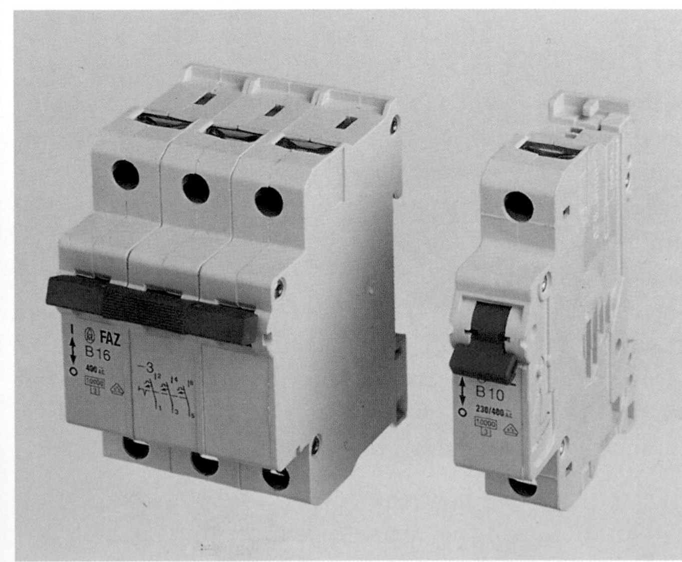

Figure 38.11 Circuit breakers

Circuit breakers, Figure 38.11, are now used instead of fuses in consumer units. They contain an electromagnet (Chapter 42) which, when the current exceeds the rated value of the circuit breaker, becomes strong enough to separate a pair of contacts and breaks the circuit. They operate much faster than fuses and have the advantage that they can be reset by pressing a button.

The **residual current circuit breaker** (RCCB), also called a **residual current device** (RCD), is an adapted circuit breaker which is used when the resistance of the earth path between the consumer and the sub-station is not small enough for a fault-current to blow the fuse or trip the circuit breaker. It works by detecting any difference between the currents in the live and neutral wires; when these become unequal due to an earth fault (i.e. some of the current returns to the sub-station via the case of the appliance and earth) it breaks the circuit before there is any danger. They have high sensitivity and a quick response.

An RCD should be plugged into a socket supplying power to a portable appliance such as an electric lawn-mower or hedge trimmer. In these cases the risk of electrocution is greater because the user is generally making a good earth connection through the feet.

h) Double insulation

Appliances such as vacuum cleaners, hairdryers and food mixers are usually double-insulated. Connection to the supply is by a two-core insulated cable, with no earth wire, and the appliance is enclosed in an insulating plastic case. Any metal attachments that the user might touch are fitted into this case so that they do not make a direct connection with the internal electrical parts, e.g. a motor. There is then no risk of a shock should a fault develop.

Paying for electricity

Electricity supply companies charge for the **electrical energy** they supply. A joule is a very small amount of energy and a larger unit, the **kilowatt-hour** (kWh) is used.

A kilowatt-hour is the electrical energy used by a 1 kW appliance in 1 hour.

$$1\text{ kWh} = 1000\text{ J/s} \times 3600\text{ s}$$
$$= 3\,600\,000\text{ J} = 3.6\text{ MJ}$$

A 3 kW electric fire working for 2 hours uses 6 kWh of electrical energy – usually called 6 'units'. Electricity meters, which are joulemeters, are marked in kWh: the latest have digital readouts like the one in Figure 38.12. At present a 'unit' costs about 8p in the UK.

Typical powers of some appliances are given in Table 38.1.

Table 38.1 Power of some appliances

video recorder	20 W	iron	1 kW
hi-fi	40 W	fire	1, 2, 3 kW
light bulbs	60, 100 W	kettle	2 kW
television	100 W	immersion heater	3 kW
fridge	150 W	cooker	8 kW

Note that the current required by an 8 kW cooker is given by

$$I = \frac{P}{V} = \frac{8000\ \text{W}}{230\ \text{V}} = 35\ \text{A}$$

This is too large a current to draw from the ring main and so a separate circuit must be used.

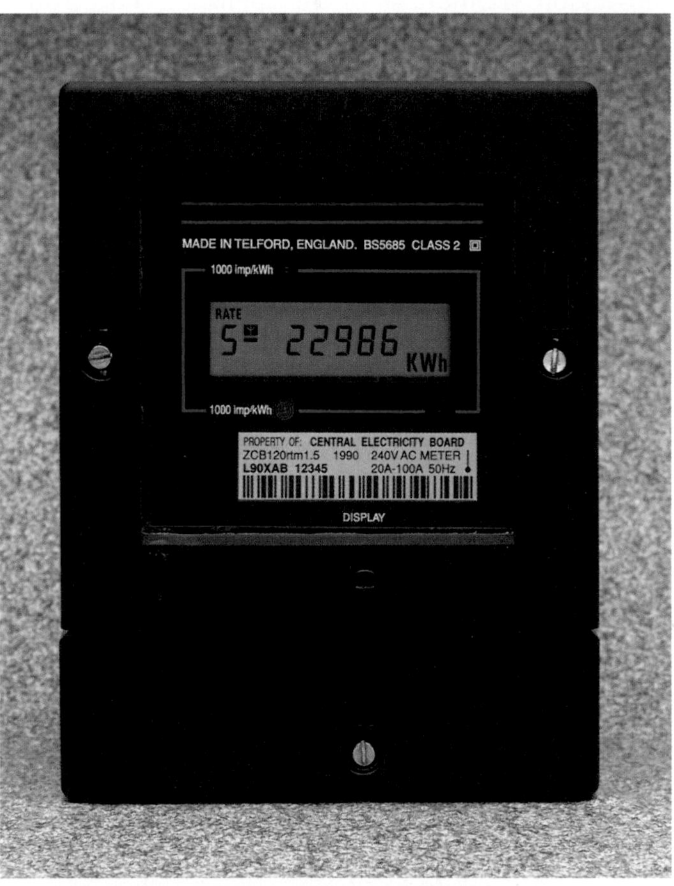

Figure 38.12 Electricity meter with digital display

Dangers of electricity

a) Electric shock

Electric shock occurs if current flows from an electric circuit through a person's body to earth. This can happen if there is **damaged insulation** or **faulty wiring**. The typical resistance of dry skin is about 10 000 Ω/m so if a person touches a wire carrying electricity at 240 V, an estimate of the current flowing through them to earth would be $I = V/R = 240/10\ 000 = 0.024$ A = 24 mA. For **wet skin**, the resistance is lowered to about 1000 Ω/m (since water is a good conductor of electricity) so the current would increase to around 240 mA.

It is **the size of the current** (not the voltage) and the **length of time** for which it acts which determine the strength of an electric shock. The path the current takes influences the effect of the shock; some parts of the body are more vulnerable than others. A current of 100 mA through the heart is likely to be fatal.

Damp conditions increase the severity of an electric shock because water lowers the resistance of the path to earth; wearing shoes with insulating rubber soles or standing on a dry insulating floor increases the resistance between a person and earth and will reduce the severity of an electric shock.

To avoid the risk of getting an electric shock:

- Switch off the electrical supply to an appliance before starting repairs.
- Use plugs that have an earth pin and a cord grip; a rubber or plastic case is preferred.
- Do not allow appliances or cables to come into contact with water, e.g. holding a hairdryer with wet hands in a bathroom can be dangerous. Keep electrical appliances well away from baths and swimming pools!
- Do not have long cables trailing across a room, under a carpet that is walked over regularly or in other situations where the insulation can become damaged. Take particular care when using electrical cutting devices (such as hedge cutters) not to cut the supply cable.

In case of an electric shock, take the following action:

1. **Switch off the supply** if the shocked person is still touching the equipment.
2. **Send for qualified medical asistance**.
3. **If breathing has stopped apply the 'kiss of life'**.
4. **If the heart has stopped try to restart it** by striking the chest three times over the heart.

b) Fire risks

If flammable material is placed too close to a hot appliance such as an electric heater, it may catch fire. Similarly if the electrical wiring in the walls of a house becomes overheated, a fire may start. Wires become hot when they carry electrical currents – the larger the current carried the hotter a particular wire will become, since the rate of production of heat equals I^2R (see p. 183).

To reduce the risk of fire through **overheated cables**, the maximum current in a circuit should be limited by taking these precautions:

- Use plugs that have the correct fuse.
- Do not attach too many appliances to a circuit.
- Don't overload circuits by using too many adapters.

- Appliances such as heaters use large amounts of power (and hence current), so do not connect them to a lighting circuit designed for low current use. (Thick wires have a lower resistance than thin wires so are used in circuits expected to carry high currents.)

Damaged insulation or faulty wiring which leads to a large current flowing to earth through flammable material can also start a fire.

The factors leading to fire or electric shock can be summarized as follows:

damaged insulation	→ electric shock and fire risk
overheated cables	→ fire risk
damp conditions	→ increased severity of electric shocks

Questions

1 How much electrical energy in **joules** does a 100 watt lamp transfer in
 a 1 second,
 b 5 seconds,
 c 1 minute?

2 a What is the power of a lamp rated at 12 V 2 A?
 b How many joules of electrical energy are transferred per second by a 6 V 0.5 A lamp?

3 The largest number of 100 W bulbs connected in parallel which can safely be run from a 230 V supply with a 5 A fuse is

 A 2 **B** 5 **C** 11 **D** 12 **E** 20

4 What is the maximum power in kilowatts of the appliance(s) that can be connected safely to a 13 A 230 V mains socket?

5 The circuits of Figure 38.13a and b show 'short circuits' between the live (L) and neutral (N) wires. In both, the fuse has blown but whereas circuit a is now safe, b is still dangerous even though the lamp is out which suggests the circuit is safe. Explain.

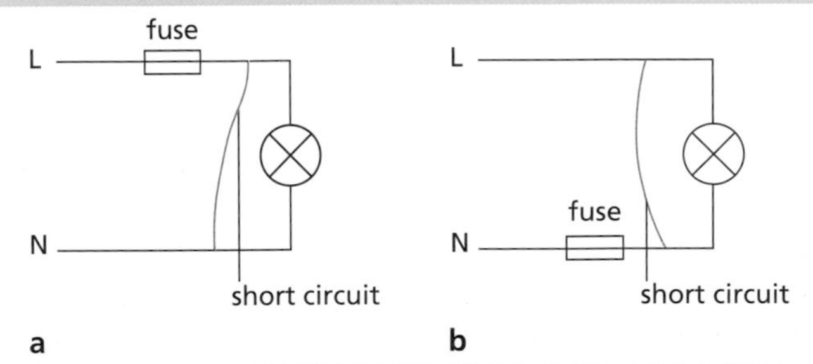

Figure 38.13

6 What steps should be taken before replacing a blown fuse in a plug?

7 What size fuse (3 A or 13 A) should be used in a plug connected to
 a a 150 W television,
 b a 900 W iron,
 c a 2 kW kettle,
 if the supply is 230 V?

8 What is the cost of heating a tank of water with a 3000 W immersion heater for 80 minutes if electricity costs 10p per kWh?

9 a Figure 38.14 shows an electric fence, designed to keep horses in a field.

Figure 38.14

When a horse touches the wire the horse receives a mild electric shock. Explain how.

 b Figure 38.15 shows how a person could receive an electric shock from a faulty electrical appliance. Using a residual circuit breaker (RCB) can help to protect the person against receiving a serious shock.

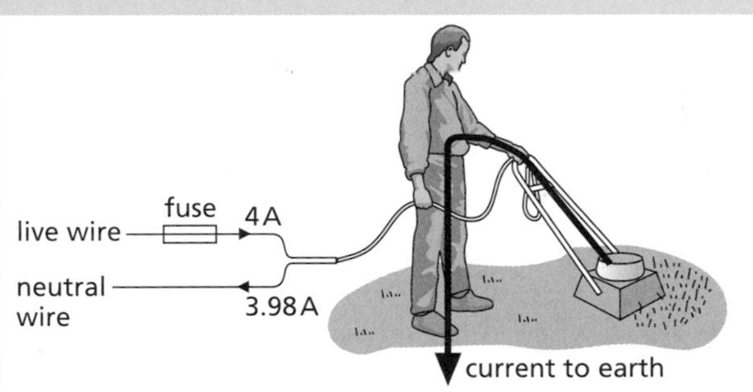

Figure 38.15

 (i) Compare the action of an RCB to that of a fuse.
 (ii) The graph in Figure 38.16 illustrates how the severity of an electric shock depends upon both the size of the current and the time for which the current flows through the body.

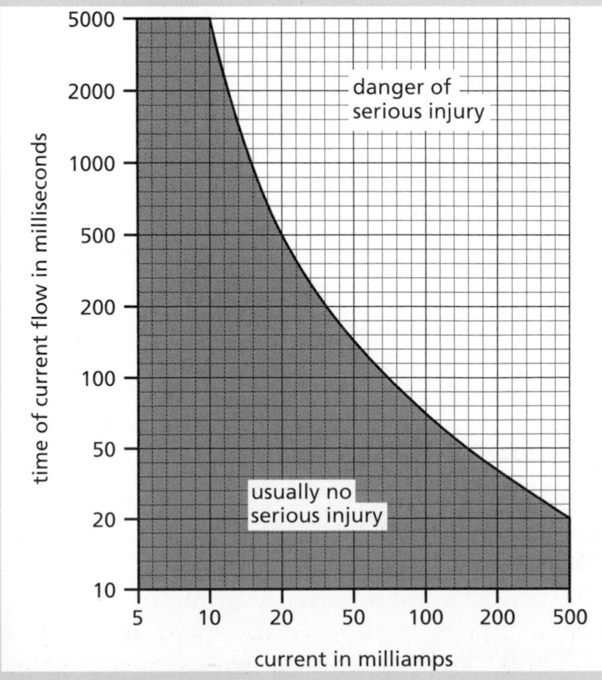

Figure 38.16

continued

Within how long must the RCB cut off the current if the person using the lawnmower is to be in no danger of serious injury?

(*AQA (SEG) Higher, June 1999*)

Checklist

After studying this chapter you should be able to

- ☐ recall the relations $E = ItV$ and $P = IV$ and use them to solve simple problems on energy transfers,
- ☐ describe experiments to measure electric power,
- ■ describe electric lamps, heating elements and fuses,
- ■ recall that a joulemeter measures electrical energy,
- ■ describe with the aid of diagrams a house wiring system and explain the functions and positions of switches, fuses, circuit breakers and earth,
- ■ state the advantages of connecting lamps in parallel in a lighting circuit,
- ■ wire a mains plug and recall the international insulation colour code,
- ■ perform calculations of the cost of electrical energy in joules and kilowatt-hours,
- ■ recall the hazards of damaged insulation, damp conditions and overheating of cables and the associated risks.

39 Electronic systems

Electronic systems	*Transistor as a switch*
Input transducers	*Practical work* Transistor switching circuits: light-operated, temperature-operated.
Output transducers	
Semiconductor diode	
Transistor	

Electronics is being used more and more in our homes, factories, offices, schools, banks, shops and hospitals. The development of semiconductor devices such as transistors and integrated circuits ('chips') has given us, among other things, automatic banking machines, laptop computers, programmable control devices, robots, computer games, digital cameras, Figure 39.1a, and heart pacemakers, Figure 39.1b.

Figure 39.1a Digital camera

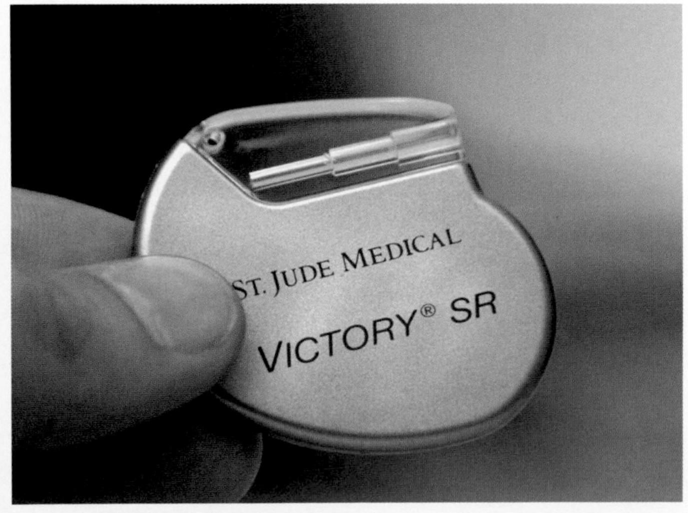

Figure 39.1b Heart pacemaker

Electronic systems

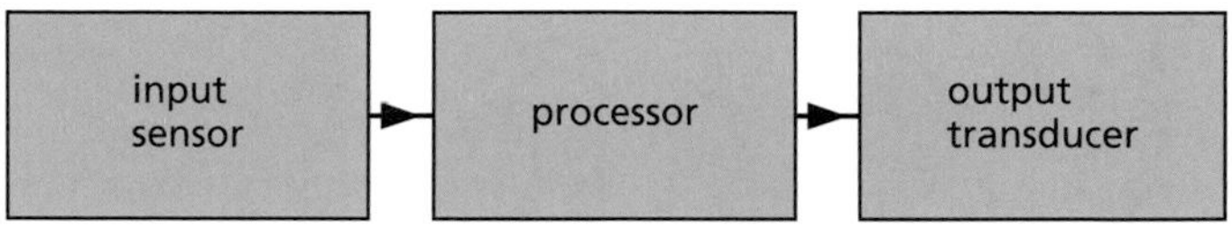

Figure 39.2 Electronic system

Any electronic system can be considered to consist of the three parts shown in the block diagram of Figure 39.2, i.e.

(i) an **input sensor** or **transducer**,
(ii) a **processor**, and
(iii) an **output transducer**.

A 'transducer' is a device for converting a non-electrical input into an electrical signal or vice versa.

The **input sensor** detects changes in the environment and converts them from their present form of energy into electrical energy. Input sensors or transducers include LDRs, thermistors, microphones and switches which respond, for instance, to pressure changes.

The **processor** decides on what action to take on the electrical signal it receives from the input sensor. It may involve an operation such as counting, amplifying, timing or storing.

The **output transducer** converts the electrical energy supplied by the processor into another form. Output transducers include lamps, LEDs, loudspeakers, motors, heaters, relays and cathode ray tubes.

In a radio, the input sensor is the aerial which sends an electrical signal to processors in the radio which, among other things, amplify the signal so that it can enable the output transducer, in the form of a loudspeaker, to produce sound.

Input transducers

a) Light dependent resistor (LDR)

The action of an LDR depends on the fact that the resistance of the semiconductor cadmium sulphide decreases as the intensity of the light falling on it increases.

An LDR and a circuit showing its action are shown in Figures 39.3a and b. Note the circuit symbol for an LDR, sometimes seen without its circle. When light from a lamp falls on the 'window' of the LDR, its resistance decreases and the increased current lights the lamp.

LDRs are used in photographic exposure meters and in series with a resistor to provide an input signal for a transistor (Figure 39.16) or other switching circuit.

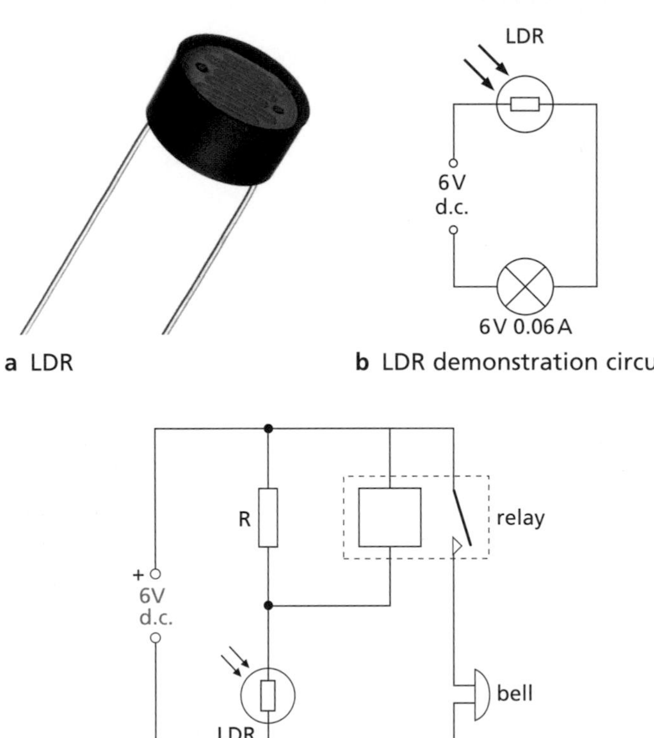

a LDR **b** LDR demonstration circuit

c Light-operated intruder alarm

Figure 39.3

Figure 39.3c shows how an LDR can be used to switch a relay (Chapter 42). The LDR forms part of a potential divider across the 6 V supply. When light falls on the LDR, the resistance of the LDR, and hence the voltage across it, decreases. There is a corresponding increase in the voltage across resistor R and the relay; when the voltage across the relay coil reaches its operating p.d., the normally open contacts close, allowing current to flow to the bell, which rings. If the light is removed, the current through the relay becomes too small to 'hold' the relay contacts open; power to the bell is cut and it stops ringing.

b) Thermistor

This contains semiconducting metallic oxides whose resistance decreases markedly when the temperature rises either due to heating the thermistor directly or to passing a current through it.

One type is shown in Figure 39.4a, and the symbol for a thermistor is shown in a circuit to demonstrate its action in Figure 39.4b. Heating the thermistor with a match lights the bulb.

A thermistor in series with a meter marked in °C can measure temperatures (Chapter 36). Used in series with a resistor it can provide an input signal to a transistor (Figure 39.18) or other switching circuit.

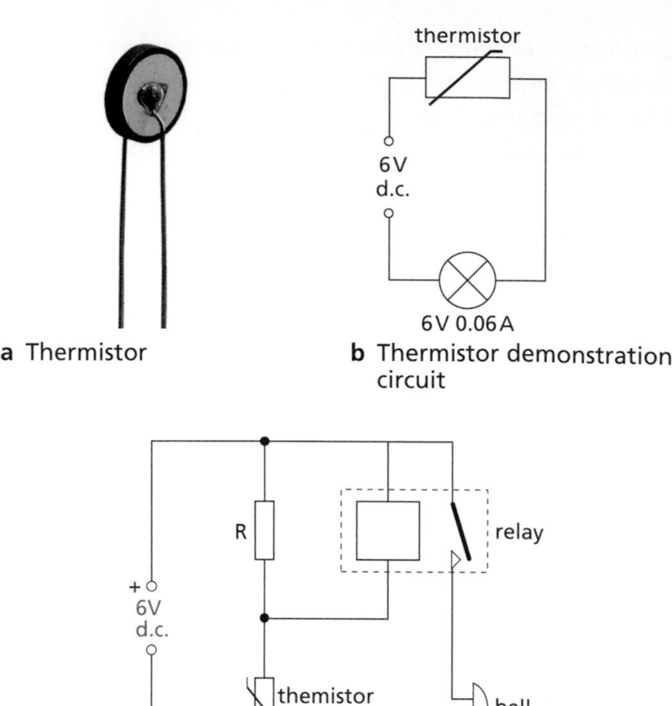

a Thermistor **b** Thermistor demonstration circuit

c High-temperature alarm

Figure 39.4

Figure 39.4c shows how a thermistor can be used to switch a relay. The thermistor forms part of a potential divider across the d.c. source. When the temperature rises, the resistance of the thermistor falls, and so does the p.d. across it. The voltage across resistor R and the relay increases. When the voltage across the relay reaches its operating p.d. the normally open contacts close, so that the circuit to the bell is completed and it rings. If a variable resistor is used in the circuit, the temperature at which the alarm sounds can be varied.

Output transducers

a) Relays

A switching circuit cannot supply much power to an appliance so a relay is often included; this allows the small current provided by the switching circuit to control the larger current needed to operate a buzzer as in a temperature-operated switch (Figure 39.18, p. 197) or other device. Relays controlled by a switching circuit can also be used to switch on the mains supply for electrical appliances in the home. In Figure 39.5 if the output of the switching circuit is 'high' (5 V), a small current flows to the relay which closes the mains switch; the relay also isolates the low voltage circuit from the high voltage mains supply.

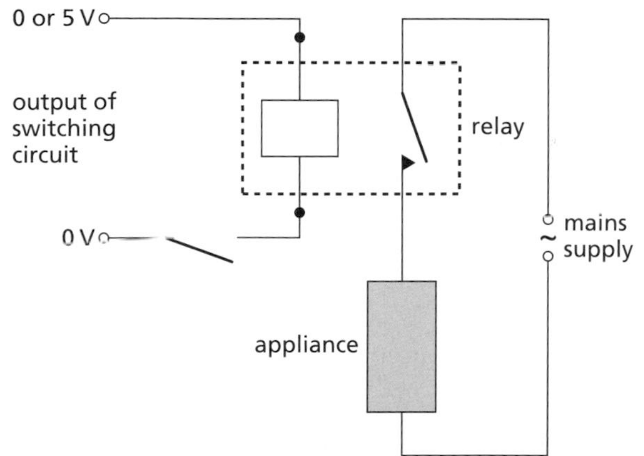

Figure 39.5 Use of a relay to switch mains supply

b) Light emitting diode (LED)

An LED, shown in Figure 39.6a, is a diode made from the semiconductor gallium arsenide phosphide. When forward biased, the current through it makes it emit red, yellow or green light. No light is emitted on reverse bias which, if it exceeds 5 V, may cause damage.

In use an LED must have a suitable resistor *R* in series with it (e.g. 300 Ω on a 5 V supply) to limit the current (typically 10 mA). Figure 39.6b shows the symbol for an LED (again the use of the circle is optional) in a demonstration circuit.

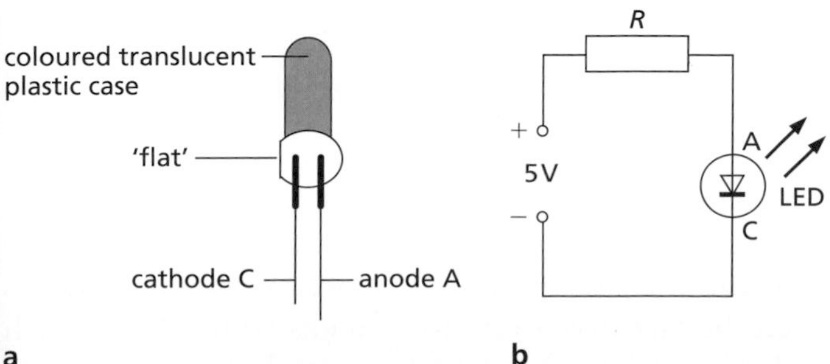

Figure 39.6 LED and demonstration circuit

LEDs are used as indicator lamps on computers, radios and other electronic equipment. Many clocks, calculators, video recorders and measuring instruments have seven-segment red or green numerical displays, Figure 39.7a. Each segment is an LED and, depending on which have a voltage across them, the display lights up the numbers 0 to 9, as in Figure 39.7b.

LEDs are small, reliable and have a long life; their operating speed is high and their current requirements very low.

Diode lasers operate in a similar way to LEDs but emit coherent laser light; they are used in optical fibre communications as transmitters.

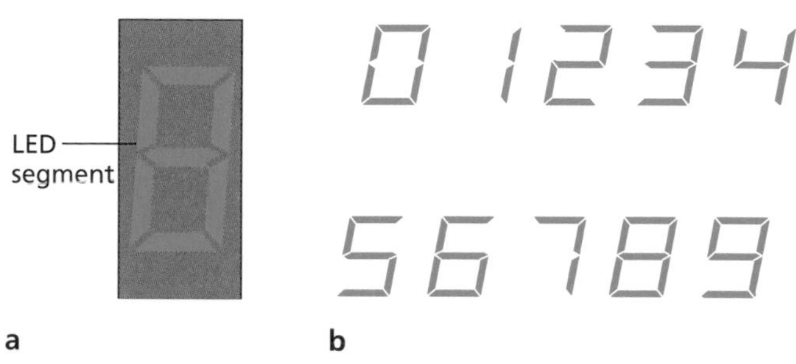

Figure 39.7 LED numerical display

Semiconductor diode

A diode is a device that lets current pass through it in one direction only. One is shown in Figure 39.8 with its symbol. (You will also come across the symbol without its outer circle.) The wire nearest the band is the **cathode** and the one at the other end is the **anode**.

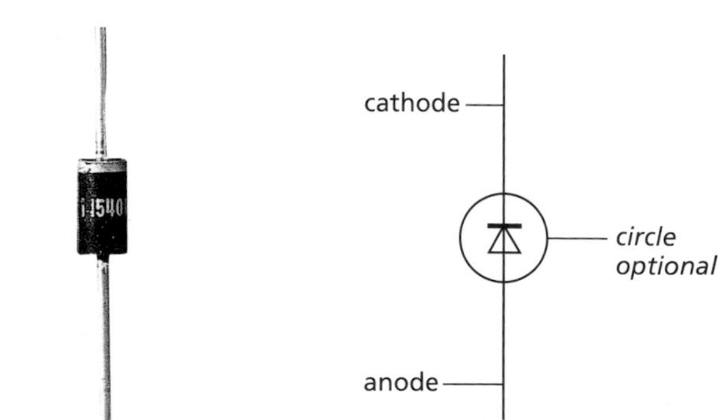

Figure 39.8 A diode and its symbol

The typical *I–V* graph is shown in Figure 36.6b (p. 176). The diode conducts when the anode goes to the + terminal of the voltage supply and the cathode to the − terminal, Figure 39.9a. It is then **forward biased**; its resistance is small and conventional current passes in the direction of the arrow on its symbol. If the connections are the other way round, it does not conduct; its resistance is large and it is **reverse biased**, Figure 39.9b.

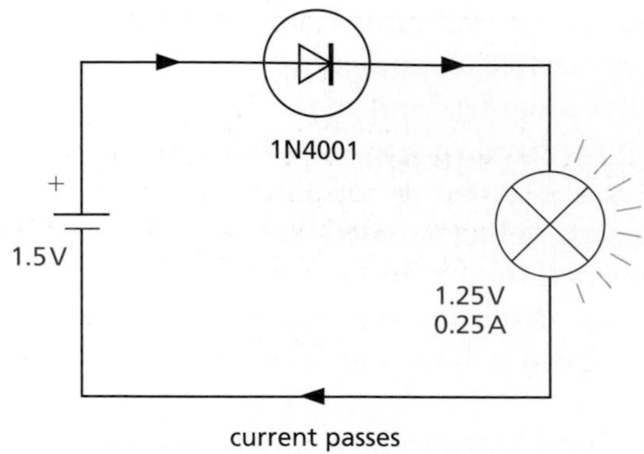

a Forward biased

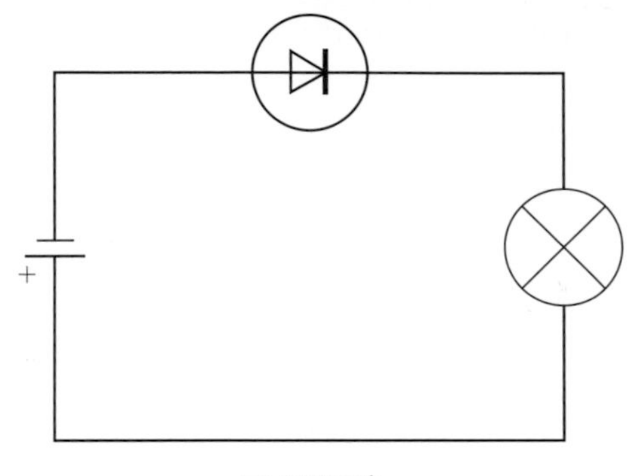

b Reverse biased

Figure 39.9 Demonstrating the action of a diode

The lamp in the circuit shows when the diode conducts by lighting up. It also acts as a resistor to limit the current when the diode is forward biased. Otherwise the diode might overheat and be damaged.

A diode is a **non-ohmic** conductor. This makes it useful as a **rectifier** for changing alternating current (a.c.) to direct current (d.c.). Figure 39.10 shows the rectified output voltage obtained from a diode when it is connected to an a.c. supply.

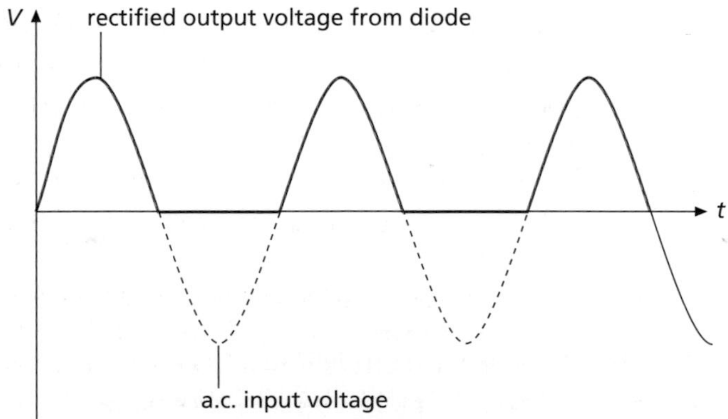

Figure 39.10 Rectification by a diode

☐ *Transistor*

Transistors are the small semiconductor devices which have revolutionized electronics. They are made both as separate components, like those in Figure 39.11a in their cases, and also as parts of **integrated circuits** (ICs) in which millions may be 'etched' on a 'chip' of silicon, Figure 39.11b.

Transistors have three connections called the **base** (B), the **collector** (C) and the **emitter** (E). In the transistor symbol shown in Figure 39.12, the arrow indicates the direction in which conventional current flows through it when C and B are connected to battery +, and E to battery −. Again, the outer circle of the symbol is not always included.

Figure 39.11a Transistor components

Figure 39.11b Integrated circuits which may each contain millions of transistors

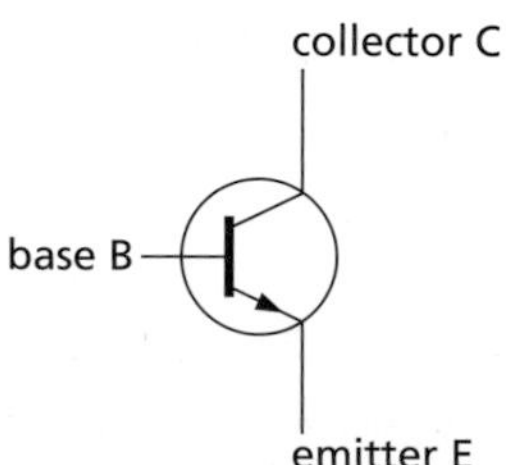

Figure 39.12 Symbol for a transistor

There are two current paths through a transistor. One is the **base–emitter path** and the other is the **collector–emitter** (via base) **path**. The transistor's usefulness arises from the fact that it can link circuits connected to each path so that the current in one controls that in the other, just like a relay.

Its action can be shown using the circuit of Figure 39.13. When **S is open**, the base current I_B is zero and neither L_1 nor L_2 lights up, showing that the collector current I_C is also zero even though the battery is correctly connected across the C–E path.

When **S is closed**, B is connected through R to battery +, and L_2 lights up but not L_1. This shows there is now collector current (which passes through L_2) and that it is much greater than the base current (which passes through L_1 but is too small to light it).

Therefore, **in a transistor the base current I_B switches on and controls the much greater collector current I_C.**

Resistor R must be in the circuit to limit the base current which would otherwise create so large a collector current as to destroy the transistor by overheating.

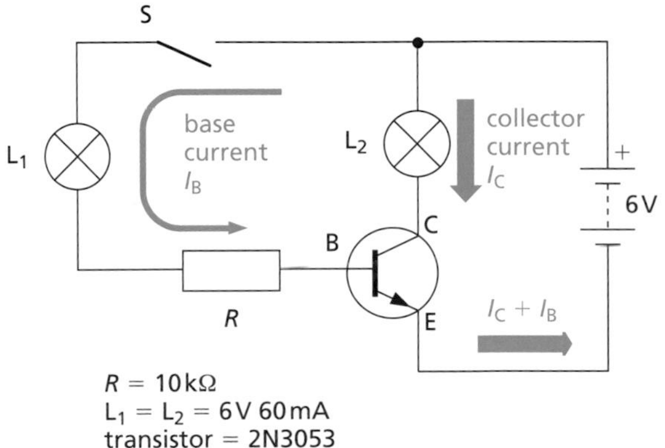

Figure 39.13 Demonstration circuit

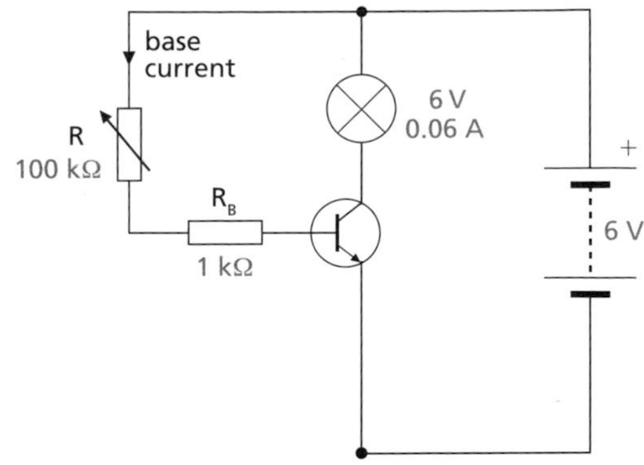

a Rheostat control

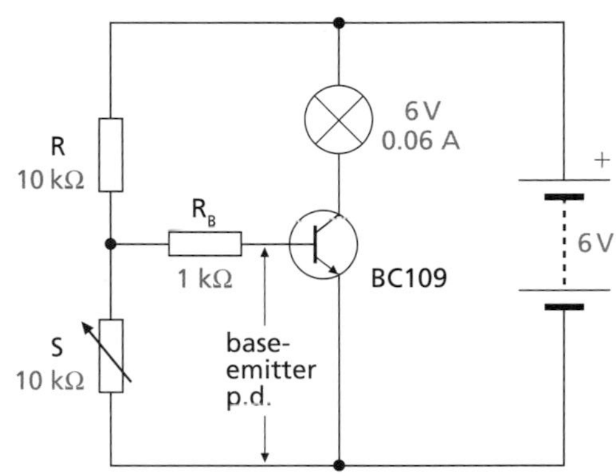

b Potential divider control

Figure 39.14 Transistor switching circuits

☐ *Transistor as a switch*

a) Advantages

Transistors have many advantages over other electrically operated switches such as relays. They are small, cheap, reliable, have no moving parts, their life is almost indefinite (in well-designed circuits) and they can switch on and off millions of times a second.

b) 'On' and 'off' states

A transistor is considered to be 'off' when the collector current is zero or very small. It is 'on' when the collector current is much larger. The resistance of the collector–emitter path is large when the transistor is 'off' (as it is for an ordinary mechanical switch) and small (ideally it should be zero) when it is 'on'.

To switch a transistor 'on' requires the base voltage (and therefore the base current) to exceed a certain minimum value (about +0.6 V).

c) Basic switching circuits

Two are shown in Figure 39.14a, b. The 'on' state is shown by the lamp in the collector circuit becoming fully lit.

Rheostat control is used in the circuit in Figure 39.14a. 'Switch-on' occurs by reducing R until the base current is large enough to make the collector current light the lamp. (The base resistor R_B is *essential* in case R is made zero and results in +6 V from the battery being applied directly to the base. This would produce very large base and collector currents and destroy the transistor by overheating.)

Potential divider control is used in the circuit in Figure 39.14b. Here 'switch-on' is obtained by adjusting the variable resistor S until the p.d. across S (which is the base–emitter p.d. and depends on the value of S compared with that of R) exceeds +0.6 V or so.

Note In a potential divider the p.ds across the resistors are in the ratio of their resistances (see Chapter 36). For example, in Figure 39.14b, if R = 10 kΩ and S = 5 kΩ then the p.d. across $R = V_R$ = 4 V and the p.d. across $S = V_S$ = 2 V, i.e. V_R/V_S = 4 V/2 V = 2/1 and $V_R + V_S$ = 6 V = p.d. across R and S in series where R/S = 10 kΩ/5 Ω = 2/1.

In general

$$\frac{V_R}{V_S} = \frac{R}{S} \quad \text{and} \quad V_S = \frac{(V_R + V_S)S}{(R + S)}$$

Also see question 2 on p. 197.

Practical work

Transistor switching circuits

The components can be mounted on a circuit board, e.g. an S-DeC, Figure 39.15a. The diagrams in Figure 39.15b show how to lengthen transistor leads and also how to make connections (without soldering) to parts that have 'tags', e.g. variable resistors.

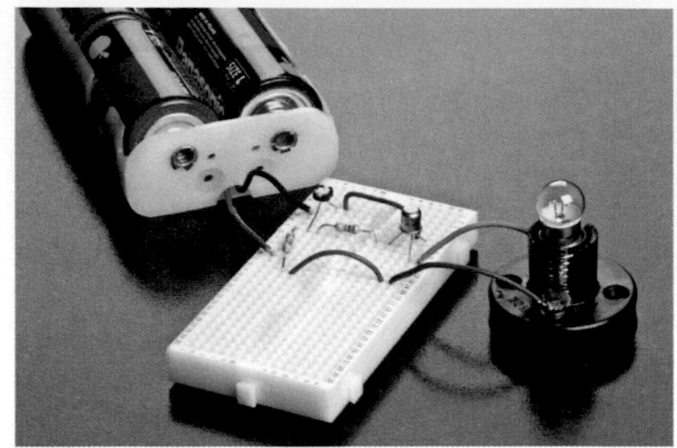

Figure 39.15a Partly built transistor switching circuit

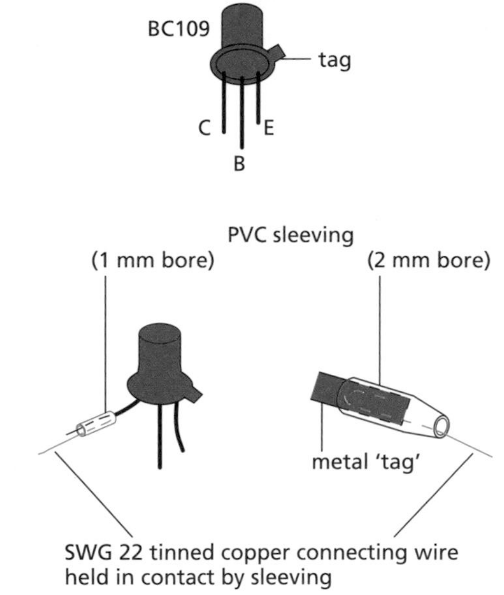

Figure 39.15b Lengthening transistor leads and making connections to tags

In many control circuits, devices such as LDRs and thermistors are used in potential divider arrangements to detect small changes of light intensity and temperature respectively. These changes then enable a transistor to act as a simple processor by controlling the current to an output transducer, e.g. a lamp or a buzzer.

a) Light-operated switch

In the circuit of Figure 39.16 the LDR is part of a potential divider. The lamp comes on when the LDR is shielded, due to more of the battery p.d. being dropped across the increased resistance of the LDR (i.e. more than 0.6 V) and less across *R*. In the dark, the base–emitter p.d. increases as does the base current and so also the collector current.

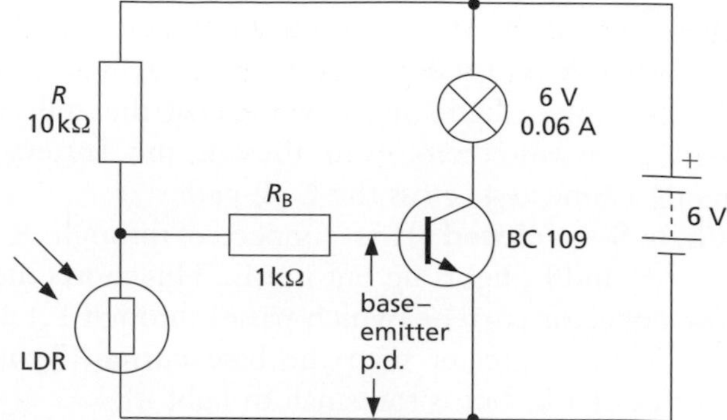

Figure 39.16 Light-operated switch

If the LDR and *R* are interchanged the lamp goes off in the dark and the circuit could act as a light-operated intruder alarm.

If a variable resistor is used for *R*, the light level at which switching occurs can be changed.

b) Temperature-operated switch

In the low-temperature operated switch of Figure 39.17, a thermistor and resistor form a potential divider across the 6 V supply. When the temperature of the thermistor falls, its resistance increases and so does the p.d. across it, i.e. the base-emitter p.d. rises. When it reaches 0.6 V, the transistor switches on and the collector current becomes large enough to operate the lamp. The circuit could act as a frost-warning device.

If the thermistor and resistor are interchanged, the circuit can be used as a high-temperature alarm, Figure 39.18.

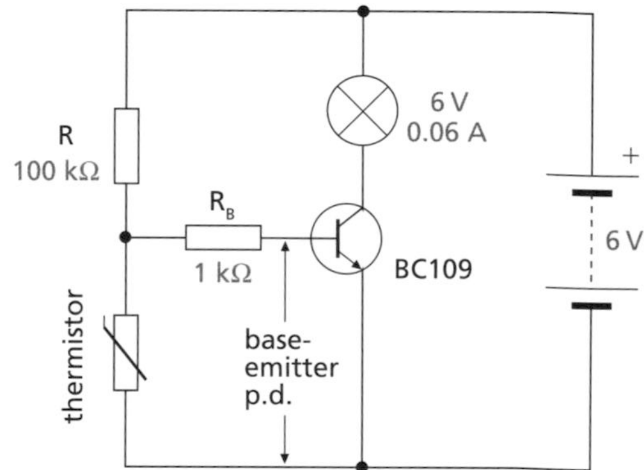

Figure 39.17 Low-temperature operated switch

When the temperature of the thermistor rises, its resistance decreases and a larger fraction of the 6 V supply is dropped across *R*, i.e. the base–emitter p.d. increases. When it exceeds 0.6 V or so the transistor switches on and collector current (too small to ring the buzzer directly) goes through the relay coil. The relay contacts close, enabling the buzzer to obtain directly from the 6 V supply the larger current it needs.

The diode D protects the transistor from damage by the large p.d. induced in the relay coil (due to its inductance) when the collector current falls to zero at switch off. The diode is forward biased by the induced p.d. (which tries to maintain the current through the relay coil) and, because of its low forward resistance (e.g. 1 Ω), offers an easy path for the current produced. To the 6 V supply the diode is reverse biased and its high resistance does not short-circuit the relay coil when the transistor is on.

If R is variable the temperature at which switching occurs can be changed.

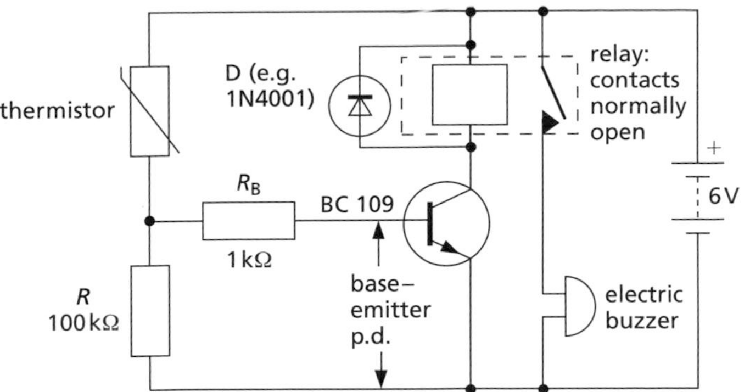

Figure 39.18 High-temperature operated switch

Questions

1 Figure 39.19a shows a lamp, a semiconductor diode and a cell connected in series. The lamp lights when the diode is connected in this direction. Say what happens to each of the lamps in b, c and d. Give reasons for your answer.

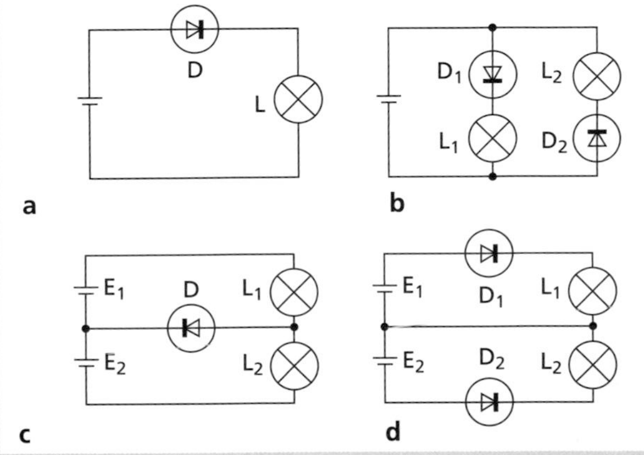

Figure 39.19

2 What are the readings V_1 and V_2 on the high-resistance voltmeters in the potential divider circuit of Figure 39.20 if

a $R_1 = R_2 = 10\ k\Omega$,

b $R_1 = 10\ k\Omega$, $R_2 = 50\ k\Omega$,

c $R_1 = 20\ k\Omega$, $R_2 = 10\ k\Omega$?

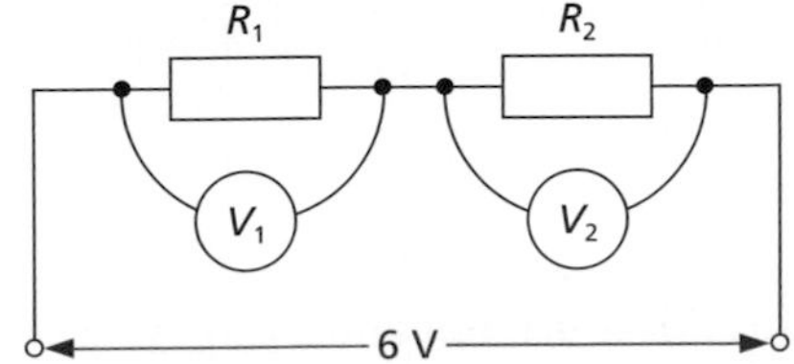

Figure 39.20

3 A simple moisture-warning circuit is shown in Figure 39.21 in which the moisture detector consists of two closely spaced copper rods.

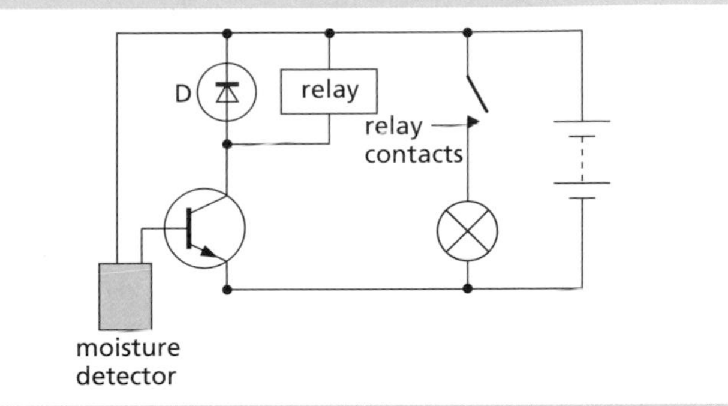

Figure 39.21

a Describe how the circuit works when the detector gets wet.

b Warning lamps are often placed in the collector circuit of a transistor. Why is a relay used here?

c What is the function of D?

4 Figure 39.22 shows a circuit that is used to switch on a lamp automatically when it starts to go dark.

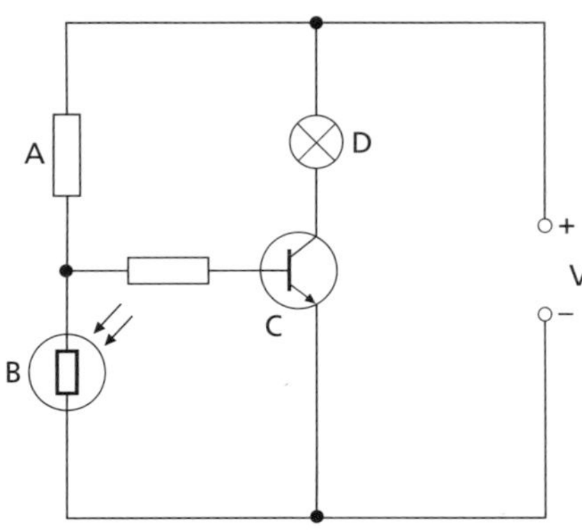

Figure 39.22

a Write down the names of the components labelled **A**, **B**, **C** and **D**.

b Which of the four components **A**, **B**, **C** or **D** acts as a switch?

c Explain why the lamp comes on as it goes dark.

(UCLES IGCSE Physics Extended, Nov 2006)

Checklist

After studying this chapter you should be able to

- recall the functions of the input sensor, processor and output transducer in an electronic system and give some examples,
- describe the action of an LDR and a thermistor and show an understanding of their use as input transducers,
- understand the use of a relay in a switching circuit,
- ☐ explain what is meant by a diode being forward biased and reverse biased and recall that a diode can produce rectified a.c.,
- ☐ describe the action of a transistor with the aid of a circuit diagram,
- ☐ describe how a transistor can be used as a switch,
- ☐ explain the operation of light- and temperature-operated transistor switching circuits with the aid of diagrams.

40 Digital electronics

Analogue and digital electronics
Logic gates
Logic gate control systems
Problems to solve
Designing logic control systems.
Electronics and society

□ *Analogue and digital electronics*

There are two main types of electronic circuits, devices or systems – analogue and digital.

In **analogue circuits**, voltages (and currents) can have any value within a certain range over which they can be varied smoothly and continuously, Figure 40.1a. They include amplifier-type circuits.

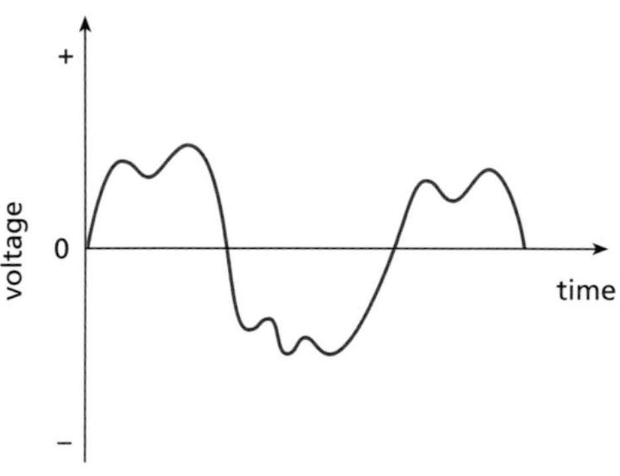

a

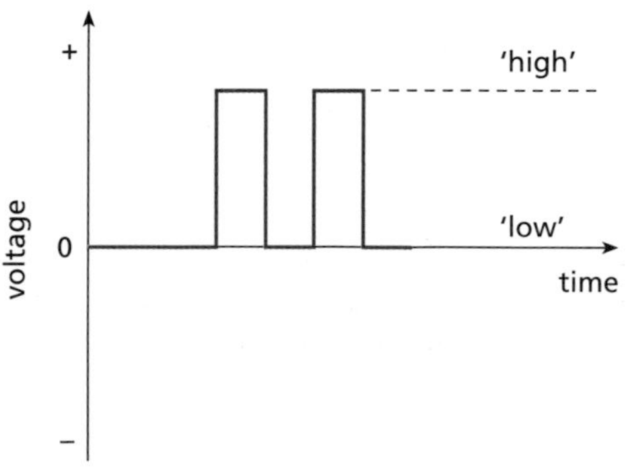

b

Figure 40.1

In **digital circuits**, voltages have only one of two values, either 'high' (e.g. 5 V) or 'low' (e.g. near 0 V), Figure 40.1b. They include switching-type circuits such as those we have considered in Chapter 39.

A **variable resistor** is an analogue device which, in a circuit with a lamp, allows the lamp to have a wide range of light levels. A **switch** is a digital device which allows a lamp to be either 'on' or 'off'.

Analogue meters display their readings by the deflection of a pointer over a continuous scale, Figure 44.4a. **Digital meters** display their readings as digits, i.e. numbers, which change by one digit at a time, Figure 44.4b (see p. 225).

□ *Logic gates*

Logic gates are switching circuits used in computers and other electronic systems. They 'open' and give a 'high' output voltage, i.e. a signal (e.g. 5 V), depending on the combination of voltages at their inputs, of which there is usually more than one.

There are five basic types, all made from transistors in integrated circuit form. The behaviour of each is described by a **truth table** showing what the output is for all possible inputs. 'High' (e.g. 5 V) and 'low' (e.g. near 0 V) outputs and inputs are represented by 1 and 0 respectively and are referred to as **logic levels** 1 and 0.

a) NOT gate or inverter

This is the simplest gate, with one input and one output. It produces a 'high' output if the input is 'low', i.e. NOT high, and vice versa. Whatever the input, the gate inverts it. The symbol and truth table are given in Figure 40.2.

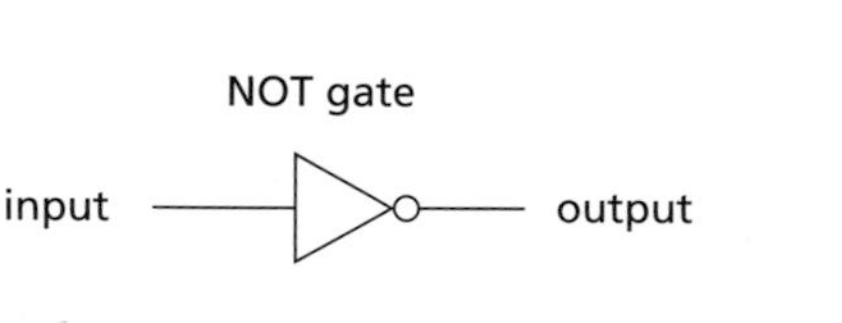

input	output
0	1
1	0

Figure 40.2 NOT gate symbol and truth table

b) OR, NOR, AND, NAND gates

All these have two or more inputs and one output. The truth tables and symbols for 2-input gates are shown in Figure 40.3. Try to remember the following.

OR: output is 1 if input A **OR** input B **OR** both are 1
NOR: output is 1 if neither input A **NOR** input B is 1
AND: output is 1 if input A **AND** input B are 1
NAND: output is 1 if input A **AND** input B are **NOT** both 1

Note from the truth tables that the outputs of the NOR and NAND gates are those of the OR and AND gates respectively inverted. They have a small circle at the output end of their symbols to show this inversion.

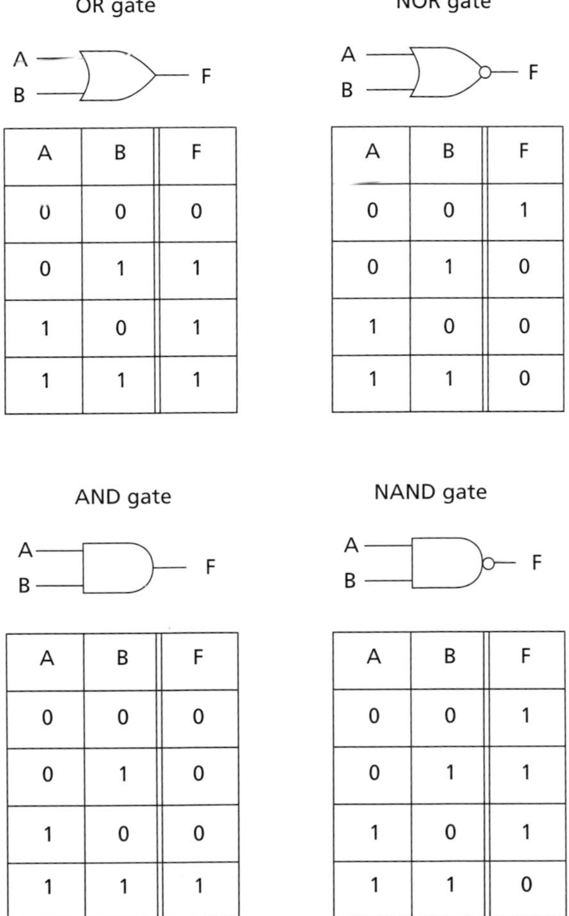

OR gate

A	B	F
0	0	0
0	1	1
1	0	1
1	1	1

NOR gate

A	B	F
0	0	1
0	1	0
1	0	0
1	1	0

AND gate

A	B	F
0	0	0
0	1	0
1	0	0
1	1	1

NAND gate

A	B	F
0	0	1
0	1	1
1	0	1
1	1	0

Figure 40.3 Symbols and truth tables for 2-input gates

c) Testing logic gates

The truth tables for the various gates can be conveniently checked by having the logic gate IC mounted on a small board with sockets for the power supply, inputs A and B and output F, Figure 40.4. A 'high' input (i.e. logic level 1) is obtained by connecting the input socket to the positive of the power supply, e.g. +5 V and a 'low' one (i.e. logic level 0) to 0 V.

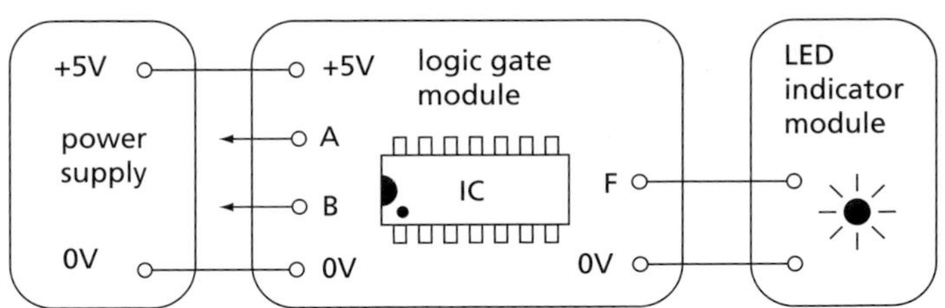

Figure 40.4 Modules for testing logic gates

The output can be detected using an indicator module containing an LED which lights up for a 1 and stays off for a 0.

□ *Logic gate control systems*

Logic gates can be used as processors in electronic control systems. Many of these can be demonstrated by connecting together commercial modules like those in Figure 40.8b (overleaf).

a) Security system

The block diagram for a simple system that might be used by a jeweller to protect an expensive clock is shown in Figure 40.5. The clock sits on a push switch which sends a 1 to the NOT gate unless the clock is lifted when a 0 is sent. In that case the output from the NOT gate is a 1 which rings the bell.

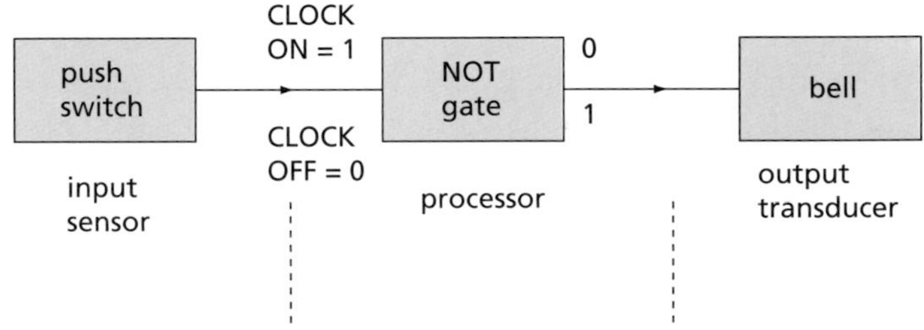

Figure 40.5 Simple alarm system

b) Safety system for a machine operator

The system could prevent a machine (e.g. an electric motor) being switched on before another switch had been operated by a protective safety guard being in the correct position. In Figure 40.6, when switches A *and* B are pressed they supply a 1 to each input of the AND gate which can then start the motor.

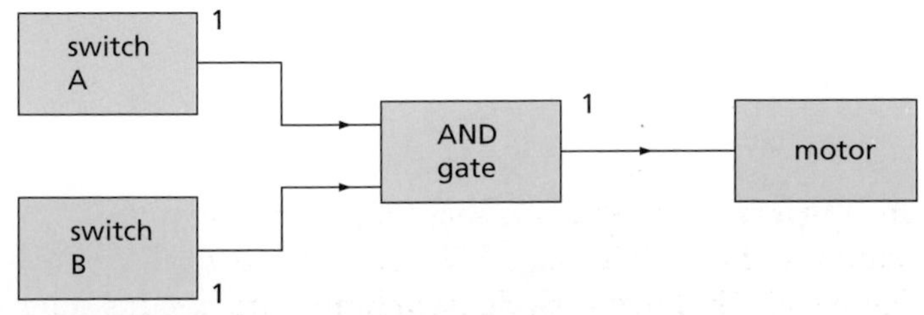

Figure 40.6 Safety system for controlling a motor

c) Heater control system

The heater control has to switch on the heating system when it is

(i) **cold**, i.e. the temperature is below a certain value and the output from the temperature sensor is 0, and
(ii) **daylight**, i.e. the light sensor output is 1.

With these outputs from the sensors applied to the processor in Figure 40.7, the AND gate has two 1 inputs. The output from the AND gate is then 1 and will turn on the heater control. Any other combination of sensor outputs produces a 0 output from the AND gate, as you can check.

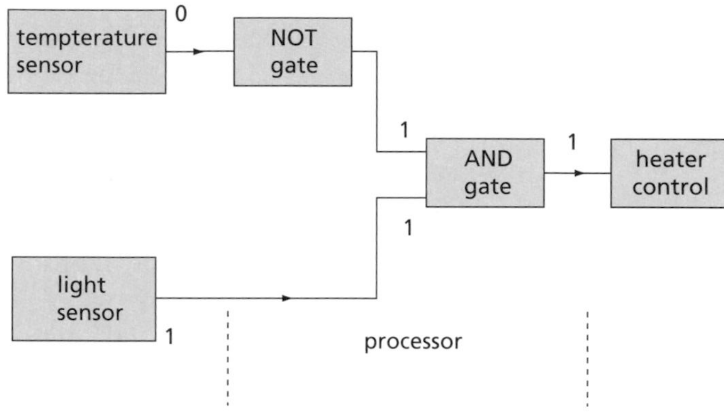

Figure 40.7 Heater control system

d) Street lights

A system is required which allows the street lights either to be turned on manually by a switch at any time, or automatically by a light sensor when it is dark. The arrangement in Figure 40.8a achieves this since the OR gate gives a 1 output when either or both of its inputs are 1.

The system can be demonstrated using the module shown in Figure 40.8b.

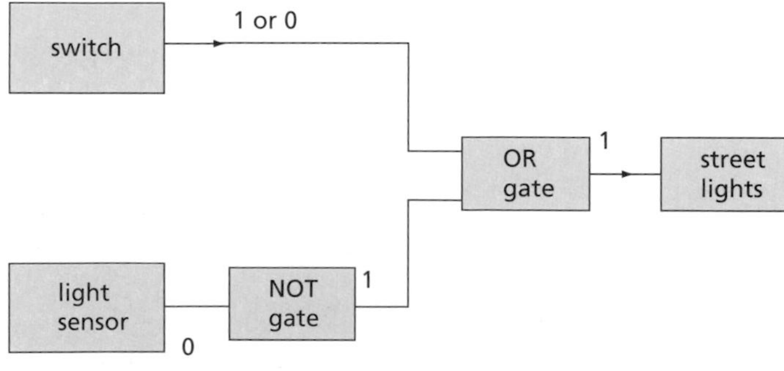

Figure 40.8a Control system with manual override

Figure 40.8b Module for demonstrating street lights

☐ *Problems to solve*

Design and draw block diagrams for logic control systems to indicate how the following jobs could be done. If possible build them using modules.

1 Allow a doorbell to work only during the **day**.
2 Give warning when the temperature of a domestic hot water system is too **high** or when a switch is pressed to **test** the alarm.
3 Switch on a bathroom heater when it is **cold** and **light**.
4 Sound an alarm when it is **cold** or a switch is **pressed**.
5 Give warning if the temperature of a room **falls** during the **day** and also allow a **test** switch to check the alarm works.
6 Give warning of **frosty** conditions at **night** to a gardener who is sometimes very tired after a hard day and may want to switch the alarm off.

F *Electronics and society*

Electronics is having an ever-increasing impact on all our lives. Work and leisure are changing as a result of the social, economic and environmental influences of new technology.

a) Reasons for the impact

Why is electronics having such a great impact? Some of the reasons are listed below.

(i) **Mass production** of large quantities of semiconductor devices (e.g. ICs) allows them to be made very cheaply.
(ii) **Miniaturization** of components means that even complex systems can be compact.
(iii) **Reliability** of electronic components is a feature of well-designed circuits. There are no moving parts to wear out and systems can be robust.
(iv) **Energy consumption** and use of natural resources is often much less than for their non-electronic counterparts, for example the transistor uses less power than a relay.

(v) **Speed of operation** can be millions of times greater than for other alternatives (e.g. mechanical devices).

(vi) **Transducers** of many different types are available for transferring information in and out of an electronic system.

To sum up, electronic systems tend to be cheaper, smaller, more reliable, less wasteful, much faster and can respond to a wider range of signals than other systems.

b) Some areas of impact

At home devices such as washing machines, burglar alarms, telephones, cookers and sewing machines contain electronic components. Central heating systems and garage doors may have automatic electronic control. For home entertainment, video recorders, CD players, interactive digital televisions or computers with Internet connections and electronic games are finding their way into more and more homes.

Medical services have benefited greatly in recent years from the use of electronic instruments and appliances. Electrocardiograph (ECG) recorders for monitoring the heart, ultrasonic scanners for checks during pregnancy, gamma ray scanners for detecting tumours, deaf aids, heart pacemakers, artificial kidneys, limbs and hands with electronic control, Figure 40.9, and 'keyhole' surgery are some examples.

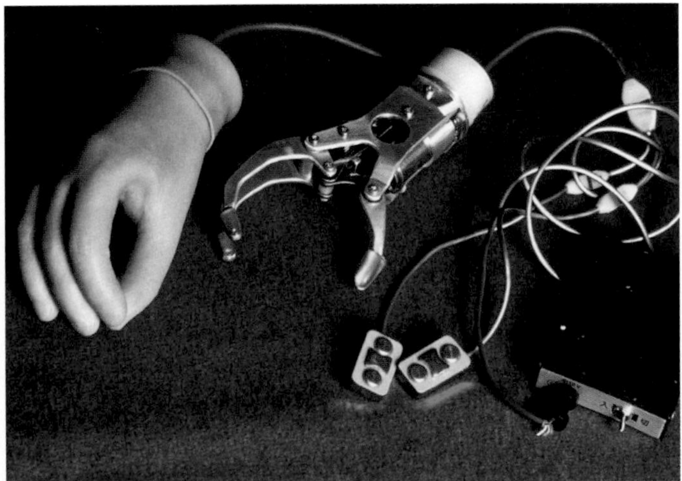

Figure 40.9 Electronically controlled artificial hands

In industry microprocessor-controlled equipment is taking over. Robots are widely used for car assembly work, and to do dull, routine, dirty jobs such as welding and paint spraying. In many cases production lines and even whole factories, e.g. sugar refineries and oil refineries, are almost entirely automated. Computer-aided design (CAD) of products is increasing, Figure 40.10, even in the clothing industry.

In offices, banks and shops computers are used for word processing, data control and communications via **e-mail**: text, numbers and pictures are transmitted by electronic means, often by high-speed digital links. Cash dispensers and other automated services at banks are a great convenience for their customers. Bar codes (like the one on the back cover of this book) on packaged products are used by shops for stock control in conjunction with a bar code reader (which uses a laser) and a data recorder connected to a computer. A similar system is operated by libraries to record the issue and return of books. Libraries provide electronic databases and Internet facilities for research.

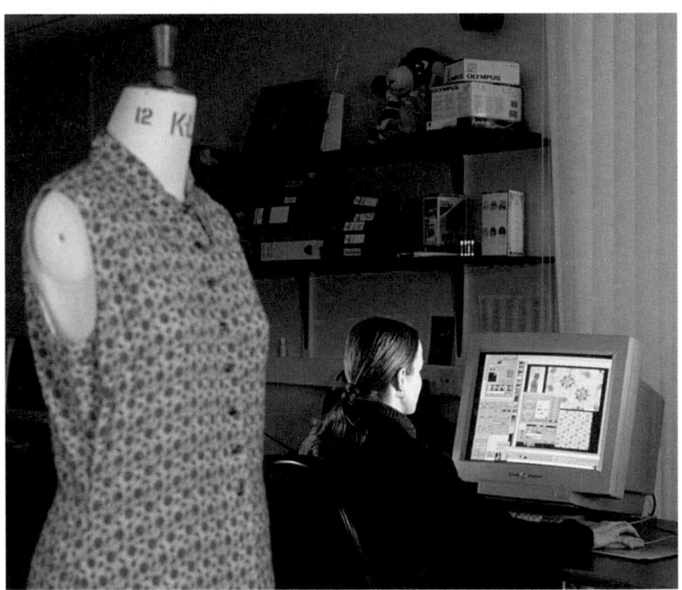

Figure 40.10 Computer-aided design of clothing

Communications have been transformed. Satellites enable events on one side of the world to be seen and heard on the other side, as they happen. Digital telephone links, mobile phones and e-mail are the order of the day.

Leisure activities have been affected by electronic developments. For some people, leisure means participating in or attending sporting activities and here the electronic scoreboard is likely to be in evidence. For the golf enthusiast, electronic machines claim to analyse 'swings' and reduce handicaps. For others, leisure means listening to music, whose production, recording and listening facilities have been transformed by the digital revolution. Electronically synthesized music has become the norm for popular recordings. The lighting and sound effects in stage shows are programmed by computer. For the cinema-goer, special effects in film production have been vastly improved by computer-generated animated images, Figure 40.11. The availability of home computers and games consoles in recent years has enabled a huge market in computer games and home-learning resources to develop.

Figure 40.11 Computer animation brings the tiger into the scene

c) Consequences of the impact

Most of the social and economic consequences of electronics are beneficial but a few cause problems.

An improved quality of life has resulted from the greater convenience and reliability of electronic systems, increased life expectancy and leisure time, and fewer dull, repetitive jobs.

Better communication has made the world a smaller place. The speed with which news can be reported to our homes by radio and television, and the influence of information systems such as teletext and the Internet enable the public to be better informed.

Databases have been developed. These are memories which can store huge amounts of information for rapid transmission from one place to another. For example, the police can obtain in seconds, by radio, details of a car they are following. Databases raise questions, however, about invasion of privacy and security.

Employment is affected by the demand for new equipment – new industry and jobs are created to make and maintain it – but when electronic systems replace mechanical ones, redundancy and/or retraining needs arise. Conditions of employment and long-term job prospects can also be affected for many people, especially certain manual and clerical workers. One industrial robot replaces four factory workers.

The public attitude to the electronics revolution is not always positive. Modern electronics is a 'hidden' technology with parts that are enclosed in a tiny package (or 'black box') and do not move. It is also a 'throwaway' technology in which the whole lot is discarded and replaced – by an expert – if a part fails, and rapid advances in design technology cause equipment to quickly become obsolete. For these reasons it may be regarded as mysterious and unfriendly – people feel they do not understand what makes it tick.

d) The future

The only certain prediction about the future is that new technologies will be developed and these, like present ones, will continue to have a considerable influence on our lives.

Today the development of 'intelligent' computers is being pursued with great vigour, and voice recognition techniques are already in use. Optical systems, which are more efficient than electronic ones, are being increasingly developed for data transmission, storage and processing of information.

Questions

1 The combined truth tables for four logic gates A, B, C, D are given below. State what kind of gate each one is.

Inputs		Outputs			
		A	B	C	D
0	0	0	0	1	1
0	1	0	1	1	0
1	0	0	1	1	0
1	1	1	1	0	0

2 What do the symbols A to E represent in Figure 40.12?

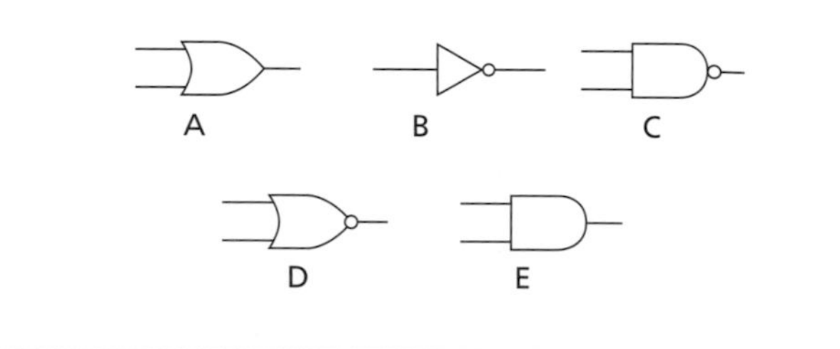

Figure 40.12

3 This question is about burglar alarms. Figure 40.13 shows part of an alarm system used to protect a CD player on display in a shop. The CD player is placed on a pressure switch. When the switch is pressed down it provides a high output (1). The alarm bell sounds if the CD player is lifted up.

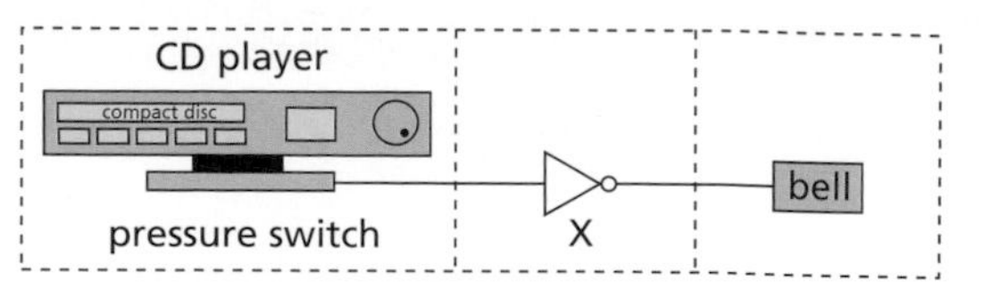

Figure 40.13

a (i) Write down the name of the logic gate labelled X in the diagram.
(ii) Write out the truth table for this logic gate, started below.

Input	Output
0	

b Someone lifts the CD player. Explain how the switch and the logic gate make the bell ring.

(*OCR Foundation, June 1999*)

4 Design and draw the block diagrams for logic control systems to:

a wake you at the crack of dawn and which you can also switch off,

b protect the contents of a drawer which you can still open without setting off the alarm.

5 a Figure 40.14 shows the symbol for one type of logic gate.

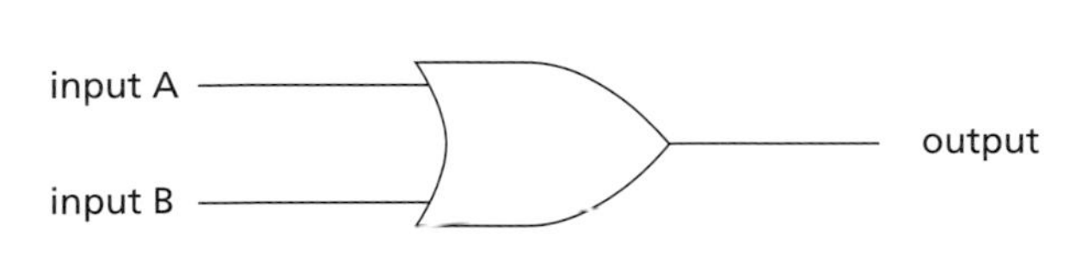

Figure 40.14

(i) What is this type of logic gate?

(ii) Copy and complete the truth table for this logic gate.

Input A	Input B	Output
0	0	
1	0	
0	1	
1	1	

b Figure 40.15 shows a *control system* which may be fitted in an automatic washing machine.

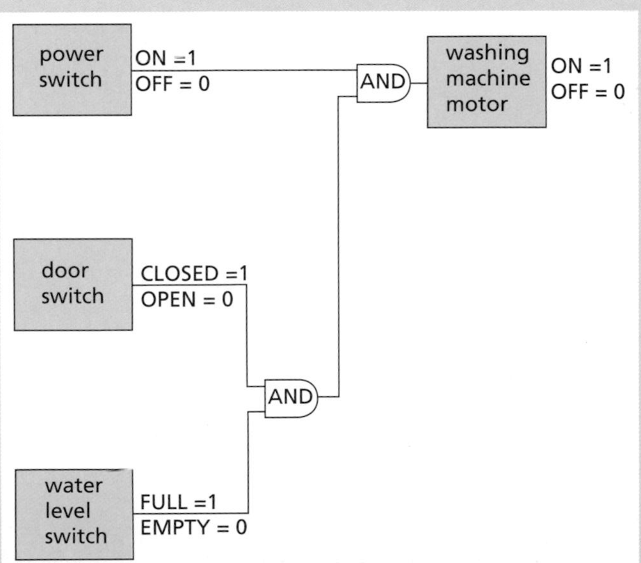

Figure 40.15

(i) Which part of the system forms the *control circuit*?

(ii) What conditions will stop the washing machine working?

c A car may be fitted with an automatic parking light which switches on when the car is parked at night. Figure 40.16 shows an incomplete system for a parking light.

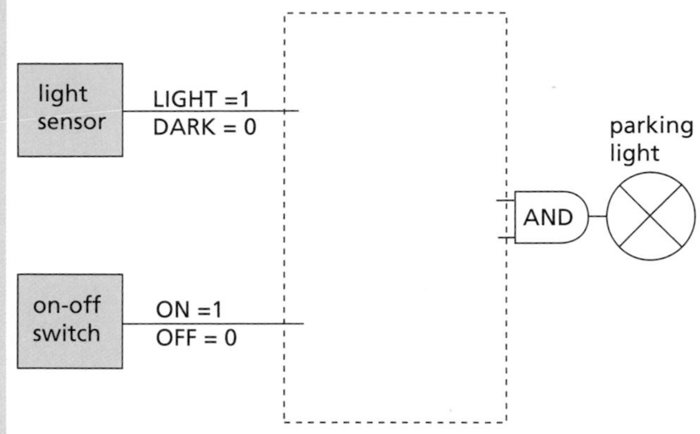

Figure 40.16

Copy and complete the diagram to show how a NOT gate can be used to make the parking light work. Use the correct symbol for a NOT gate.

(*AQA (SEG) Foundation, Summer 1999*)

6 a A student is to design an electrical circuit which will indicate when the temperature inside an incubator is 40 °C.

(i) Name a suitable temperature sensor for the student to use.

(ii) Draw the circuit symbol for the temperature sensor you named above.

b Before designing the circuit the student measured the resistance of the sensor at different temperatures. The results are shown in the table.

Temperature in °C	Resistance of sensor in ohms
0	3000
10	2000
20	1400
30	1000
40	700
50	500

(i) Draw a graph of these results, with resistance on the vertical axis (0–3000 Ω) and temperature on the horizontal axis (0–50 °C).

(ii) Is the change in the resistance of the sensor digital or analogue? Give a reason for your answer.

c The student uses the sensor in the circuit shown in Figure 40.17. The LED comes on when the temperature reaches 40 °C.

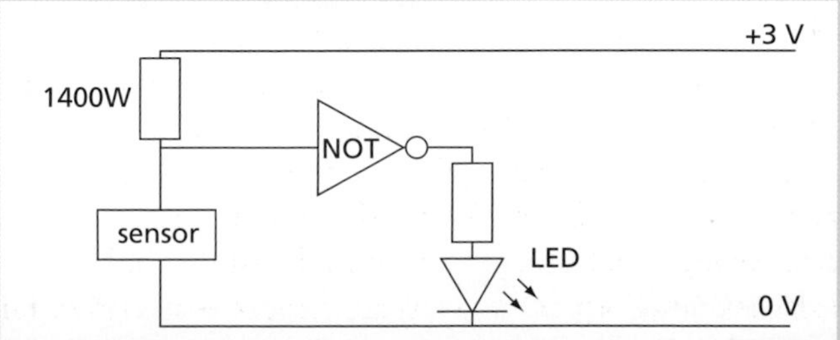

Figure 40.17

continued

(i) Using the information given in part b, what is the resistance of the sensor at 40 °C?
(ii) Use this equation

total resistance of resistors in series = $\text{resistance}_1 + \text{resistance}_2$

to work out the combined resistance of the sensor and resistor at 40 °C.
(iii) What is the voltage across the sensor at 40 °C?
(iv) The graph in Figure 40.18 shows how the output from the NOT gate changes with the input to the NOT gate.

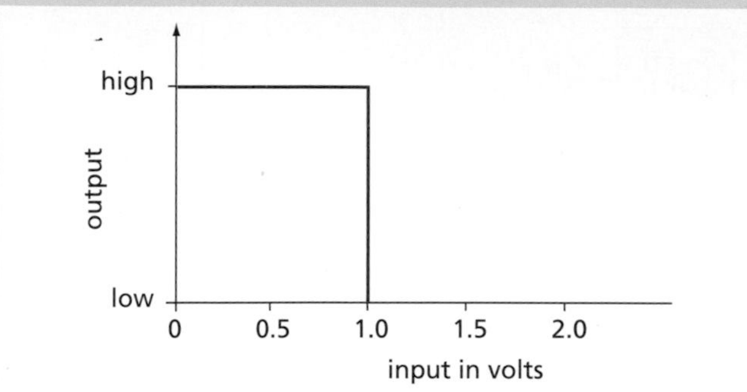

Figure 40.18

Explain why the LED remains ON when the temperature inside the incubator goes above 40 °C.

(*SEG Foundation, Summer 1998*)

7 Figure 40.19 shows a circuit which can be used as an automatic switch.

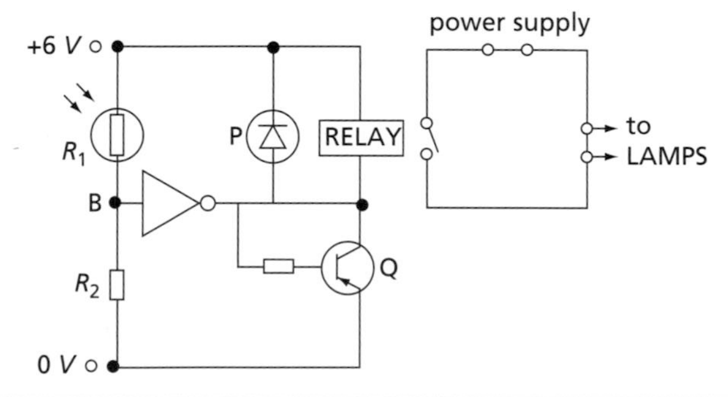

Figure 40.19

a Name the components P, Q and R_1.

Use the information in Figure 40.20 for parts b and c.

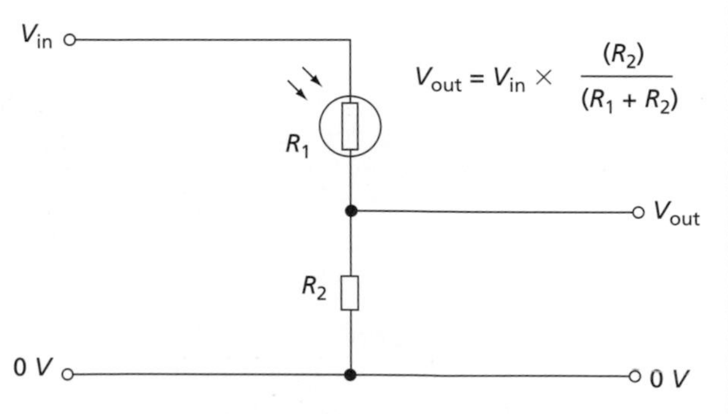

Figure 40.20

b The resistance of R_2 = 2000 Ω. V_{in} is 6 V.
(i) In daylight the resistance of R_1 = 500 Ω. Calculate the voltage across R_2.
(ii) In daylight the lamps will be OFF. Explain why.

c In the dark the resistance of R_1 is 198 000 Ω. Calculate the voltage across R_2.

(*AQA (NEAB) Higher, June 1999*)

Checklist

After studying this chapter you should be able to

- ☐ explain and use the terms **analogue** and **digital**,
- ☐ state that logic gates are switching circuits containing transistors and other components,
- ☐ describe the action of **NOT, OR, NOR, AND** and **NAND** logic gates and recall their truth tables,
- ☐ design and draw block diagrams of logic control systems for given requirements.

Electricity
Additional questions

Static electricity; electric current; potential difference; resistance

1 In the process of electrostatic induction

A a conductor is rubbed with an insulator
B a charge is produced by friction
C negative and positive charges are separated
D a positive charge induces a positive charge
E electrons are 'sprayed' into an object.

2 a (i) Copper is an electrical conductor. What is meant by a *conductor*?
(ii) Ebonite, glass and polythene are electrical insulators. What is meant by an *insulator*?

b Polythene is easily given a negative charge by rubbing it with a dry woollen cloth.
(i) The diagram shows a charged polythene rod being held close to a suspended charged polythene rod.
Complete the phrase,

'like charges . . . . . .'.

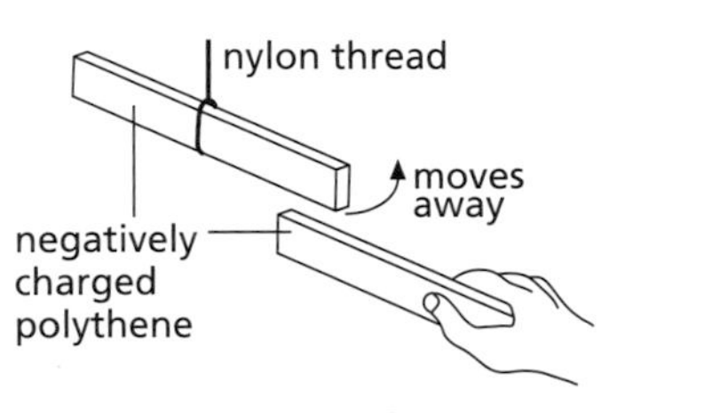

(ii) The diagram below shows rod X being held near the suspended charged polythene rod.
Which of the following might correctly describe rod X?

positively charged glass
negatively charged ebonite
uncharged copper
negatively charged polythene

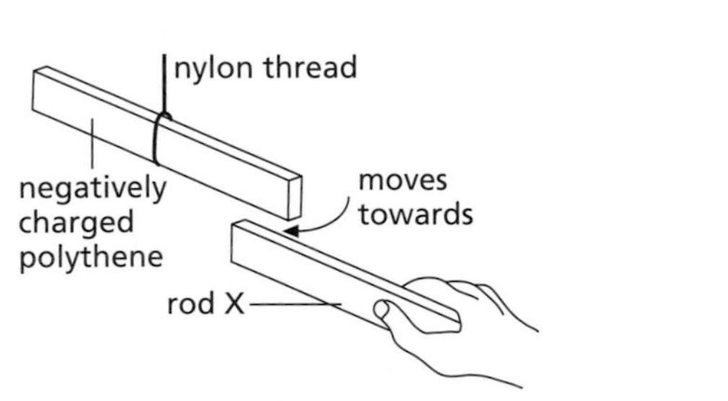

(*UCLES IGCSE Physics Core, Nov 2000*)

3 If two resistors of 25 Ω and 15 Ω are joined together in series and then placed in parallel with a 40 Ω resistor, the effective resistance in ohms of the combination is

A 0.1 **B** 10 **C** 20 **D** 40 **E** 400

4 V_1, V_2, V_3 are the p.ds across the 1 Ω, 2 Ω, and 3 Ω resistors in the following diagram and the current is 5 A. Which one of the columns **A** to **E** shows the correct values of V_1, V_2 and V_3 measured in volts?

	A	B	C	D	E
V_1	1.0	5	0.2	5.0	4.0
V_2	2.0	10	0.4	2.5	3.0
V_3	3.0	15	0.6	1.6	2.0

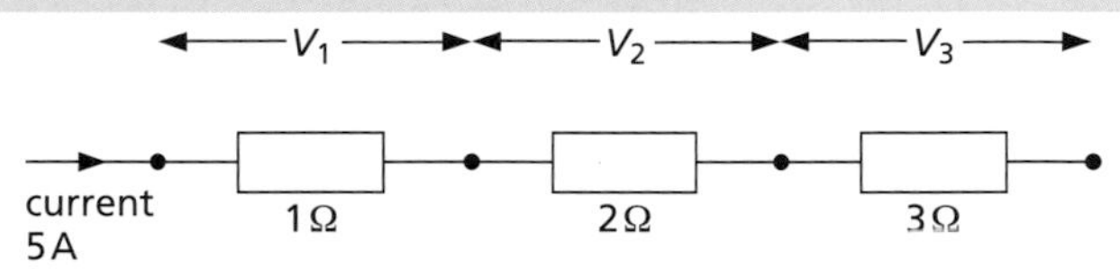

5 a The graph below illustrates how the p.d. across the ends of a conductor is related to the current flowing through it.
(i) What law may be deduced from the graph?
(ii) What is the resistance of the conductor?

b Draw diagrams to show how six 2 V lamps could be lit to normal brightness when using a
(i) 2 V supply,
(ii) 6 V supply,
(iii) 12 V supply.

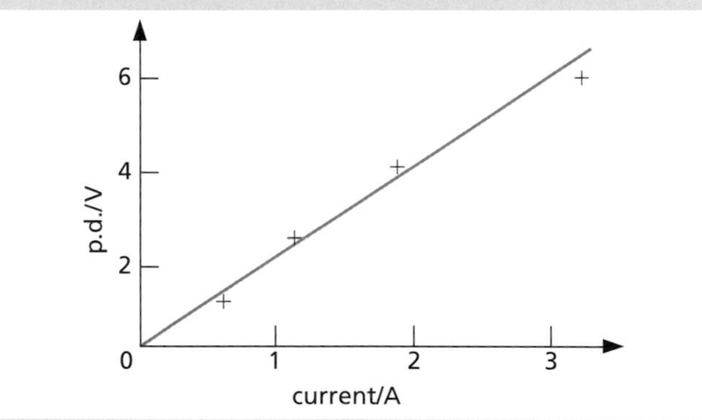

6 When a 4 Ω resistor is connected across the terminals of a 12 V battery, the number of coulombs passing through the resistor per second is

A 0.3 **B** 3 **C** 4 **D** 12 **E** 48

7 The diagram below shows a 2 V cell connected to an arrangement of resistors, part in series and part in parallel.
a What is the total resistance of the two 4 Ω resistors in parallel?
b What is the current flowing in the 1 Ω resistor?
c What is the current in one of the 4 Ω resistors?

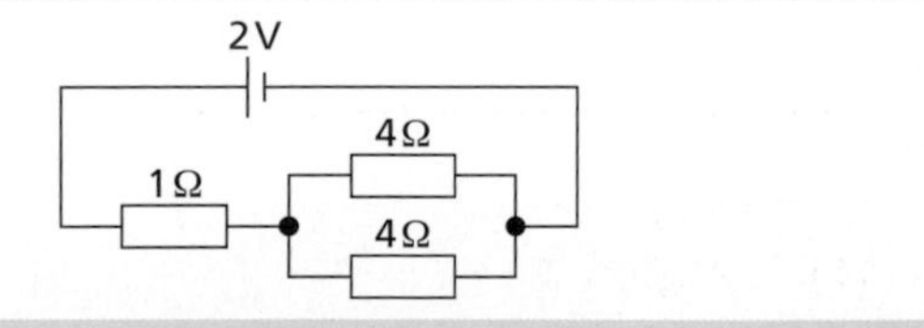

continued

8 A 4 Ω coil and a 2 Ω coil are connected in parallel. What is their combined resistance? A total current of 3 A passes through the coils. What current flows through the 2 Ω coil?

9 The circuit shown below is connected up.

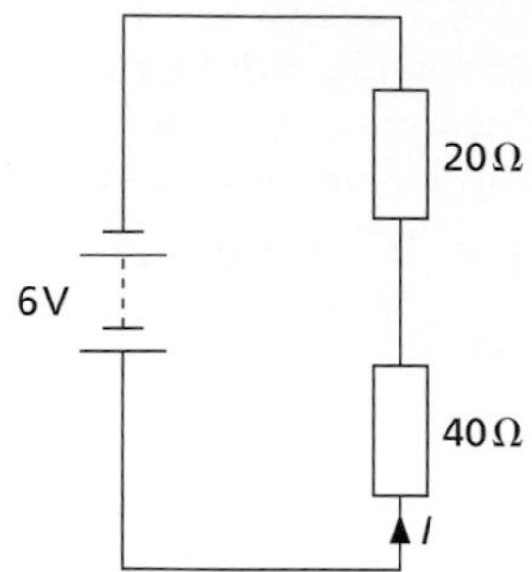

a Calculate the combined resistance of the two resistors in the diagram.

b (i) State the relationship between resistance, p.d. and current by completing the following equation.

resistance = ______________________

(ii) Calculate the current, I, in the circuit. State the unit in your answer.

c Use your answer to **b**(ii) to calculate the p.d. across the 40 Ω resistor. State the unit in your answer.

d The circuit is now used as a potential divider, as shown below.

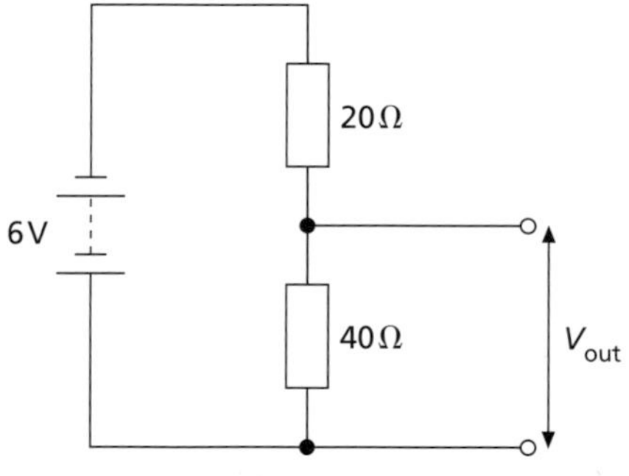

Use your answer to part **c** to state the value of V_{out}, the output voltage of the potential divider.

(*UCLES IGCSE Physics Core, Nov 2000*)

10 Sam is investigating how the resistance of a lamp changes as she alters the current through it. She uses this circuit.

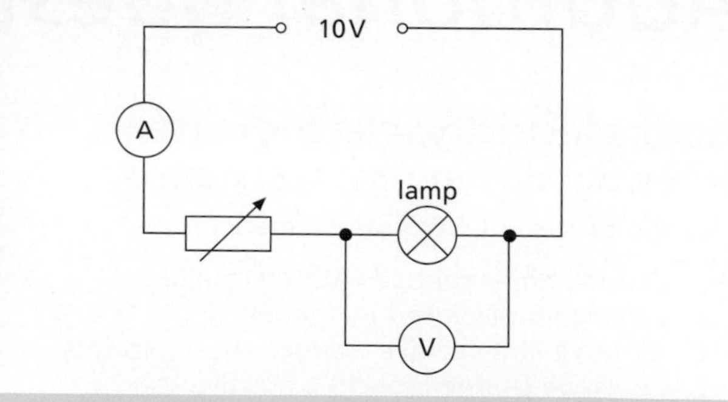

a She adjusts the setting of the variable resistor. Explain how this affects the current.

b She records the values of the voltage across the lamp as the current changes. She plots this graph.

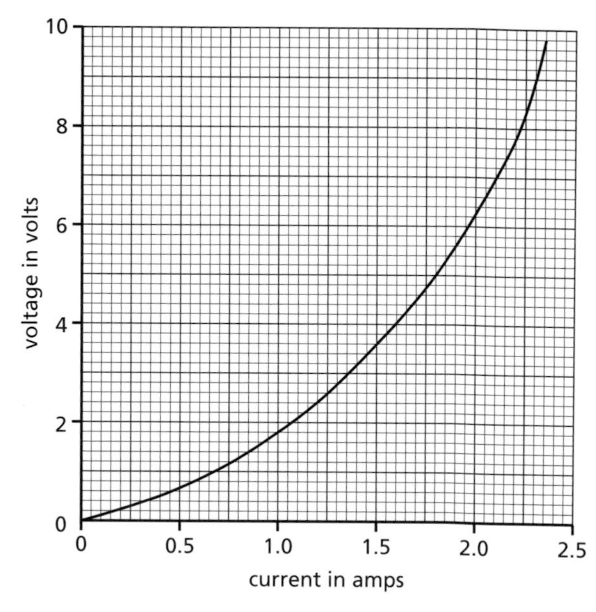

(i) Use the graph to find the value of the current when the voltage is 4.0 V.

(ii) Calculate the resistance of the lamp when the voltage is 4.0 V. You *must* show how you work out your answer.

c How can you tell from the graph that the resistance of the lamp increases between 4.0 V and 8.0 V?

(*OCR Higher, June 1999*)

11 What is the resistance of a wire of length 300 m and cross-section area 1.0 mm^2 made of material of resistivity 1.0×10^{-7} Ω m?

Electric power

12 a Below is a list of wattages of various appliances. State which is most likely to be the correct one for each of the appliances named.

60 W 250 W 850 W 2 kW 3.5 kW

(i) kettle
(ii) table lamp
(iii) iron

b What current will be taken by a 920 W appliance if the supply voltage is 230 V?

13 The diagram shows four designs of switch.

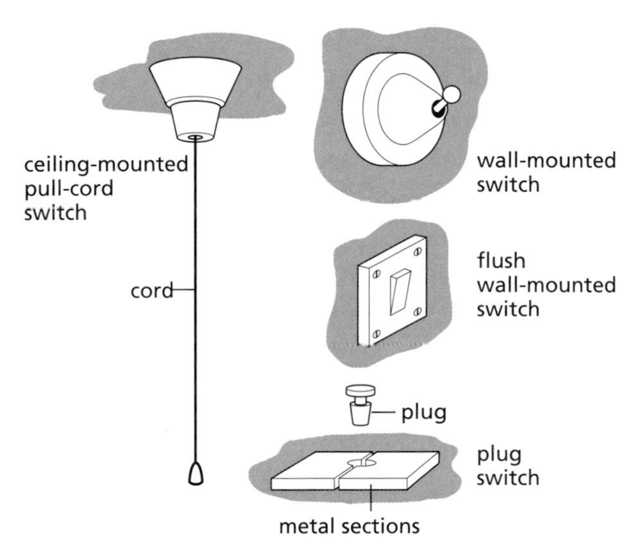

a Which switch is the only one which would be safe to use in a steamy washroom?
b Give a reason for your choice.

(*UCLES IGCSE Physics Core, Nov 1998*)

Electronic systems; digital electronics

14 A certain light-dependent resistor (LDR) has a resistance of 10 MΩ in the dark, a resistance of 200 Ω when bright light shines on it.

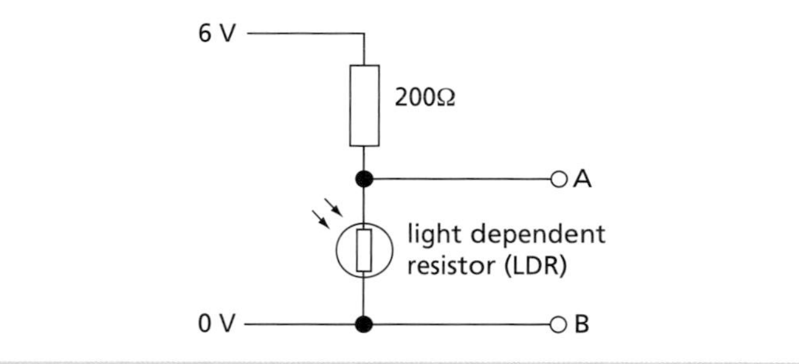

The diagram shows the LDR connected with a 200 Ω resistor in a potential divider circuit. What is the potential difference between A and B when

a the LDR is in the dark?
A just below 6 V **B** 3 V **C** just above 0 V
b bright light shines on the LDR?
A just below 6 V **B** 3 V **C** just above 0 V

(*UCLES IGCSE Physics Core, May 1998*)

15 a What is stored in a charged capacitor?
The components in the diagram below are connected as a time-delay circuit.

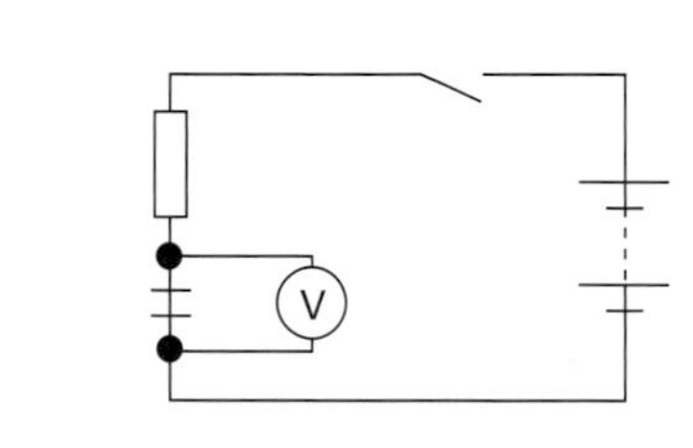

b Describe what happens to the voltmeter after the switch is closed.
c Give one example of an electronic device which make use of a time-delay circuit.

(*UCLES IGCSE Physics Core, May 1997*)

16 The diagram shows an electric lamp being powered by an alternating current supply.

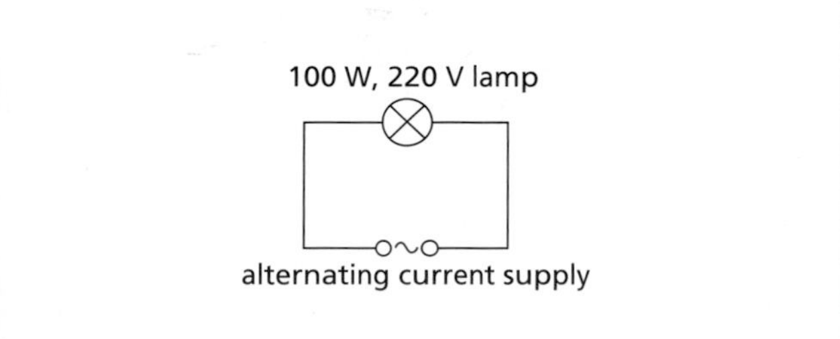

a Explain the meaning of 100 W, 220 V.
b Calculate the current carried by the lamp when it is lit at its normal brightness. Write down the equation that you use and show your working.

continued

c Calculate the resistance of the lamp when it is lit at its normal brightness. Write down the equation that you use and show your working.

d (i) Copy the circuit diagram and add the symbol for a device which could be used to change the alternating current into a direct current.
(ii) State the name of this device by labelling it on your drawing.

(*UCLES IGCSE Physical Science Extended, Nov 2000*)

17 a Draw the symbol for a NOR gate. Label the inputs and the outputs

b State whether the output of a NOR gate will be high (ON) or low (OFF) when
(i) one input is high and one input is low,
(ii) both inputs are high.

c A digital circuit made from three NOT gates and one NAND gate is shown.

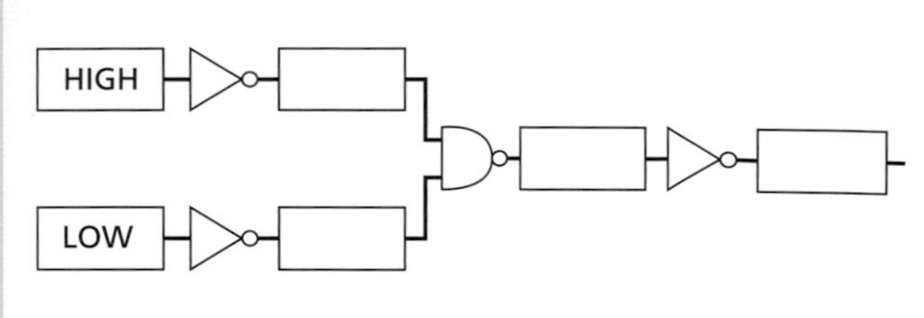

(i) write HIGH or LOW in each of the boxes.
(ii) State the effect on the output of changing both of the inputs.

(*UCLES IGCSE Physics Extended, Nov 2005*)

Electromagnetic effects

41 Magnetic fields

Properties of magnets
Magnetization of iron and steel
Magnetic fields
Earth's magnetic field
Practical work
Plotting lines of force.

Properties of magnets

a) Magnetic materials

Magnets only attract strongly certain materials such as iron, steel, nickel, cobalt, which are called 'ferro-magnetics'.

b) Magnetic poles

The poles are the places in a magnet to which magnetic materials are attracted, e.g. iron filings. They are near the ends of a bar magnet and occur in pairs of equal strength.

c) North and south poles

If a magnet is supported so that it can swing in a horizontal plane it comes to rest with one pole, the north-seeking or N pole, always pointing roughly towards the Earth's north pole. A magnet can therefore be used as a compass.

d) Law of magnetic poles

If the N pole of a magnet is brought near the N pole of another magnet repulsion occurs. Two S (south-seeking) poles also repel. By contrast, N and S poles always attract. The law of magnetic poles summarizes these facts and states:

Like poles repel, unlike poles attract.

The force between magnetic poles decreases as their separation increases.

Magnetization of iron and steel

Chains of small iron nails and steel paper clips can be hung from a magnet, Figure 41.1. Each nail or clip magnetizes the one below it and the unlike poles so formed attract.

If the iron chain is removed by pulling the top nail away from the magnet, the chain collapses, showing that **magnetism induced in iron is temporary**. When the same is done with the steel chain, it does not collapse; **magnetism induced in steel is permanent**.

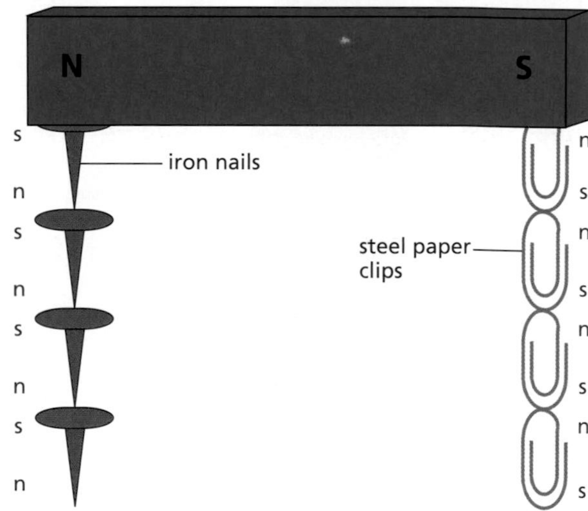

Figure 41.1 Investigating the magnetization of iron and steel

Magnetic materials like iron which magnetize easily but readily lose their magnetism (are easily demagnetized) are said to be **soft**. Those like steel which are harder to magnetize than iron but stay magnetized are **hard**. Both types have their uses; very hard ones are used to make permanent magnets. Solenoids can be used to magnetize and demagnetize magnetic materials (p. 215); dropping or heating a magnet also causes demagnetization.

Magnetic fields

The space surrounding a magnet where it produces a magnetic force is called a **magnetic field**. The force around a bar magnet can be detected and shown to vary in direction using the apparatus in Figure 41.2. If the floating magnet is released near the N pole of the bar magnet, it is repelled to the S pole and moves along a curved path known as a **line of force** or a **field line**. It moves in the opposite direction if its south pole is uppermost.

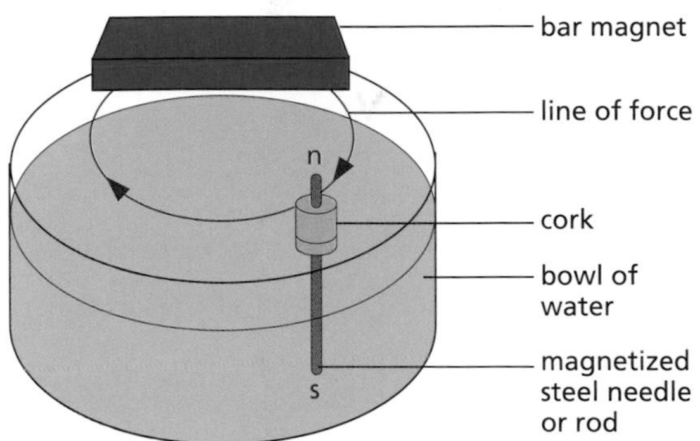

Figure 41.2 Detecting magnetic force

It is useful to consider that a magnetic field has a direction and to represent the field by lines of force. It has been decided that **the direction of the field at any point should be the direction of the force on a N pole**. To show the direction, arrows are put on the lines of force and point away from a N pole towards a S pole. The magnetic field is stronger in regions where the field lines are close together than where they are further apart.

Practical work

Plotting lines of force

a) Plotting compass method

A plotting compass is a small pivoted magnet in a glass case with non-magnetic metal walls, Figure 41.3a.

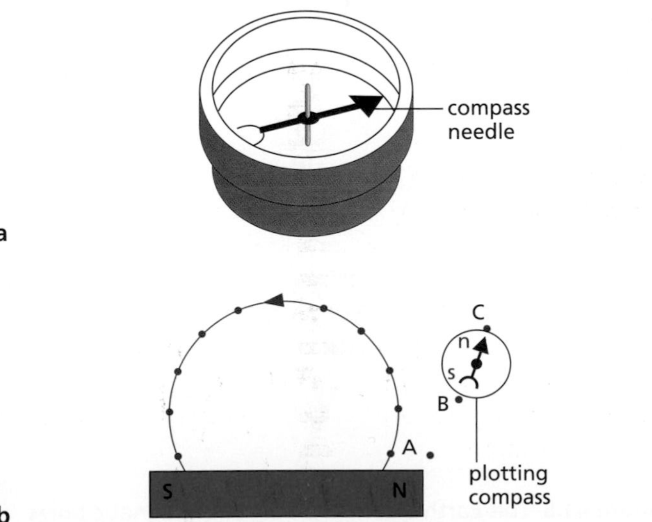

Figure 41.3

Lay a bar magnet on a sheet of paper. Place the plotting compass at a point such as A, Figure 41.3b, near one pole of the magnet. Mark the position of the poles n, s of the compass by pencil dots B, A. Move the compass so that pole s is exactly over B, mark the new position of n by dot C.

Continue this process until the S pole of the bar magnet is reached. Join the dots to give one line of force and show its direction by putting an arrow on it. Plot other lines by starting at different points round the magnet.

A typical field pattern is shown in Figure 41.4.

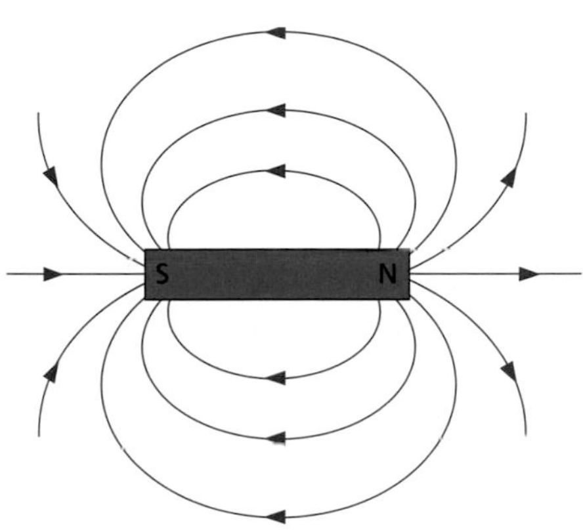

Figure 41.4 Magnetic field lines around a bar magnet

The combined field due to two neighbouring magnets can also be plotted to give patterns like those in Figures 41.5a, b. In part a, where two like poles are facing each other, the point × is called a **neutral point**. At × the field due to one magnet cancels out that due to the other and there are no lines of force.

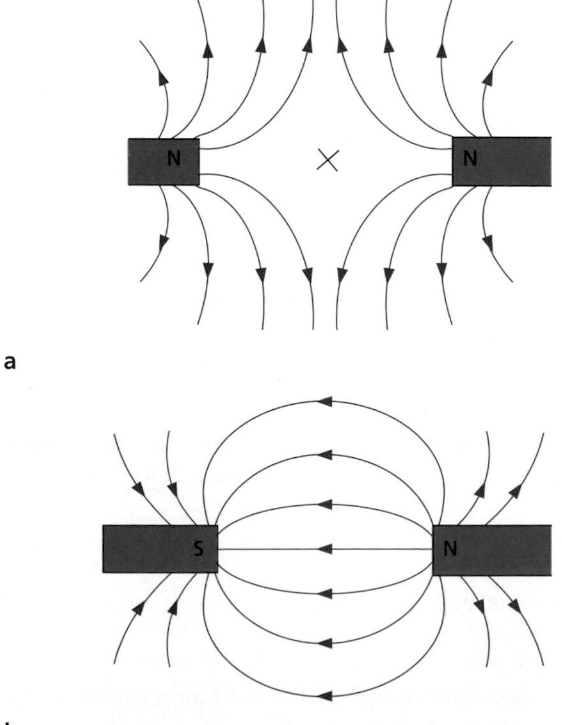

Figure 41.5 Field lines due to two neighbouring magnets

b) Iron filings method

Place a sheet of paper *on top of* a bar magnet and sprinkle iron filings *thinly and evenly* on to the paper from a 'pepper pot'.

Tap the paper gently with a pencil and the filings should form patterns of the lines of force. Each filing turns in the direction of the field when the paper is tapped.

This method is quick but no use for weak fields. Figures 41.6a, b show typical patterns with two magnets. Why are they different? What combination of poles would give the observed patterns?

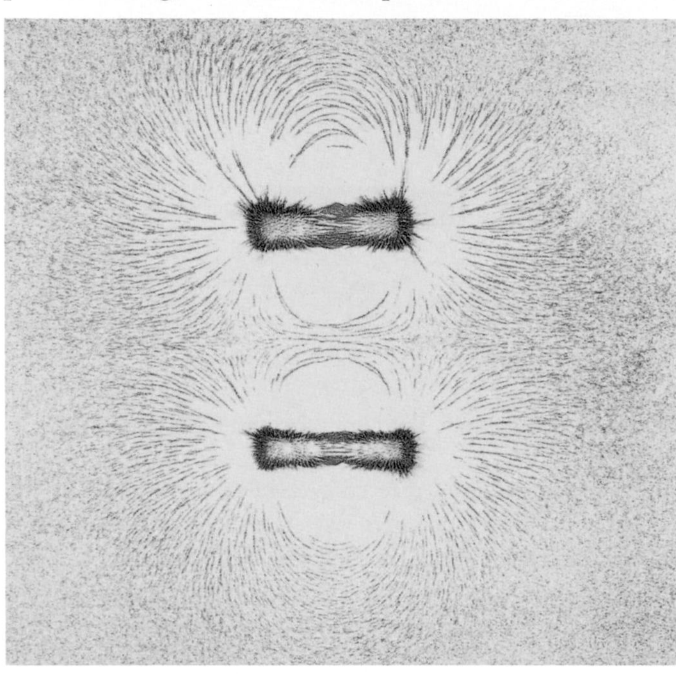

a

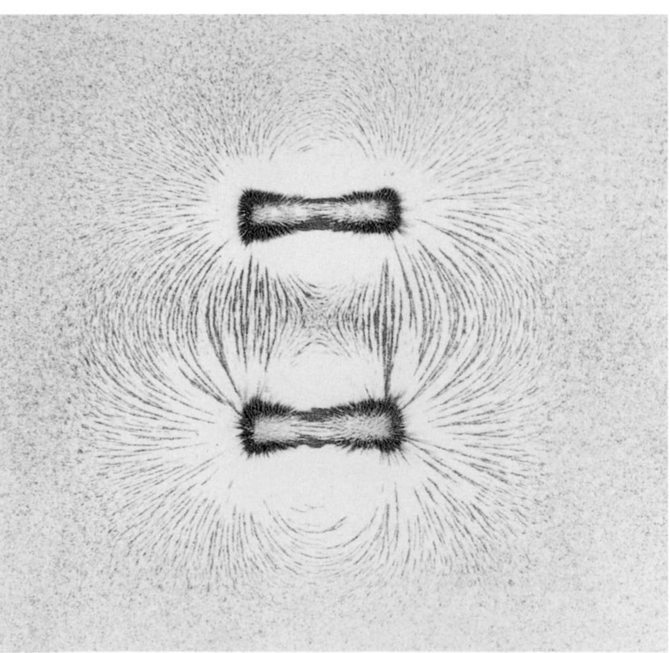

b

Figure 41.6 Field lines round two bar magnets shown by iron filings

F Earth's magnetic field

If lines of force are plotted on a sheet of paper with no magnets nearby, a set of parallel straight lines is obtained. They run roughly from S to N geographically, Figure 41.7, and represent a small part of the Earth's magnetic field in a horizontal plane.

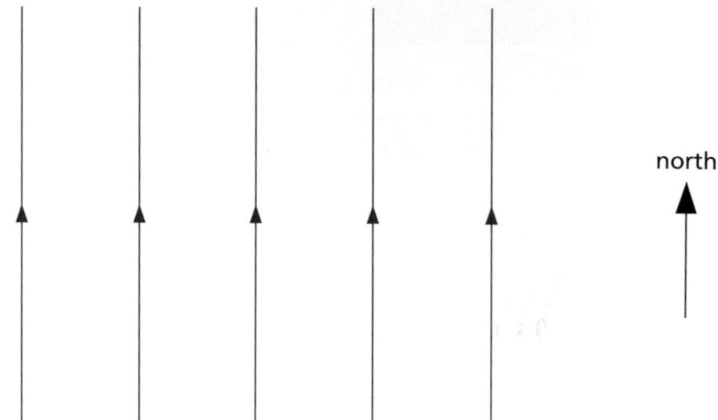

Figure 41.7 Lines of force due to the Earth's field

At most places on the Earth's surface a magnetic compass points slightly east or west of true north, i.e. the Earth's geographical and magnetic north poles do not coincide. The angle between magnetic north and true north is called the **declination**, Figure 41.8. In London at present (2008) it is 6 °W of N and is decreasing. By about the year 2140 it should be 0 °.

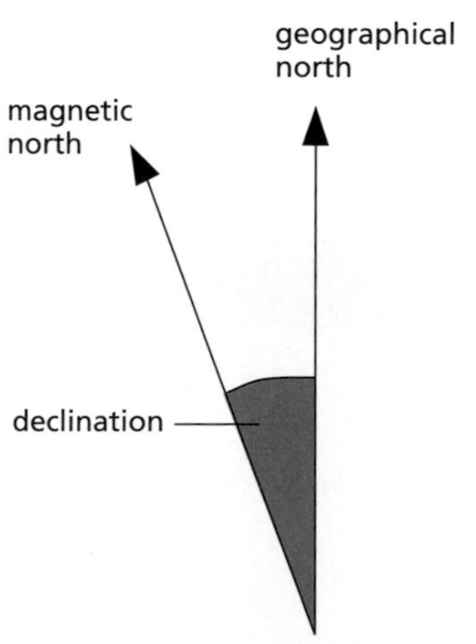

Figure 41.8 The Earth's geographical and magnetic poles do not coincide

Questions

1 A magnet attracts

A plastics B any metal C iron and steel
D aluminium E carbon

2 Copy Figure 41.9 which shows a plotting compass and a magnet. Label the N pole of the magnet and draw the field line on which the compass lies.

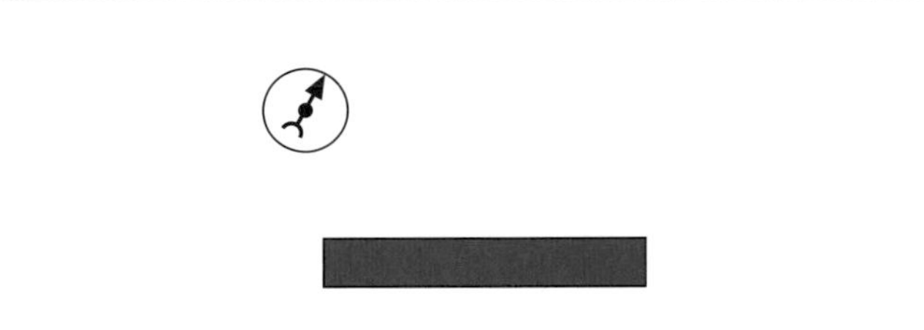

Figure 41.9

3 a Figure 41.10 shows the magnetic field between two magnets.

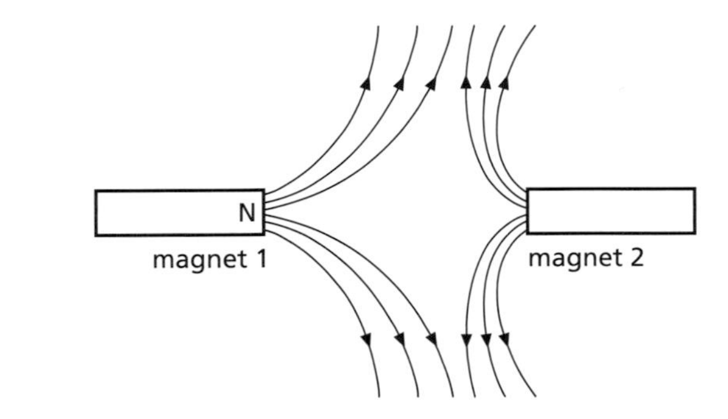

Figure 41.10

(i) Copy the diagram and label the other poles of the magnets.
(ii) Which is the weaker magnet?

b A child has accidentally swallowed some small metal objects. These objects have stuck in the child's throat. A doctor uses the tool in Figure 41.11 to remove them.
(i) What happens to the fixed iron tip when the permanent magnet is moved towards it?
(ii) Why is it an advantage to make the tip out of iron?
(iii) The tool is pushed down the child's throat and guided to the metal object. Explain why the sheath needs to be flexible.
(iv) The doctor tries to use the tool to remove a steel paper clip and an aluminium washer from the child's throat. Which object can the doctor remove? Explain your answer.

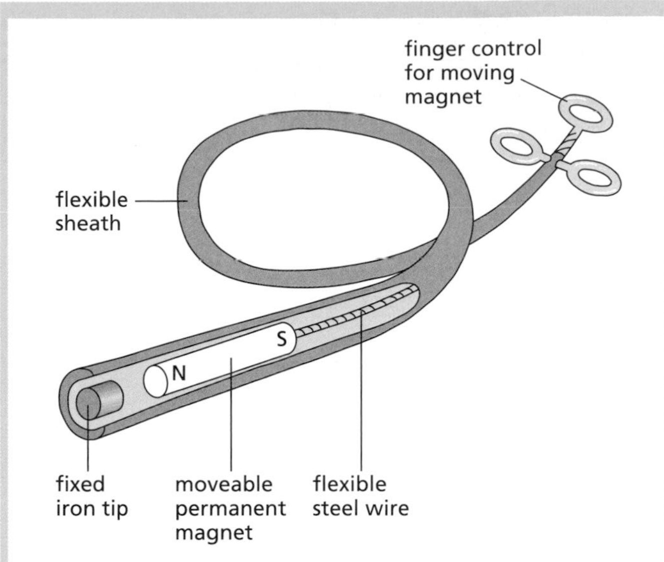

Figure 41.11

(*London Foundation, June 1998*)

Checklist

After studying this chapter you should be able to

- state the properties of magnets,
- explain what is meant by **soft** and **hard** magnetic materials,
- recall that a magnetic field is the region round a magnet where a magnetic force is exerted and is represented by lines of force whose direction at any point is the direction of the force on a N pole,
- map magnetic fields (by the plotting compass and iron filings methods) round (a) one magnet, (b) two magnets,
- recall that at a neutral point the field due to one magnet cancels that due to another.

42 Electromagnets

Oersted's discovery	*Electric bell*
Field due to a straight wire	*Relay, reed switch and circuit breaker*
Field due to a circular coil	*Telephone*
Field due to a solenoid	*Practical work* Simple electromagnet.
Magnetization and demagnetization	
Electromagnets	

Oersted's discovery

In 1819 Oersted accidentally discovered the magnetic effect of an electric current. His experiment can be repeated by holding a wire over and parallel to a compass needle which is pointing N and S, Figure 42.1. The needle moves when the current is switched on. Reversing the current causes the needle to move in the opposite direction.

Evidently around a wire carrying a current there is a magnetic field. As with the field due to a permanent magnet, we represent the field due to a current by field lines or lines of force. Arrows on the lines show the direction of the field, i.e. the direction in which a N pole points.

Different field patterns are given by differently shaped conductors.

current direction
compass needle
S
N
movement of needle
low voltage high current supply

Figure 42.1 An electric current produces a magnetic effect

Field due to a straight wire

If a straight vertical wire passes through the centre of a piece of card held horizontally and a current is passed through the wire, Figure 42.2, iron filings sprinkled on the card set in concentric circles when the card is gently tapped.

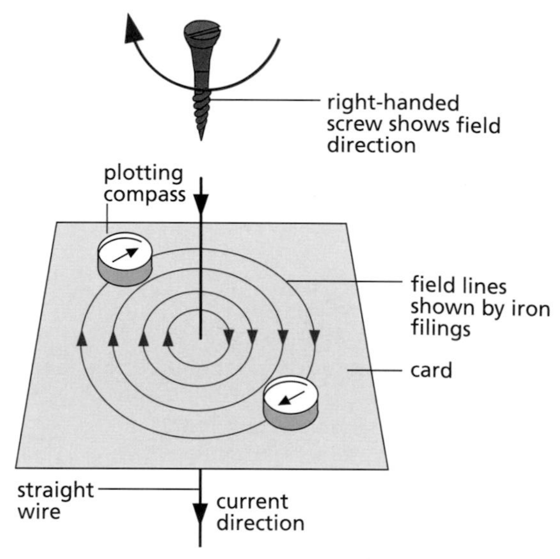

Figure 42.2 Field due to a straight wire

Plotting compasses placed on the card settle along the field lines and show the direction of the field at different points. When the current direction is reversed, the compasses point in the opposite direction showing that the direction of the field reverses when the current reverses.

If the current direction is known, the direction of the field can be predicted by the **right-hand screw rule**:

> If a right-handed screw moves forwards in the direction of the current (conventional), the direction of rotation of the screw gives the direction of the field.

Field due to a circular coil

The field pattern is shown in Figure 42.3. At the centre of the coil the field lines are straight and at right angles to the plane of the coil. The right-hand screw rule again gives the direction of the field at any point.

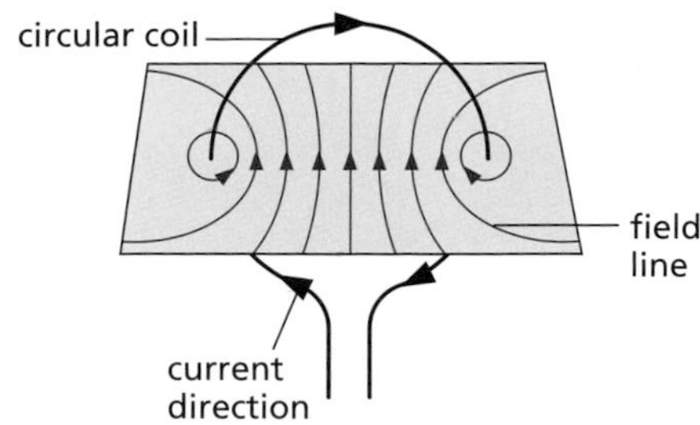

Figure 42.3 Field due to a circular coil

Field due to a solenoid

A solenoid is a long cylindrical coil. It produces a field similar to that of a bar magnet; in Figure 42.4a, end A behaves like a N pole and end B like a S pole. The polarity can be found as before by applying the right-hand screw rule to a short length of one turn of the solenoid. Alternatively the **right-hand grip rule** can be used. This states that if the fingers of the right hand grip the solenoid in the direction of the current (conventional), the thumb points to the N pole, Figure 42.4b. Figure 42.4c shows how to link the end-on view of the current direction in the solenoid to the polarity.

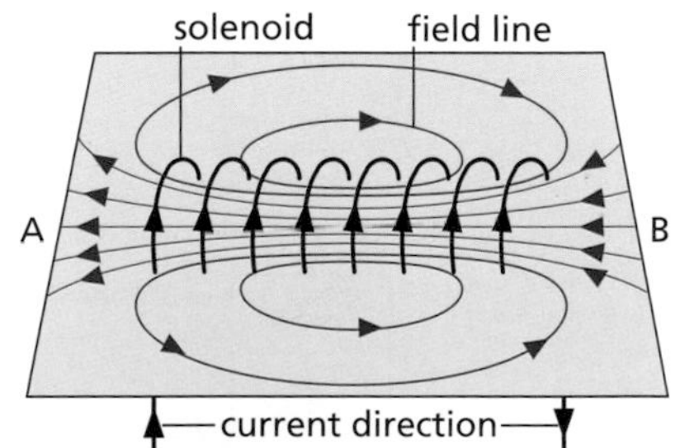

a Field due to a solenoid

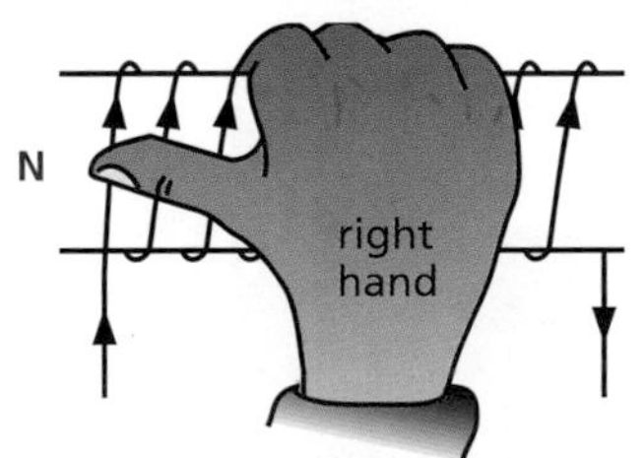

b The right-hand grip rule

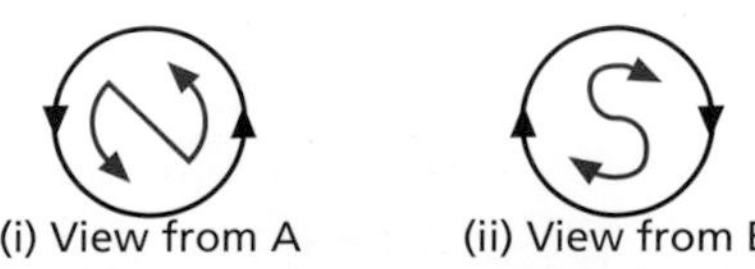

c End-on views

Figure 42.4

Inside the solenoid in Figure 42.4a, the field lines are closer together than they are outside the solenoid. This indicates that the magnetic field is stronger inside a solenoid than outside it.

The field inside a solenoid can be made very strong if it has a large number of turns or a large current. Permanent magnets can be made by allowing molten ferromagnetic metal to solidify in such fields.

Magnetization and demagnetization

A ferromagnetic material can be magnetized by placing it inside a solenoid and gradually increasing the current. This increases the magnetic field strength in the solenoid (the density of the field lines increases), and the material becomes magnetized. Reversing the direction of current flow reverses the direction of the magnetic field and reverses the polarity of the magnetization. A magnet can be demagnetized by placing it inside a solenoid through which the current is repeatedly reversed and reduced.

Practical work

Simple electromagnet

An electromagnet is a coil of wire wound on a soft iron core. A 5 cm iron nail and 3 m of PVC-covered copper wire (SWG 26) are needed.

(a) Leave about 25 cm at one end of the wire (for connecting to the circuit) and then wind about 50 cm as a single layer on the nail. **Keep the turns close together and always wind in the same direction.** Connect the circuit of Figure 42.5, setting the rheostat at its maximum resistance.
Find the number of paper clips the electromagnet can support when known currents between 0.2 A and 2.0 A pass through it. Record the results in a table. How does the 'strength' of the electromagnet depend on the current?

(b) Add another two layers of wire to the nail, winding in the *same direction* as the first layer. Repeat the experiment. What can you say about the 'strength' of an electromagnet and the number of turns of wire?

(c) Place the electromagnet on the bench and under a sheet of paper. Sprinkle iron filings on the paper, tap it gently and observe the field pattern. How does it compare with that given by a bar magnet?

(d) Use the right-hand screw (or grip) rule to predict which end of the electromagnet is a N pole. Check with a plotting compass.

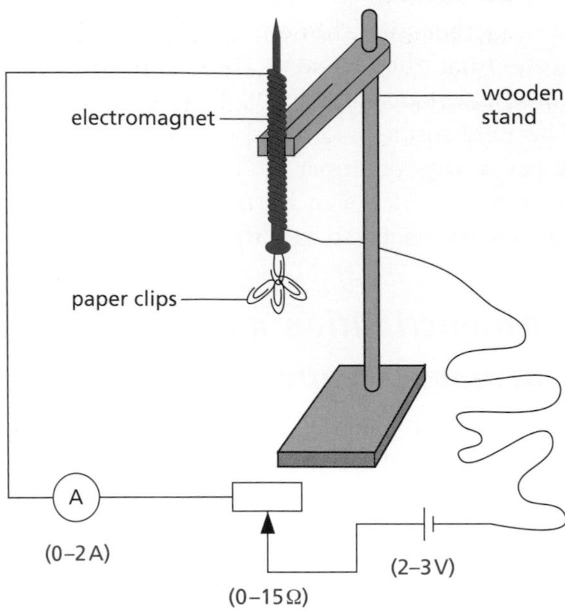

Figure 42.5

Electromagnets

The magnetism of an electromagnet is *temporary* and can be switched on and off, unlike that of a permanent magnet. It has a core of soft iron which is magnetized only when current flows in the surrounding coil.

The strength of an electromagnet increases if

(i) the current in the coil increases,
(ii) the number of turns on the coil increases,
(iii) the poles are moved closer together.

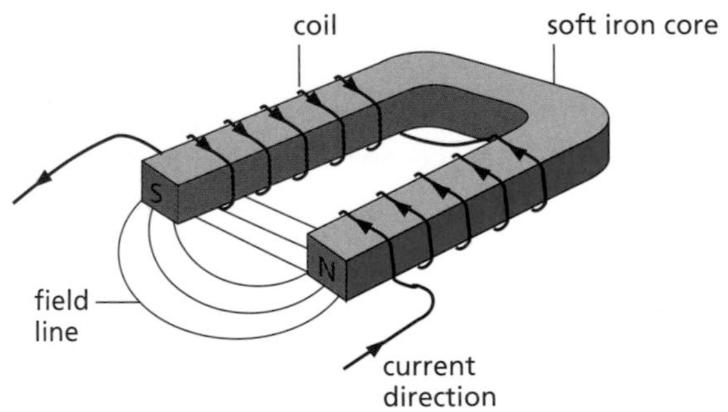

Figure 42.6 C-core or horseshoe electromagnet

In C-core (or horseshoe) electromagnets condition (iii) is achieved, Figure 42.6. Note that the coil is wound in *opposite* directions on each limb of the core.

As well as being used as a crane to lift iron objects, scrap iron, etc., Figure 42.7, an electromagnet is an essential part of many electrical devices.

Figure 42.7 Electromagnet being used to lift scrap metal

F Electric bell

When the circuit in Figure 42.8 is completed, current flows in the coils of the electromagnet which becomes magnetized and attracts the soft iron bar (the armature).

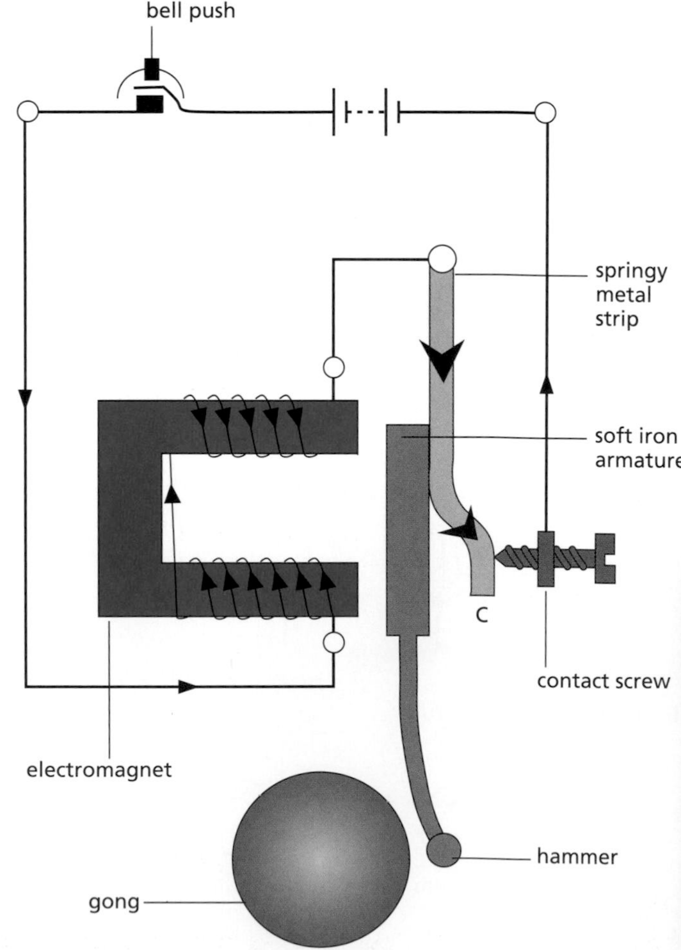

Figure 42.8 Electric bell

The hammer hits the gong but the circuit is now broken at the point C of the contact screw.

The electromagnet loses its magnetism (becomes demagnetized) and no longer attracts the armature. The springy metal strip is then able to pull the armature back, remaking contact at C and so completing the circuit again. This cycle is repeated so long as the bell push is depressed and continuous ringing occurs.

Relay, reed switch and circuit breaker

a) Relay

A relay is a switch based on the principle of an electromagnet. It is useful if we want one circuit to control another, especially if the current and power are larger in the second circuit (see question 2, p. 237). Figure 42.9 shows a typical relay. When a current is in the coil from the circuit connected to AB, the soft iron core is magnetized and attracts the L-shaped iron armature. This rocks on its pivot and closes the contacts at C in the circuit connected to DE. The relay is then 'energized' or 'on'.

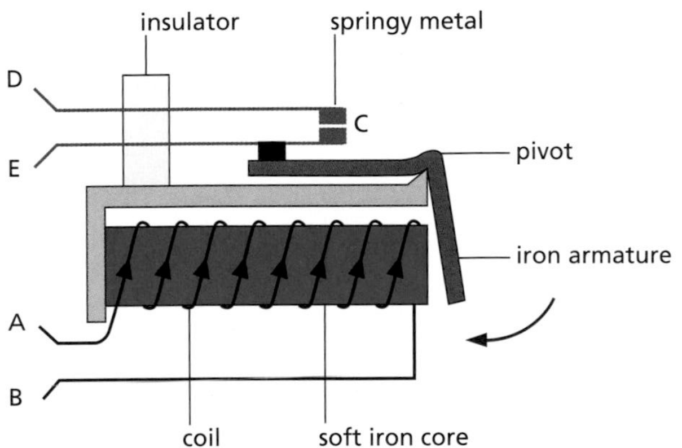

Figure 42.9 Relay

The current needed to operate a relay is called the **pull-on** current, and the **drop-off** current is the smaller current in the coil when the relay just stops working. If the coil resistance R of a relay is 185 Ω and its operating p.d. V is 12 V, the pull-on current $I = V/R = 12/185 = 0.065$ A $=$ 65 mA. The symbols for relays with normally open and normally closed contacts are given in Figure 42.10a and b.

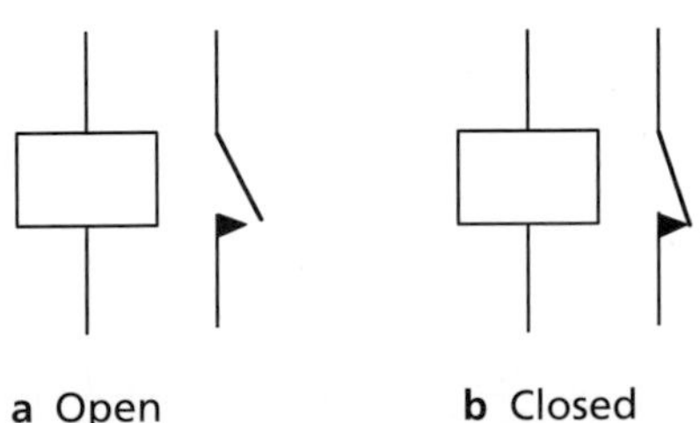

Figure 42.10 Symbols for a relay

Some examples of the use of relays in circuits appear in Chapter 39.

b) Reed switch

One such switch is shown in Figure 42.11a. When current flows in the coil, the magnetic field produced magnetizes the strips (called 'reeds') of magnetic material. The ends become opposite poles and one reed is attracted to the other so completing the circuit connected to AB. The reeds separate when the current in the coil is switched off. This type of reed switch is sometimes called a 'reed relay'.

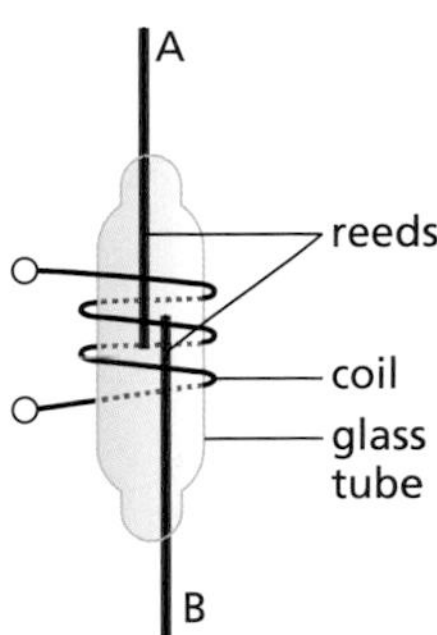

a Reed switch

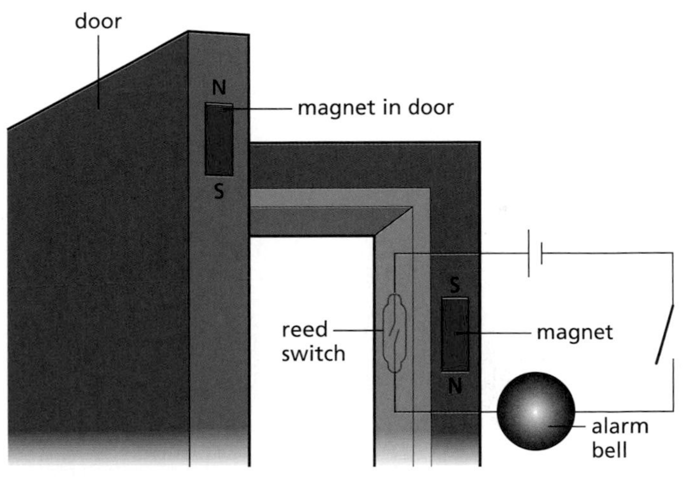

b Burglar alarm activated by a reed switch

Figure 42.11

Reed switches are also operated by permanent magnets. Figure 42.11b shows the use of a normally open reed switch as a burglar alarm. How does it work?

c) Circuit breaker

A circuit breaker (p. 187) acts in a similar way to a normally closed relay; when the current in the electromagnet exceeds a critical value the contact points are separated and the circuit is broken. In the design shown in question 4 on p. 219, when the iron bolt is attracted far enough towards the electromagnet, the plunger is released and the push switch opens, breaking contact to the rest of the circuit.

F Telephone

A telephone contains a microphone at the speaking end and a receiver at the listening end.

a) Carbon microphone

When someone speaks into a carbon microphone, Figure 42.12, sound waves cause the diaphragm to move backwards and forwards. This varies the pressure on the carbon granules between the movable carbon dome which is attached to the diaphragm and the fixed carbon cup at the back. When the pressure increases, the granules are squeezed closer together and their electrical resistance decreases. A decrease of pressure has the opposite effect. The current passing through the microphone varies in a similar way to the sound wave variations.

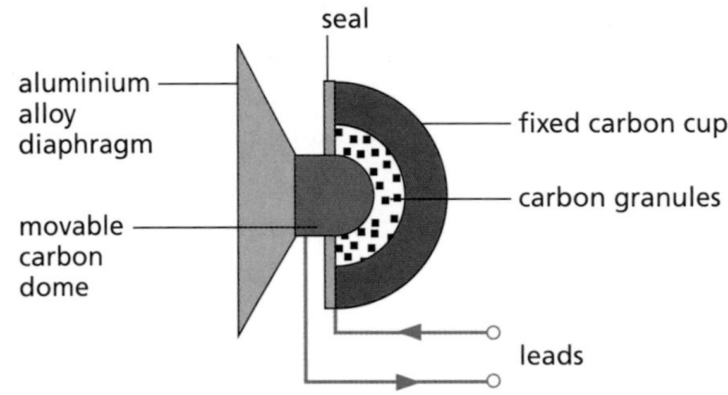

Figure 42.12 Carbon microphone

b) Receiver

The coils are wound in opposite directions on the *two* S poles of a magnet, Figure 42.13. If the current goes round one in a clockwise direction, it goes round the other anticlockwise, so making one S pole stronger and the other weaker. This causes the iron armature to rock on its pivot towards the stronger S pole. When the current reverses, the armature rocks the other way due to the S pole which was the stronger before becoming the weaker. These armature movements are passed on to the diaphragm, making it vibrate and produce sound of the same frequency as the alternating current in the coil (received from the microphone).

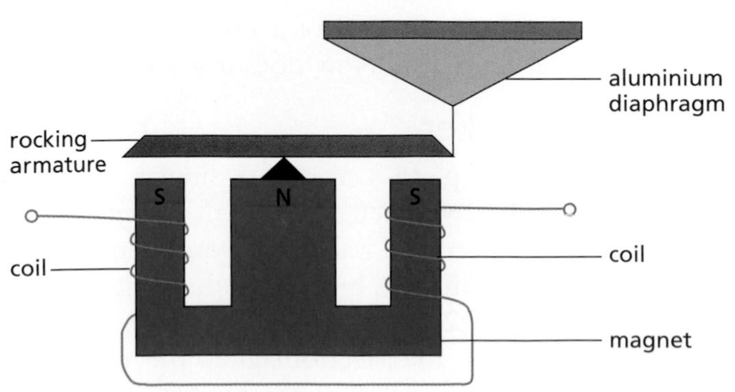

Figure 42.13 Telephone receiver

Questions

1 The vertical wire in Figure 42.14 is at right angles to the card. In what direction will a plotting compass at A point when
a there is no current in the wire,
b current flows upwards?

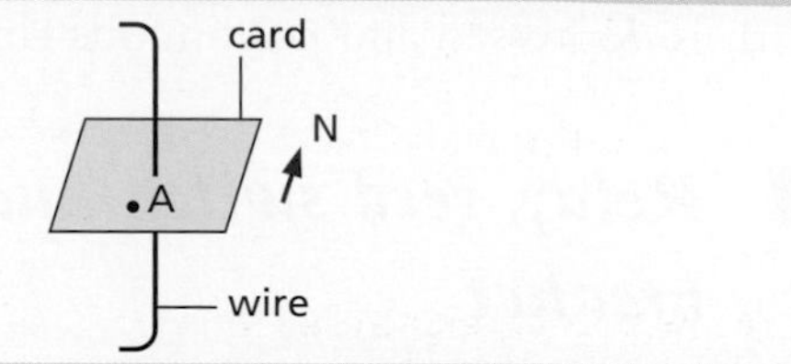

Figure 42.14

2 Figure 42.15 shows a solenoid wound on a core of soft iron. Will the end A be a N pole or S pole when the current flows in the direction shown?

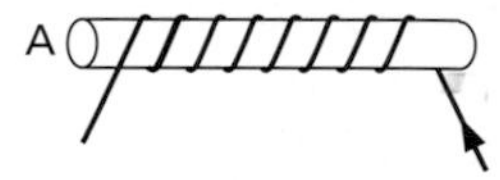

Figure 42.15

3 Figure 42.16 shows an electromagnet being used to lift some weights.

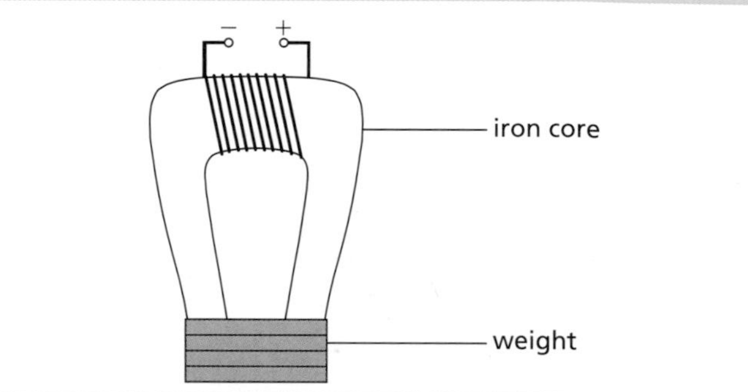

Figure 42.16

The graph in Figure 42.17 shows how the weight lifted by the electromagnet depends on the current in the coil.

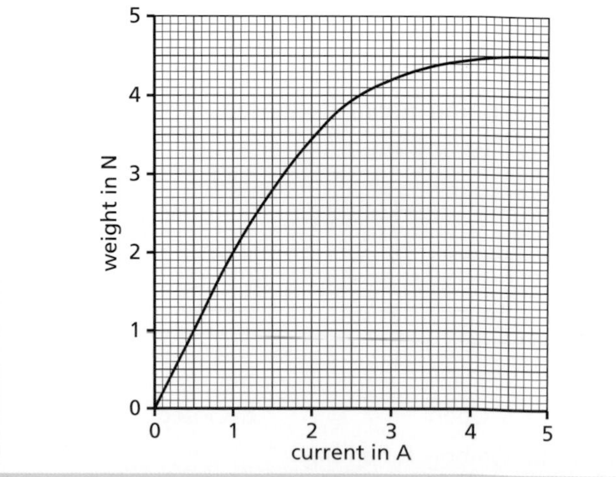

Figure 42.17

a (i) What is the heaviest weight that the electromagnet can lift?
(ii) Suggest how the electromagnet can be changed so that it will lift a heavier weight.

b The electromagnet is used to pick up iron cans filled with fruit juice. Each can weighs 2 N.
(i) What current is needed for the electromagnet to lift one can?
(ii) When this current is doubled, is the electromagnet able to lift two cans at the same time? Use data from the graph to give the reason for your answer.

(*London Foundation, June 1999*)

4 A fault in an electrical circuit can cause too great a current to flow. Some circuits are switched off by a circuit breaker.

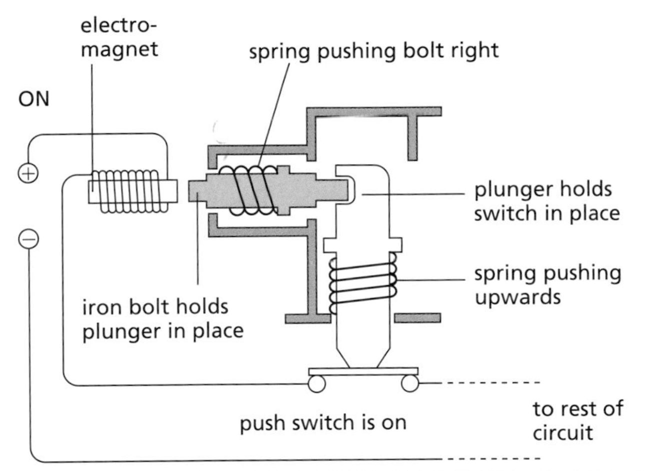

Figure 42.18

One type of circuit breaker is shown in Figure 42.18. A normal current is flowing. Explain, in detail, what happens when a current which is bigger than normal flows.

(*AQA (NEAB) Foundation, June 1999*)

Checklist

After studying this chapter you should be able to

- describe and draw sketches of the magnetic fields round current-carrying, straight and circular conductors and solenoids,
- recall the right-hand screw and right-hand grip rules for relating current direction and magnetic field direction,
- □ describe the effect on the magnetic field of changing the magnitude and direction of the current in a solenoid,
- □ identify regions of different magnetic field strength around a solenoid,
- make a simple electromagnet,
- describe uses of electromagnets,
- explain the action of an electric bell, a relay, a reed switch and a circuit breaker.

43 Electric motors

The motor effect	***Moving-coil loudspeaker***
Fleming's left-hand rule	***Practical work***
Simple d.c. electric motor	A model motor.
Practical motors	

Electric motors form the heart of a whole host of electrical devices ranging from domestic appliances such as vacuum cleaners and washing machines to electric trains and lifts. In a car the windscreen wipers are usually driven by one and the engine is started by another.

The motor effect

A wire carrying a current in a magnetic field experiences a force. If the wire can move it does so.

a) Demonstration

In Figure 43.1 the flexible wire is loosely supported in the strong magnetic field of a C-shaped magnet (permanent or electro). When the switch is closed, current flows in the wire which jumps upwards as shown. If either the direction of the current or the direction of the field is reversed, the wire moves downwards. **The force increases if the strength of the field increases and if the current increases.**

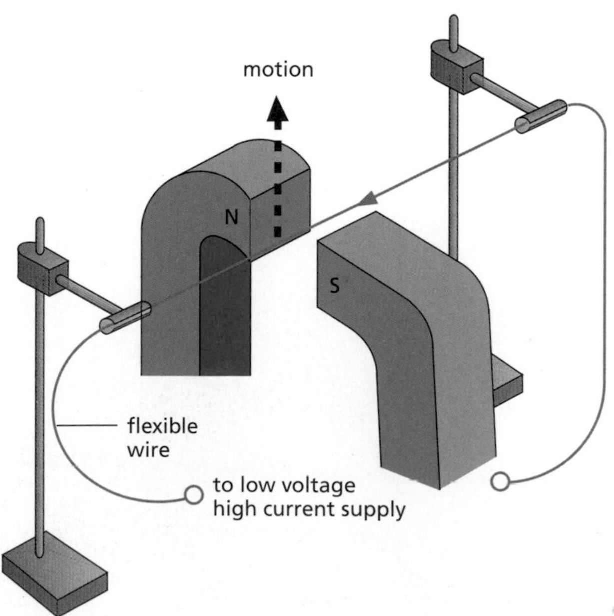

Figure 43.1 A wire carrying a current in a magnetic field experiences a force

b) Explanation

Figure 43.2a is a side view of the magnetic field lines due to the wire and the magnet. Those due to the wire are circles and we will assume their direction is as shown. The dotted lines represent the field lines of the magnet and their direction is towards the right.

The resultant field obtained by combining both fields is shown in Figure 43.2b. There are more lines below than above the wire since both fields act in the same direction below but in opposition above. If we *suppose* the lines are like stretched elastic, those below will try to straighten out and in so doing will exert an upward force on the wire.

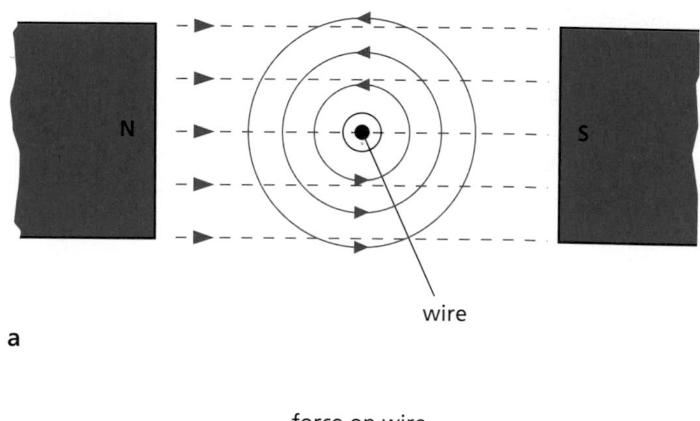

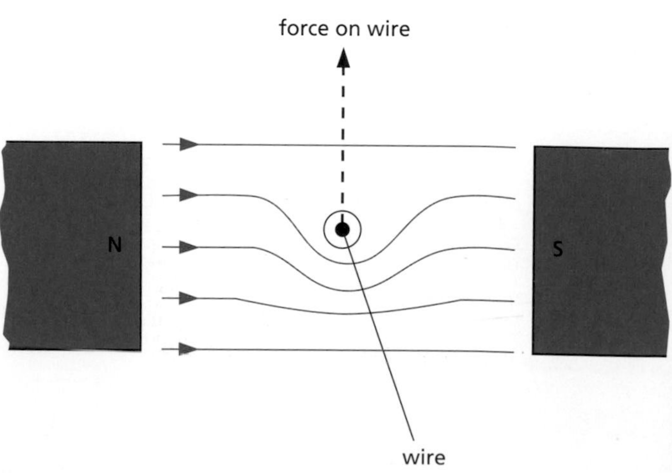

Figure 43.2

☐ *Fleming's left-hand rule*

The direction of the force or thrust on the wire can be found by this rule which is also called the 'motor rule', Figure 43.3.

> Hold the thumb and first two fingers of the left hand at right angles to each other with the **F**irst finger pointing in the direction of the **F**ield and the se**C**ond finger in the direction of the **C**urrent, then the **Th**umb points in the direction of the **Th**rust.

If the wire is not at right angles to the field, the force is smaller and is zero if the wire is parallel to the field.

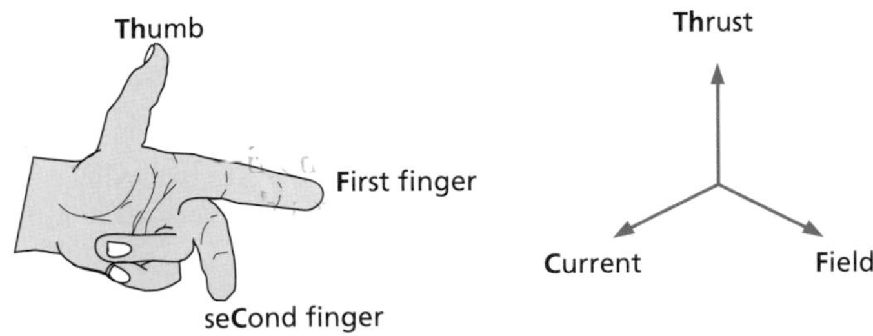

Figure 43.3 Fleming's left-hand (motor) rule

Simple d.c. electric motor

A simple motor to work from direct current (d.c.) consists of a rectangular coil of wire mounted on an axle which can rotate between the poles of a C-shaped magnet, Figure 43.4. Each end of the coil is connected to half of a split ring of copper, called a **commutator**, which rotates with the coil. Two carbon blocks, the **brushes**, are pressed lightly against the commutator by springs. The brushes are connected to an electrical supply.

If Fleming's left-hand rule is applied to the coil in the position shown, we find that side **ab** experiences an upward force and side **cd** a downward force. (No forces act on **ad** and **bc** since they are parallel to the field.) These two forces form a **couple** which rotates the coil in a clockwise direction until it is vertical.

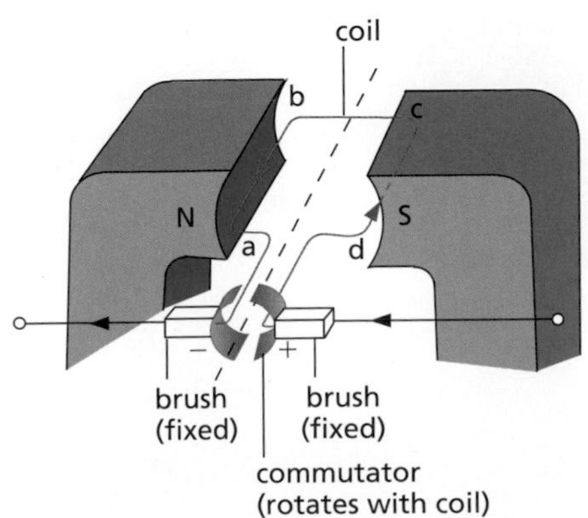

Figure 43.4 Simple d.c. motor

The brushes are then in line with the gaps in the commutator and the current stops. However, because of its inertia, the coil overshoots the vertical and the commutator halves change contact from one brush to the other. This reverses the current through the coil and so also the directions of the forces on its sides. Side **ab** is on the right now, acted on by a downward force, while **cd** is on the left with an upward force. The coil thus carries on rotating clockwise.

The more turns there are on the coil, or the larger the current through it, the greater is the couple on the coil and the faster it turns.

Practical motors

Practical motors have:

(a) **a coil of many turns wound on a soft iron cylinder or core** which rotates with the coil. This makes it more powerful. The coil and core together are called the **armature**.

(b) **several coils** each in a slot in the core and each having a pair of commutator segments. This gives increased power and smoother running. The motor of an electric drill is shown in Figure 43.5.

(c) **an electromagnet** (usually) to produce the field in which the armature rotates.

Most electric motors used in industry are **induction motors**. They work off a.c. (alternating current) on a different principle from the d.c. motor.

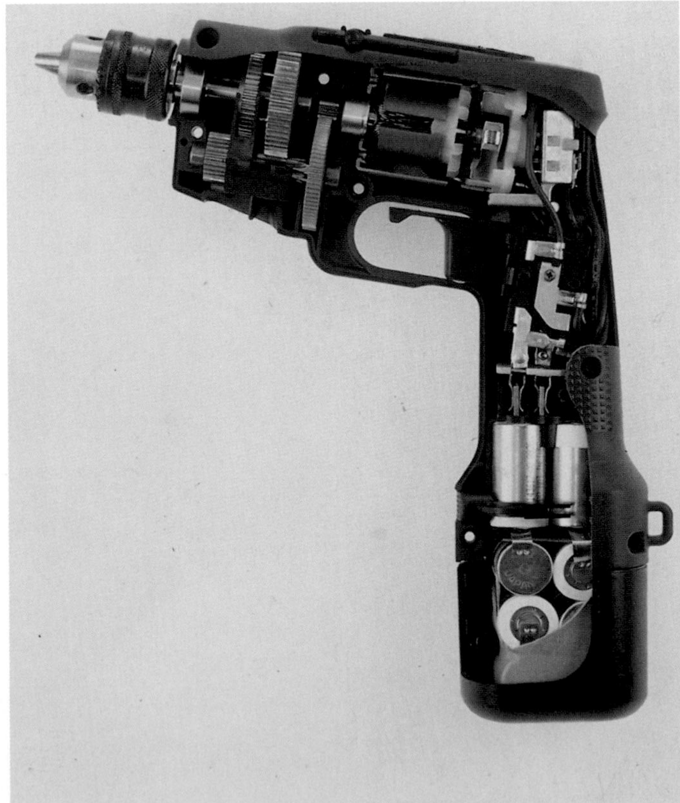

Figure 43.5 Motor inside an electric drill

Practical work

A model motor

The motor shown in Figure 43.6 is made from a kit.

1 Wrap Sellotape round one end of the metal tube which passes through the wooden block.
2 Cut two rings off a piece of narrow rubber tubing; slip them on to the Sellotaped end of the metal tube.
3 Remove the insulation from one end of a 1.5 metre length of SWG 26 PVC-covered copper wire and fix it under both rubber rings so that it is held tight against the Sellotape. This forms one end of the coil.
4 Wind 10 turns of the wire in the slot in the wooden block and finish off the second end of the coil by removing the PVC and fixing this too under the rings but on the *opposite* side of the tube from the first end. The bare ends act as the **commutator**.
5 Push the axle through the metal tube of the wooden base so that the block spins freely.
6 Arrange two 0.5 metre lengths of wire to act as **brushes** and leads to the supply, as shown. Adjust the brushes so that they are vertical and each touches one bare end of the coil when the plane of the coil is horizontal. **The motor will not work if this is not so.**
7 Slide the base into the magnet with *opposite poles facing*. Connect to a 3 V battery (or other low voltage d.c. supply) and a slight push of the coil should set it spinning at high speed.

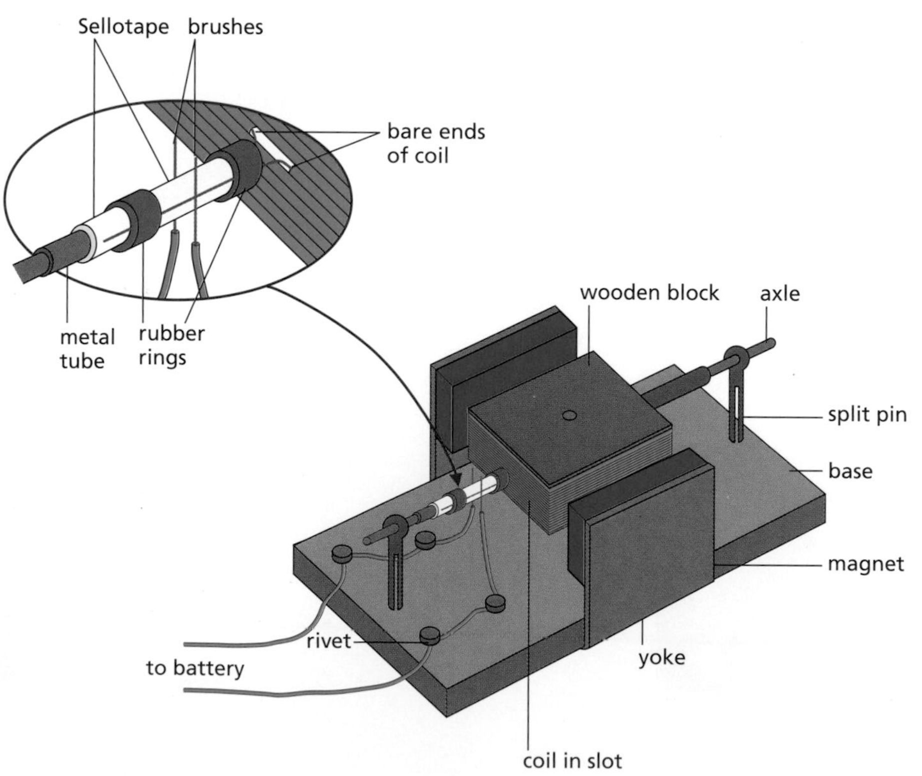

Figure 43.6 A model motor

F Moving-coil loudspeaker

Varying currents from a radio, disc player, etc. pass through a short cylindrical coil whose turns are at right angles to the magnetic field of a magnet with a central pole and a surrounding ring pole, Figure 43.7a.

A force acts on the coil which, according to Fleming's left-hand rule, makes it move in and out. A paper cone attached to the coil moves with it and sets up sound waves in the surrounding air, Figure 43.7b.

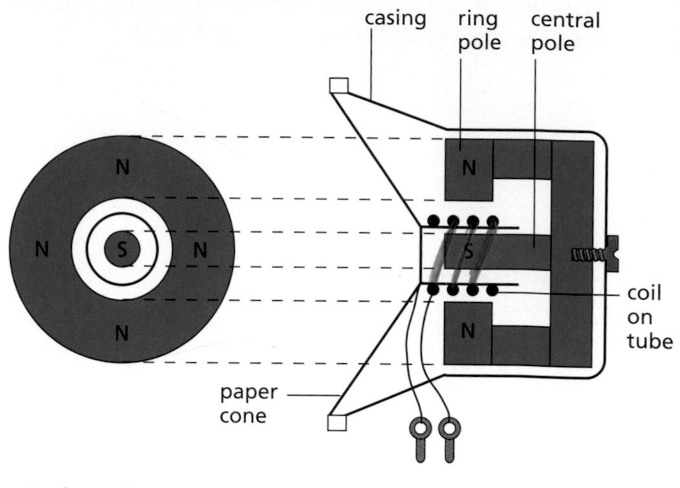

Figure 43.7 Moving-coil loudspeaker

Questions

1 A current flows in a wire running between the N and S poles of a magnet lying horizontally as shown in Figure 43.8. The force on the wire due to the magnet is directed

A from N to S
B from S to N
C opposite to the current direction
D in the direction of the current
E vertically upwards.

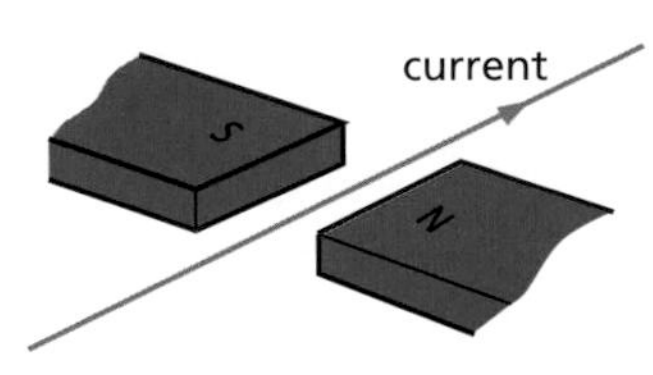

Figure 43.8

2 In the simple electric motor of Figure 43.9, the coil rotates anticlockwise as seen by the eye from the position X when current flows in the coil. Is the current flowing clockwise or anticlockwise around the coil when viewed from above?

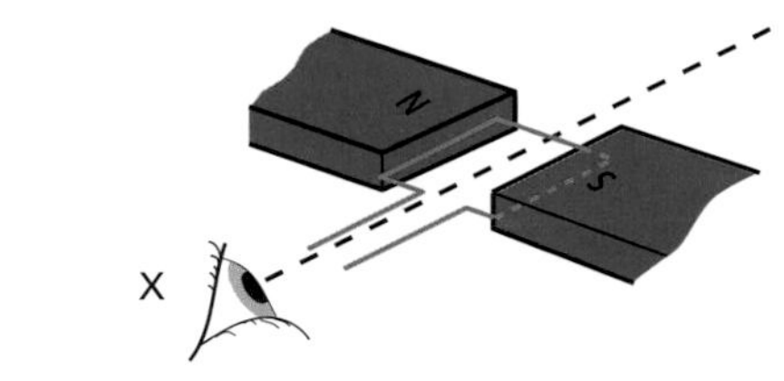

Figure 43.9

3 An electric motor is a device which transfers

A mechanical energy to electrical energy
B heat energy to electrical energy
C electrical energy to heat only
D heat to mechanical energy
E electrical energy to mechanical energy and heat.

Checklist

After studying this chapter you should be able to

- describe a demonstration to show that a force acts on a current-carrying conductor in a magnetic field and recall that it increases with the strength of the field and the size of the current,
- draw the resultant field pattern for a current-carrying conductor which is at right angles to a uniform magnetic field,
- explain why a rectangular current-carrying coil experiences a couple in a uniform magnetic field,
- draw a diagram of a simple d.c. electric motor and explain how it works,
- describe a practical d.c. motor.

44 Electric meters

Moving-coil galvanometer
Ammeters and shunts
Voltmeters and multipliers
Multimeters
Reading a voltmeter

Moving-coil galvanometer

A galvanometer detects small currents or small p.ds, often of the order of milliamperes (mA) or millivolts (mV).

In the moving-coil **pointer-type** meter, a coil is pivoted between the poles of a permanent magnet, Figure 44.1a. Current enters and leaves the coil by hair springs above and below it. When current flows, a couple acts on the coil (as in an electric motor), causing it to rotate until stopped by the springs. The greater the current, the greater the deflection which is shown by a pointer attached to the coil.

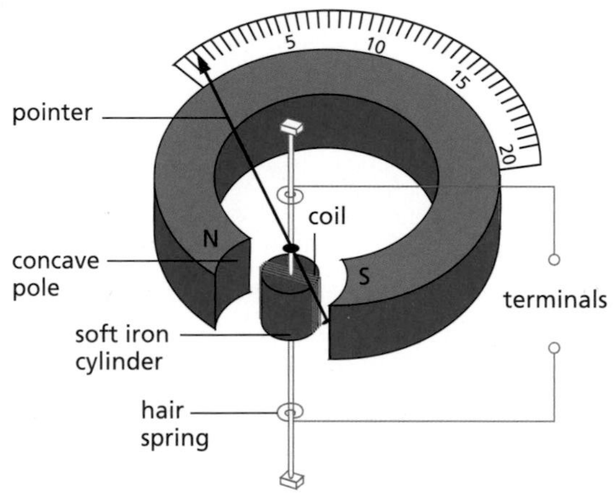

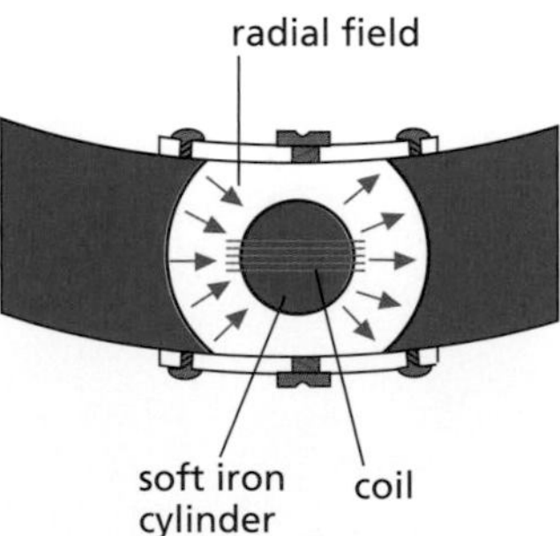

b View from above

Figure 44.1 Moving-coil pointer-type galvanometer

The soft iron cylinder at the centre of the coil is fixed and along with the concave poles of the magnet it produces a **radial** field, Figure 44.1b, i.e. the field lines are directed to the centre of the cylinder. The scale on the meter is then even or linear, i.e. all divisions are the same size.

The sensitivity of a galvanometer is increased by having

(i) more turns on the coil,
(ii) a stronger magnet,
(iii) weaker hair springs or a wire suspension,
(iv) as a pointer, a long beam of light reflected from a mirror on the coil.

The last two are used in **light-beam** meters which have a full-scale deflection of a few microamperes (μA). ($1\ \mu A = 10^{-6}$ A)

Ammeters and shunts

An ammeter is a galvanometer having a known low resistance, called a **shunt**, in parallel with it to take most of the current, Figure 44.2. An ammeter is placed in *series* in a circuit and must have a *low resistance* otherwise it changes the current to be measured.

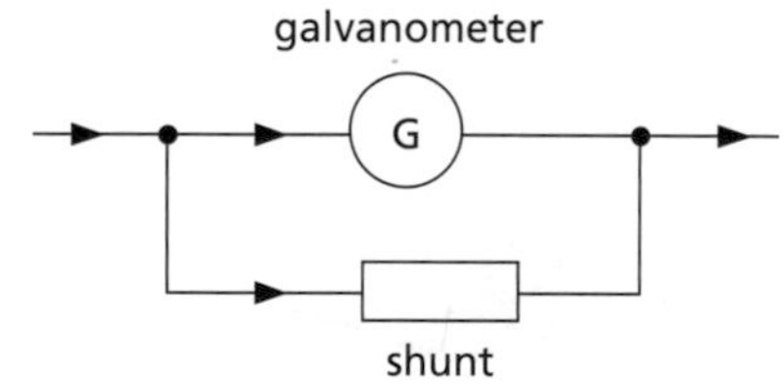

Figure 44.2 An ammeter

Voltmeters and multipliers

A voltmeter is a galvanometer having a known high resistance, called a **multiplier**, in series with it, Figure 44.3. A voltmeter is placed in *parallel* with the part of the circuit across which the p.d. is to be measured and must have a *high resistance* – otherwise the total resistance of the whole circuit is reduced so changing the current and the p.d. required.

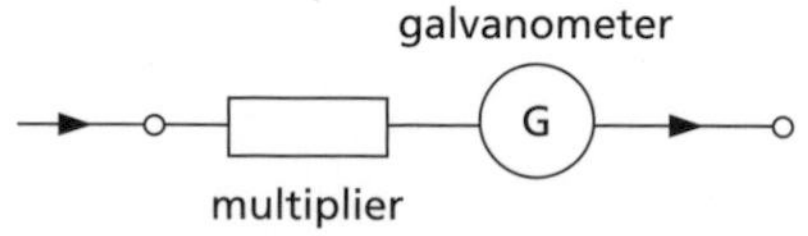

Figure 44.3 A voltmeter

Multimeters

These can have analogue or digital displays (see Figures 44.4a and b) and can be used to measure a.c. or d.c. currents or voltages and also resistance. The required function is first selected, say a.c. current, and then a suitable range chosen. For example if a current of a few milliamps is expected, the 10 mA range might be selected and the value of the current (in mA) read from the display; if the reading is off-scale, the sensitivity should be reduced by changing to the higher, perhaps 100 mA, range.

For the measurement of resistance, the resistance function is chosen and the appropriate range selected. The terminals are first short-circuited to check the zero of resistance, then the unknown resistance is disconnected from any circuit and reconnected across the terminals of the meter in place of the short circuit.

Analogue multimeters are adaptions of moving-coil galvanometers. Digital multimeters are constructed from integrated circuits. On the voltage setting they have a very high input resistance (10 MΩ), i.e. they affect most circuits very little and so give very accurate readings.

Figure 44.4a Analogue multimeter

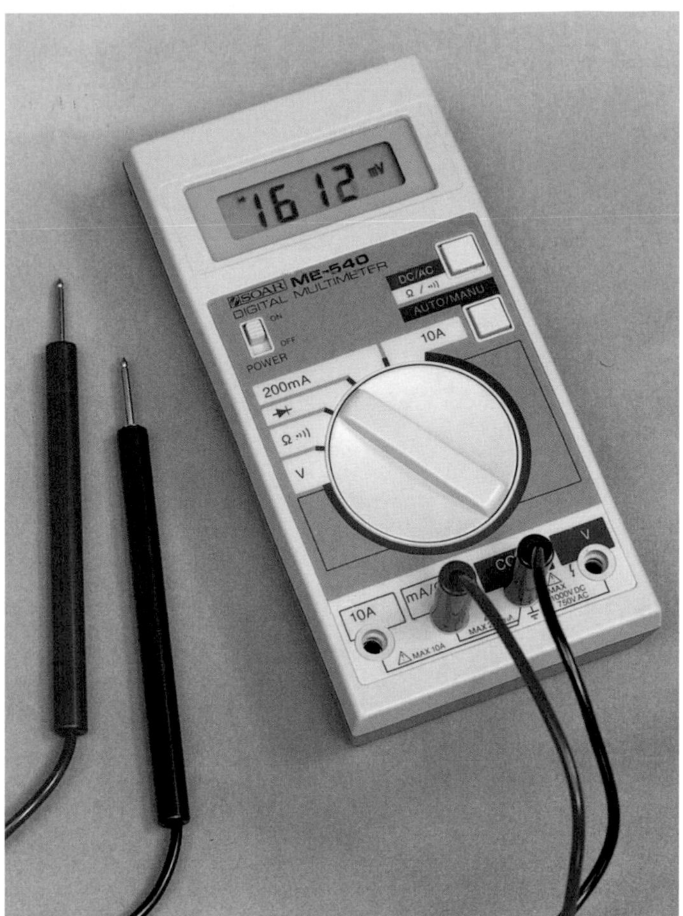

Figure 44.4b Digital multimeter

Reading a voltmeter

The face of an analogue voltmeter is represented in Figure 44.5. The voltmeter has two scales. The 0–5 scale has a **full-scale deflection** of 5.0 V. Each small division on the 0–5 scale represents 0.1 V. This voltmeter scale can be read to the nearest 0.1 V. However the human eye is very good at judging a half division, so we are able to estimate the voltmeter reading to the nearest 0.05 V with considerable precision.

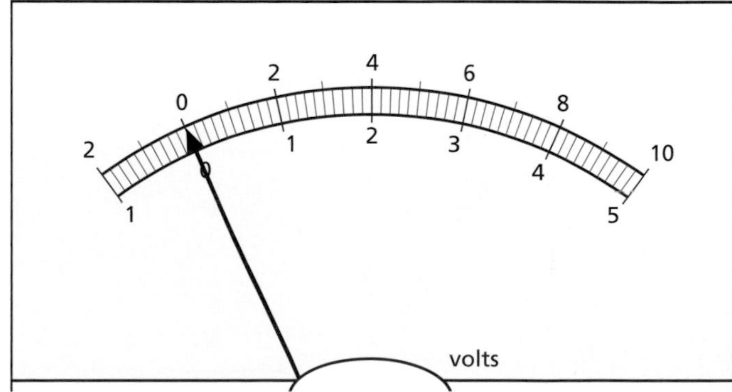

Figure 44.5 An analogue voltmeter scale

Every measuring instrument has a calibrated scale. When you write an account of an experiment (see p. xii, *scientific enquiry*) you should include details about each scale that you use.

Questions

1 What does a galvanometer do?

2 Why should the resistance of
 a an ammeter be very small,
 b a voltmeter be very large?

3 The scales of a voltmeter are shown in Figure 44.6.

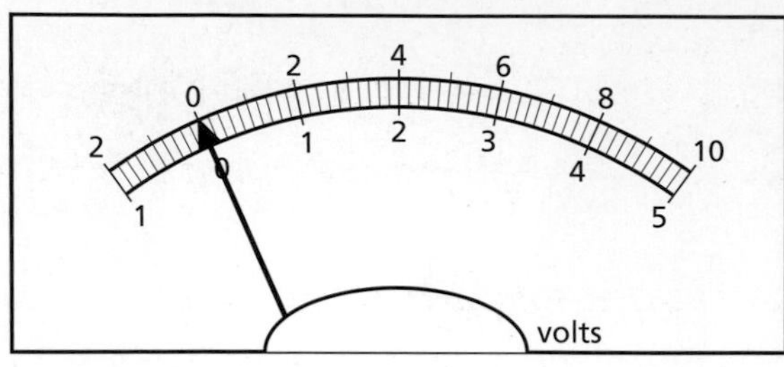

Figure 44.6

 a What are the two ranges available when using the voltmeter?
 b What do the small divisions between the numbers 3 and 4 represent?
 c Which scale would you use to measure a voltage of 4.6 V?
 d When the voltmeter reads 4.0 V where should you position your eye to make the reading?
 e When making the reading for 4.0 V an observer's eye is over the 0 V mark. Explain why the value obtained by this observer is *higher* than 4.0 V.

Checklist

After studying this chapter you should be able to

- draw a diagram of a simple moving-coil galvanometer and explain how it works,
- explain how a moving-coil galvanometer can be modified for use as (a) an ammeter and (b) a voltmeter,
- explain why (a) an ammeter should have a very low resistance and (b) a voltmeter should have a very high resistance.

45 Generators

Electromagnetic induction	*Simple d.c. generator (dynamo)*
Faraday's law	*Practical generators*
Lenz's law	*Applications of electromagnetic induction*
Simple a.c. generator (alternator)	

An electric current creates a magnetic field. The reverse effect of producing electricity from magnetism was discovered in 1831 by Faraday and is called **electromagnetic induction**. It led to the construction of generators for producing electrical energy in power stations.

Electromagnetic induction

Two ways of investigating the effect follow.

a) Straight wire and U-shaped magnet

First the wire is held at rest between the poles of the magnet. It is then moved in each of the six directions shown in Figure 45.1 and the meter observed. Only *when it is moving upwards* (direction 1) or *downwards* (direction 2) is there a deflection on the meter, indicating an induced current in the wire. The deflection is in opposite directions in these two cases and only lasts while the wire is in motion.

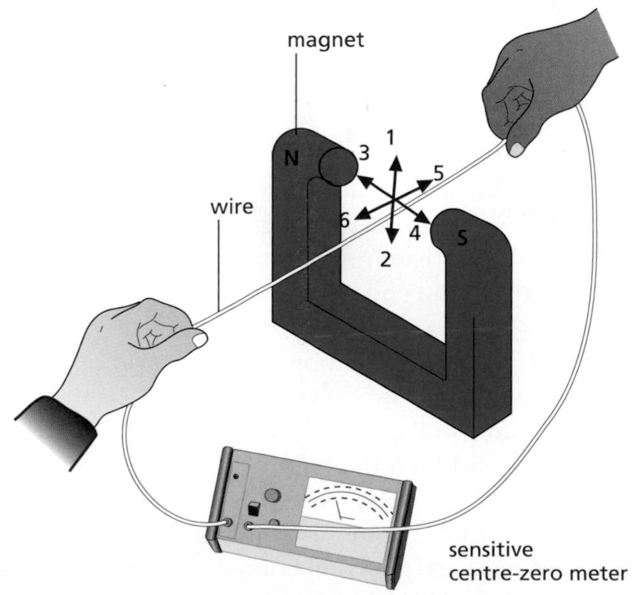

Figure 45.1 A current is induced in the wire when it is moved up or down between the magnet poles

b) Bar magnet and coil

The magnet is pushed into the coil one pole first, Figure 45.2, then held still inside it. It is next withdrawn. The meter shows that current is induced in the coil in one direction as the magnet *moves in* and in the opposite direction as it is *removed*. There is no deflection when the magnet is at rest. The results are the same if the coil is moved instead of the magnet, i.e. only **relative motion** is needed.

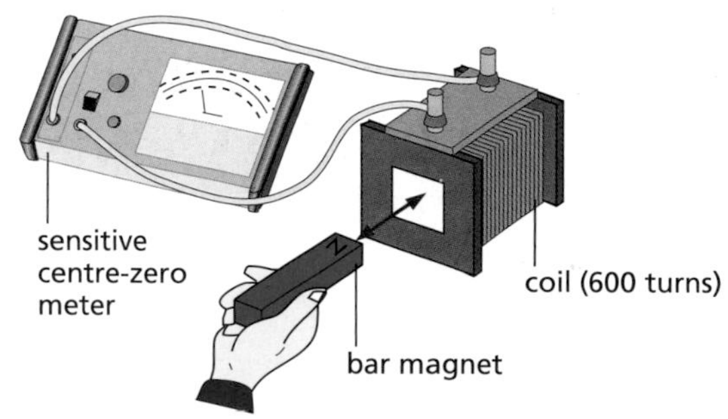

Figure 45.2 A current is induced in the coil when the magnet is moved in or out

Faraday's law

To 'explain' electromagnetic induction Faraday suggested that a voltage is induced in a conductor whenever it 'cuts' magnetic field lines, i.e. moves *across* them, but not when it moves along them or is at rest. If the conductor forms part of a complete circuit, an induced current is also produced.

Faraday found, and it can be shown with apparatus like that in Figure 45.2, that the induced p.d. or voltage increases with increases of

(i) the speed of motion of the magnet or coil,
(ii) the number of turns on the coil,
(iii) the strength of the magnet.

These facts led him to state a law:

> The size of the induced p.d. is directly proportional to the rate at which the conductor cuts magnetic field lines.

□ *Lenz's law*

The direction of the induced current can be found by a law due to the Russian scientist, Lenz.

> The direction of the induced current is such as to oppose the change causing it.

In Figure 45.3a the magnet approaches the coil, north pole first. According to Lenz's law the induced current should flow in a direction that makes the coil behave like a magnet with its top a north pole. The downward motion of the magnet will then be opposed since like poles repel.

When the magnet is withdrawn, the top of the coil should become a south pole, Figure 45.3b, and attract the north pole of the magnet, so hindering its removal. The induced current is thus in the opposite direction to that when the magnet approaches.

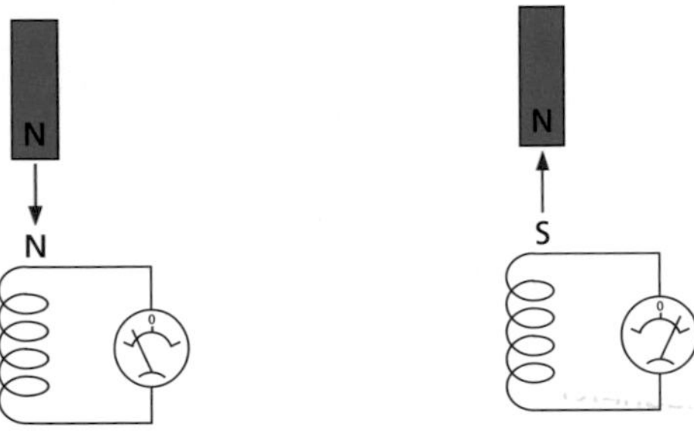

Figure 45.3 The induced current opposes the motion of the magnet

Lenz's law is an example of the principle of conservation of energy. If the currents caused opposite poles to those that they do, electrical energy would be created from nothing. As it is, mechanical energy is provided, by whoever moves the magnet, to overcome the forces that arise.

For a straight wire moving at right angles to a magnetic field a more useful form of Lenz's law is **Fleming's right-hand rule** (the 'dynamo rule'), Figure 45.4.

> Hold the thumb and first two fingers of the right hand at right angles to each other with the **F**irst finger pointing in the direction of the **F**ield and the thu**M**b in the direction of **M**otion of the wire, then the se**C**ond finger points in the direction of the induced **C**urrent.

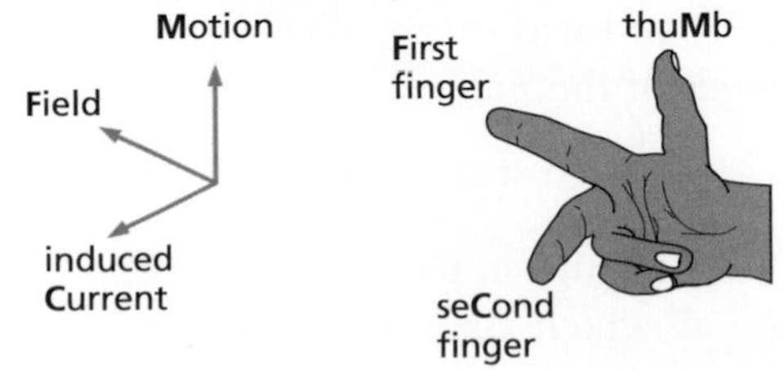

Figure 45.4 Fleming's right-hand (dynamo) rule

■ *Simple a.c. generator (alternator)*

The simplest alternating current (a.c.) generator consists of a rectangular coil between the poles of a C-shaped magnet, Figure 45.5a. The ends of the coil are joined to two **slip rings** on the axle and against which carbon **brushes** press.

When the coil is rotated it cuts the field lines and a voltage is induced in it. Figure 45.5b shows how the voltage varies over one complete rotation.

As the coil moves through the vertical position with **ab** uppermost, **ab** and **cd** are moving along the lines (**bc** and **da** do so always) and no cutting occurs. The induced voltage is zero.

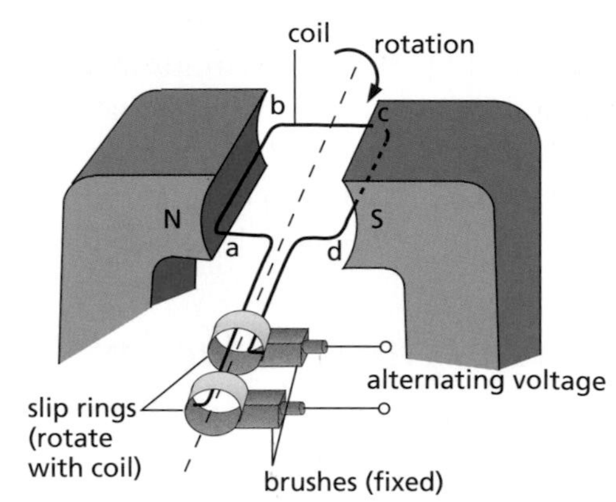

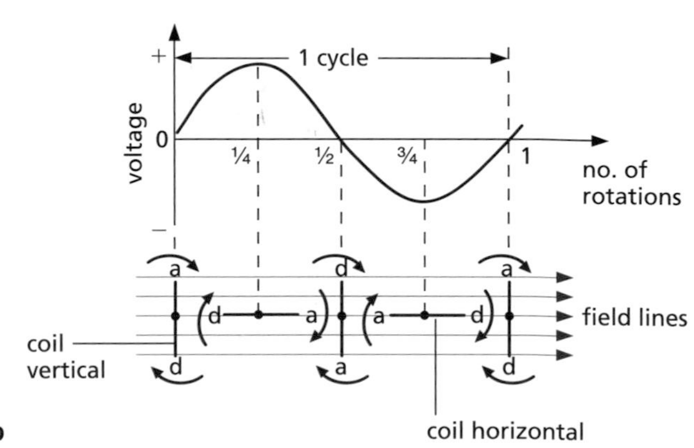

Figure 45.5 A simple a.c. generator and its output

During the first quarter rotation the p.d. increases to a maximum when the coil is horizontal. Sides **ab** and **dc** are then cutting the lines at the greatest rate.

In the second quarter rotation the p.d. decreases again and is zero when the coil is vertical with **dc** uppermost. After this, the direction of the p.d. reverses because, during the next half rotation, the motion of **ab** is directed upwards and **dc** downwards.

An alternating voltage is generated which acts first in one direction and then the other; it would cause a.c. to flow in a circuit connected to the brushes. The **frequency** of an a.c. is the number of complete cycles it makes each second and is measured in **hertz** (Hz), i.e. 1 cycle per second = 1 Hz. If the coil rotates twice per second, the a.c. has frequency 2 Hz. The mains supply is a.c. of frequency 50 Hz.

F Simple d.c. generator (dynamo)

An a.c. generator becomes a direct current (d.c.) one if the slip rings are replaced by a **commutator** (like that in a d.c. motor), Figure 45.6a.

The brushes are arranged so that as the coil goes through the vertical, changeover of contact occurs from one half of the split ring of the commutator to the other. In this position the voltage induced in the coil reverses and so one brush is always positive and the other negative.

The voltage at the brushes is shown in Figure 45.6b; although varying in value, it never changes direction and would produce a direct current (d.c.) in an external circuit.

In construction the simple d.c. dynamo is the same as the simple d.c. motor and one can be used as the other. When an electric motor is working it acts as a dynamo and creates a voltage which opposes the applied voltage. The current in the coil is therefore much less once the motor is running.

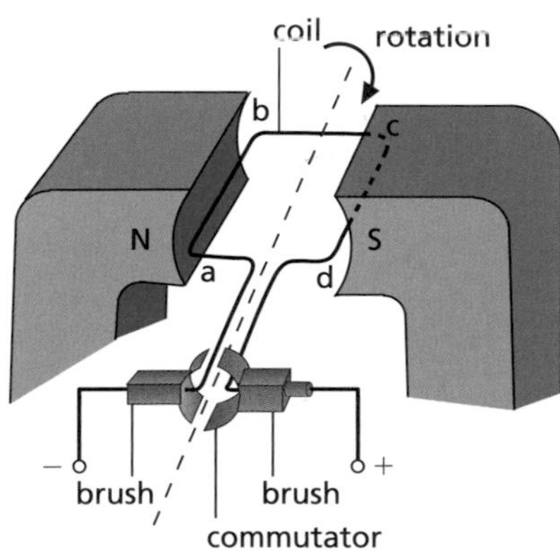

a

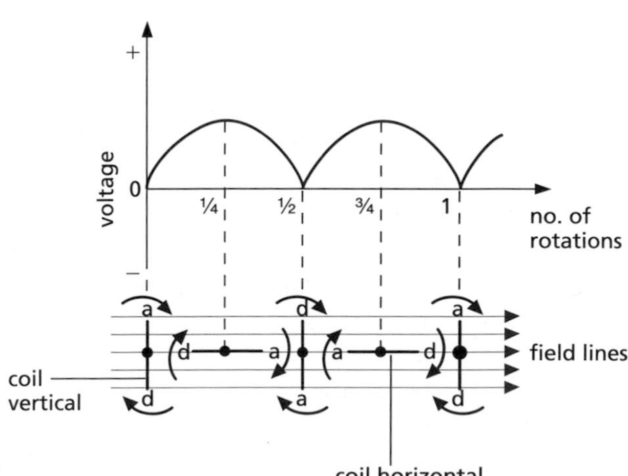

b

Figure 45.6 A simple d.c. generator and its output

Practical generators

In actual generators several coils are wound in evenly spaced slots in a soft iron cylinder and electromagnets usually replace permanent magnets.

a) Power stations

In power station alternators the electromagnets rotate (the **rotor**, Figure 45.7a) while the coils and their iron core are at rest (the **stator**, Figure 45.7b). The large p.ds and currents (e.g. 25 kV at several thousand amps) induced in the stator are led away through stationary cables, otherwise they would quickly destroy the slip rings by sparking. Instead the relatively small d.c. required by the rotor is fed via the slip rings from a small dynamo (the **exciter**) which is driven by the same turbine as the rotor.

a Rotor (electromagnets)

b Stator (induction coils)

Figure 45.7 The rotor and stator of a power station alternator

In a thermal power station (Chapter 17), the turbine is rotated by high-pressure steam obtained by heating water in a coal- or oil-fired boiler or in a nuclear reactor (or by hot gas in a gas-fired power station). A block diagram of a thermal power station is shown in Figure 45.8. The energy transfer diagram was given in Figure 17.7, p. 79.

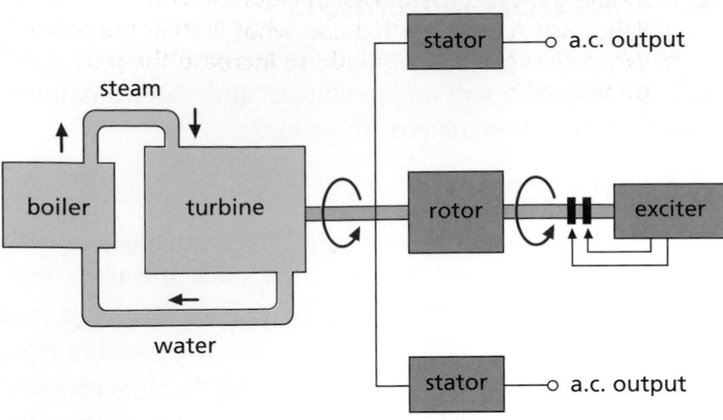

Figure 45.8 Block diagram of a thermal power station

b) Cars

Most cars are now fitted with alternators because they give a greater output than dynamos at low engine speeds.

c) Bicycles

The rotor of a bicycle dynamo is a permanent magnet and the voltage is induced in the coil which is at rest, Figure 45.9.

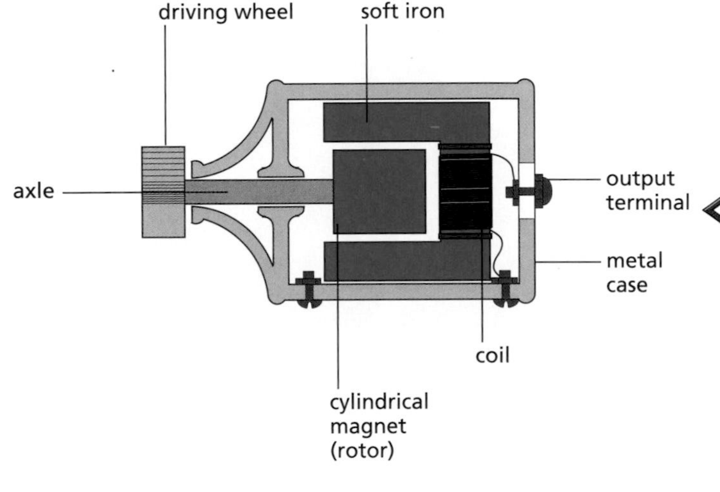

Figure 45.9 Bicycle dynamo

F *Applications of electromagnetic induction*

a) Moving-coil microphone

The moving-coil loudspeaker shown in Figure 43.7 (p. 222) can be operated in reverse mode as a microphone. When sound is incident on the paper cone it vibrates causing the attached coil to move in and out between the poles of the magnet. A varying electric current, representative of the sound, is then induced in the coil by electromagnetic induction.

b) Magnetic recording

Magnetic tapes or disks are used to record information in sound systems and computers. In the recording head shown in Figure 45.10, the tape becomes magnetized when it passes over the gap in the pole piece of the electromagnet and retains a magnetic record of the electrical signal applied to the coil from a microphone or computer. In playback mode, the varying magnetization on the moving tape or disk induces a corresponding electrical signal in the coil as a result of electromagnetic induction.

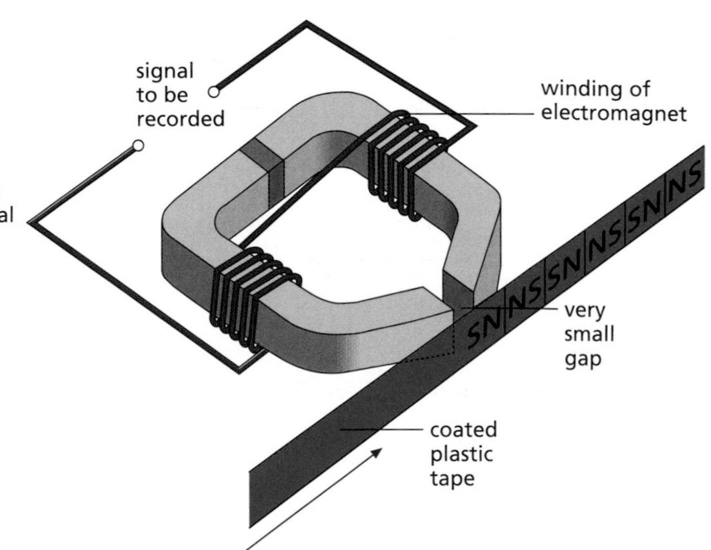

Figure 45.10 Magnetic recording or playback head

Questions

1 a Figure 45.11 shows a magnet being moved into a coil of wire. The reading on the meter is shown in the diagram.

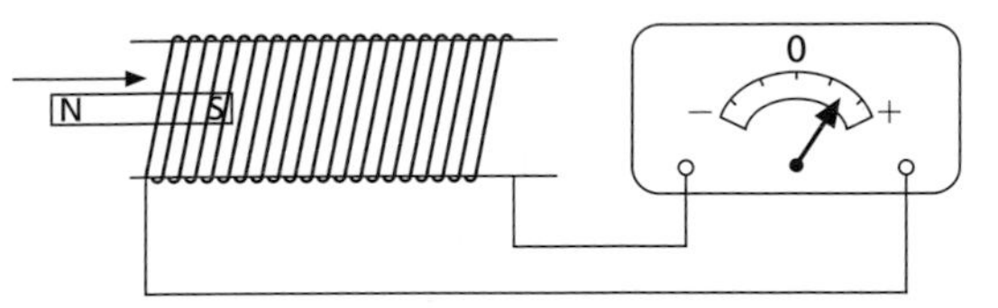

Figure 45.11

Copy the meter scale and draw the meter reading which you would expect to get in each of the following cases.

(i) The magnet is at rest inside the coil, Figure 45.12.

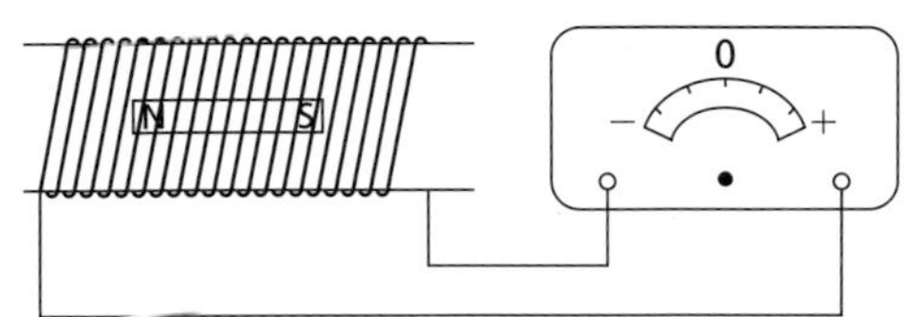

Figure 45.12

(ii) The magnet is moved out of the coil, Figure 45.13.

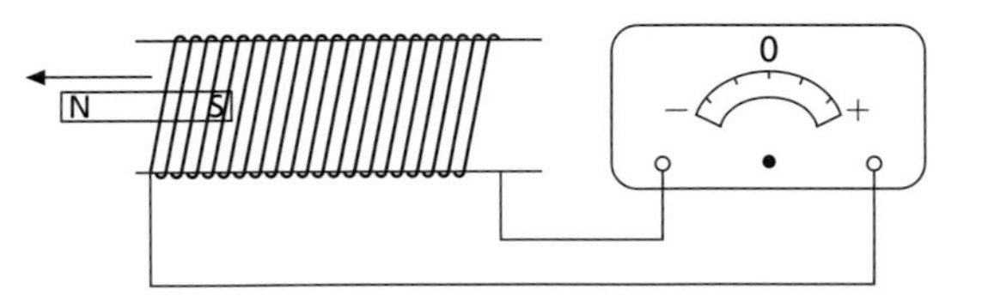

Figure 45.13

b Figure 45.14 shows a bicycle dynamo and part of the wheel.

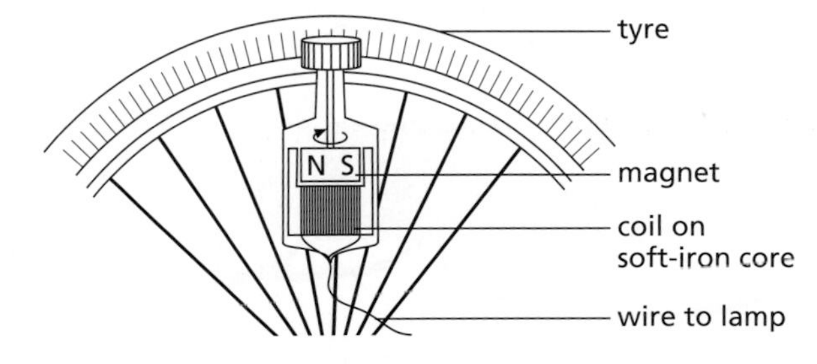

Figure 45.14

Explain, as fully as you can, why a current flows through the bicycle lamp when the wheel of the bicycle turns.

(*NEAB Foundation, June 1998*)

2 A simple generator is shown in Figure 45.15.

a What are A and B called and what is their purpose?

b What changes can be made to increase the p.d. generated?

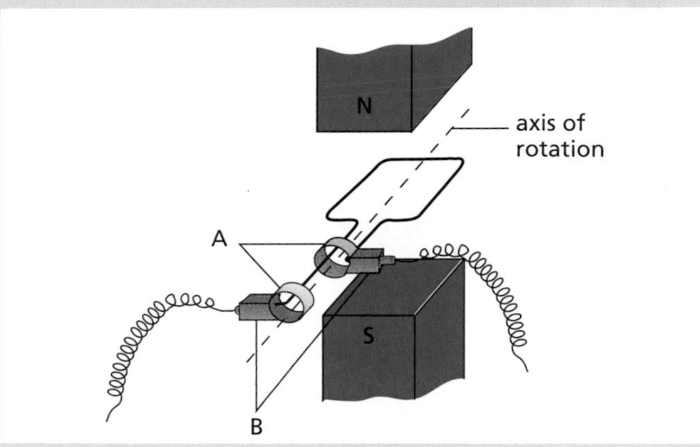

Figure 45.15

■ *Checklist*

After studying this chapter you should be able to

- ■ describe experiments to show electromagnetic induction,
- □ recall Faraday's explanation of electromagnetic induction,
- □ predict the direction of the induced current using Lenz's law or Fleming's right-hand rule,
- ■ draw a diagram of a simple a.c. generator and sketch a graph of its output,
- ■ recall some applications of electromagnetic induction.

46 Transformers

Mutual induction
Transformer equation
Energy losses in a transformer
Transmission of electrical power
Applications of eddy currents
Practical work
Mutual induction with a.c.

Mutual induction

When the current in a coil is switched on or off or changed, a voltage is induced in a neighbouring coil. The effect, called **mutual induction**, is an example of electromagnetic induction and can be shown with the arrangement of Figure 46.1. Coil A is the **primary** and coil B the **secondary**.

Switching on the current in the primary sets up a magnetic field and as its field lines 'grow' outwards from the primary they 'cut' the secondary. A p.d. is induced in the secondary until the current in the primary reaches its steady value. When the current is switched off in the primary, the magnetic field dies away and we can imagine the field lines cutting the secondary as they collapse, again inducing a p.d. in it. Changing the primary current by *quickly* altering the rheostat has the same effect.

The induced p.d. is increased by having a soft iron rod in the coils or, better still, by using coils wound on a complete iron ring. More field lines then cut the secondary due to the magnetization of the iron.

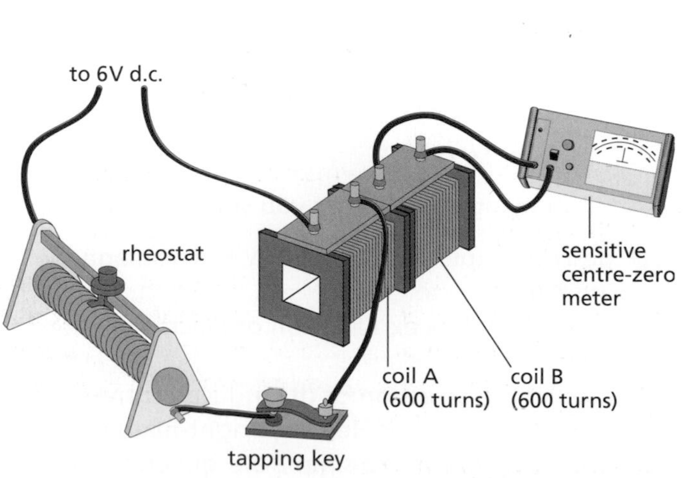

Figure 46.1 A changing current in a primary coil (A) induces a current in a secondary coil (B)

Practical work

Mutual induction with a.c.

An alternating current is changing all the time and if it flows in a primary coil, an alternating voltage and current are induced in a secondary coil.

Connect the circuit of Figure 46.2. The 1 V high current power unit supplies a.c. to the primary and the lamp detects the secondary current.

Find the effect on the brightness of the lamp of

(i) pulling the C-cores apart slightly,
(ii) increasing the secondary turns to 15,
(iii) decreasing the secondary turns to 5.

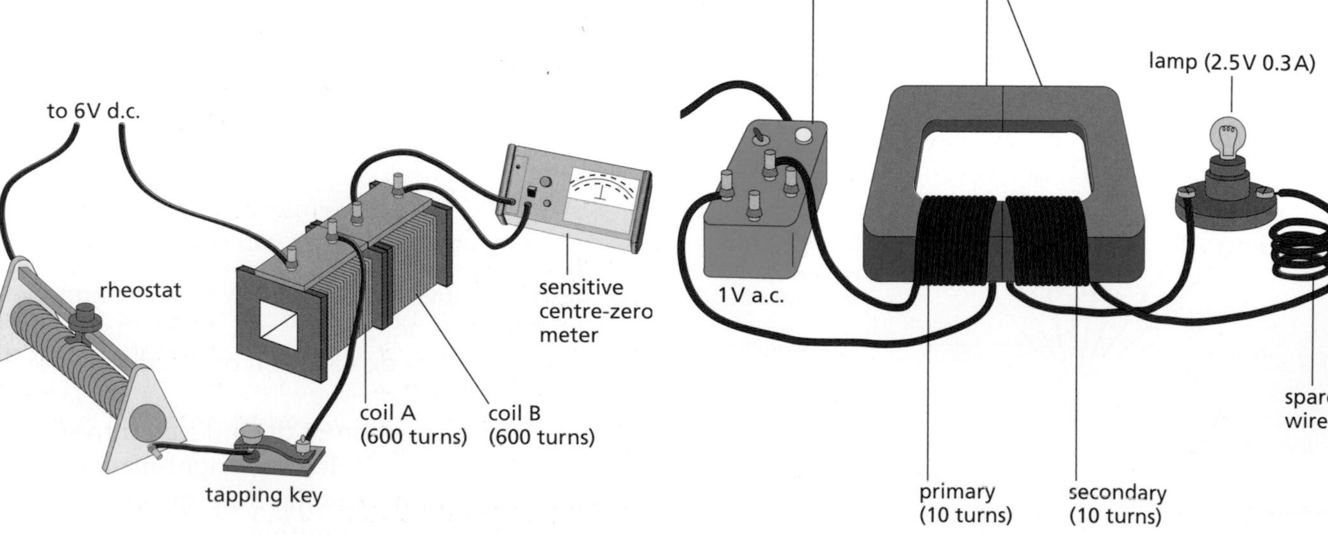

Figure 46.2

Transformer equation

A **transformer** transforms (changes) an *alternating* voltage from one value to another of greater or smaller value. It has a primary coil and a secondary coil wound on a complete soft iron core, either one on top of the other, Figure 46.3a, or on separate limbs of the core, Figure 46.3b.

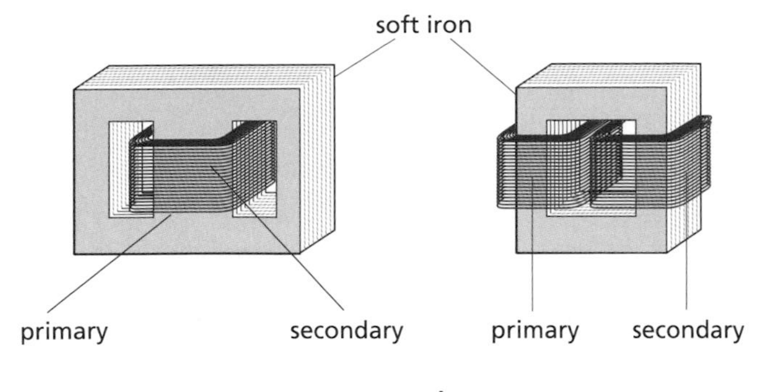

Figure 46.3 Primary and secondary coils of a transformer

An alternating voltage applied to the primary induces an alternating voltage in the secondary. The value of the secondary voltage can be shown, for a transformer in which all the field lines cut the secondary, to be given by

$$\frac{\text{secondary voltage}}{\text{primary voltage}} = \frac{\text{secondary turns}}{\text{primary turns}}$$

In symbols

$$\frac{V_s}{V_p} = \frac{N_s}{N_p}$$

A 'step-up' transformer has more turns on the secondary than the primary and V_s is greater than V_p, Figure 46.4a. For example, if the secondary has twice as many turns as the primary, V_s is about twice V_p. In a 'step-down' transformer there are fewer turns on the secondary than the primary and V_s is less than V_p, Figure 46.4b.

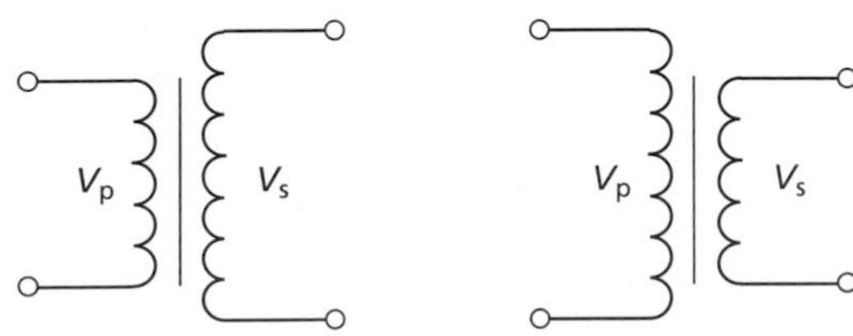

a Step-up: $V_s > V_p$ **b** Step-down: $V_p > V_s$

Figure 46.4 Symbols for a transformer

Energy losses in a transformer

If the p.d. is stepped up in a transformer the current is stepped down in proportion. This must be so if we assume that all the electrical energy given to the primary appears in the secondary, i.e. that energy is conserved and the transformer is 100% efficient or 'ideal' (many approach this). Then

$$\text{power in primary} = \text{power in secondary}$$
$$V_p \times I_p = V_s \times I_s$$

where I_p and I_s are the primary and secondary currents respectively.

$$\therefore \qquad \frac{I_s}{I_p} = \frac{V_p}{V_s}$$

So, for the ideal transformer, if the p.d. is doubled the current is halved. In practice, it is more than halved due to small energy losses in the transformer arising from the following three causes.

a) Resistance of windings

The windings of copper wire have some resistance and heat is produced by the current in them. Large transformers like those in Figure 46.5 have to be oil-cooled to prevent overheating.

Figure 46.5 Step-up transformers at a power station

b) Eddy currents

The iron core is in the changing magnetic field of the primary, and currents, called eddy currents, are induced in it which cause heating. These are reduced by using a **laminated** core made of sheets, insulated from each other to have a high resistance.

c) Leakage of field lines

All the field lines produced by the primary may not cut the secondary, especially if the core has an air gap or is badly designed.

Worked example

A transformer steps down the mains supply from 230 V to 10 V to operate an answering machine.

a What is the turns ratio of the transformer windings?

b How many turns are on the primary if the secondary has 100 turns?

c What is the current in the primary if the transformer is 100% efficient and the current in the answering machine is 2 A?

a Primary voltage = V_p = 230 V

Secondary voltage = V_s = 10 V

$$\text{Turns ratio} = N_s/N_p = V_s/V_p = 10\ \text{V}/230\ \text{V} = 1/23$$

b Secondary turns = N_s = 100

From **a**, $$\frac{N_s}{N_p} = \frac{1}{23}$$

$$\therefore\quad N_p = 23\,N_s = 23 \times 100 = 2300\ \text{turns}$$

c Efficiency = 100%

$\therefore$ power in primary = power in secondary

$$V_p \times I_p = V_s \times I_s$$

$$\therefore\quad I_p = \frac{V_s \times I_s}{V_p} = \frac{10\ \text{V} \times 2\ \text{A}}{230\ \text{V}}$$

$$= 2/23\ \text{A} \qquad = 0.09\ \text{A}$$

Note In this ideal transformer the current is stepped up in the same ratio as the voltage is stepped down.

Transmission of electrical power

a) Grid system

The Grid is a network of cables throughout Britain, mostly supported on pylons, which connect over 100 power stations to consumers. In the largest modern stations, electricity is generated at 25 000 V (25 kilovolts = 25 kV) and stepped up at once in a transformer to 275 or 400 kV to be sent over long distances on the Supergrid. Later, the p.d. is reduced by substation transformers for distribution to local users, Figure 46.6.

At the National Control Centre engineers direct the flow and re-route it when breakdown occurs. This makes the supply more reliable, and cuts costs by enabling smaller, less efficient stations to be shut down at off-peak periods.

b) Use of high alternating p.ds

The efficiency with which transformers step alternating p.ds up and down accounts for the use of a.c. rather than d.c. in power transmission. High voltages are used in the transmission of electric power to reduce the amount of energy 'lost' as heat.

Power cables have resistance, and so electrical energy is transferred to heat during the transmission of electricity from the power station to the user. The power 'lost' as heat in cables of resistance R is I^2R, so I should be kept low to reduce energy loss. Since power = IV, if 400 000 W of electrical power has to be sent through cables, it can be done either as 1 A at 400 000 V or as 1000 A at 400 V. Less energy will be transferred to heat if the power is transmitted at the lower current

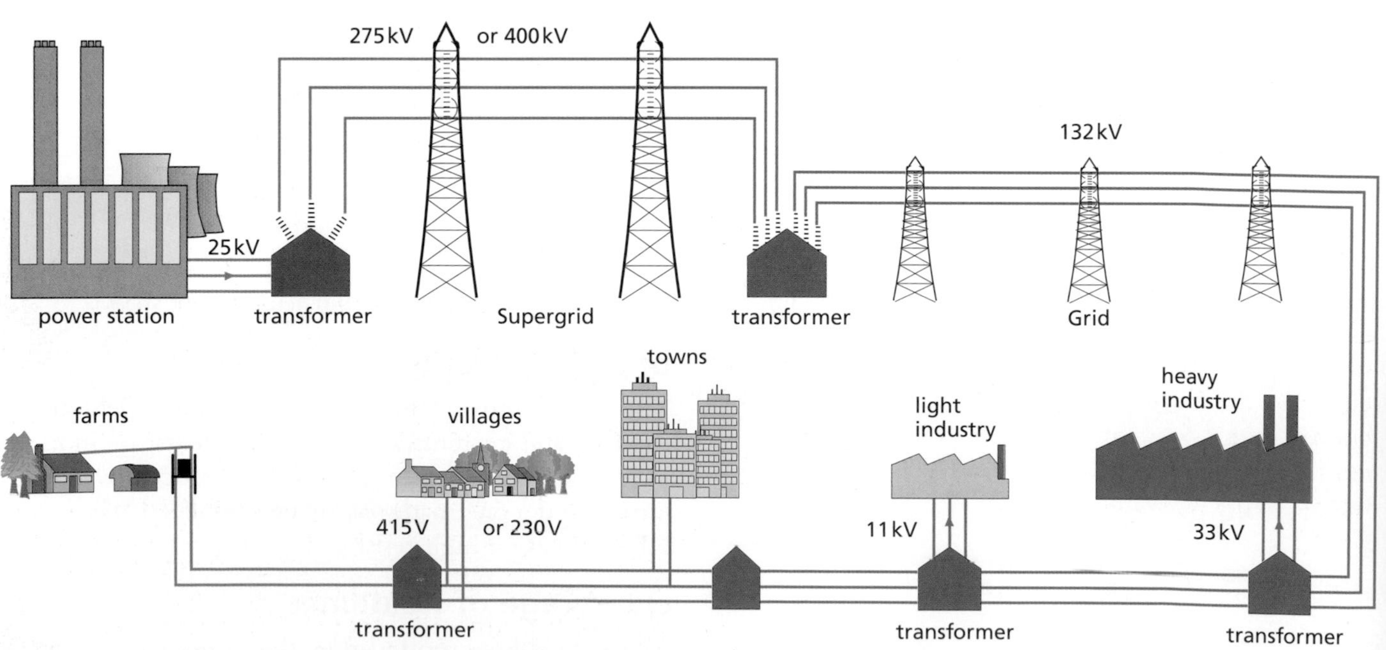

Figure 46.6 The National Grid transmission system in Britain

and higher voltage, i.e. 1 A at 400 000 V. High p.ds require good insulation but are readily produced by a.c. generators.

F Applications of eddy currents

Eddy currents are the currents induced in a piece of metal when it cuts magnetic field lines. They can be quite large due to the low resistance of the metal. They have their uses as well as their disadvantages.

a) Car speedometer

The action depends on the eddy currents induced in a thick aluminium disc when a permanent magnet, near it but *not touching it*, is rotated by a cable driven from the gearbox of the car, Figure 46.7. The eddy currents in the disc make it rotate in an attempt to reduce the relative motion between it and the magnet (see Chapter 45). The extent to which the disc can turn however is controlled by a spring. The faster the magnet rotates the more the disc turns before it is stopped by the spring. A pointer fixed to the disc moves over a scale marked in mph (or km/h) and gives the speed of the car.

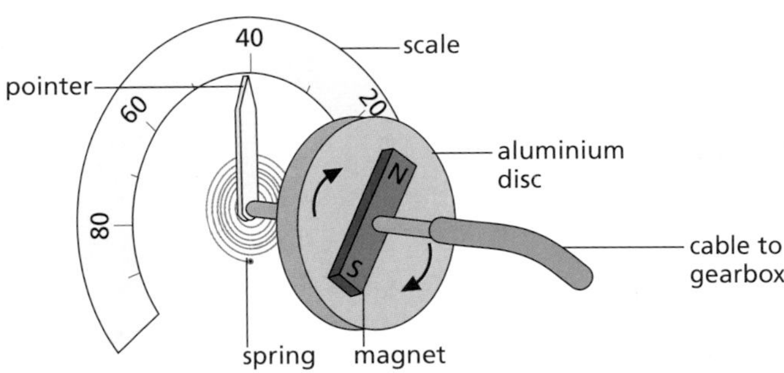

Figure 46.7 Car speedometer

b) Metal detector

The metal detector shown in Figure 46.8 consists of a large primary coil (A) through which an a.c. current is passed and a smaller secondary coil (B). When the detector is swept over a buried metal object (such as a nail, coin or pipe) the fluctuating magnetic field lines associated with the alternating current in coil A 'cut' the hidden metal and induce eddy currents in it. The changing magnetic field lines associated with these eddy currents cut the secondary coil B in turn and induce a current which can be used to operate an alarm. The coils are set at right angles to each other so that their magnetic fields do not interact.

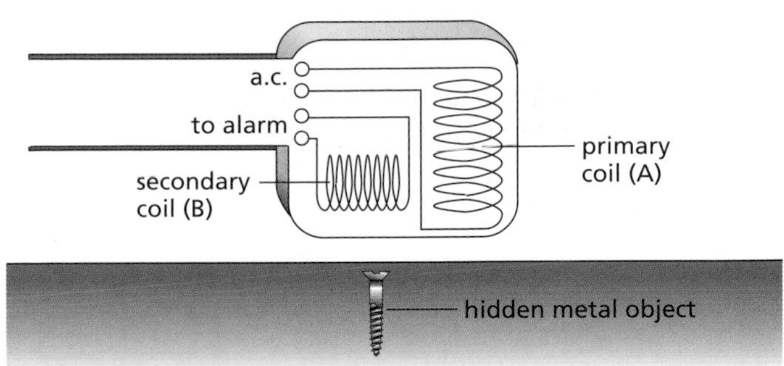

Figure 46.8 Metal detector

Questions

1 Two coils of wire, A and B, are placed near one another, Figure 46.9. Coil A is connected to a switch and battery. Coil B is connected to a centre-reading moving-coil galvanometer.

a If the switch connected to coil A were closed for a few seconds and then opened, the galvanometer connected to coil B would be affected. Explain and describe, step by step, what would actually happen.

b What changes would you expect if a bundle of soft iron wires were placed through the centre of the coils? Give a reason for your answer.

c What would happen if more turns of wire were wound on the coil B?

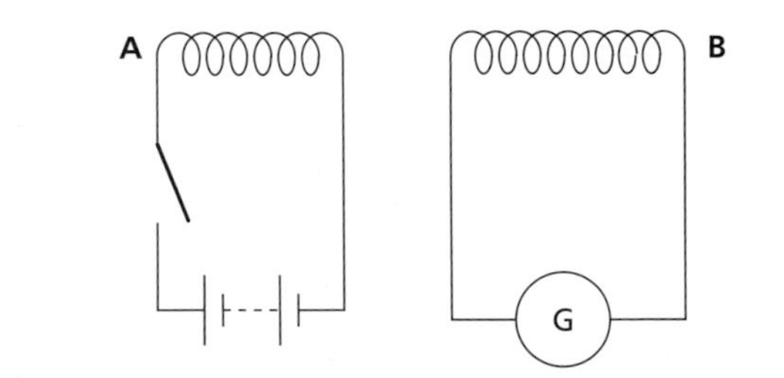

Figure 46.9

2 The main function of a step-down transformer is to

A decrease current
B decrease voltage
C change a.c. to d.c.
D change d.c. to a.c.
E decrease the resistance of a circuit.

3 a Calculate the number of turns on the secondary of a step-down transformer, which would enable a 12 V bulb to be used with a 230 V a.c. mains power, if there are 460 turns on the primary.

b What current will flow in the secondary when the primary current is 0.10 A? Assume there are no energy losses.

continued

4 a Figure 46.10 represents a simple transformer used to light a 12 V lamp. When the power supply is switched on the lamp is very dim.

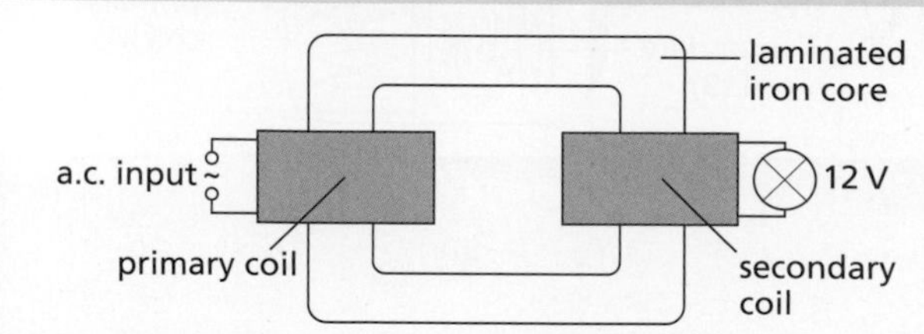

Figure 46.10

(i) Give *one* way to increase the voltage at the lamp without changing the power supply.
(ii) What is meant by the iron core being *laminated*?

b Electrical energy is distributed around the country by a network of high voltage cables, Figure 46.11.

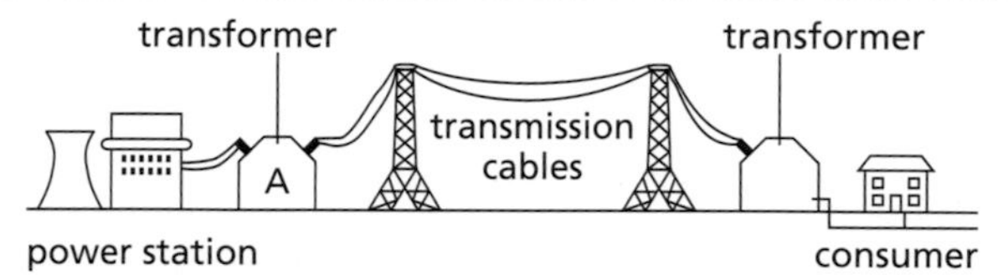

Figure 46.11

(i) For the system to work the power is generated and distributed using alternating current rather than direct current. Why?
(ii) Transformers are an essential part of the distribution system. Explain why.
(iii) The transmission cables are suspended high above the ground. Why?

c The power station generates 100 MW of power at a voltage of 25 kV. Transformer A, which links the power station to the transmission cables, has 44 000 turns in its 275 kV secondary coil.
(i) Write down the equation which links the number of turns in each transformer coil to the voltage across each transformer coil.
(ii) Calculate the number of turns in the primary coil of transformer A. Show clearly how you work out your answer.

d Figure 46.12 shows how the cost of transmitting the electricity along the cables depends upon the thickness of the cable.

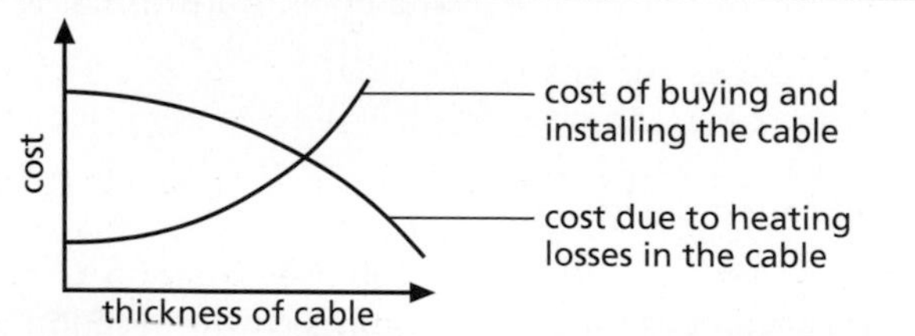

Figure 46.12

(i) Why does the cost due to the heating losses go down as the cable is made thicker?
(ii) By what process is most heat energy lost from the cables?

(AQA (SEG) Higher, Summer 1999)

Checklist

After studying this chapter you should be able to

- ■ explain the principle of the transformer,
- ■ recall the transformer equation $V_s/V_p = N_s/N_p$ and use it to solve problems,
- □ recall that for an ideal transformer $V_p \times I_p = V_s \times I_s$ and use the relation to solve problems,
- □ recall the causes of energy losses in practical transformers,
- ■ explain why high voltage a.c. is used for transmitting electrical power.

Electromagnetic effects *Additional questions*

Magnetic fields; electromagnets

1 a Two bar magnets are held close together.

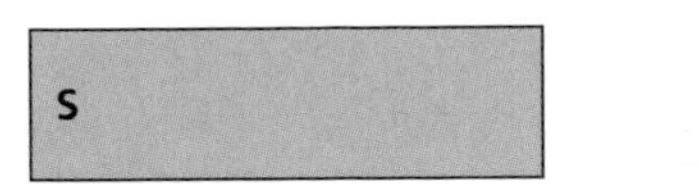

When released the magnets *push away from* each other.
(i) Copy the diagram above and use the letter N to label the north pole of each magnet.
(ii) Write down *one* word which describes what happens to the magnets as they are released.

b The diagram shows an electromagnet. Copy the diagram and draw the shape of the magnetic field of the electromagnet.

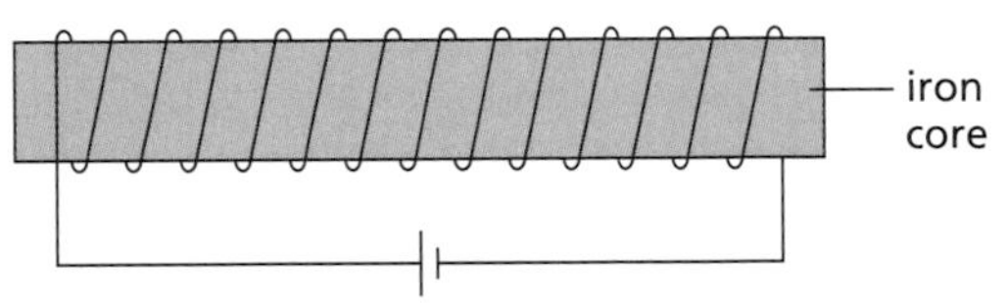

c The diagram shows five electromagnets, **L**, **M**, **N**, **O** and **P**. Each electromagnet is able to pick up a different number of paper clips.

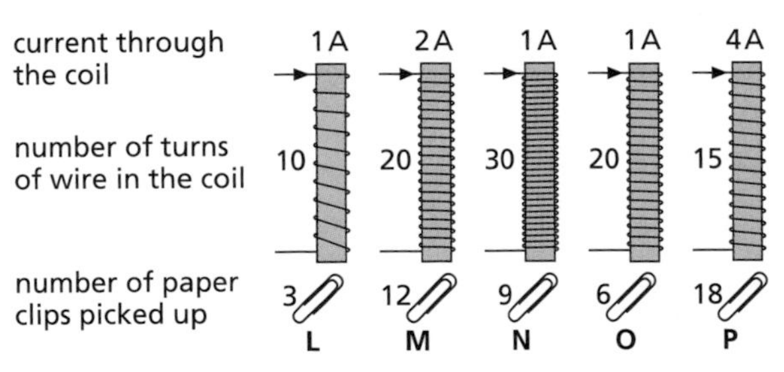

(i) Which *three* electromagnets should you compare if you want to find out how the strength of an electromagnet depends upon the number of turns of wire in the coil?
(ii) What is the connection between the strength of an electromagnet and the number of turns of wire in the coil?

(AQA (SEG) Foundation, June 1999)

2 Part of the electrical system of a car is shown in the following diagram.

a Why are connections made to the car body?
b There are *two* circuits in parallel with the battery. What are they?
c Why is wire A thicker than wire B?
d Why is a relay used?

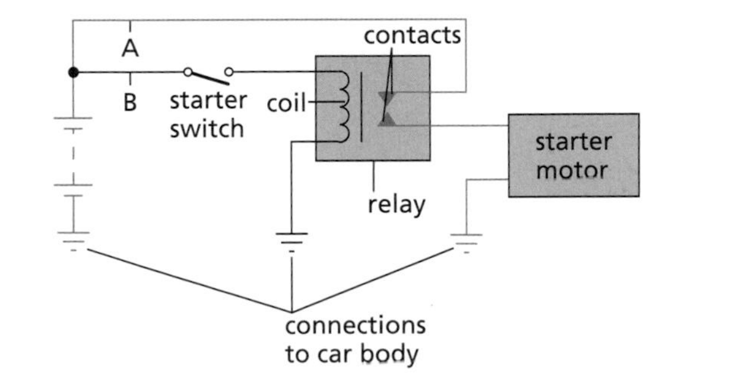

Electric motors; generators; transformers

3 Describe the deflections observed on the sensitive, centre-zero galvanometer G when the copper rod XY is connected to its terminals and is made to vibrate up and down (as shown by the arrows), between the poles of a U-shaped magnet so that it is at right angles to the magnetic field.

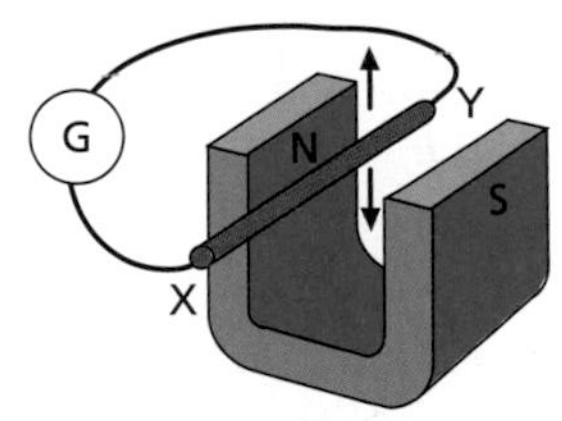

Explain what is happening.

4 A transformer has 1000 turns on the primary coil. The voltage applied to the primary coil is 230 V a.c. How many turns are on the secondary coil if the output voltage is 46 V a.c.?

A 20 **B** 200 **C** 2000 **D** 4000 **E** 8000

5 a Draw a labelled diagram of the essential components of a simple motor. Explain how continuous rotation is produced and show how the direction of rotation is related to the direction of the current.

b State what would happen to the direction of rotation of the motor you have described if
(i) the current were reversed,
(ii) the magnetic field were reversed,
(iii) both current and field were reversed simultaneously.

6 This question is about electromagnetism.

a Michael is investigating how a short length of copper wire can be made to move in a magnetic field. He uses this apparatus.

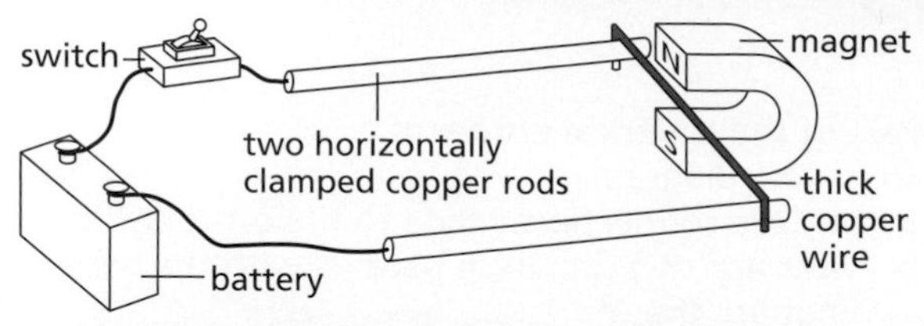

He places the magnet so that the wire is midway between the poles. He writes down these observations.

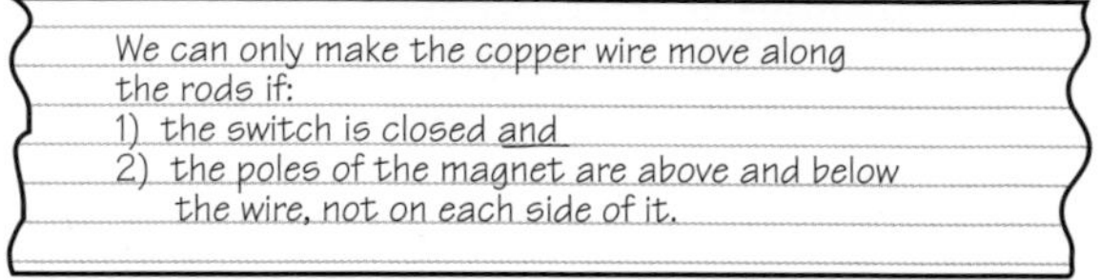
We can only make the copper wire move along the rods if:
1) the switch is closed and
2) the poles of the magnet are above and below the wire, not on each side of it.

Explain these *two* observations. Use your ideas about electromagnetism.

b The diagram shows a model generator.

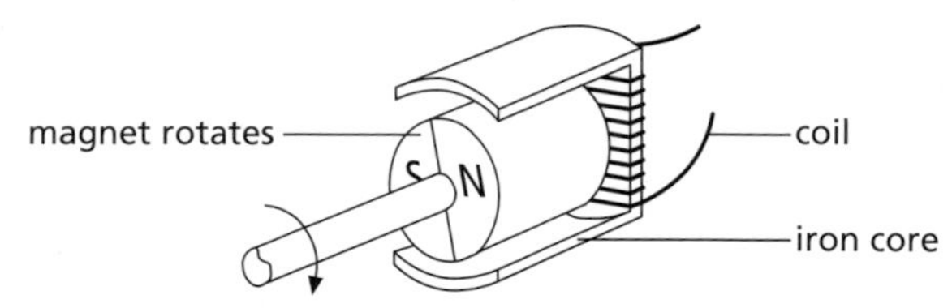

(i) What happens in the coil of wire when the magnet rotates?
(ii) Why does this happen?

c The ends of the coil are connected to a cathode ray oscilloscope (CRO). The diagram shows the trace on the screen as the magnet rotates.

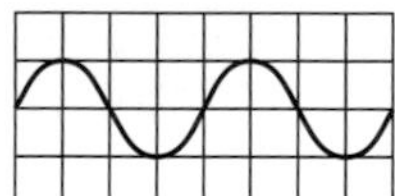

Copy the diagram and draw new traces for each of the following changes. (Assume the settings of the oscilloscope remain the same.)
(i) The magnet rotates at the same speed but in the opposite direction.
(ii) The magnet rotates at the same speed, in the same direction as the original, but the number of turns of the coil is doubled.
(iii) The magnet rotates at twice the speed, in the same direction, with the original number of turns of the coil.

d Explain why **iron** is used as the core in the model generator.

e The output from a power station generator is connected to a step-up transformer. The transformer is connected to transmission lines.

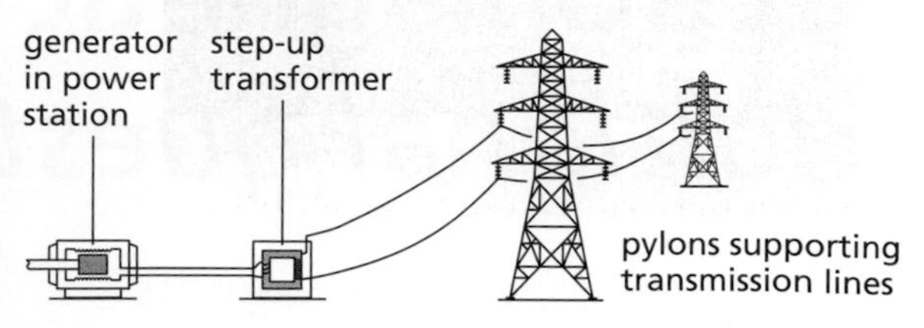

Explain why a step-up transformer is needed. Use your ideas about power losses in transmission.

(OCR Higher, June 1999)

7 The figure below shows the basic parts of a transformer.

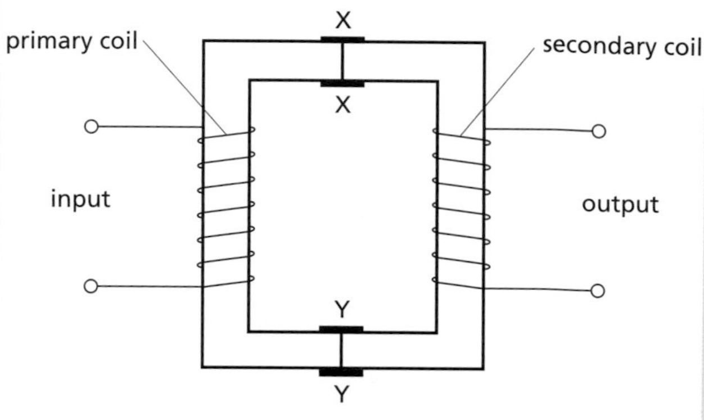

a Use ideas of electromagnetic induction to explain how the input voltage is transformed into an output voltage. Use the three questions below to help you with your answer.
(i) What happens in the primary coil?
(ii) What happens in the core?
(iii)What happens in the secondary coil?

b State what is needed to make the output voltage higher than the input voltage.

c The core of this transformer splits along XX and YY. Explain why the transformer would not work if the two halves of the core were separated by about 30 cm.

d A 100% efficient transformer is used to step up the voltage of a supply from 100 V to 200 V. A resistor is connected to the output. The current in the primary coil is 0.4 A.
Calculate the current in the secondary coil.

(UCLES IGCSE Physics Extended, Nov 2005)

Electrons and atoms

47 Electrons

Thermionic emission	*Uses of the CRO*
Cathode rays	*X-rays*
Deflection of an electron beam	*Photoelectric effect*
Cathode ray oscilloscope (CRO)	*Waves or particles?*

The discovery of the electron was a landmark in physics and led to great technological advances.

Thermionic emission

The evacuated bulb in Figure 47.1 contains a small coil of wire, the **filament**, and a metal plate called the **anode** because it is connected to the positive of the 400 V d.c. power supply. The negative of the supply is joined to the filament which is also called the **cathode**. The filament is heated by current from a 6 V supply (a.c. or d.c.).

With the circuit as shown, the meter deflects, indicating current flow in the circuit containing the gap between anode and cathode. The current stops if *either* the 400 V supply is reversed to make the anode negative, *or* the filament is not heated.

This demonstration shows that negative charges, in the form of electrons, escape from the filament when it is hot because they have enough energy to get free from the metal surface. The process is known as **thermionic emission** and the bulb as a thermionic diode (since it has two electrodes). There is a certain minimum **threshold energy** (depending on the metal) which the electrons must have to escape. Also, the higher the temperature of the metal, the greater the number of electrons emitted. The electrons are attracted to the anode if it is positive and are able to reach it because there is a vacuum in the bulb.

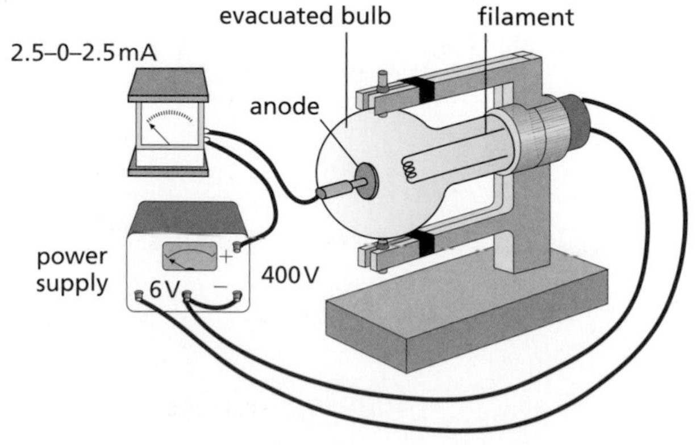

Figure 47.1 Demonstrating thermionic emission

Cathode rays

Beams of electrons moving at high speed are called **cathode rays**. Their properties can be studied using the 'Maltese cross tube', Figure 47.2.

Electrons emitted by the hot cathode are accelerated towards the anode but most pass through the hole in it and travel on along the tube. Those that miss the cross cause the screen to fluoresce with green or blue light and cast a shadow of the cross on it. The cathode rays evidently travel in straight lines.

If the N pole of a magnet is brought up to the neck of the tube, the rays (and the fluorescent shadow) can be shown to move upwards. The rays are clearly deflected by a magnetic field and, using Fleming's left-hand rule (Chapter 43), we see that they behave like conventional current (positive charge flow) travelling from anode to cathode.

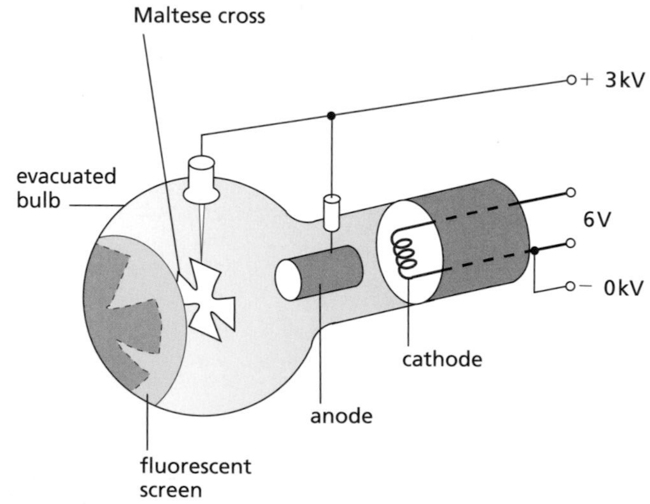

Figure 47.2 Maltese cross tube

There is also an optical shadow of the cross, due to light emitted by the cathode. This is unaffected by the magnet.

Deflection of an electron beam

a) By a magnetic field

In Figure 47.3 the evenly spaced crosses represent a uniform magnetic field (i.e. one of the same strength throughout the area shown) acting into and perpendicular to the paper. An electron beam entering the field at right angles to the field experiences a force due to the motor effect (Chapter 43), whose direction is given by Fleming's left-hand rule. This indicates that the force acts inwards at right angles to the direction of the beam and makes it follow a **circular** path as shown (the beam being treated as conventional current in the opposite direction).

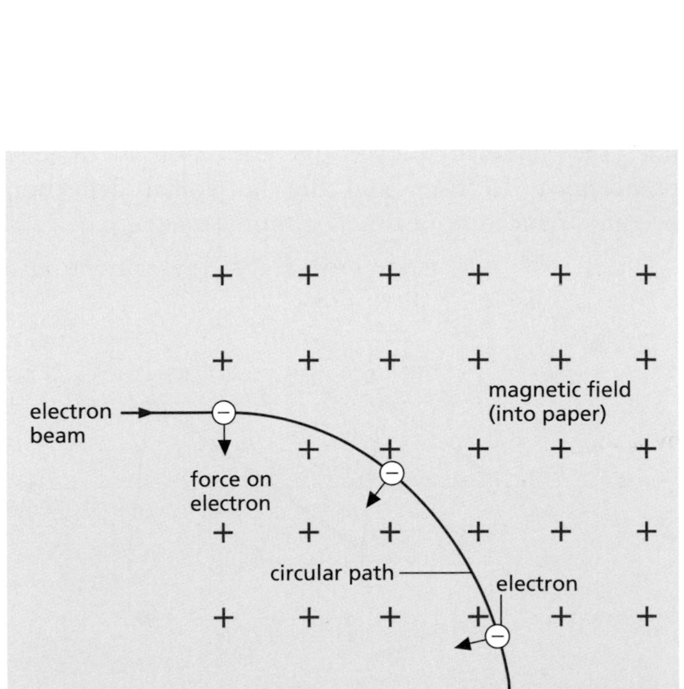

Figure 47.3 Path of an electron beam at right angles to a magnetic field

b) By an electric field

An electric field is a region where an electric charge experiences a force due to other charges (see p. 164). In Figure 47.4 the two metal plates behave like a capacitor that has been charged by connection to a voltage supply. If the charge is evenly spread over the plates, a uniform electric field is created between them and is represented by parallel, equally spaced lines; the arrows indicate the direction in which a positive charge would move.

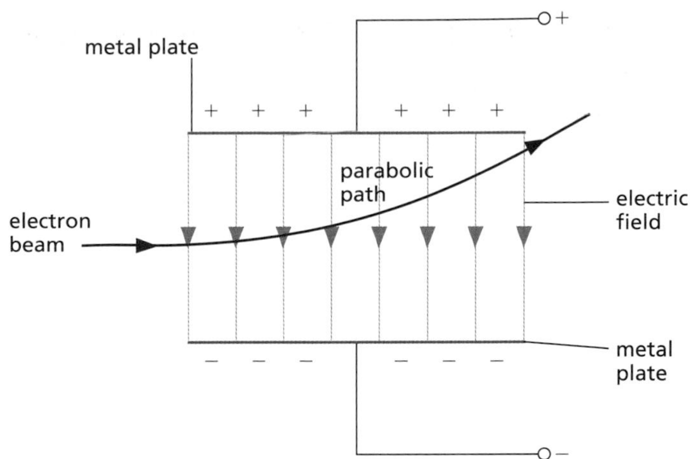

Figure 47.4 Path of an electron beam incident perpendicular to an electric field

If a beam of electrons enters the field perpendicular to it, the negatively charged beam is attracted towards the positively charged plate and follows a **parabolic** path, as shown. In fact its behaviour is not unlike that of a projectile (Chapter 21) in which the horizontal and vertical motions can be treated separately.

c) Demonstration

The deflection tube in Figure 47.5 can be used to show the deflection of an electron beam in electric and magnetic fields. Electrons from a hot cathode strike a fluorescent screen S set at an angle. A p.d. applied across two horizontal metal plates Y_1Y_2 creates a *vertical* electric field which deflects the rays upwards if Y_1 is positive (as shown) and downwards if it is negative.

When current flows in the two coils X_1X_2 (in series) outside the tube, a *horizontal* magnetic field is produced across the tube. It can be used instead of a magnet to deflect the rays, or to cancel the deflection due to an electric field.

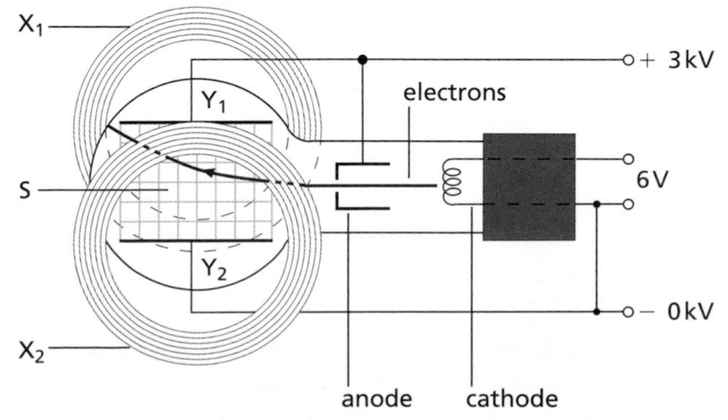

Figure 47.5 Deflection tube

☐ *Cathode ray oscilloscope (CRO)*

The CRO is one of the most important scientific instruments ever to be developed. It contains a cathode ray tube which has three main parts, Figure 47.6.

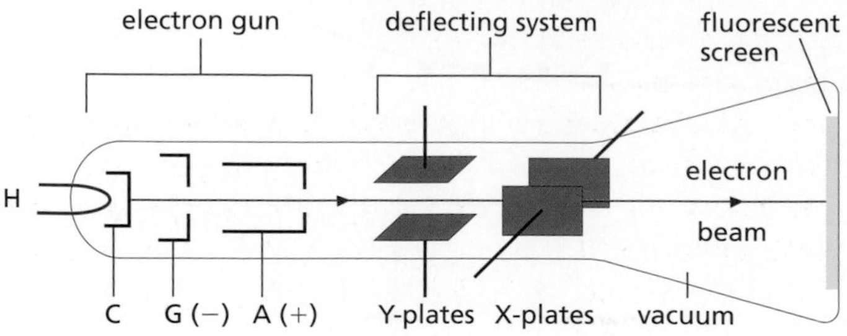

Figure 47.6 Main parts of a CRO

a) Electron gun

This consists of a **heater** H, a **cathode** C, another electrode called the **grid** G and two or three **anodes** A. G is at a negative voltage with respect to C and controls the number of electrons passing through its central hole from C to A; it is the **brilliance** or **brightness** control. The anodes are at high positive voltages relative to C; they accelerate the electrons along the highly evacuated tube and also **focus** them into a narrow beam.

b) Fluorescent screen

A bright spot of light is produced on the screen where the beam hits it.

c) Deflecting system

Beyond A are two pairs of deflecting plates to which p.ds can be applied. The **Y-plates** are horizontal but create a vertical electric field which deflects the beam vertically. The **X-plates** are vertical and deflect the beam horizontally.

The p.d. to create the electric field between the Y-plates is applied to the **Y-input** terminals (often marked 'high' and 'low') on the front of the CRO. The input is usually amplified by an amount that depends on the setting of the **Y-amp gain** control, before it is applied to the Y-plates. It can then be made large enough to give a suitable vertical deflection of the beam.

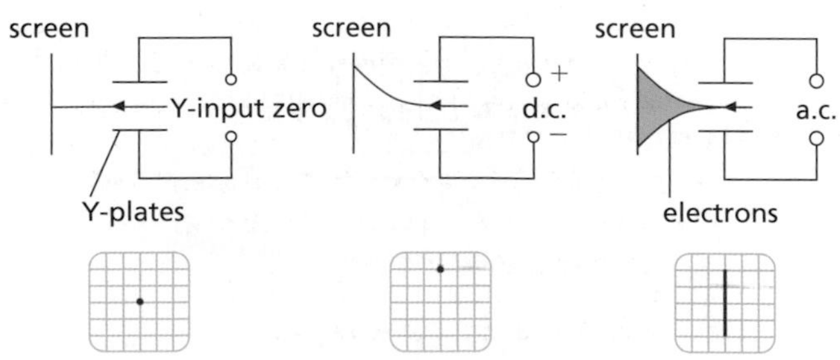

Figure 47.7 Deflection of the electron beam

In Figure 47.7a the p.d. between the Y-plates is zero as is the deflection. In b the d.c. input p.d. makes the upper plate positive and it attracts the beam of negatively charged electrons upwards. In c the 50 Hz a.c. input makes the beam move up and down so rapidly that it produces a continuous vertical line (whose length increases if the Y-amp gain is turned up).

The p.d. applied to the X-plates is also via an amplifier, the X-amplifier, and can either be from an external source connected to the **X-input** terminal or, more commonly, from the **time base** circuit in the CRO.

The time base deflects the beam horizontally in the X-direction and makes the spot sweep across the screen from left to right at a steady speed determined by the setting of the time base controls (usually 'coarse' and 'fine'). It must then make the spot 'fly' back very rapidly to its starting point, ready for the next sweep. The p.d. from the time base should therefore have a sawtooth waveform like that in Figure 47.8. Since AB is a straight line, the distance moved by the spot is directly proportional to time and the horizontal deflection becomes a measure of time, i.e. a time axis or base.

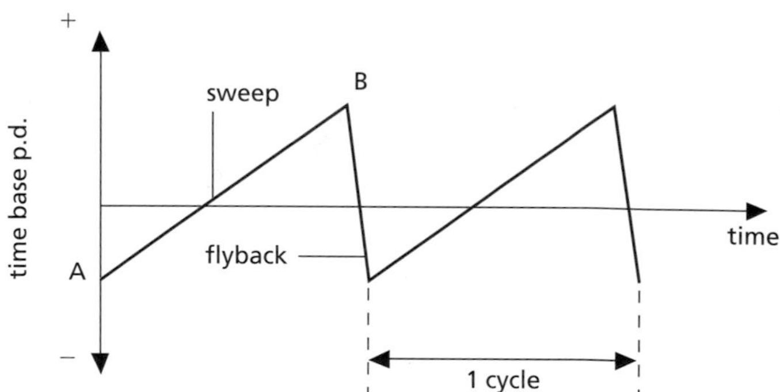

Figure 47.8 Time base waveform

In Figures 47.9a, b, c, the time base is on. For trace a the Y-input p.d. is zero, for b the Y-input is d.c. which makes the upper Y-plate positive. In both cases the spot traces out a horizontal line which appears to be continuous if the flyback is fast enough. For trace c the Y-input is a.c., i.e. the Y-plates are alternately positive and negative and the spot moves accordingly.

Figure 47.9 Deflection of the spot with time base on

□ *Uses of the CRO*

A small CRO is shown in Figure 47.10.

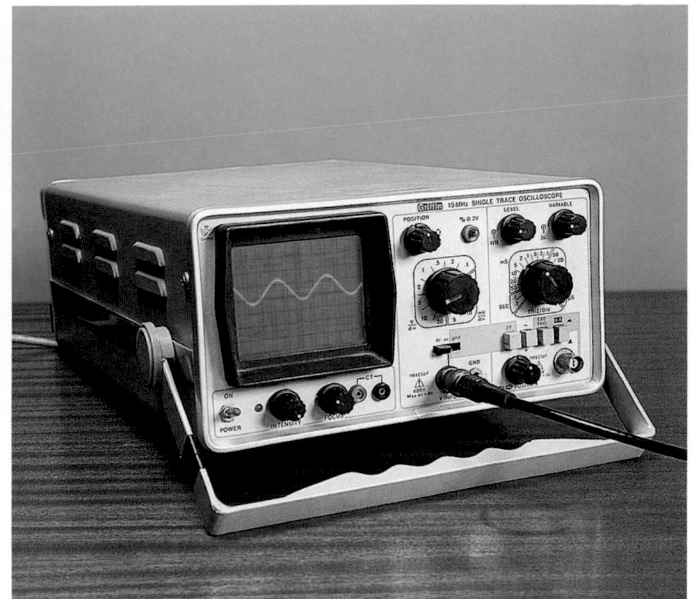

Figure 47.10 Single-beam CRO

a) Practical points

The **brilliance** or **intensity** control, which is sometimes the **on/off** switch as well, should be as low as possible when there is just a spot on the screen. Otherwise screen 'burn' occurs which damages the fluorescent material. If possible it is best to defocus the spot when not in use, or draw it into a line by running the time base.

When preparing the CRO for use, set the **brilliance**, **focus**, **X-** and **Y-shift** controls (which allow the spot to be moved 'manually' over the screen in the X and Y directions respectively) to their mid-positions. The **time base** and **Y-amp gain** controls can then be adjusted to suit the input.

When the **a.c./d.c. selector** switch is in the 'd.c.' (or 'direct') position, both d.c. and a.c. can pass to the Y-input. In the 'a.c.' (or 'via C') position, a capacitor blocks d.c. in the input but allows a.c. to pass.

b) Measuring p.ds

A CRO can be used as a d.c./a.c. voltmeter if the p.d. to be measured is connected across the Y-input terminals; **the deflection of the spot is proportional to the p.d.**

For example, if the **Y-amp gain** control is on, say, 1 V/div, a deflection of 1 vertical division on the screen graticule (like graph paper with squares for measuring deflections) would be given by a 1 V d.c. input. A line 1 division long (time base off) would be produced by an a.c. input of 1 V peak-to-peak, i.e. peak p.d. = 0.5 V.

c) Displaying waveforms

In this widely used role, the time base is on and the CRO acts as a 'graph-plotter' to show the waveform, i.e. the variation with time, of the p.d. applied to its Y-input. The displays in Figures 47.11a and b are of alternating p.ds with sine waveforms. For trace a, the time base frequency *equals* that of the input and one complete wave is obtained. For b it is *half* that of the input and two waves are formed. If the traces are obtained with the Y-amp gain control on, say, 0.5 V/div, the peak-to-peak voltage of the a.c. = 3 divs × 0.5 V/div = 1.5 V and the peak p.d. = 0.75 V.

Sound waveforms can be displayed if a microphone is connected to the Y-input terminals (see Chapter 9).

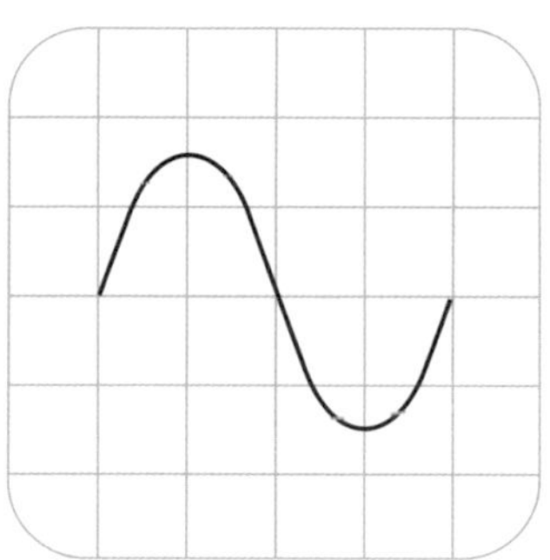

a

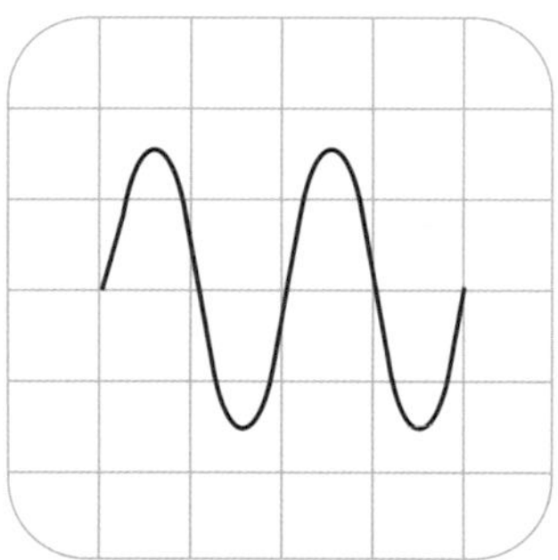

b

Figure 47.11 Alternating p.d. waveforms on the CRO

d) Measuring time intervals and frequency

These can be measured if the CRO has a calibrated time base. For example, when the time base is set on 10 ms/div, the spot takes 10 milliseconds to move 1 division horizontally across the screen graticule. If this is the time base setting for the waveform in Figure 47.11b then since one complete wave occupies two horizontal divisions, we can say

$$\begin{aligned}\text{time for 1 complete wave} &= 2 \text{ divs} \times 10 \text{ ms/div}\\ &= 20 \text{ ms}\\ &= 20/1000 = 1/50 \text{ s}\end{aligned}$$

∴ number of complete waves per second = 50

∴ frequency of a.c. applied to Y-input = 50 Hz

F X-rays

X-rays are produced when high-speed electrons are stopped by matter.

a) Production

In an X-ray tube, Figure 47.12, electrons from a hot filament are accelerated across a vacuum to the anode by a large p.d. (up to 100 kV). The anode is a copper block with a 'target' of a high-melting-point metal such as tungsten on which the electrons are focused by the electric field between the anode and the concave cathode. The tube has a lead shield with a small exit for the X-rays.

The work done in transferring a charge Q through a p.d. V is

$$E = Q \times V \qquad \text{(see p. 171)}$$

This will equal the k.e. of the electrons reaching the anode if Q = charge on an electron = 1.6×10^{-19} C and V is the accelerating p.d. Less than 1% of the k.e. of the electrons becomes X-ray energy; the rest heats the anode which has to be cooled.

High p.ds give short wavelength, very penetrating (**hard**) X-rays. Less penetrating (**soft**) rays, of longer wavelength, are obtained with lower p.ds. The absorption of X-rays by matter is greatest by materials of high density having a large number of outer electrons in their atoms, i.e. of high atomic number (Chapter 49). A more intense beam of rays is produced if the rate of emission of electrons is raised by increasing the filament current.

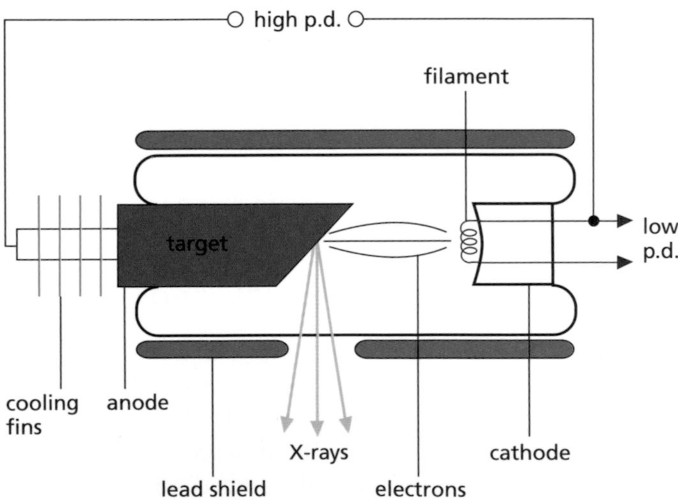

Figure 47.12 X-ray tube

b) Properties and nature

X-rays

(i) readily penetrate matter – up to 1 mm of lead,
(ii) are not deflected by electric or magnetic fields,
(iii) ionize a gas, making it a conductor, e.g. a charged electroscope discharges when X-rays pass through the surrounding air,
(iv) affect a photographic film,
(v) cause fluorescence,
(vi) give interference and diffraction effects.

These facts (and others) suggest that X-rays are electromagnetic waves of very short wavelength.

c) Uses

These were considered earlier (Chapter 8).

F Photoelectric effect

Electrons are emitted by certain metals when electromagnetic radiation of small enough wavelength falls on them. The effect is called **photoelectric emission** and is given by zinc exposed to ultraviolet.

The photoelectric effect only occurs for a given metal if the frequency of the incident electromagnetic radiation exceeds a certain **threshold frequency**. We can explain this by assuming that

(i) all electromagnetic radiation is emitted and absorbed as packets of energy, called **photons**, and
(ii) the energy of a photon is directly proportional to its frequency.

Ultraviolet photons would therefore have more energy than light photons since UV has a higher frequency than light. The behaviour of zinc (and most other substances) in not giving photoelectric emission with light but with UV would therefore be explained: a photon of light has less than the minimum energy required to emit an electron.

The absorption of a photon by an atom results in the electron gaining energy and the photon disappearing. If the photon has more than the minimum amount of energy required to enable an electron to escape, the excess appears as k.e. of the emitted electron.

energy of photon = energy needed for electron to escape + k.e. of electron

The photoelectric effect is the process by which X-ray photons are absorbed by matter; in effect it causes **ionization** (Chapter 48) since electrons are ejected and positive ions remain. Photons not absorbed pass through with unchanged energy.

F Waves or particles?

The wave theory of electromagnetic radiation can account for properties such as interference, diffraction and polarization which the photon theory cannot. On the other hand it does not explain the photoelectric effect which the photon theory does.

It would seem that electromagnetic radiation has a dual nature and has to be regarded as waves on some occasions and as 'particles' (photons) on others.

Questions

1 a In Figure 47.13a, to which terminals on the power supply must plates A and B be connected to deflect the cathode rays downwards?

b In Figure 47.13b, in which direction will the cathode rays be deflected?

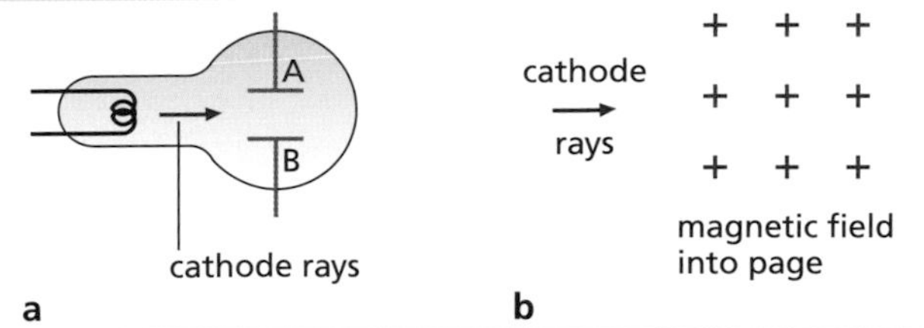

Figure 47.13

2 In an oscilloscope, a beam of electrons hits a screen. The screen has a fluorescent coating.

a The purpose of the fluorescent coating is to emit

A infra-red radiation when electrons hit it
B light when electrons hit it
C microwaves when electrons hit it
D ultraviolet radiation when electrons hit it

b The beam of electrons is produced by an electron gun. Figure 47.14 shows the principle of an electron gun.

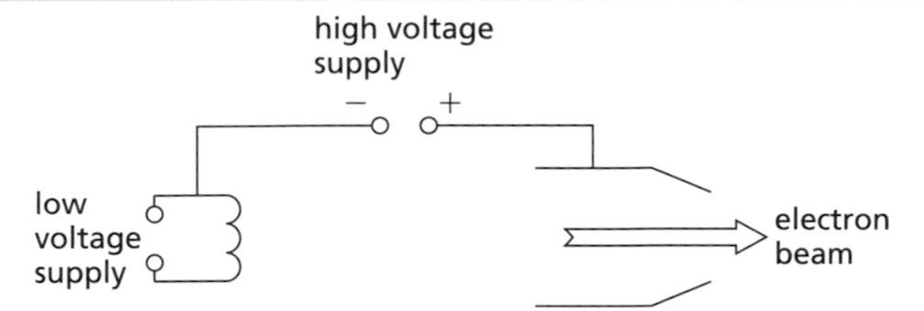

Figure 47.14

(i) Copy the diagram and label the cathode (the part that emits electrons).
(ii) Label the anode (the part that accelerates the electrons).
(iii) How can the speed of the electrons in the beam be changed?
(iv) The positive and negative connections to the high voltage supply are swapped over. State and explain what happens to the electron beam.

c An alternating voltage is applied to the input of an oscilloscope. Figure 47.15 shows the appearance of the screen. The sensitivity is set to 3 volts per cm.

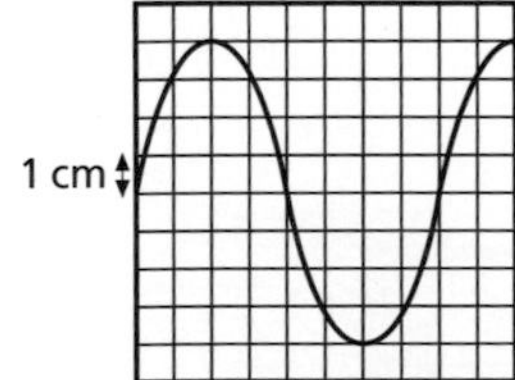

Figure 47.15

(i) What is the maximum voltage?
(ii) What is the minimum voltage?
(iii) The frequency of the alternating voltage is doubled, but the value of the voltage is unchanged. On a similar grid draw the new oscilloscope display.

(*London Foundation, June 1998*)

Checklist

After studying this chapter you should be able to

- explain the terms **thermionic emission** and **cathode rays**,
- describe experiments to show that cathode rays are deflected by magnetic and electric fields,
- □ describe with the aid of diagrams the jobs done in a CRO by the electron gun, the X- and Y-plates, the fluorescent screen and the timebase,
- □ describe how the CRO is used to display waveforms.

48 Radioactivity

Ionizing effect of radiation
Geiger–Müller (GM) tube
Alpha, beta, gamma rays
Particle tracks
Radioactive decay: half-life
Uses of radioactivity
Dangers and safety

The discovery of radioactivity in 1896 by the French scientist Becquerel was accidental. He found that uranium compounds emitted radiation which (i) affected a photographic plate even when it was wrapped in black paper and (ii) ionized a gas. Soon afterwards Marie Curie discovered the radioactive element radium. We now know that radioactivity arises from unstable nuclei (Chapter 49) which may occur naturally or be produced in reactors. Radioactive materials are widely used in industry, medicine and research.

We are all exposed to natural **background radiation** caused partly by radioactive materials in rocks, the air and our bodies, and partly by cosmic rays from outer space (see p. 251).

Ionizing effect of radiation

A charged electroscope discharges when a lighted match or a radium source (**held in forceps**) is brought near the cap, Figures 48.1a and b.

In the first case the flame knocks electrons out of surrounding air molecules leaving them as positively charged **ions**, i.e. air molecules which have lost one or more electrons, Figure 48.2; in the second case radiation causes the same effect, called **ionization**. The positive ions are attracted to the cap if it is negatively charged; if it is positively charged the electrons are attracted. As a result in either case the charge on the electroscope is neutralized, i.e. it loses its charge.

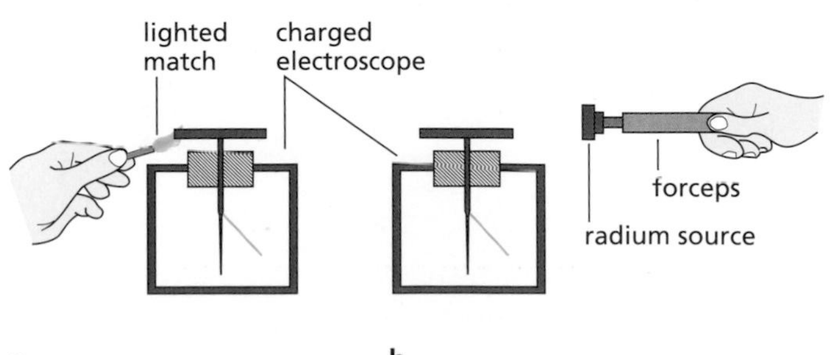

Figure 48.1

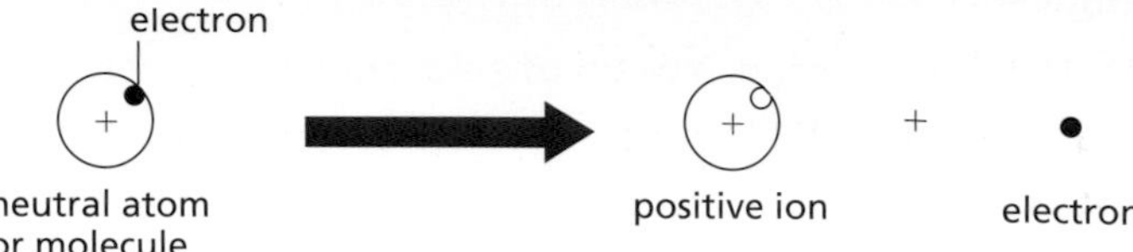

Figure 48.2 Ionization

Geiger–Müller (GM) tube

The ionizing effect is used to detect radiation.

When radiation enters a GM tube, Figure 48.3, either through a thin end-window made of mica, or, if it is very penetrating, through the wall, it creates argon ions and electrons. These are accelerated towards the electrodes and cause more ionization by colliding with other argon atoms.

On reaching the electrodes, the ions produce a current pulse which is amplified and fed either to a **scaler** or a **ratemeter**. A scaler counts the pulses and shows the total received in a certain time. A ratemeter gives the counts per second (or minute), or **count-rate**, directly. It usually has a loudspeaker which gives a 'click' for each pulse.

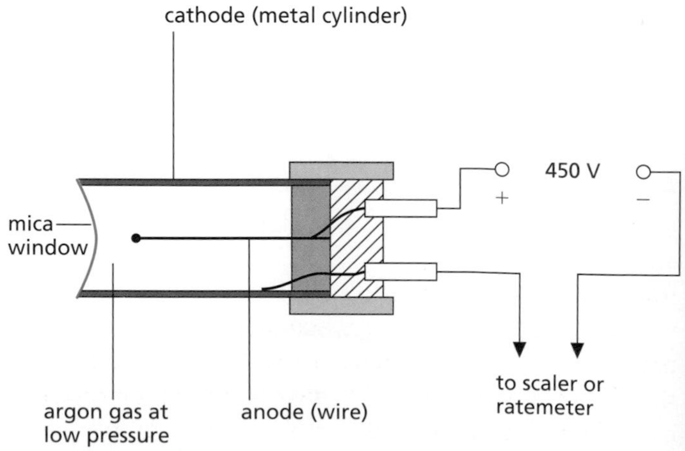

Figure 48.3 GM tube

Alpha, beta and gamma rays

Experiments to study the penetrating power, ionizing ability and behaviour of radiation in magnetic and electric fields show that a radioactive substance emits one or more of three types of radiation – called alpha (α), beta (β^- or β^+) and gamma (γ) rays.

Penetrating power can be investigated as in Figure 48.4 by observing the effect on the count-rate of placing one of the following in turn between the GM tube and the lead sheet:

(i) a sheet of thick paper (the radium source, lead and tube must be close together for this part),
(ii) a sheet of aluminium 2 mm thick,
(iii) a sheet of lead 2 cm thick.

Radium (Ra-226) emits α-rays, β^--rays and γ-rays. Other sources can be tried, e.g. americium, strontium and cobalt.

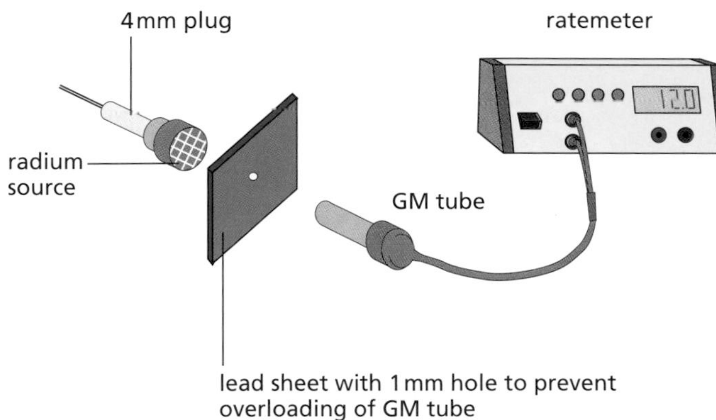

Figure 48.4 Investigating the penetrating power of radiation

a) Alpha rays

These are stopped by a thick sheet of paper and have a range in air of only a few centimetres since they cause intense ionization in a gas due to frequent collisions with gas molecules. They are deflected by electric and *strong* magnetic fields in a direction and by an amount which suggests they are helium atoms minus two electrons, i.e. **helium ions with a double positive charge**. From a particular substance, they are all emitted with the same speed (about 1/20th of that of light).

Americium (Am-241) is a pure α source.

b) Beta rays

These are stopped by a few millimetres of aluminium and some have a range in air of several metres. Their ionizing power is much less than that of α-particles. As well as being deflected by electric fields, they are more easily deflected by magnetic fields. Measurements show that β^--rays are streams of **high-energy electrons**, like cathode rays, emitted with a range of speeds up to that of light.

Strontium (Sr-90) emits β^--rays only.

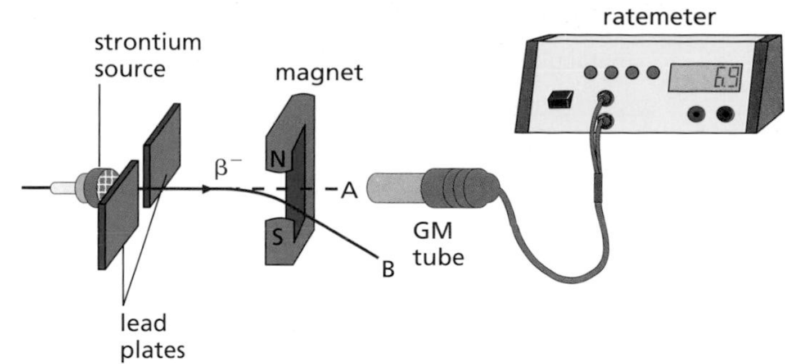

Figure 48.5 Demonstrating magnetic deflection of β^--particles

The magnetic deflection of β^--particles can be shown as in Figure 48.5. With the GM tube at A and without the magnet, the count-rate is noted. Inserting the magnet reduces the count-rate but it increases again when the GM tube is moved sideways to B.

c) Gamma rays

These are the most penetrating and are stopped only by many centimetres of lead. They ionize a gas even less than β-particles and are not deflected by electric and magnetic fields. They give interference and diffraction effects and are **electromagnetic radiation** travelling at the speed of light. Their wavelengths are those of very short X-rays, from which they differ only because they arise in atomic nuclei whereas X-rays come from energy changes in the electrons outside the nucleus.

Cobalt (Co-60) emits γ-rays and β^--rays but can be covered with aluminium to provide pure γ-rays.

In a collision, α-particles, with their relatively large mass and charge, have more of a chance of knocking an electron from an atom and causing ionization than the lighter β-particles. γ-rays, which have no charge, are even less likely to produce ionization.

A GM tube detects β-particles and γ-rays and energetic α-particles; a charged electroscope detects α only. All three types of rays cause fluorescence.

The behaviour of the three kinds of radiation in a magnetic field is summarized in Figure 48.6a. The deflections (not to scale) are found from Fleming's

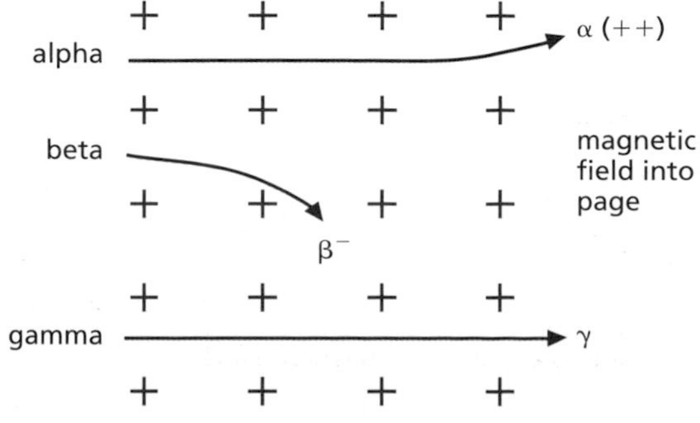

Figure 48.6a Deflection of α-rays, β^--rays and γ-rays in a magnetic field

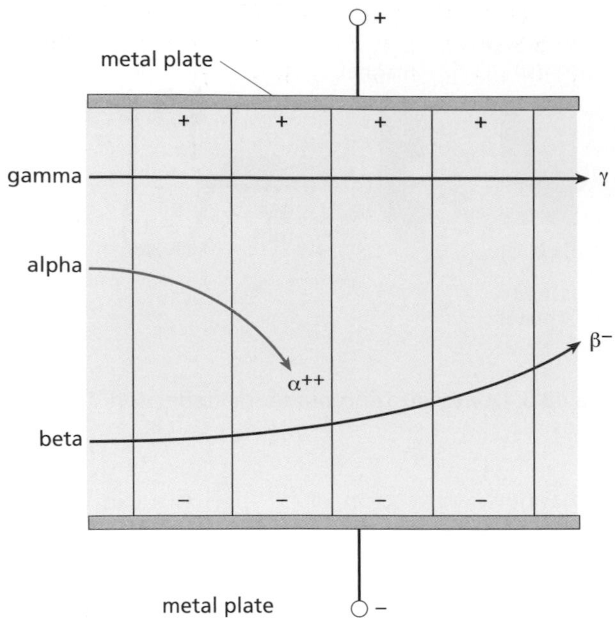

Figure 48.6b Deflection of α-rays, β^--rays and γ-rays in a uniform electric field

left-hand rule, taking negative charge moving to the right as equivalent to positive (conventional) current to the left.

Figure 48.6b shows the behaviour of α-rays, β^--rays and γ-rays in a uniform electric field. α-rays are attracted towards the negatively charged metal plate, β-rays are attracted towards the positively charged plate, and γ-rays pass through undeflected.

Particle tracks

The paths of particles of radiation were first shown up by the ionization they produced in devices called cloud chambers. When air containing vapour, e.g. alcohol, is cooled enough, saturation occurs. If ionizing radiation passes through the air, further cooling causes the saturated vapour to condense on the ions created. The resulting white line of tiny liquid drops shows up as a track when illuminated.

In a **diffusion cloud chamber**, α-particles gave straight, thick tracks, Figure 48.7a. Very fast β-particles produced thin, straight tracks while slower ones gave short, twisted, thicker tracks, Figure 48.7b. γ-rays eject electrons from air molecules; the ejected electrons behaved like β^--particles in the cloud chamber and produced their own tracks spreading out from the γ-rays.

The **bubble chamber**, in which the radiation leaves a trail of bubbles in liquid hydrogen, has now replaced the cloud chamber in research work. The higher density of atoms in the liquid gives better defined tracks, Figure 48.8, than obtained in a cloud chamber. A magnetic field is usually applied across the bubble chamber which causes charged particles to move in circular paths; the sign of the charge can be deduced from the way the path curves.

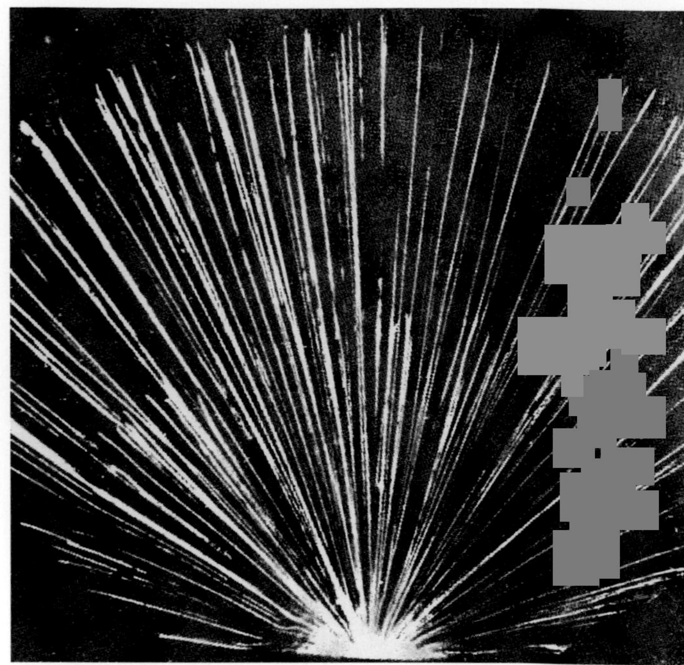

a α-particles

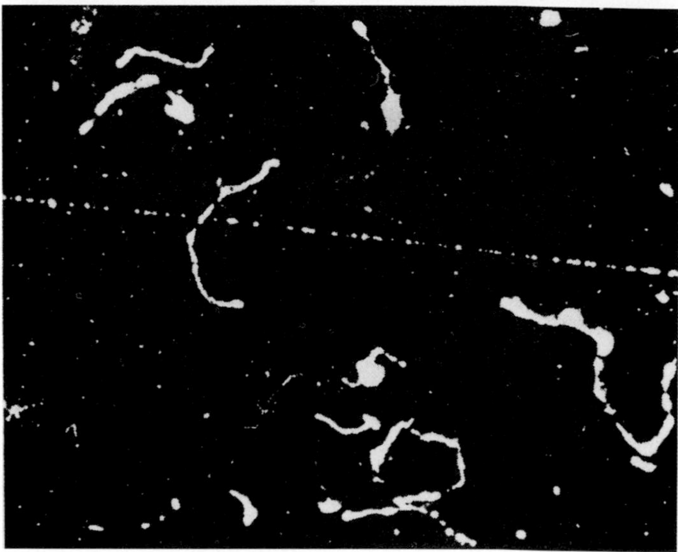

b Fast and slow β-particles

Figure 48.7 Tracks in a cloud chamber

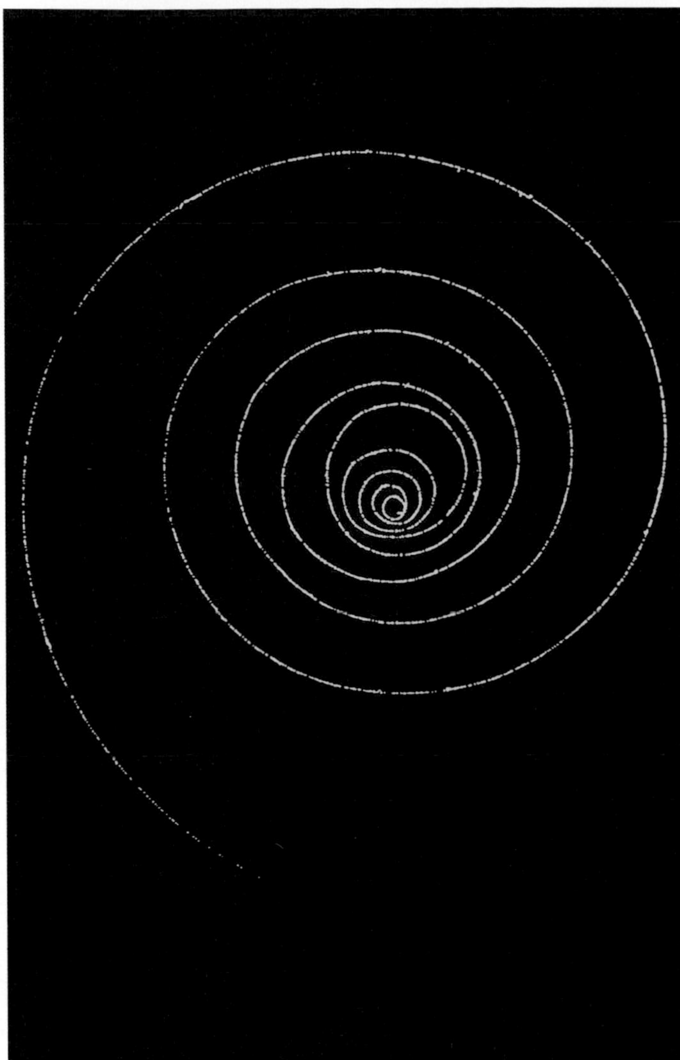

Figure 48.8 Charged particle track in a bubble chamber

Radioactive decay: half-life

Radioactive atoms have unstable nuclei and 'decay' into atoms of different elements with more stable nuclei when they emit α- or β-particles. These changes are spontaneous and cannot be controlled; also, it does not matter whether the material is pure or combined chemically with something else.

a) Half-life

The **rate of decay** is unaffected by temperature but every radioactive element has its own definite decay rate, expressed by its **half-life**. This is the **average time for half the atoms in a given sample to decay**. It is difficult to know when a substance has lost all its radioactivity, but the time for its activity to fall to half its value can be found more easily.

b) Decay curve

The average number of disintegrations (i.e. decaying atoms) per second of a sample is its **activity**. If it is measured at different times (e.g. by finding the count-rate using a GM tube and ratemeter), a decay curve of activity against time can be plotted. The ideal one in Figure 48.9 shows that the activity decreases by the *same* fraction in successive equal time intervals. It falls from 80 to 40 disintegrations per second in 10 minutes, from 40 to 20 in the next 10 minutes, from 20 to 10 in the third 10 minutes and so on. The half-life is 10 minutes.

Half-lives vary from millionths of a second to millions of years. For radium it is 1600 years.

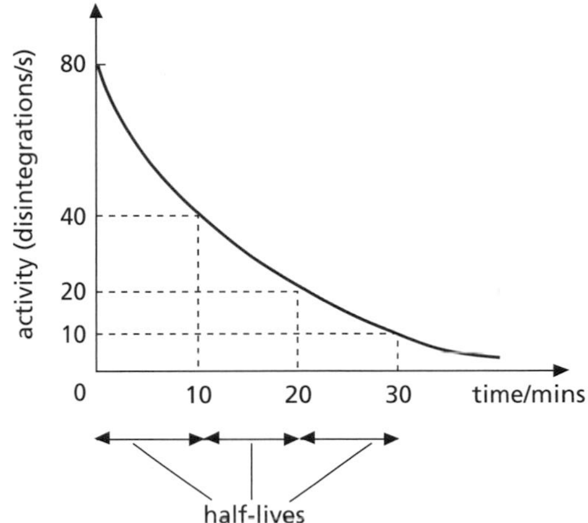

Figure 48.9 Decay curve

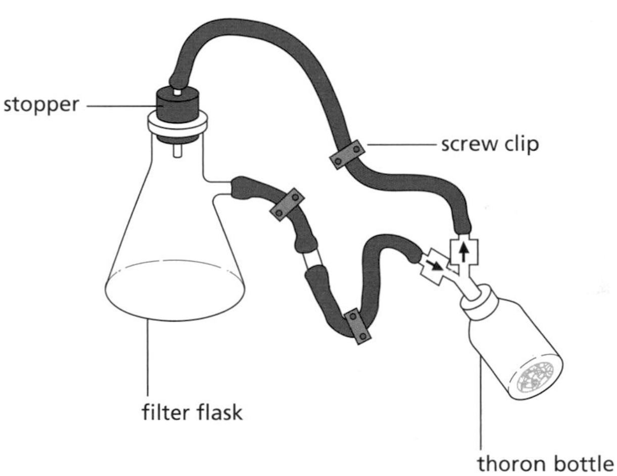

a

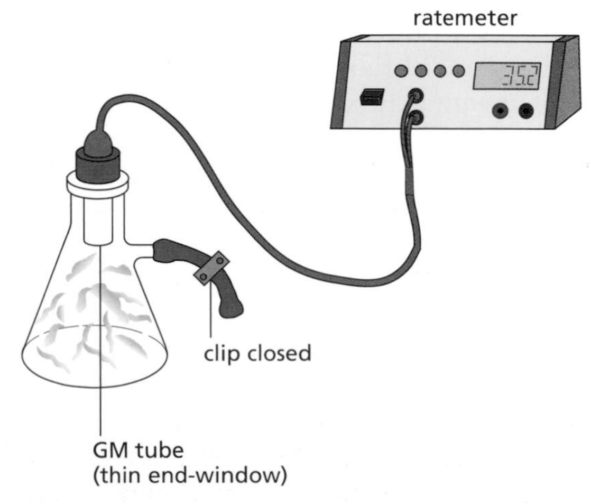

b

Figure 48.10

c) Experiment

The half-life of the α-emitting gas **thoron** can be found as in Figure 48.10. The thoron bottle is squeezed three or four times to transfer some thoron to the flask, Figure 48.10a. The clips are then closed, the bottle removed and the stopper replaced by a GM tube so that it seals the top, Figure 48.10b.

When the ratemeter reading has reached its maximum and started to fall, the count-rate is noted every 15 s for 2 minutes and then every 60 s for the next few minutes. (The GM tube is left in the flask for at least 1 hour until the radioactivity has decayed.)

A measure of the background radiation is obtained by recording the counts for a period (say 10 minutes) at a position well away from the thoron equipment. The count-rates in the thoron decay experiment are then corrected by subtracting the average background count-rate from each reading (as in question 2, p. 252). A graph of the corrected count-rate against time is plotted and the half-life (52 s) estimated from it.

d) Random nature

During the previous experiment it becomes evident that the count-rate varies irregularly: the loudspeaker of the ratemeter 'clicks' erratically, not at a steady rate. This is because radioactive decay is a **random** process, in that it is a matter of pure chance whether or not a particular atom will decay during a certain period of time. All we can say is that about half the atoms in a sample will decay during the half-life. We cannot say which atoms these will be, nor can we influence the process in any way. Radioactive emissions occur randomly over space and time.

Uses of radioactivity

Radioactive substances, called **radioisotopes**, are now made in nuclear reactors and have many uses.

a) Thickness gauge

If a radioisotope is placed on one side of a moving sheet of material and a GM tube on the other, the count-rate decreases if the thickness increases. This technique is used to control automatically the thickness of paper, plastic and metal sheets during manufacture, Figure 48.11. Because of their range, β-emitters are suitable sources for monitoring the thickness of thin sheets but γ-emitters would be needed for thicker materials.

Flaws in a material can be detected in a similar way; the count-rate will increase where a flaw is present.

Figure 48.11 Quality control in the manufacture of paper using a radioactive gauge

b) Tracers

The progress of a small amount of a weak radioisotope injected into a system can be 'traced' by a GM tube or other detector. The method is used in medicine to detect brain tumours and internal bleeding, in agriculture to study the uptake of fertilizers by plants, and in industry to measure fluid flow in pipes.

A tracer should be chosen whose half-life matches the time needed for the experiment; the activity of the source is then low after it has been used and so will not pose an ongoing radiation threat. For medical purposes, where short exposures are preferable, the time needed to transfer the source from the production site to the patient also needs to be considered.

c) Radiotherapy

Gamma rays from strong cobalt radioisotopes are used in the treatment of cancer.

d) Sterilization

Gamma rays are used to sterilize medical instruments by killing bacteria. They are also used to 'irradiate' certain foods, again killing bacteria to preserve the food for longer. They are safe to use as no radioactive material goes into the food.

e) Archaeology

A radioisotope of carbon present in the air, carbon-14, is taken in by living plants and trees. When a tree dies no fresh carbon is taken in and the carbon-14 starts to decay with a half-life of 5700 years. By measuring the residual activity of carbon-containing material such as wood, linen or charcoal, the age of archaeological remains can be estimated within the range 1000 to 50 000 years, Figure 48.12. See *Worked example* (b), on p. 252.

The ages of rocks have been estimated in a similar way by measuring the ratio of the number of atoms of a radioactive element to those of its decay product in a sample. See *Worked example* (c), on p.252.

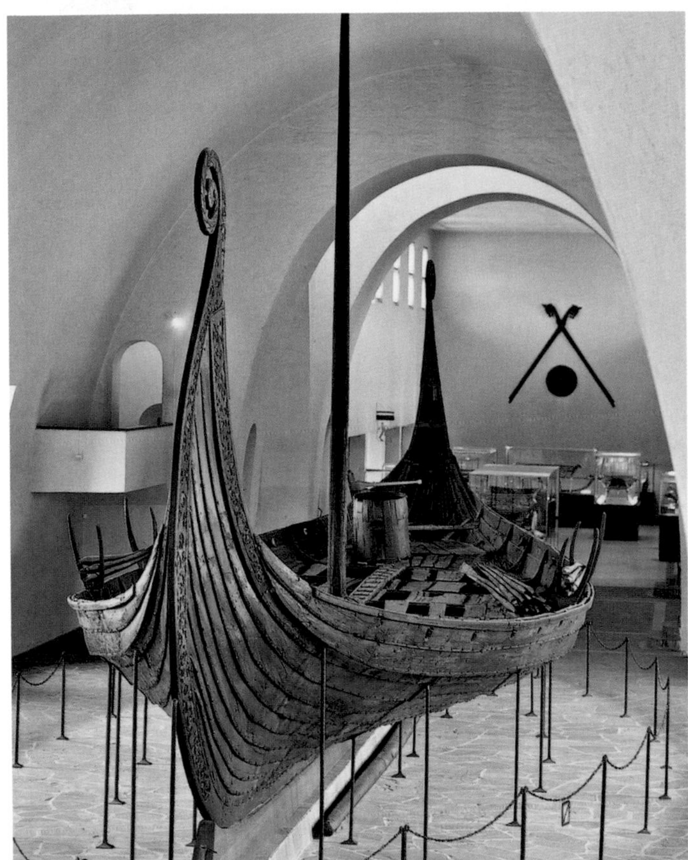

Figure 48.12 The year of construction of this Viking ship has been estimated by radiocarbon techniques to be AD 800

Dangers and safety

We are continually being exposed to radiation from a range of sources, both natural ('background') and artificial, as indicated in Figure 48.13.

(i) Cosmic rays (high-energy particles from outer space) are mostly absorbed by the atmosphere and produce radioactivity in the air we breathe, but some reach the Earth's surface.

(ii) Numerous homes, particularly in Scotland, are built from granite rocks which emit radioactive radon gas; this can collect in basements or well-insulated rooms if the ventilation is poor.

(iii) Radioactive potassium-40 is present in food and is absorbed by our bodies.

(iv) Various radioisotopes are used in certain medical procedures.

(v) Radiation is produced in the emissions from nuclear power stations and in fall-out from the testing of nuclear bombs; the latter produce strontium isotopes with long half-lives which are absorbed by bone.

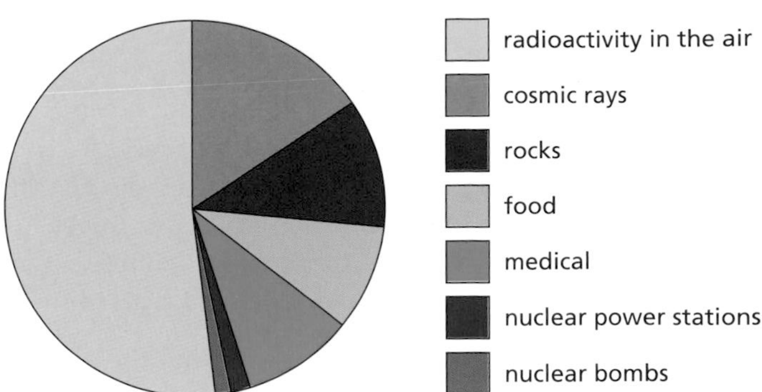

Figure 48.13 Radiation sources

We cannot avoid exposure to radiation in small doses but large doses can be dangerous to our health. The ionizing effect produced by radiation causes damage to cells and tissues in our bodies and can also lead to the mutation of genes. The danger from α-particles is small, unless the source enters the body, but β- and γ-rays can cause radiation burns (i.e. redness and sores on the skin) and delayed effects such as eye cataracts and cancer. Large exposures may lead to radiation sickness and death. The symbol used to warn of the presence of radioactive material is shown in Figure 48.14.

Figure 48.14 Radiation hazard sign

The increasing use of radioisotopes in medicine and industry has made it important to find ways of disposing of radioactive waste safely. One method is to enclose the waste in steel containers which are then buried in concrete bunkers; possible leakage is a cause of public concern, as water supplies could be contaminated allowing radioactive material to enter the food chain.

The weak sources used at school **should always be lifted with forceps, never held near the eyes and should be kept in their boxes when not in use**. In industry sources are handled by long tongs and transported in thick lead containers. Workers are protected by lead and concrete walls, and wear radiation dose badges which keep a check on the amount of radiation they have been exposed to over a period (usually one month). The badge contains several windows which allow different types of radiation to fall onto a photographic film (see question 5, p. 253); when the film is developed it is darkest where the exposure to radiation was greatest.

Worked examples

(a) A radioactive source has a half-life of 20 minutes. What fraction is left after 1 hour?

After 20 minutes, fraction left = 1/2
After 40 minutes, fraction left = 1/2 × 1/2 = 1/4
After 60 minutes, fraction left = 1/2 × 1/4 = 1/8

(b) Carbon-14 has a half-life of 5700 years. A 10 g sample of wood cut recently from a living tree has an activity of 160 counts/minute. A piece of charcoal taken from a prehistoric campsite also weighs 10 g but has an activity of 40 counts/minute. Estimate the age of the charcoal.

After 1 × 5700 years the activity will be 160/2 = 80 counts per minute
After 2 × 5700 years the activity will be 80/2 = 40 counts per minute
The age of the charcoal is 2 × 5700 = 11 400 years

(c) The ratio of the number of atoms of argon-40 to potassium-40 in a sample of radioactive rock is analysed to be 1 : 3. Assuming that there was no potassium in the rock originally and that argon-40 decays to potassium-40 with a half-life of 1500 million years, estimate the age of the rock.

Assume there were N atoms of argon-40 in the rock when it was formed.

After 1 × 1500 million years there will be $N/2$ atoms of argon left and $(N - N/2) = N/2$ atoms of potassium formed, giving an Ar : K ratio of 1 : 1.

After 2 × 1500 = 3000 million years there would be $(N/2)/2 = N/4$ argon atoms left and $(N - N/4) = 3N/4$ potassium atoms formed, giving an Ar : K ratio of 1 : 3 as measured.

The rock must be about 3000 million years old.

Questions

1 Which type of radiation from radioactive materials
- **a** has a positive charge?
- **b** is the most penetrating?
- **c** is easily deflected by a magnetic field?
- **d** consists of waves?
- **e** causes the most intense ionization?
- **f** has the shortest range in air?
- **g** has a negative charge?
- **h** is not deflected by an electric field?

2 Kate's teacher wants to find how much beta radiation passes through different thicknesses of aluminium, Figure 48.15.

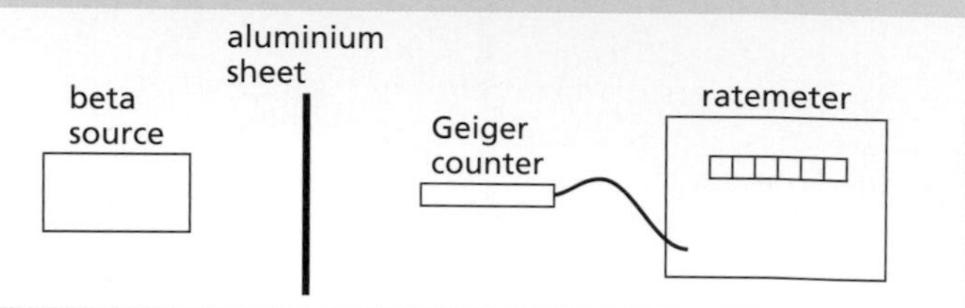

Figure 48.15

First he measures background radiation. It gives a reading of 60 counts per minute on the ratemeter.
- **a** Suggest *two* possible sources of background radiation.
- **b** Write down *two* safety precautions that he should take when using the beta source.

He now records the count rate for different thicknesses of aluminium. The table shows the results.

Thicknesses of aluminium in mm	1.0	2.0	3.0	4.0	5.0	6.0	7.0	8.0
Actual ratemeter reading in counts per minute	560	310	180	120	90	75	60	60
Corrected count rate in counts per minute	500	250						

- **c** Finish the table. There are *six* gaps.
- **d** (i) Plot the points on a grid like that in Figure 48.16.
  (ii) Finish the graph by drawing the best curve.

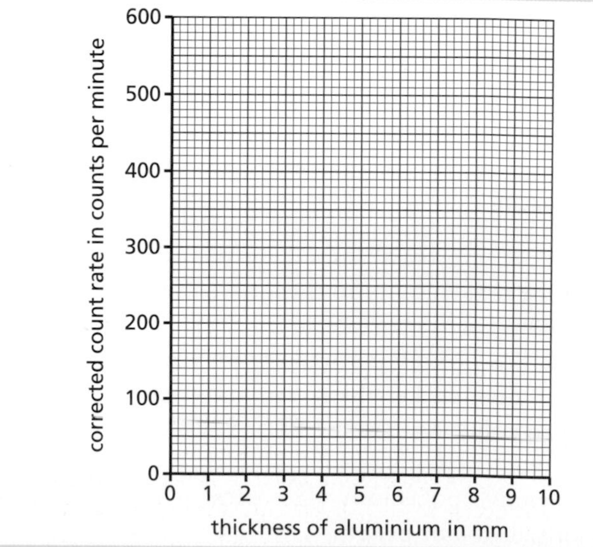

Figure 48.16

e Aluminium is rolled into sheets twenty millimetres thick in a rolling mill. A radioactive source and a detector are used to check the thickness of the sheet as it leaves the rollers, Figure 48.17.

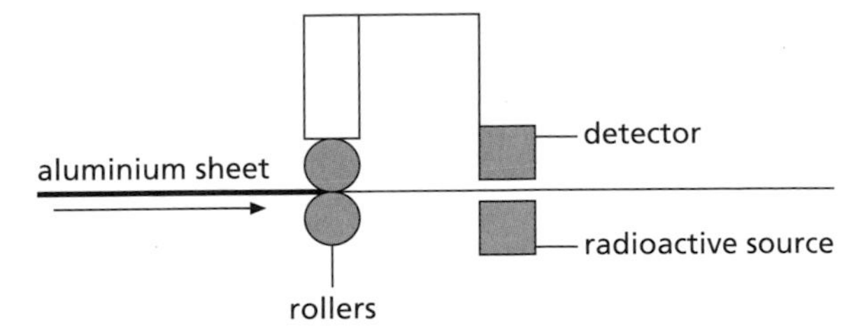

Figure 48.17

(i) Why is beta radiation *not* suitable for checking twenty millimetre sheet?
(ii) Suggest *one* type of radiation which could be used to check the thickness of twenty millimetre sheet.

(*OCR Foundation, June 1999*)

3 In an experiment to find the half-life of radioactive iodine, the count-rate falls from 200 counts per second to 25 counts per second in 75 minutes. What is its half-life?

4 If the half-life of a radioactive gas is 2 minutes, then after 8 minutes the activity will have fallen to a fraction of its initial value. This fraction is

A $\frac{1}{4}$ **B** $\frac{1}{6}$ **C** $\frac{1}{8}$ **D** $\frac{1}{16}$ **E** $\frac{1}{32}$

5 Figure 48.18 shows a film badge worn by people who work with radioactive materials. The badge has been opened. The badge is used to measure the amount of radiation the workers have been exposed to.

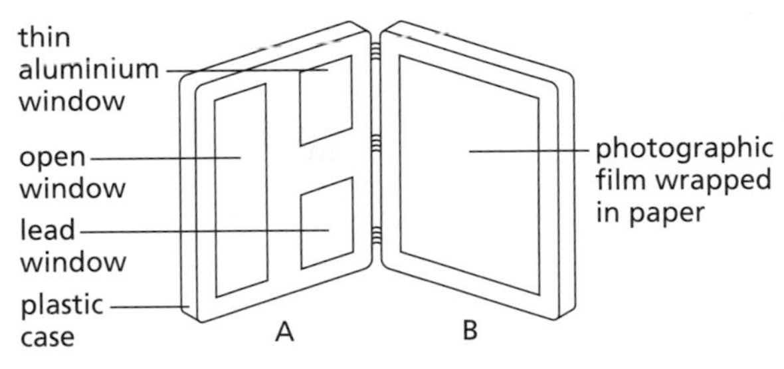

Figure 48.18

The detector is a piece of photographic film wrapped in paper inside part B of the badge. Part A has 'windows' as shown.

a Complete the sentences below.

When the badge is closed
(i) ...... radiation and ...... radiation can pass through the open window and affect the film.
(ii) most of the ...... radiation will pass through the lead window and affect the film.

b Other detectors of radiation use a gas which is ionised by the radiation.
(i) Explain what is meant by *ionised*.
(ii) Write down *one* use of ionising radiation.

c Uranium-238 has a very long half-life. It decays via a series of short-lived radioisotopes to produce the stable isotope lead-204. Explain, in detail, what is meant by the following terms.
(i) *half-life*
(ii) *radioisotopes*

d The relative proportions of uranium-238 and lead-204 in a sample of igneous rock can be used to date the rock. A rock sample contains three times as many lead atoms as uranium atoms.
(i) What fraction of the original uranium was left in the rock? Assume that there was no lead in the original rock.
(ii) The half-life of uranium-238 is 4500 million years. Calculate the age of the rock.

(*AQA (NEAB) Higher, June 1999*)

6 The radioactive isotope, carbon-14, decays by beta (β) particle emission.

a What is a beta (β) particle?

b Plants absorb carbon-14 from the atmosphere. The graph in Figure 48.19 shows the decay curve for 1 g of carbon-14 taken from a flax plant.

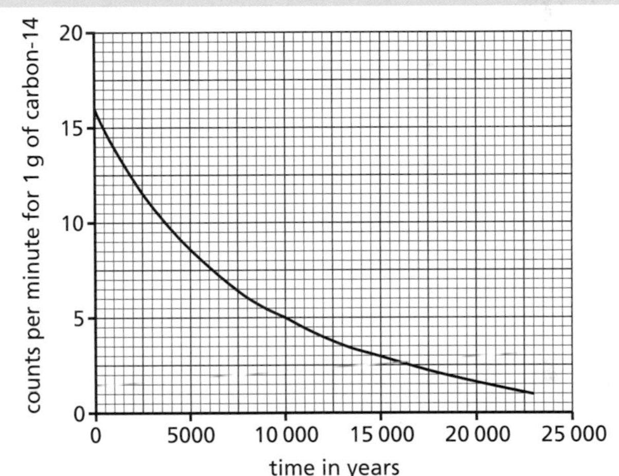

Figure 48.19

Use the graph to find the half-life of carbon-14.

c Linen is a cloth made from the flax plant. A recent exhibition included part of a linen shirt, believed to have belonged to St. Thomas à Becket, who died in 1162. Extracting carbon-14 from the cloth would allow the age of the shirt to be verified. If 1 g of carbon-14 extracted from the cloth were to give 870 counts in 1 hour, would it be possible for the shirt to have once belonged to St. Thomas à Becket? You must show clearly the steps used and reason for your decision.

(*AQA (SEG) Higher, June 1999*)

Checklist

After studying this chapter you should be able to

- recall that the radiation emitted by a radioactive substance can be detected by its ionizing effect,
- explain the principle of operation of a Geiger–Müller tube and a diffusion cloud chamber,
- recall the nature of α-, β- and γ-rays,
- describe experiments to compare the range and penetrating power of α-, β- and γ-rays in different materials,
- recall the ionizing abilities of α-, β- and γ-rays and relate them to their ranges,
- □ predict how α-, β- and γ-rays will be deflected in magnetic and electric fields,
- define the term **half-life**,
- describe an experiment from which a radioactive decay curve can be obtained,
- show from a graph that radioactive decay processes have a constant half-life,
- solve simple problems on half-life,
- recall that radioactivity is (a) a random process, (b) due to nuclear instability, (c) independent of external conditions,
- recall some uses of radioactivity,
- describe sources of radiation,
- discuss the dangers of radioactivity and safety precautions necessary.

49 Atomic structure

Nuclear atom
Protons and neutrons
Isotopes and nuclides
Radioactive decay
Nuclear stability
Models of the atom
Nuclear energy

The discoveries of the electron and of radioactivity seemed to indicate that atoms contained negatively and positively charged particles and were not indivisible as was previously thought. The questions then were 'How are the particles arranged inside an atom?', and 'How many are there in the atom of each element?'

An early theory, called the 'plum-pudding' model, regarded the atom as a positively charged sphere in which the negative electrons were distributed all over it (like currants in a pudding) and in sufficient numbers to make the atom electrically neutral. Doubts arose about this model.

Nuclear atom

While investigating radioactivity, the physicist Rutherford noticed that not only could α-particles pass straight through very thin metal foil as if it weren't there but also that some were deflected from their initial direction. With the help of Geiger (of GM tube fame) and Marsden, Rutherford investigated this in detail at Manchester University using the arrangement in Figure 49.1. The fate of the α-particles after striking the gold foil was detected by the scintillations (flashes of light) they produced on a glass screen coated with zinc sulfide and fixed to a rotatable microscope.

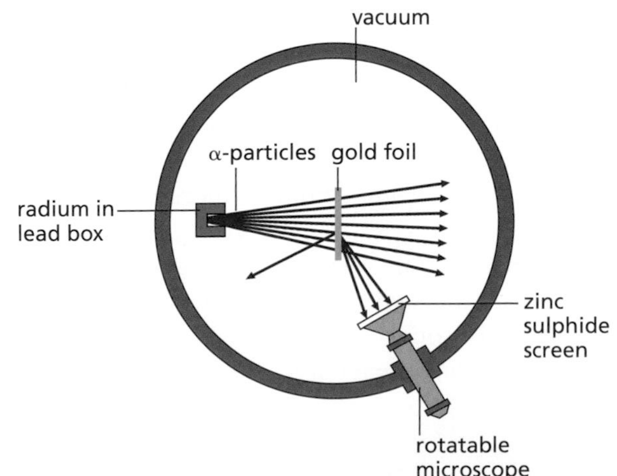

Figure 49.1 Geiger and Marsden's scattering experiment

They found that most of the α-particles were undeflected, some were scattered by appreciable angles and a few (about 1 in 8000) surprisingly 'bounced' back. To explain these results Rutherford proposed in 1911 a 'nuclear' model of the atom in which **all the positive charge and most of the mass of an atom** formed a dense core or **nucleus**, of very small size compared with the whole atom. The electrons surrounded the nucleus some distance away.

He derived a formula for the number of α-particles deflected at various angles, assuming that the electrostatic force of repulsion between the positive charge on an α-particle and the positive charge on the nucleus of a gold atom obeyed an inverse square law (i.e. the force increases four times if the separation is halved). Geiger and Marsden's experimental results completely confirmed Rutherford's formula and supported the view that an atom is mostly empty space. In fact the nucleus and electrons occupy about one million millionth of the volume of an atom. Putting it another way, the nucleus is like a sugar lump in a very large hall and the electrons a swarm of flies.

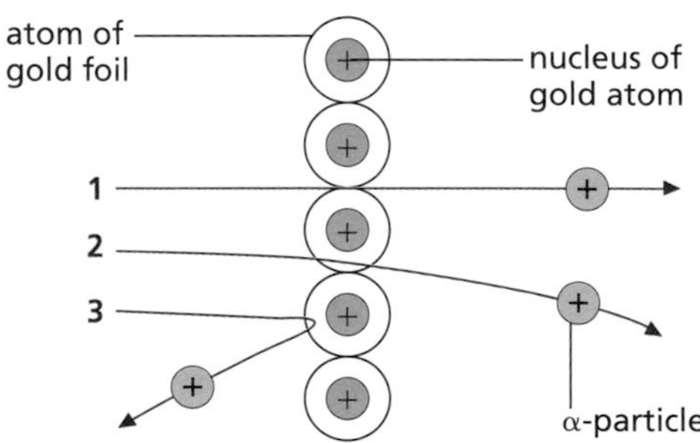

Figure 49.2 Electrostatic scattering of α-particles

Figure 49.2 shows the paths of three α-particles.

1 is clear of all nuclei and passes straight through the gold atoms.
2 suffers some deflection.
3 approaches a gold nucleus so closely as to be violently repelled by it and 'rebounds', appearing to have had a head-on 'collision'.

Protons and neutrons

We now believe as a result of other experiments, in some of which α and other high-speed particles were used as 'atomic probes', that atoms contain three basic particles – protons, neutrons and electrons.

A **proton** is a hydrogen atom minus an electron, i.e. a positive hydrogen ion. Its charge is equal in size but opposite in sign to that of an electron but its mass is about 2000 times greater.

A **neutron** is uncharged with almost the same mass as a proton.

Protons and neutrons are in the nucleus and are called **nucleons**. Together they account for the mass of the nucleus (and most of that of the atom); the protons account for its positive charge. These facts are summarized in Table 49.1.

Table 49.1

Particle	Relative mass	Charge	Location
proton	1836	+e	in nucleus
neutron	1839	0	in nucleus
electron	1	−e	outside nucleus

In a neutral atom the number of protons equals the number of electrons surrounding the nucleus. Table 49.2 shows the particles in some atoms. Hydrogen is simplest with 1 proton and 1 electron. Next is the inert gas helium with 2 protons, 2 neutrons and 2 electrons. The soft white metal lithium has 3 protons and 4 neutrons.

Table 49.2

	Hydrogen	Helium	Lithium	Oxygen	Copper
protons	1	2	3	8	29
neutrons	0	2	4	8	34
electrons	1	2	3	8	29

The atomic or proton number Z of an atom is the number of protons in the nucleus.

It is also the number of electrons in the atom. The electrons determine the chemical properties of an atom and when the elements are arranged in order of atomic number in the Periodic Table, they fall into chemical families.

In general

$$A = Z + N$$

where N is the **neutron number** of the element.

Atomic **nuclei** are represented by symbols. Hydrogen is written ${}^{1}_{1}\text{H}$, helium ${}^{4}_{2}\text{He}$, lithium ${}^{7}_{3}\text{Li}$ and in general atom X is written ${}^{A}_{Z}\text{X}$ where A is the nucleon number and Z the proton number.

The mass or nucleon number A of an atom is the number of nucleons in the nucleus.

Isotopes and nuclides

Isotopes of an element are atoms which have the same number of protons but different numbers of neutrons. That is, their proton numbers are the same but not their nucleon numbers.

Isotopes have identical chemical properties since they have the same number of electrons and occupy the same place in the Periodic Table. (In Greek, *isos* means same and *topos* means place.)

Few elements consist of identical atoms; most are mixtures of isotopes. Chlorine has two isotopes; one has 17 protons and 18 neutrons (i.e. $Z = 17$, $A = 35$) and is written ${}^{35}_{17}\text{Cl}$, the other has 17 protons and 20 neutrons (i.e. $Z = 17$, $A = 37$) and is written ${}^{37}_{17}\text{Cl}$. They are present in ordinary chlorine in the ratio of three atoms of ${}^{35}_{17}\text{Cl}$ to one atom of ${}^{37}_{17}\text{Cl}$, giving chlorine an average atomic mass of 35.5.

Hydrogen has three isotopes: ${}^{1}_{1}\text{H}$ with 1 proton, **deuterium** ${}^{2}_{1}\text{D}$ with 1 proton and 1 neutron and **tritium** ${}^{3}_{1}\text{T}$ with 1 proton and 2 neutrons. Ordinary hydrogen contains 99.99 per cent of ${}^{1}_{1}\text{H}$ atoms. Water made from deuterium is called 'heavy water' (D_2O); it has a density of 1.108 g/cm^3, it freezes at 3.8 °C and boils at 101.4 °C.

Each form of an element is called a **nuclide**. Nuclides with the same Z but different A are isotopes. Radioactive isotopes are termed radioisotopes or radionuclides; their nuclei are unstable.

Radioactive decay

The emission of an α or a β-particle from an unstable nucleus produces an atom of a different element, which may itself be unstable. After a series of changes a stable end-element is formed.

a) Alpha decay

An α-particle is a helium nucleus having 2 protons and 2 neutrons and when an atom decays by α emission, its nucleon number decreases by 4 and its proton number by 2. For example, when radium of nucleon number 226 and proton number 88 emits an α-particle, it decays to radon of nucleon number 222 and proton number 86. We can write:

$$^{226}_{88}\text{Ra} \rightarrow {}^{222}_{86}\text{Rn} + {}^{4}_{2}\text{He}$$

The values of A and Z must balance on both sides of the equation since nucleons and charge are conserved.

b) Beta decay

In β^- decay a neutron changes to a proton and an electron. The proton remains in the nucleus and the electron is emitted as a β^--particle. The new nucleus has the same nucleon number, but its proton number increases by one since it has one more proton. Radioactive carbon, called carbon-14, decays by β^- emission to nitrogen:

$$^{14}_{6}\text{C} \rightarrow {}^{14}_{7}\text{N} + {}^{0}_{-1}\text{e}$$

A particle called an antineutrino (ν^-), with no charge and negligible mass, is also emitted in β^- decay.

Positrons – subatomic particles with the same mass as an electron but with opposite (positive) charge – are emitted in some decay processes as β^+-particles. Their tracks can be seen in bubble chamber photographs. The symbol for a positron is $_{+1}^{0}\text{e}$. In β^+ decay a proton in a nucleus is converted to a neutron and a positron, for example in the reaction:

$$^{64}_{29}\text{Cu} \rightarrow {}^{64}_{28}\text{Ni} + {}^{0}_{+1}\text{e}$$

A neutrino (ν) is also emitted in β^+ decay. Neutrinos are emitted from the Sun in large numbers, but they rarely interact with matter so are very difficult to detect. Antineutrinos and positrons are the 'antiparticles' of neutrinos and electrons respectively. If a particle and its antiparticle collide they annihilate each other, producing energy in the form of γ-rays.

c) Gamma emission

After emitting an α-particle, β^-- or β^+-rays, some nuclei are left in an 'excited' state. Rearrangement of the protons and neutrons occurs and a burst of γ-rays is released.

F Nuclear stability

The stability of a nucleus depends on both the number of protons (Z) and the number of neutrons (N) it contains. Figure 49.3 is a plot of N against Z for all known nuclides; the continuous line indicates the stable nuclides and the shaded regions either side of this line are the regions of unstable nuclides.
It is found that for **stable nuclides**:

(i) $N = Z$ for the lightest nuclides,
(ii) $N > Z$ for the heaviest nuclides,
(iii) most nuclides have *even* N and Z, implying that the α-particle combination of two neutrons and two protons is likely to be particularly stable.

For **unstable nuclides**:

(i) disintegration tends to produce new nuclides nearer the stability line and continues until a stable nuclide is formed,
(ii) a nuclide above the stability line decays by β^- emission (a neutron changes to a proton and electron) so that the N/Z ratio decreases,
(iii) a nuclide below the stability line decays by β^+ emission (a proton changes to a neutron and positron) so that the N/Z ratio increases,
(iv) nuclei with more than 82 protons usually emit an α-particle when they decay.

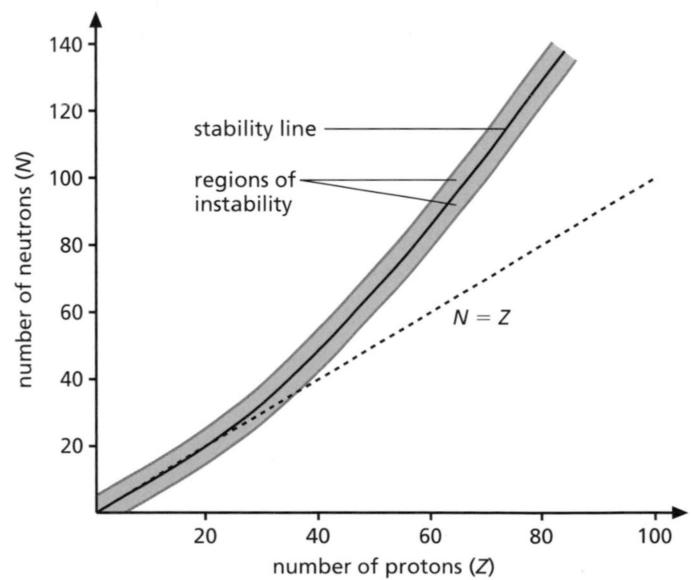

Figure 49.3 Stability of nuclei

Models of the atom

a) Rutherford–Bohr model

Shortly after Rutherford proposed his nuclear model of the atom, Bohr, a Danish physicist, developed it to explain how an atom emits light. He suggested that the electrons circled the nucleus at high speed being kept in **certain orbits** by the electrostatic attraction of the nucleus for them. He pictured atoms as miniature solar systems. Figure 49.4 shows the model for three elements.

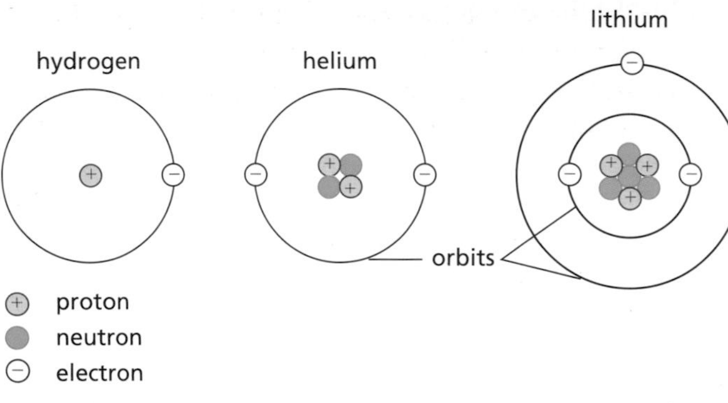

Figure 49.4 Electron orbits

Normally the electrons remain in their orbits but if the atom is given energy, e.g. by being heated, electrons may jump to an outer orbit. The atom is then said to be **excited**. Very soon afterwards the electrons return to an inner orbit and, as they do, they emit energy in the form of bursts of electromagnetic radiation (called **photons**), e.g. as infrared light, ultraviolet or X-rays, Figure 49.5. The wavelength of the radiation emitted depends on the two orbits between which the electrons jump. If an atom gains enough energy for an electron to escape altogether, the atom becomes an ion and the energy needed to achieve this is called the **ionization energy** of the atom.

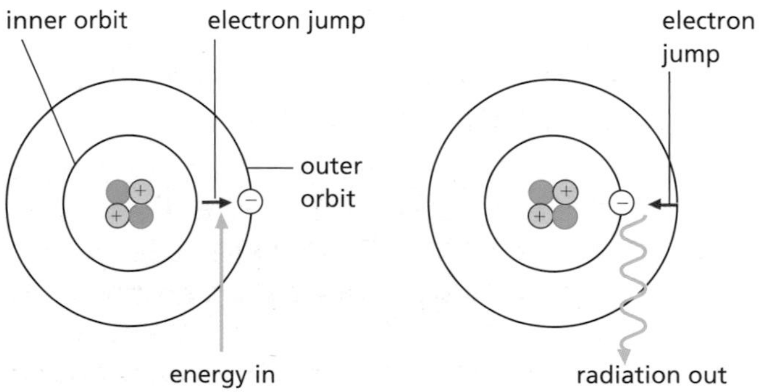

Figure 49.5 Bohr's explanation of energy changes in an atom

b) Schrödinger model

Although it remains useful for some purposes, the Rutherford–Bohr model was replaced by a **mathematical model** developed by Erwin Schrödinger, which is not easy to picture. The best we can do, without using advanced mathematics, is to say that the atom consists of a nucleus surrounded by a hazy cloud of electrons. Regions of the atom where the mathematics predicts electrons are more likely to be found are represented by denser shading, Figure 49.6.

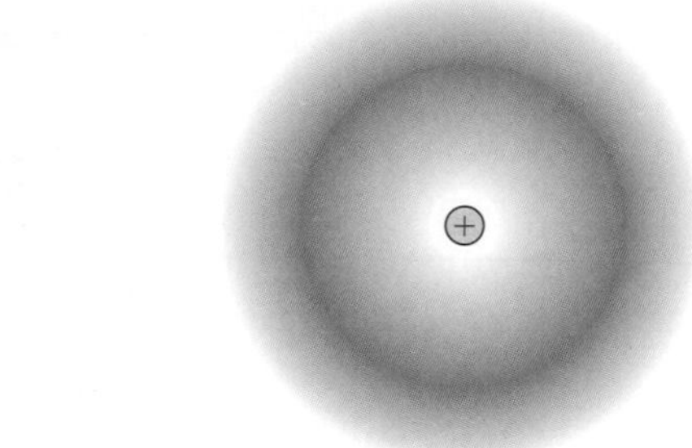

Figure 49.6 Electron cloud

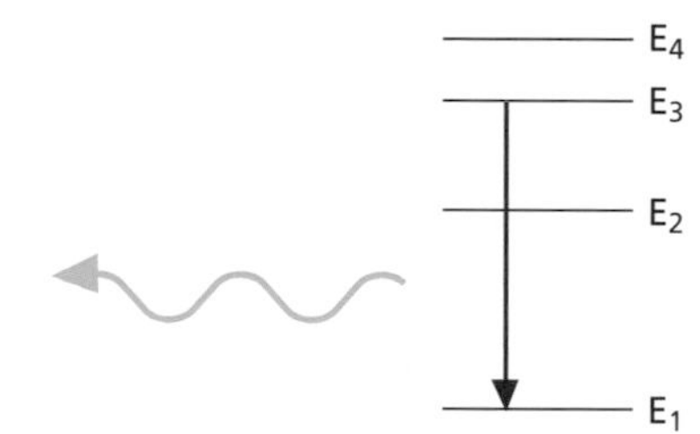

Figure 49.7 Energy levels of an atom

This theory does away with the idea of electrons moving in definite orbits and replaces them by **energy levels** that are different for each element. When an electron 'jumps' from one level, say E_3 in Figure 49.7, to a lower one E_1, a photon of electromagnetic radiation is emitted with energy equal to the difference in energy of the two levels. The frequency (and wavelength) of the radiation emitted by an atom is thus dependent on the arrangement of energy levels. For an atom emitting visible light, the resulting spectrum (produced for example by a prism) is a series of coloured **lines** which is unique to each element. Sodium vapour in a gas discharge tube (such as a yellow street light) gives two adjacent yellow–orange lines, Figure 49.8a. Light from the Sun is due to energy changes in many different atoms and the resulting spectrum is a **continuous** one with all colours (see Figure 49.8b).

Figure 49.8a Line spectrum due to energy changes in sodium

Figure 49.8b A continuous spectrum

☐ Nuclear energy

a) $E = mc^2$

Einstein predicted that if the energy of a body changes by an amount E, its mass changes by an amount m given by the equation

$$E = mc^2$$

where c is the speed of light (3×10^8 m/s). The implication is that any reaction in which there is a decrease of mass, called a **mass defect**, is a source of energy. The energy and mass changes in physical and chemical changes are very small; those in some nuclear reactions, e.g. radioactive decay, are millions of times greater. It appears that mass (matter) is a very concentrated form of energy.

b) Fission

The heavy metal uranium is a mixture of isotopes of which $^{235}_{92}U$, called uranium-235, is the most important. Some atoms of this isotope decay quite naturally, emitting high-speed neutrons. If one of these hits the nucleus of a neighbouring uranium-235 atom (being uncharged the neutron is not repelled by the nucleus), this may break (**fission**) into two nearly equal radioactive nuclei, often of barium and krypton, with the production of two or three more neutrons:

$$^{235}_{92}\text{U} + \underset{\text{neutron}}{^{1}_{0}\text{n}} \rightarrow \underbrace{^{144}_{56}\text{Ba} + ^{90}_{36}\text{Kr}}_{\text{fission fragments}} + \underset{\text{neutrons}}{2^{1}_{0}\text{n}}$$

The mass defect is large and appears mostly as k.e. of the fission fragments. These fly apart at great speed, colliding with surrounding atoms and raising their average k.e., i.e. their temperature, so producing heat.

If the fission neutrons split other uranium-235 nuclei, a **chain reaction** is set up, Figure 49.11. In practice some fission neutrons are lost by escaping from the surface of the uranium before this happens. The ratio of those causing fission to those escaping increases as the mass of uranium-235 increases. This must exceed a certain **critical** value to sustain the chain reaction.

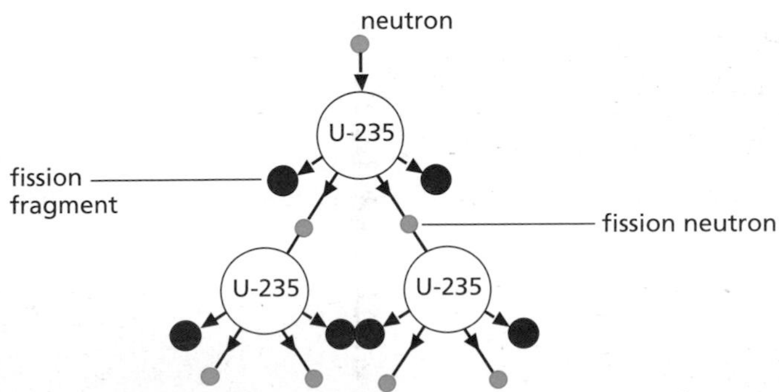

Figure 49.11 Chain reaction

c) Nuclear reactor

In a nuclear power station heat from a nuclear reactor produces the steam for the turbines. Figure 49.10 is a simplified diagram of one type of reactor.

The chain reaction occurs at a steady rate which is controlled by inserting or withdrawing neutron-absorbing rods of boron among the uranium rods. The graphite core is called the **moderator** and slows down the fission neutrons; fission of uranium-235 occurs more readily with slow than with fast neutrons. Carbon dioxide gas is pumped through the core and carries off heat to the **heat exchanger** where steam is produced. The concrete shield gives workers protection from γ-rays and escaping neutrons. The radioactive fission fragments must be removed periodically if the nuclear fuel is to be used efficiently.

In an **atomic bomb** an increasing uncontrolled chain reaction occurs when two pieces of uranium-235 come together and exceed the critical mass.

d) Fusion

The union of light nuclei into heavier ones can also lead to a loss of mass and, as a result, the release of energy. Such a reaction has been achieved in the 'hydrogen bomb'. At present, research is being done on the controlled fusion of isotopes of hydrogen (deuterium and tritium) to give helium.

$$\underset{\text{deuterium}}{^{2}_{1}\text{H}} + \underset{\text{tritium}}{^{3}_{1}\text{H}} \rightarrow \underset{\text{helium}}{^{4}_{2}\text{He}} + \underset{\text{neutron}}{^{1}_{0}\text{n}}$$

Fusion can only occur if the reacting nuclei have enough energy to overcome their mutual electrostatic repulsion. This can happen if they are raised to a very high temperature (over 100 million °C) so that they collide at very high speeds. If fusion occurs, the energy released is enough to keep the reaction going; since heat is required, it is called **thermonuclear** fusion.

The source of the Sun's energy is nuclear fusion. The temperature in the Sun is high enough for the conversion of hydrogen into helium to occur in a sequence of thermonuclear fusion reactions known as the 'hydrogen burning' sequence.

$$^{1}_{1}\text{H} + ^{1}_{1}\text{H} \rightarrow ^{2}_{1}\text{H} + \text{positron } (^{0}_{1}\text{e}) + \text{neutrino } (\nu)$$

$$^{1}_{1}\text{H} + ^{2}_{1}\text{H} \rightarrow ^{3}_{2}\text{He} + \gamma\text{-ray}$$

$$^{3}_{2}\text{He} + ^{3}_{2}\text{He} \rightarrow ^{4}_{2}\text{He} + ^{1}_{1}\text{H} + ^{1}_{1}\text{H}$$

Each of these fusion reactions results in a loss of mass and a release of energy. Overall, tremendous amounts of energy are created that help to maintain the very high temperature of the Sun.

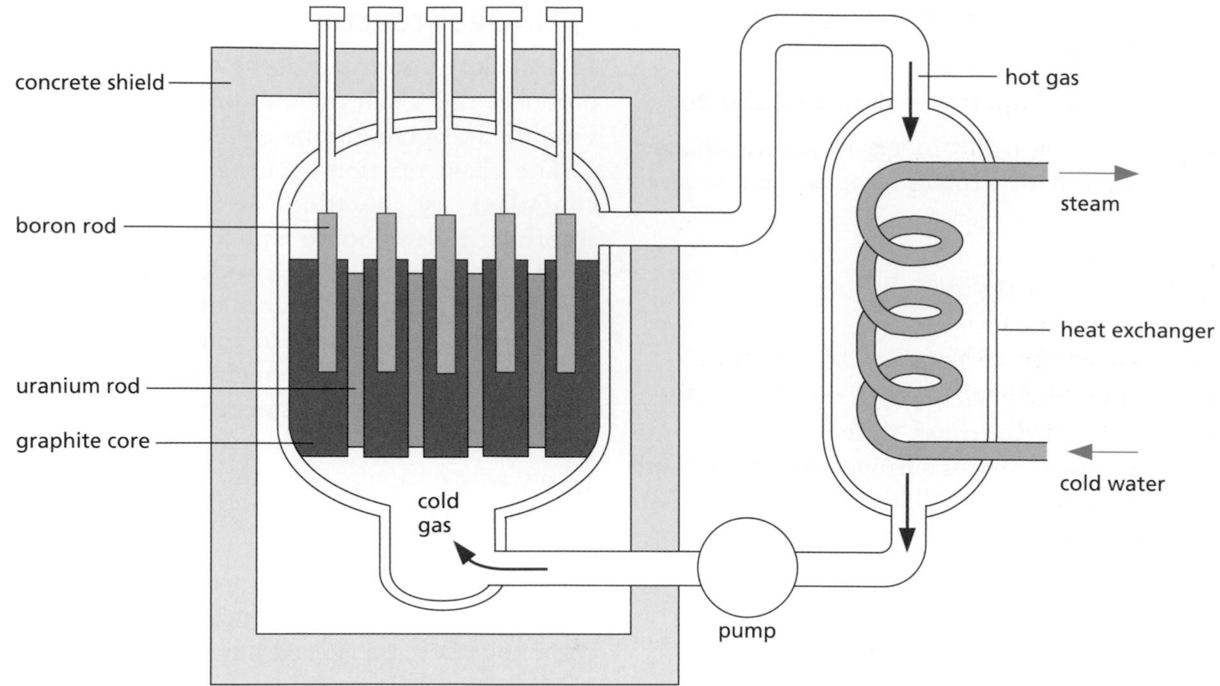

Figure 49.10 Nuclear reactor

Questions

1 a One nuclide is written as $^{210}_{84}\text{Po}$.
(i) Which figure is the proton number (atomic number)?
(ii) Which figure is the nucleon number (mass number)?
(iii) Which figure gives the number of protons in the nucleus?
(iv) How can you find the number of neutrons in the nucleus?

b An α-particle can be written as $^{4}_{2}\alpha$. Polonium $^{210}_{84}\text{Po}$ decays into lead (Pb) by emitting an α-particle. Copy and complete the nuclear equation below, by writing the correct numbers in the boxes.

$$^{210}_{84}\text{Po} \longrightarrow {}^{\square}_{\square}\text{Pb} + {}^{4}_{2}\alpha$$

(*UCLES IGCSE Physics Core, Nov 2000*)

2 The table below contains some information about uranium-238.

Proton number $Z = 92$
Nucleon number $A = 238$
Decays by emitting α-particle

a State how many electrons there are in a neutral atom of uranium-238.
b State where in the atom the electrons are to be found.
c State how many neutrons there are in an atom of uranium-238.
d State where in the atom the neutrons are to be found.
e State what happens to the number of protons in an atom of uranium-238 when an α-particle is emitted.

(*UCLES IGCSE Physics Core, Nov 2006*)

3 The diagrams in Figure 49.11 represent three atoms, **A**, **B** and **C**.

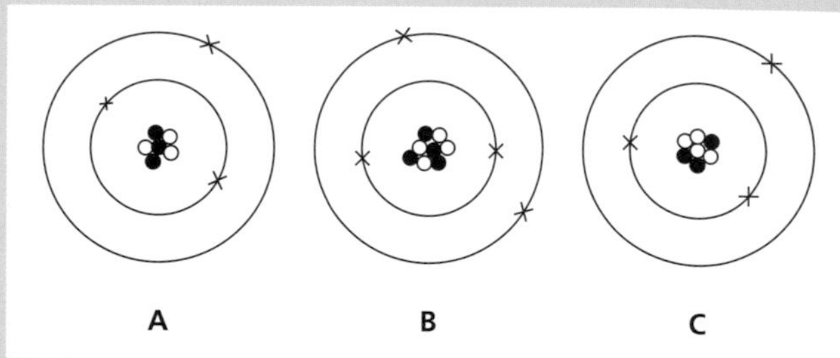

Figure 49.11

a Two of the atoms are from the *same* element.
(i) Which of **A**, **B** and **C** is an atom of a different element?
(ii) Give *one* reason for your answer.
b Two of these atoms are isotopes of the same element.
(i) Which two are isotopes of the same element?
(ii) Explain your answer.
c Which of the particles ○, X and ●, shown in the diagrams
(i) has a positive charge?
(ii) has no charge?
(iii) has the smallest mass?
d Using the same symbols as those used in the atom diagrams, draw an alpha particle.

(*AQA (NEAB) Higher, June 1999*)

Checklist

After studying this chapter you should be able to

- ■ describe how Rutherford and Bohr contributed to views about the structure of the atom,
- □ describe the Geiger–Marsden experiment which established the nuclear model of the atom,
- ■ recall the charge, relative mass and location in the atom of protons, neutrons and electrons,
- ■ define the terms **proton number** (Z), **neutron number** (N) and **nucleon number** (A) and use the equation $A = Z + N$,
- ■ explain the terms **isotope** and **nuclide** and use symbols to represent them, e.g. $^{35}_{17}Cl$,
- ■ write equations for radioactive decay and interpret them,
- □ connect the release of energy in a nuclear reaction with a change of mass according to the equation $E = mc^2$,
- □ outline the process of *fission*,
- □ outline the process of *fusion*.

Electrons and atoms
Additional questions

Electrons

1 The diagram shows a simplified diagram of the front of a cathode ray oscilloscope (c.r.o.).

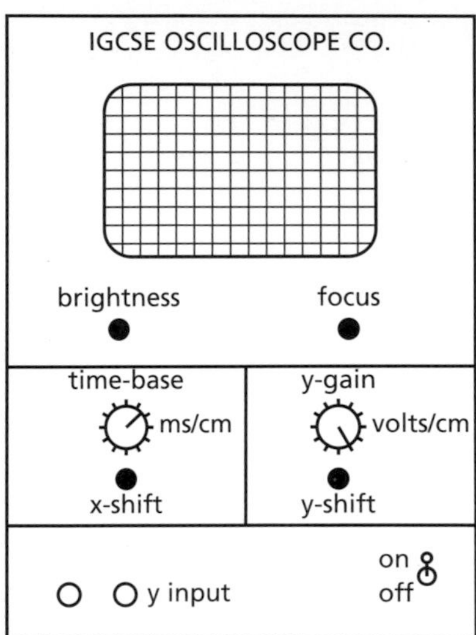

a When the oscilloscope is switched on, a bright spot is seen across the centre of the screen.
(i) Describe what causes this bright spot.
(ii) The spot is rather blurred. What control should be adjusted to make it sharper?
(iii) Which control would be switched on to turn the spot into a horizontal line?
(iv) Describe what happens inside the oscilloscope to turn the spot into a horizontal line.

b You have an alternating p.d. whose waveform you wish to display on the screen.
(i) Where would you connect this alternating p.d. to the oscilloscope?
(ii) The diagram below shows what the trace on the screen might look like.

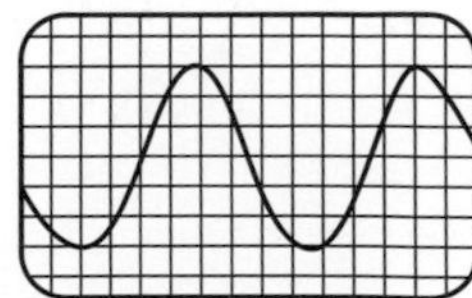

1 What change would you see on the screen if you adjusted the *x*-shift control?

2 What change would you see on the screen if you adjusted the *y*-shift control?

(UCLES IGCSE Physics Core, May 2003)

2 A tube for producing cathode rays is shown below. The tube contains various parts.

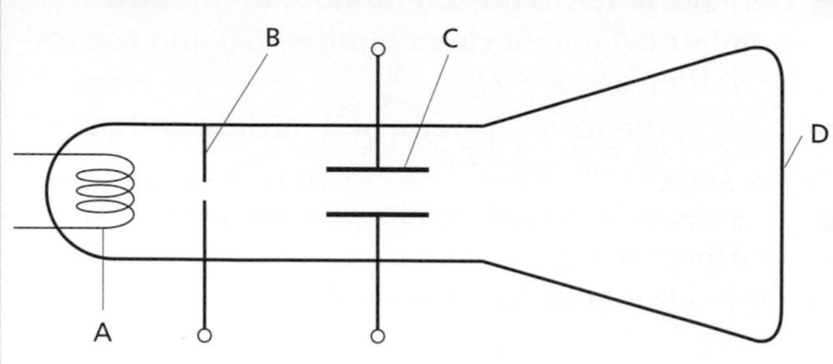

A spot is formed on the screen by the cathode rays.

a What do cathode rays consist of?

b Which part, A, B, C or D, must be heated to create the cathode rays?

c (i) Which part, A, B, C or D, is coated with fluorescent material?
(ii) What is the purpose of the fluorescent material?

d A potential difference is applied between the two halves of part C. What effect does this have on the cathode rays?

e Explain why there needs to be a vacuum inside the tube.

(UCLES IGCSE Physics Core, Nov 2005)

3 a The diagram illustrates a cathode-ray tube.

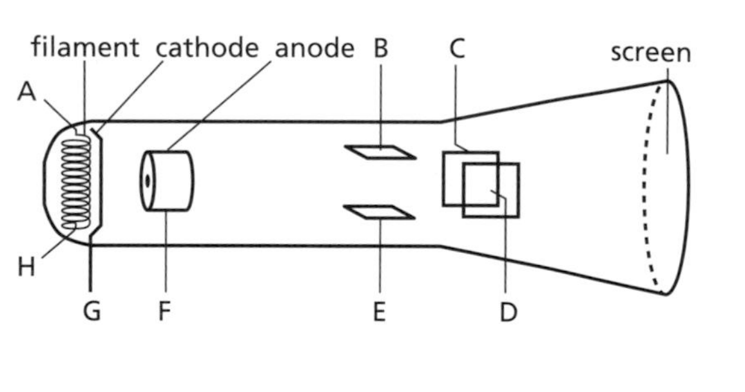

(i) Between which two points would you connect a low potential difference in order to heat the cathode?
(ii) Between which two points would you connect a high potential difference in order to produce cathode rays?
(iii) Between which two points would you connect a potential difference in order to deflect the cathode rays upwards?

b When the time base of a cathode-ray oscilloscope is turned on, there is a horizontal trace across the screen, as shown on page 263.

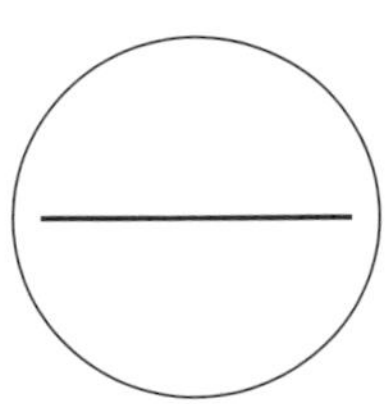

(i) An alternating potential difference of constant frequency and constant amplitude is connected to the Y-input of the oscilloscope. Sketch the trace which might be obtained.
(ii) The time base is switched off but the alternating potential difference is left connected. *Describe* what would be seen on the screen.

c A microphone is connected to another cathode-ray oscilloscope, with the time base switched to a suitable setting. First, a lady with a high-pitched voice sings into the microphone. Then a man with a low-pitched voice sings into the microphone. Describe how the traces seen on the screen would differ.

(*UCLES IGCSE Physics Core, May 1999*)

4 a The diagram shows how a beam of electrons would be deflected by an electric field produced between two metal plates. The connections of the source of high potential difference are not shown.

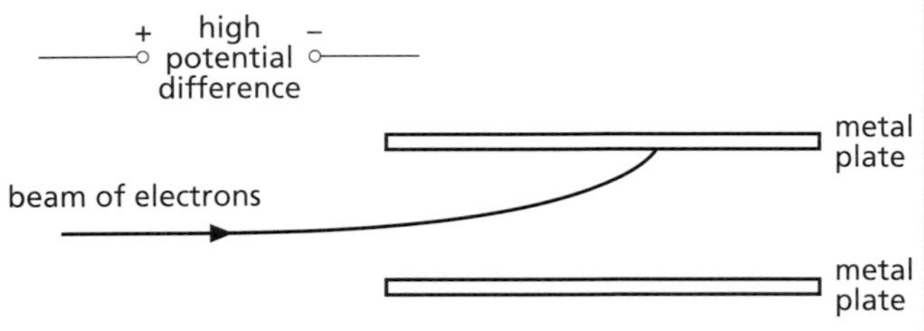

(i) Copy the diagram and draw in the missing connections.
(ii) Explain why the beam of electrons is deflected in the direction shown. In your answer, consider all the charges involved and their effect on each other.

b The deflection of a beam of electrons by an electric field is used in cathode-ray oscilloscopes.
(i) What makes the electron beam move backwards and forwards across the screen?
(ii) What makes the electron beam move up and down the screen?

c An a.c. waveform is displayed so that two full waves appear on the screen of a cathode ray oscilloscope. Sketch the waveform.

(*UCLES IGCSE Physics Extended, Nov 2006*)

5 An electron, charge *e* and mass *m*, is accelerated in a cathode ray tube by a p.d. of 1000 V. Calculate
a the kinetic energy gained by the electron,
b the speed it acquires.

($e = 1.6 \times 10^{-19}$ C, $m = 9.1 \times 10^{-31}$ kg)

Radioactivity; atomic structure

6 In some medical conditions, a radioactive isotope of iodine is injected into the bloodstream.
a Explain what is meant by the term *isotope*.
b The isotope emits β-particles. Explain what is meant by a *β-particle*.
c The half-life of the isotope is 8 days.
(i) A sample of 48 micrograms of the isotope is injected into the patient. Calculate the mass of the isotope which remains after 32 days. Show your working.
(ii) Explain why an isotope with a relatively short half-life is used.

(*UCLES IGCSE Physical Science Core, May/June 2000*)

7 Atomic nuclei contain both protons and neutrons. ${}^{11}_{6}C$, ${}^{12}_{6}C$ and ${}^{14}_{6}C$ are all forms of carbon. The nucleus of ${}^{11}_{6}C$ consists of 6 protons and 5 neutrons.

a (i) Write down the number of protons and neutrons in ${}^{12}_{6}C$:
${}^{12}_{6}C$ contains ...... protons and ...... neutrons.
(ii) Explain why ${}^{14}_{6}C$ is described as 'neutron-rich'.
(iii) What do the nuclei of ${}^{11}_{6}C$, ${}^{12}_{6}C$ and ${}^{14}_{6}C$ all have in common?

b The following graph can be used to show the numbers of neutrons and protons in the lighter nuclei. The crosses represent the nuclei of ${}^{12}_{6}C$ and ${}^{14}_{6}C$.

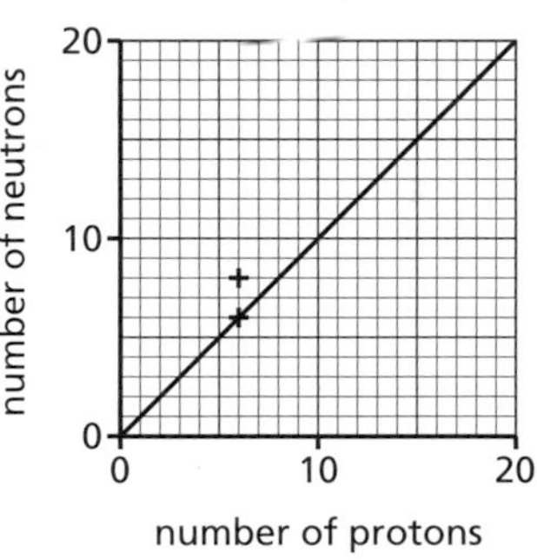

For a nucleus to be stable, a plot of its number of neutrons against its number of protons must lie on or close to the diagonal line.
(i) Copy the graph and mark X to show the position of ${}^{11}_{6}C$.
(ii) What is the condition for a nucleus with less than 10 protons to be stable?
(iii) Which form of carbon is the most stable? Explain your answer.

c The diagram below represents the decay of a nucleus of ${}^{14}_{6}C$.

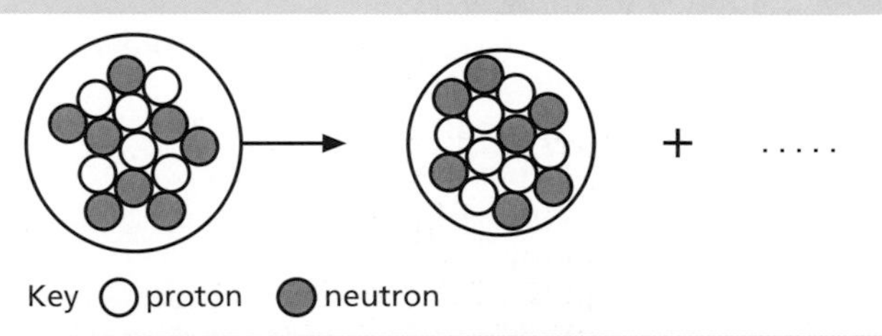

(i) Use the diagram to write down what happens to the numbers of protons and neutrons when ${}^{14}_{6}C$ decays.
(ii) What particle is missing from the right hand side of the diagram?

(*London Foundation, June 1998*)

continued

8 a State what is meant by
(i) the *half-life* of a radioactive substance,
(ii) *background radiation.*

b In a certain laboratory, the background radiation level is 25 counts/minute. A graph of the count-rate measured by a detector placed a short distance from a radioactive source in the laboratory is shown below.

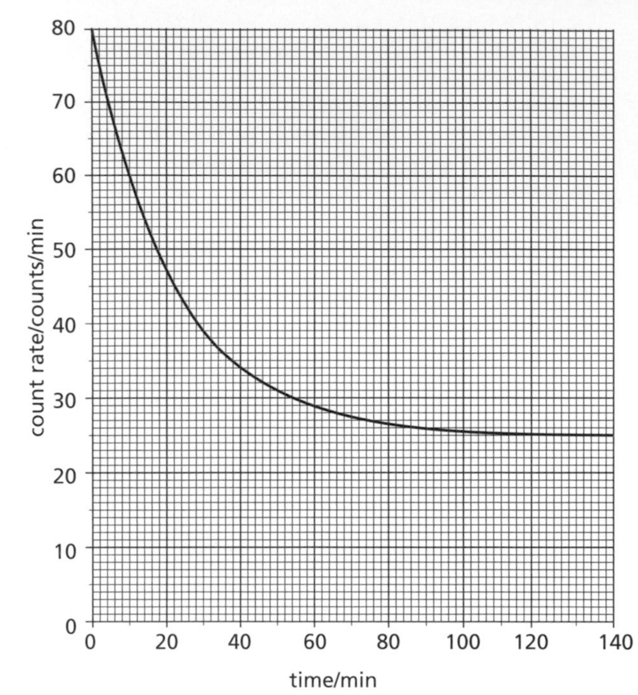

(i) At zero time, the measured count-rate of the source and background together is 80 counts/minute. Calculate the count-rate due to the source alone.
(ii) After one half-life has elapsed, what is the count-rate
1 due to the source alone,
2 measured by the detector?
(iii) Use the graph to find the half-life of the source.
(iv) Why does the graph **not** drop below the 25 counts/minute line?
(v) On the graph, sketch the curve that might be obtained for a source with a shorter half-line.

(*UCLES IGCSE Physics Core, Nov 2005*)

9 a α-particles, β-particles and γ-rays are known as ionising radiations.
(i) Describe what happens when gases are ionised by ionising radiations.
(ii) Suggest why α-particles are considered better ionisers of gas than β-particles.

b (i) Suggest two practical applications of radioactive isotopes.
(ii) For one of the applications that you have suggested, describe how it works, or draw a labelled diagram to illustrate it in use.

(*UCLES IGCSE Physics Extended, Nov 2005*)

10 The diagram shows the variation in background radiation in England, Scotland and Wales.

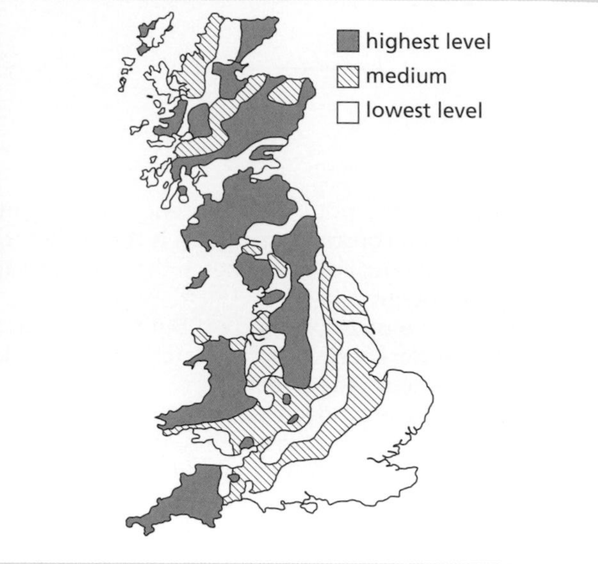

a The background radiation is calculated by finding the average value of a large number of readings. Suggest why this method is used.

b The high levels of radiation in some parts of Britain are caused by radon gas escaping from underground rocks such as granite.
Radium-224 ($^{224}_{88}$Ra) decays to form radon-220 ($^{220}_{86}$Rn).
(i) What particle is emitted when radium-224 decays?
(ii) Radon-220 then decays to polonium by emitting an alpha particle. Copy and complete the decay equation for radon-220.

$$^{220}_{86}\text{Rn} \rightarrow \ \text{Po} \ + \ \text{He}$$

(iii) The half-lives of these isotopes are given in the table.

Isotope	Half-life
radium-224	3.6 days
radon-220	52 seconds

A sample of radium-224 decays at the rate of 360 nuclei per second. The number of radon-220 nuclei is growing at less than 360 per second. Suggest a reason for this.
(iv) Radon-220 has a short half-life and it emits the least penetrative of the three main types of radioactive emission. Explain why the presence of radon gas in buildings is a health hazard.

(*London Higher, June 1999*)

Revision questions

Light and sight

1 In the diagram a ray of light is shown reflected at a plane mirror. What is
  a the angle of incidence,
  b the angle the reflected ray makes *with the mirror*?

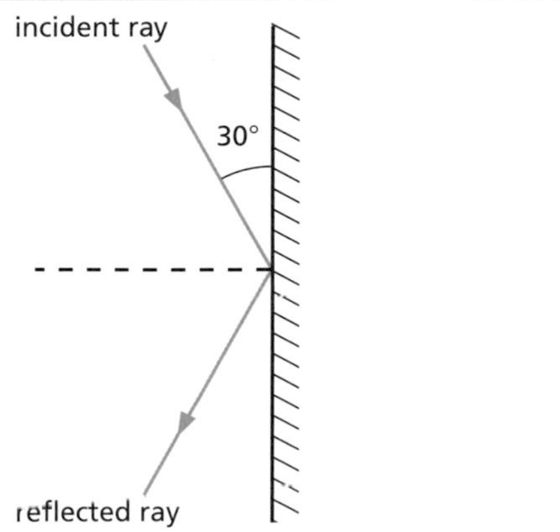

2 In the diagram below a ray of light IO changes direction as it enters glass from air.
  a What name is given to this effect?
  b Which line is the normal?
  c Is the ray bent towards or away from the normal in the glass?

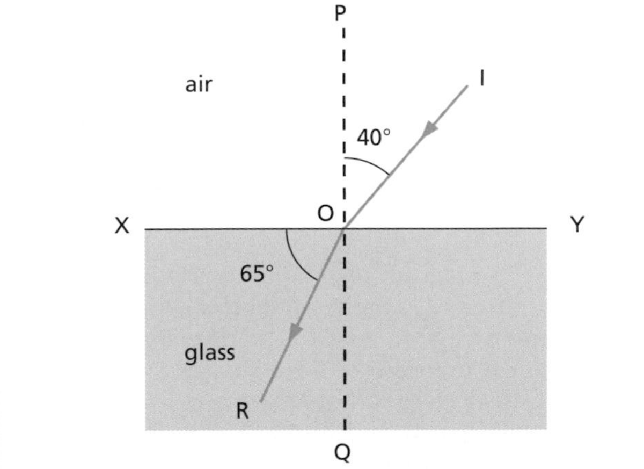

  d What is the value of the angle of incidence in air?
  e What is the value of the angle of refraction in glass?

3 In the diagram below which of the rays **A** to **E** is most likely to represent the ray emerging from the parallel-sided sheet of glass?

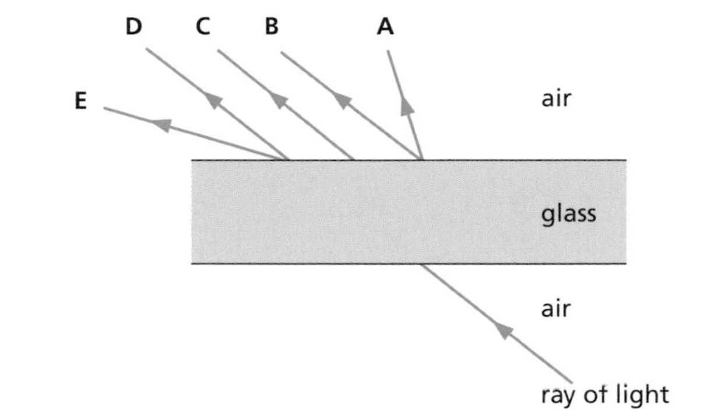

4 At night, the light beam from a torch is shone into a swimming pool along the line TSA. Instead of striking the bottom of the pool at A, the beam travels to B, as shown below.

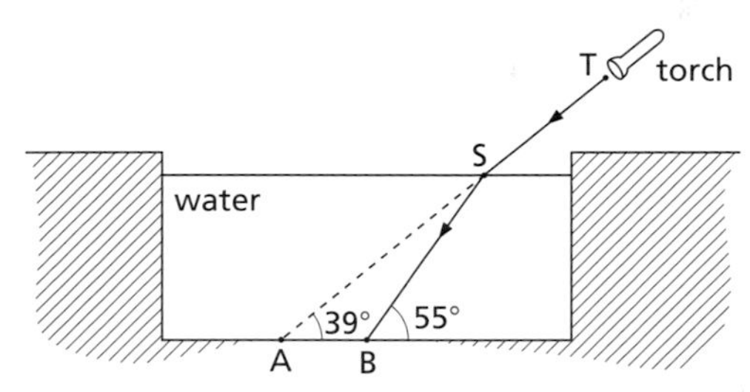

  a At S, the direction of the beam changes. State the name we use to describe this change.
  b (i) Copy the diagram above and draw the normal to the surface at S.
  (ii) Clearly mark and label the angle of incidence.
  c The diagram below shows the same pool and the same points A, B, S and T. The critical angle for the water is 50°.

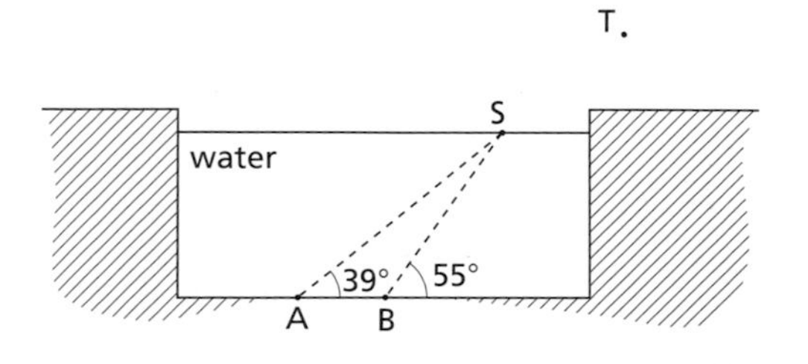

  (i) A beam of light is directed up from B to S. On a copy of the diagram above, carefully draw the path of the ray from B to S and then out into the air.
  (ii)1. A beam of light is directed up from A to S. Describe what happens to the beam at S.
  2. Explain why this happens.

(*UCLES IGCSE Physics Core, May/June 2000*)

5 A woman stands so that she is 1.0 m from a mirror mounted on a wall, as shown below.

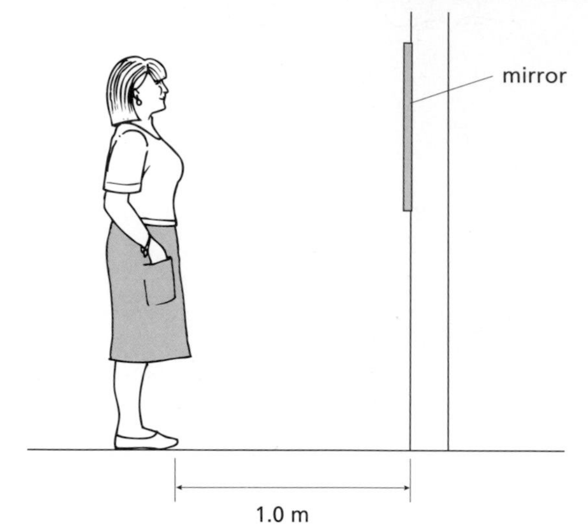

a Copy the diagram and carefully draw
(i) a clear dot to show the position of the image of her eye,
(ii) the normal to the mirror at the bottom edge of the mirror,
(iii) a ray from her toes to the bottom edge of the mirror and then reflected from the mirror.

b Explain why the woman cannot see the reflection of her toes.

c (i) How far is the woman from her image?
(ii) How far must the woman walk, and in what direction, before the distance between her and her image is 6.0 m?

(*UCLES IGCSE Physics Core, Nov 2006*)

6 The diagram shows a ray of light, from the top of an object PQ, passing through two glass prisms.

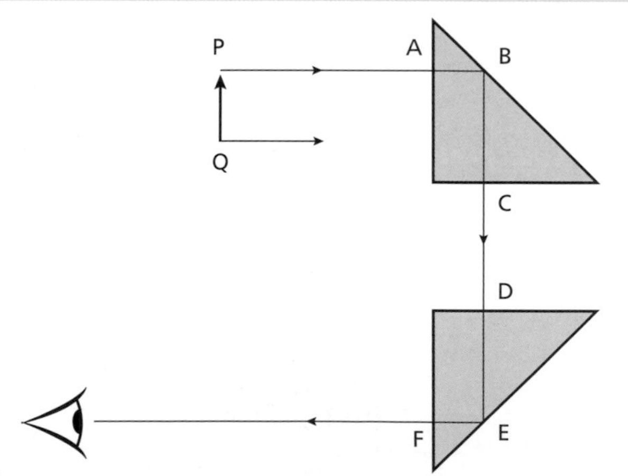

a Copy the sketch and complete the path through the two prisms of the ray shown leaving Q.

b A person looking into the lower prism, at the position indicated by the eye symbol, sees an image of PQ. State the properties of this image.

c Explain why there is no change in direction of the ray from P at points A, C, D and F.

d The speed of light as it travels from P to A is 3×10^8 m/s and the refractive index of the prism glass is 1.5. Calculate the speed of light in the prism.

e Explain why the ray AB reflects through 90° at B and does not pass out of the prism at B.

(*UCLES IGCSE Physics Extended, Nov 2006*)

7 A narrow beam of white light is shown passing through a glass prism and forming a spectrum on a screen.

a What is the effect called?

b Which colour of light appears at (i) A, (ii) B?

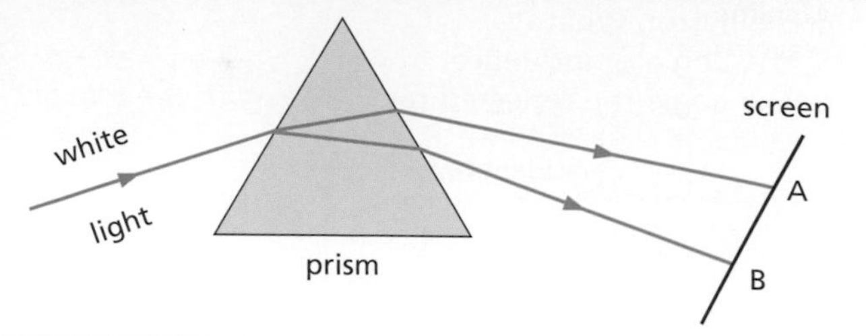

8 When using a magnifying glass to see a small object
1 an upright image is seen
2 the object should be less than one focal length away
3 a real image is seen.

Which statement(s) is (are) correct?

A 1, 2, 3 **B** 1, 2 **C** 2, 3 **D** 1 **E** 3

9 a The sketch shows two rays of light from a point O on an object. These rays are incident on a plane mirror.

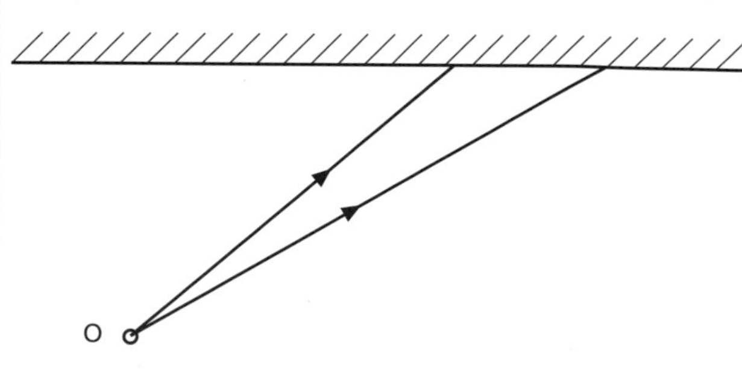

(i) Copy the diagram and continue the paths of the two rays after they reach the mirror. Hence locate the image of the object O. Label the image I.
(ii) Describe the nature of the image I.

b The diagram below is drawn to scale. It shows an object PQ and a convex lens.

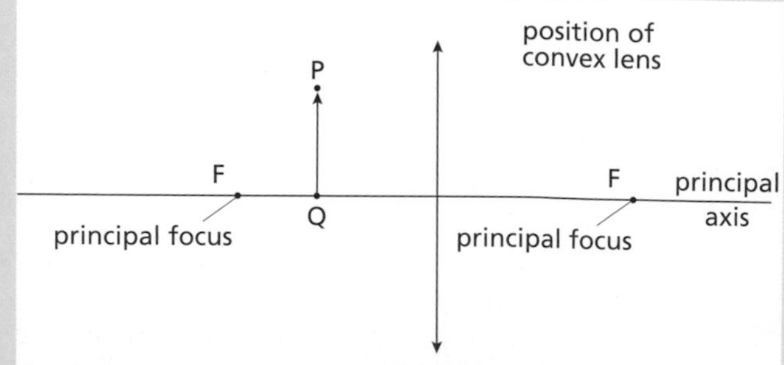

(i) Copy the diagram and draw two rays from the top of the object P that pass through the lens. Use

these rays to locate the top of the image. Label this point T.
(ii) Draw an eye sumbol to show the position from which the image T should be viewed.

(*UCLES IGCSE Physics Extended, Nov 2005*)

Waves and sound

10 In the transverse wave shown below distances are in centimetres. Which pair of entries **A** to **E** is correct?

	A	B	C	D	E
Amplitude	2	4	4	8	8
Wavelength	4	4	8	8	12

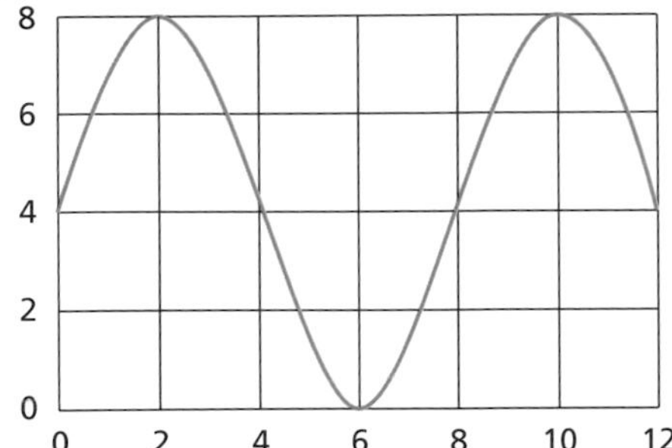

11 When a water wave goes from deep to shallow water, the changes (if any) in its speed, wavelength and frequency are

	Speed	**Wavelength**	**Frequency**
A	greater	greater	the same
B	greater	less	less
C	the same	less	greater
D	less	the same	less
E	less	less	the same

12 When the straight water waves in the diagram pass through the narrow gap in the barrier they are diffracted. What changes (if any) occur in
- **a** the shape of the waves,
- **b** the speed of the waves,
- **c** the wavelength?

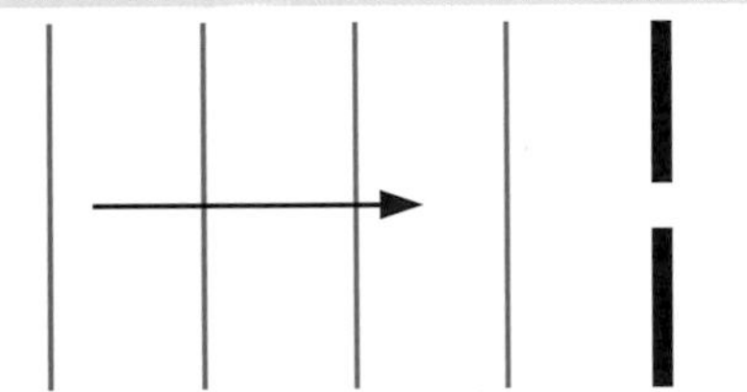

13 Here is a list of different types of waves.

gamma (γ)
infrared
radio
sound
ultraviolet
visible
X-rays

- **a** Which one of these is the only one which is *not* part of the electromagnetic spectrum?
- **b** Which one of these makes us feel warm when the Sun shines?
- **c** Which one of these do doctors use to detect broken bones?
- **d** (i) On the Moon, two astronauts cannot hear each other, even when they shout, unless they have their radios switched on.

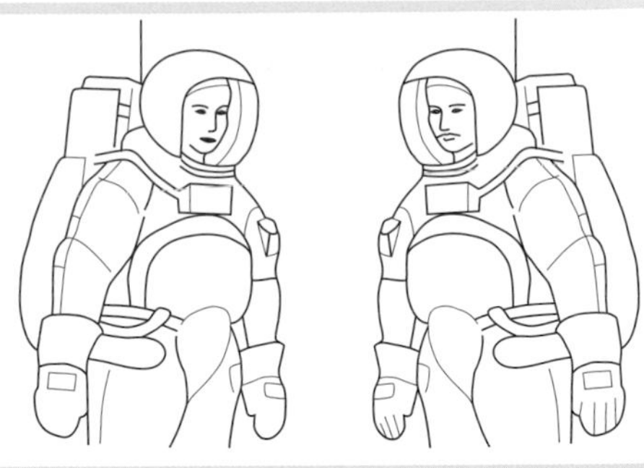

1. Why cannot they hear each other even when they shout?
2. Why can they hear each other using their radios?

(ii) Which type of wave is used to carry messages from the astronauts to mission control on Earth?

(*UCLES IGCSE Physics Core, May/June 2000*)

14 The diagram below shows the complete electromagnetic spectrum.

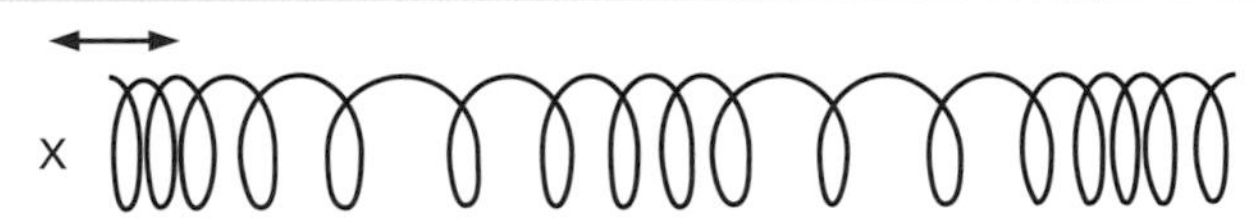

- **a** Name the radiation found at
  (i) **A**,
  (ii) **B**.
- **b** State which of the radiations marked on the diagram would have
  (i) the lowest frequency,
  (ii) the shortest wavelength.

15 The wave travelling along the spring in the diagram is produced by someone moving end X of the spring to and fro in the directions shown by the arrows.
- **a** Is the wave longitudinal or transverse?
- **b** What is the region called where the coils of the spring are (i) closer together, (ii) farther apart, than normal?

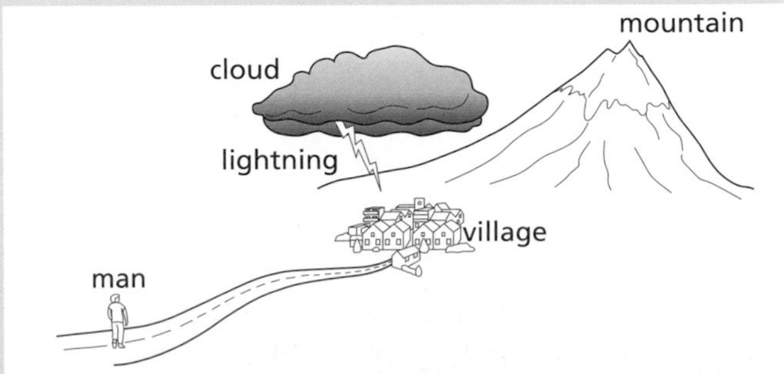

16 A man is watching a thunderstorm which is directly over a village. Some distance behind the village is a mountain.

- **a** Thunder is created at the same time as the lightning flash but, after the man sees a lightning flash, he has to wait a short time before he hears the thunder. Why is there this delay?

b When he listens carefully, the man realises that, for each lightning flash, he can hear a loud sound of thunder followed by a quieter one.
(i) After studying the diagram above, explain why he hears two sounds for each lightning flash.
(ii) Suggest why the second sound is quieter.

c The man measures the time between seeing a flash of lightning over the village, and hearing the first sound of thunder. The time is 4 s. The speed of sound in air is 330 m/s. How far away is the village?

(*UCLES IGCSE Physics Core, May/June 2000*)

17 If a note played on a piano has the same pitch as one played on a guitar, they have the same

A frequency **B** amplitude **C** quality
D loudness **E** harmonics.

18 The waveforms of two notes P and Q are shown below. Which one of the statements **A** to **E** is true?

A P has a higher pitch than Q and is not so loud.
B P has a higher pitch than Q and is louder.
C P and Q have the same pitch and loudness.
D P has a lower pitch than Q and is not so loud.
E P has a lower pitch than Q and is louder.

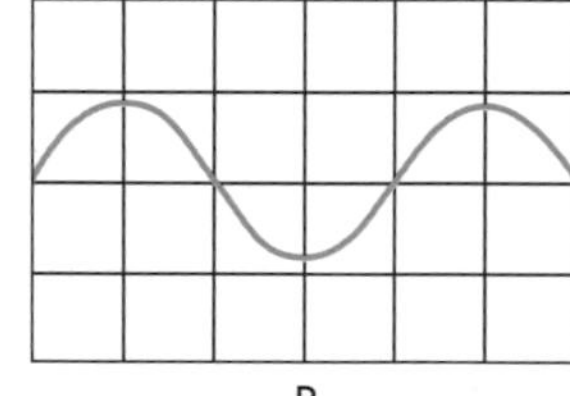
P

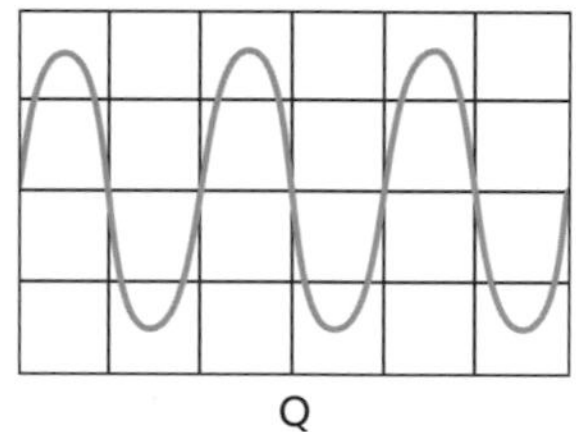
Q

19 Examples of transverse waves are

1 water waves in a ripple tank
2 all electromagnetic waves
3 sound waves.

Which statement(s) is (are) correct?

A 1, 2, 3 **B** 1, 2 **C** 2, 3 **D** 1 **E** 3

20 **a** The diagram shows a view from above of a person standing at the edge of a pond, dipping the end of a stick up and down in the water. Some of the wavefronts that spread out are shown.

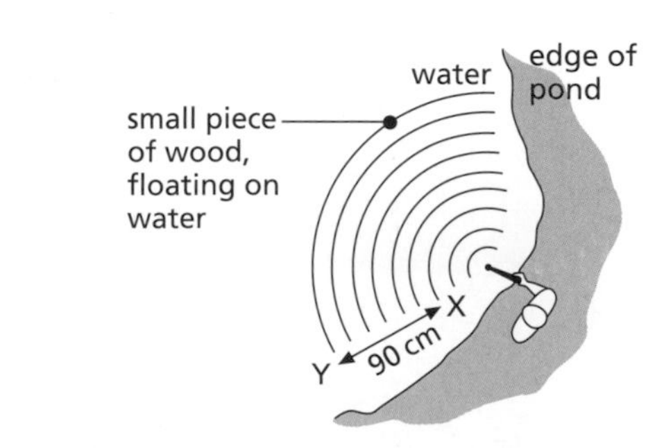

(i) How many wavelengths are there between X and Y?
(ii) The distance from X to Y is 90 cm. Calculate the wavelength of the waves.
(iii) The speed of the waves is affected by the depth of the water.

1. Describe the shape of the wavefronts, as seen from above.
2. What does the shape of the wavefronts tell you about the depth of the pond? Give a reason for your answer.

(iv) The diagram below shows a sideways view of the water surface just before the first wave reaches the floating piece of wood.

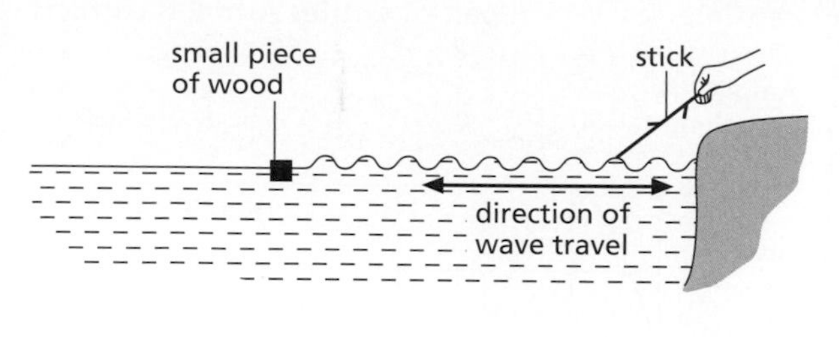

Describe how the piece of wood moves after the waves reach it. You may copy the diagram and add to it, if it helps you to answer the question.

b An underwater loudspeaker, placed in the pond in part **a**, sends out sound waves through the water, as shown below.

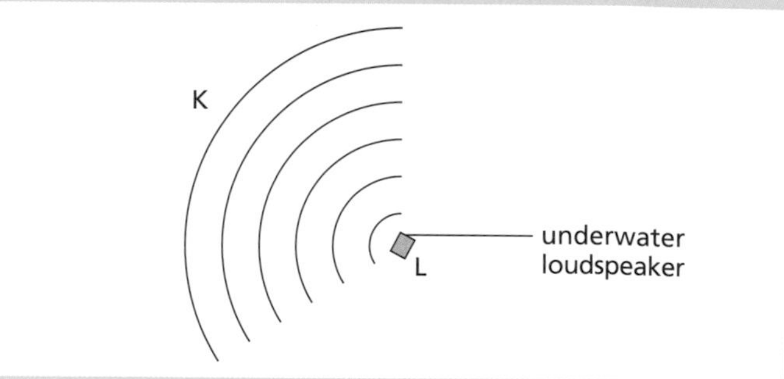

(i) What is the difference between the nature of these sound waves and the water waves in **a**? Write the appropriate words to fill the gaps in the following sentences.

'Water waves are ..... waves.'
'Sound waves are ..... waves.'

(ii) The diagram below shows a sideways view along the line KL.

K M • underwater loudspeaker L

The dot labelled M represents a water molecule on the line KL. Describe how the molecule moves when the loudspeaker is working. You may copy the diagram and add to it, if it helps you to answer the question.

(*UCLES IGCSE Physics Core, Nov 2000*)

21 The diagram shows a series of wavefronts of light travelling in air towards a glass block.

a Copy the diagram and complete the wavefronts, **A**, **B** and **C**, to show them entering the glass block.

b Before they enter the block, the wavelength of the waves is 6×10^{-7} m and their speed is 3×10^{8} m/s.

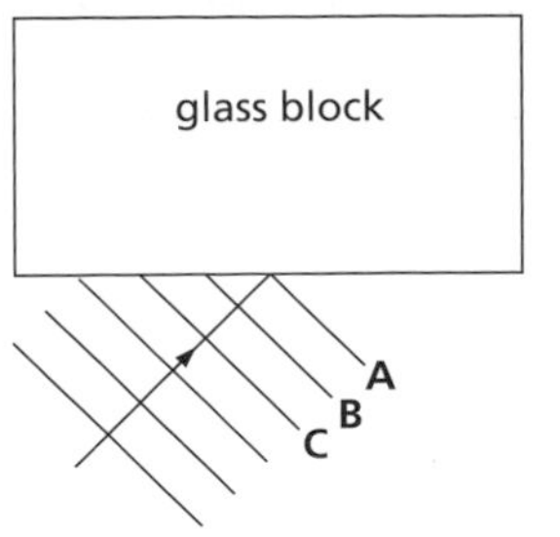

Calculate the frequency of the waves. State the equation that you use and show all your working.

C Without doing any more calculations, state what happens (i) to the *frequency* and (ii) to the *speed* of the waves as they enter the block.

(*UCLES IGCSE Physical Science Extended, May/June 2000*)

Matter and measurements

22 The basic SI units of mass, length and time are

	Mass	Length	Time
A	kilogram	kilometre	second
B	gram	centimetre	minute
C	kilogram	centimetre	second
D	gram	centimetre	second
E	kilogram	metre	second

23 Density can be calculated from the expression

A mass/volume
B mass × volume
C volume/mass
D weight/area
E area × weight

24 Read the sentences below and then answer the questions which follow.

'When potatoes are bought in a market, the weight of a bag full of potatoes is affected by the density of the potatoes. A lady fills her bag when she buys 5 kg of large potatoes. A man buys 5 kg of small potatoes. He puts them in a bag of the same size as the lady's, but his bag is not filled.'

a Which word in these sentences describes a quantity which is a force?

b What does the 5 kg measure?

the density of the potatoes
the mass of the potatoes
the volume of the potatoes
the weight of the potatoes

c Suggest one reason why the man's 5 kg of potatoes occupies less volume than the lady's potatoes.

(*UCLES IGCSE Physics Core, May/June 2000*)

25 Which of the following properties are the same for an object on Earth and on the Moon?

1 weight **2** mass **3** density

Use the answer code:

A 1, 2, 3 **B** 1, 2 **C** 2, 3 **D** 1 **E** 3

26 Here are the approximate densities of some metals.

platinum	21 000 kg/m^3	(21 g/cm^3)
gold	19 000 kg/m^3	(19 g/cm^3)
lead	11 000 kg/m^3	(11 g/cm^3)
brass	9 000 kg/m^3	(9 g/cm^3)
iron	8 000 kg/m^3	(8 g/cm^3)
aluminium	3 000 kg/m^3	(3 g/cm^3)

A person sees a coin offered for sale in an antiques market.

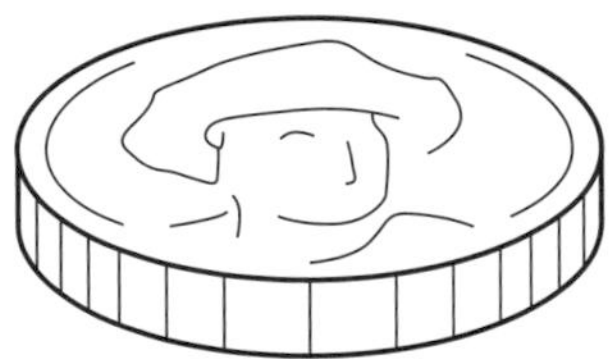

The market trader says that the coin is made of gold. After buying the coin, the person finds that its volume is 1.4 cm^3 and its mass is 12.6 g.

a Write down the equation which enables you to calculate density.

b Calculate the density of the metal from which the coin is made.

c Is the coin made of gold?

d If not, use the list above to suggest what it might be made from.

e If a country wanted to keep its coinage the same but of as low a mass as possible, which of the metals in the list should it choose?

(*UCLES IGCSE Physics Core, Nov 2000*)

27 In an experiment, forces are applied to a spring as shown in Figure a. The results of this experiment are shown in Figure b.

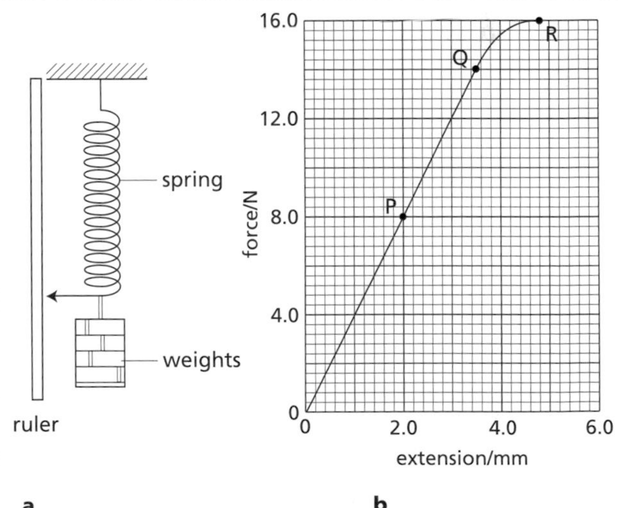

a What is the name given to the point marked Q on the graph?

b For the part OP of the graph, the spring obeys Hooke's Law. State what this means.

c The spring is stretched until the force and extension are shown by the point R on the graph. Compare how the spring stretches, as shown by the part of the graph OQ, with that shown by QR.

a The part OP of the graph shows the spring stretching according to the expression

$$F = kx$$

Use values from the graph to calculate the value of k.

(*UCLES IGCSE Physics Extended, November 2006*)

28 a The smallest division marked on a metre rule is 1 mm. A student measures a length with the ruler and records it as 0.835 m. Is he justified in giving three significant figures?

b The SI unit of density is

A kg m **B** kg/m^2 **C** kg m^3 **D** kg/m **E** kg/m^3

Forces and pressure

29 a An experiment is performed to find the centre of mass of a flat card.
The diagram below shows the card freely suspended at point **A**. A string is also fixed at **A**. The string has a weight attached to it.

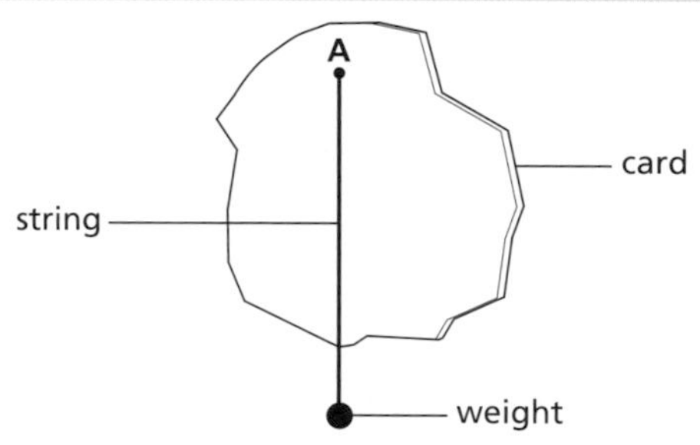

(i) Copy the diagram and mark with **X** a possible position of the centre of mass of the card.
(ii) Describe what you would do to find the exact position of the centre of mass of the card.

b The diagrams below show two motor vehicles. The centre of mass of **A** is much lower than that of **B**.

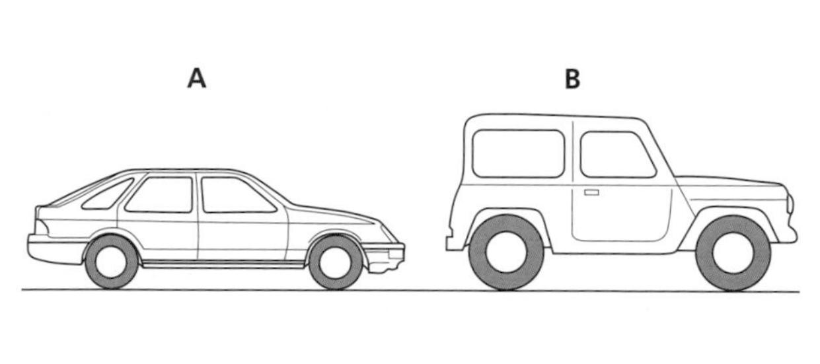

Explain the advantage of having a low centre of mass.

c Car **A** has a mass of 800 kg. Calculate its weight. (g = 10 N/kg)

(*UCLES IGCSE Physical Science Core, May 1999*)

30 The work done by a force is

1 calculated by multiplying the force by the distance moved in the direction of the force
2 measured in joules
3 the amount of the energy changed.

Which statement(s) is (are) correct?

A 1, 2, 3 **B** 1, 2 **C** 2, 3 **D** 1 **E** 3

31 The main energy change occurring in the device named is

	Device	Main energy change
1	electric lamp	electrical to heat and light
2	battery	chemical to electrical
3	pile driver	k.e. to p.e.

Which statement(s) is (are) correct?

A 1, 2, 3 **B** 1, 2 **C** 2, 3 **D** 1 **E** 3

32 Two workers, **A** and **B**, are lifting boxes of food in a store-room. The boxes all weigh the same and are lifted from the floor on to the same shelf.

A is able to lift 10 boxes in 2 minutes.
B takes longer than 2 minutes to lift 10 boxes.

a How does the total work done by **A** compare with the total work done by **B**?
b How does the power of **A** compare with the power of **B**?
c (i) Which form of energy in their bodies do the workers transform in order to do the work lifting the boxes?
(ii) From what did they obtain this supply of energy?
d The boxes have more energy when they are on the shelf than when they were on the floor. Which form of energy has increased?
e One of the boxes falls off the shelf and crashes to the ground. Describe the energy changes as the box falls and hits the ground.

(*UCLES IGCSE Physics Core, May 2001*)

33 a State the two factors on which the turning effect of a force depends.

b Forces F_1 and F_2 are applied vertically downwards at the ends of a beam resting on a pivot P. The beam has weight W.

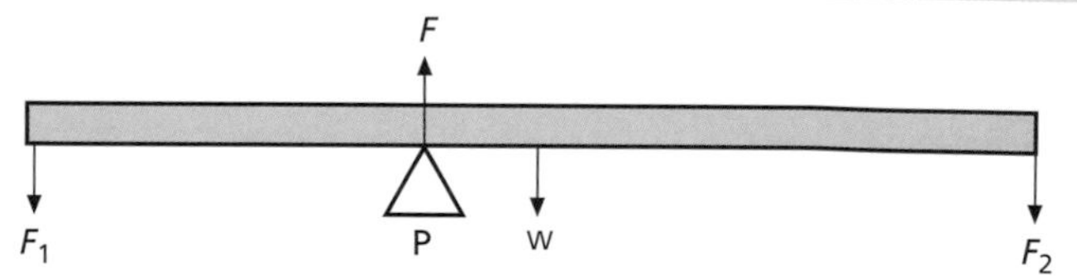

(i) Complete the statements about the two requirements for the beam to be in equilibrium.
1 There must be no resultant . . .
2 There must be no resultant . . .
(ii) The beam is in equilibrium. F is the force exerted on the beam by the pivot P. Complete the following equation about the forces on the beam.

$$F =$$

(iii) Which one of the four forces on the beam does **not** exert a moment about P?

(*UCLES IGCSE Physics Core, Nov 2006*)

34 a Two horizontal strings are attached to a soft rubber ball, as shown below.

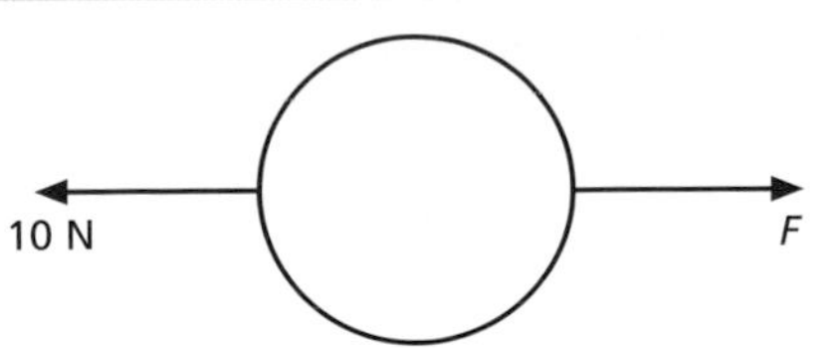

A force of 10 N pulls on one string.
(i) The ball does not move. What is the value of the force F on the other string?
(ii) What change to the rubber ball do the two forces cause?

b A garden pot containing soil weighs a total of 360 N. The pot rests on three equally spaced blocks, so that surplus water can drain out of the holes in the base of the pot. The soil is uniformly distributed in the pot, as shown below.

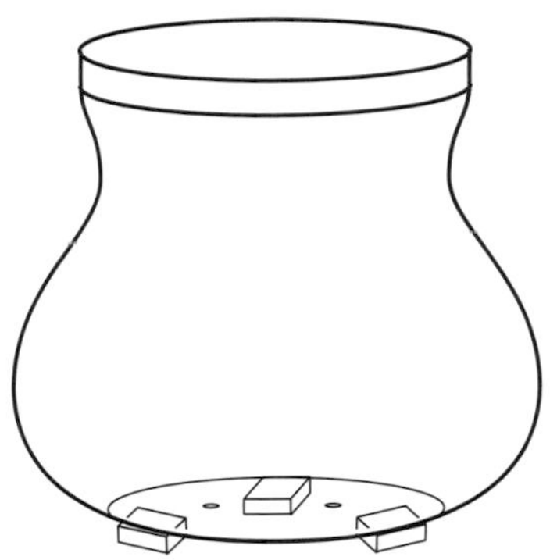

(i) What is the force exerted by each block on the pot?
(ii) State the direction of these forces.
(iii) The gardener finds that the blocks sink into the ground, but he must have the pot up on blocks to allow the drainage. What can he do to reduce the sinking of the pot?

(UCLES IGCSE Physics Core, May 2003)

35 The diagram shows the arm of a crane when it is lifting a heavy box.

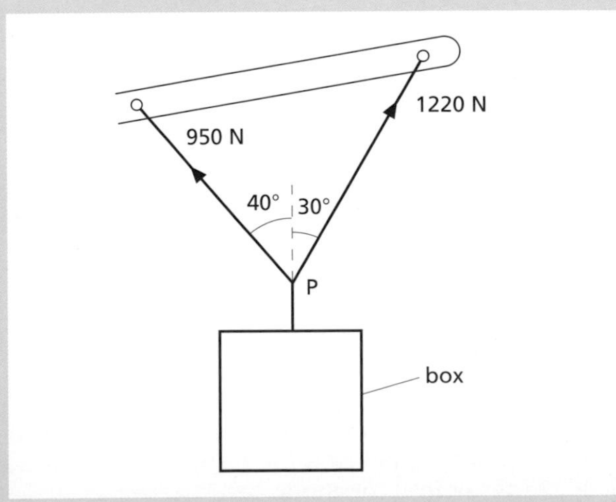

a By the use of a scale diagram (**not** calculation) of the forces acting at P, find the weight of the box.

b Another box of weight 1500 N is raised vertically by 3.0 m.
(i) Calculate the work done on the box.
(ii) The crane takes 2.5 s to raise this box 3.0 m. Calculate the power output of the crane.

(UCLES IGCSE Physics Extended, May 2003)

36 The diagram shows a motorbike, travelling at constant speed, along a level track.

a Copy and complete the following sentence, by filling in the blank spaces.
The combustion of fuel in the motorbike engine converts the ..... energy of the fuel into ..... energy.

b The motorbike engine develops a power of 15 kW. Explain what is meant by the term *power*.

c The motorbike reaches the hill. State and explain what must happen to the power output of the engine, if the motorbike is to continue at the same speed.

(UCLES IGCSE Physical Science Core, May/June 2000)

37 The efficiency of a machine which raises a load of 200 N through 2 m when an effort of 100 N moves 8 m is

A 0.5% **B**5% **C** 50% **D** 60% **E** 80%

38 Which one of the following statements is *not* true?

A Pressure is the force acting on unit area.
B Pressure is calculated from force/area.
C The SI unit of pressure is the pascal (Pa) which equals 1 newton per square metre (1 N/m^2).
D The greater the area over which a force acts the greater is the pressure.
E Force = pressure × area.

39 If the piston in the diagram below is pulled out of the cylinder from position X to position Y, without changing the temperature of the air enclosed, the air pressure in the cylinder is

A reduced to a quarter **B** reduced to a third
C the same **D** trebled
E quadrupled.

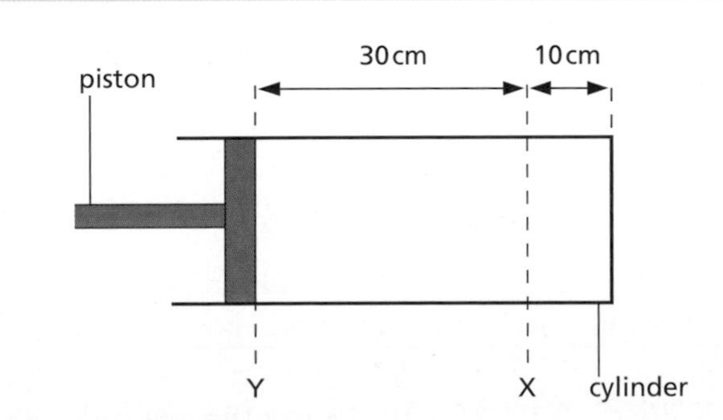

Motion and energy

40 An insect lands on a 30 cm ruler and walks along the edge, as shown in the diagram.

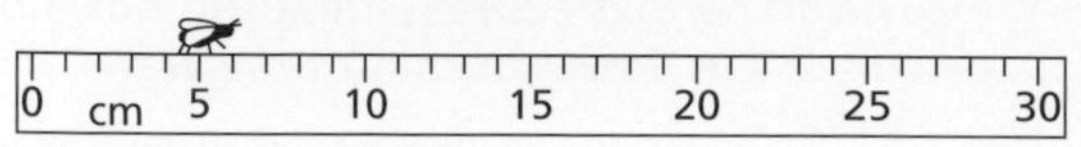

A child measures the time the insect takes to walk from the 5 cm mark to the 25 cm mark. It takes 50 s to do this.
What is the average speed, in cm/s, of the insect?

(*UCLES IGCSE Physics Core, May 1999*)

41 a The diagram shows the speed/time graph for a motorcycle.

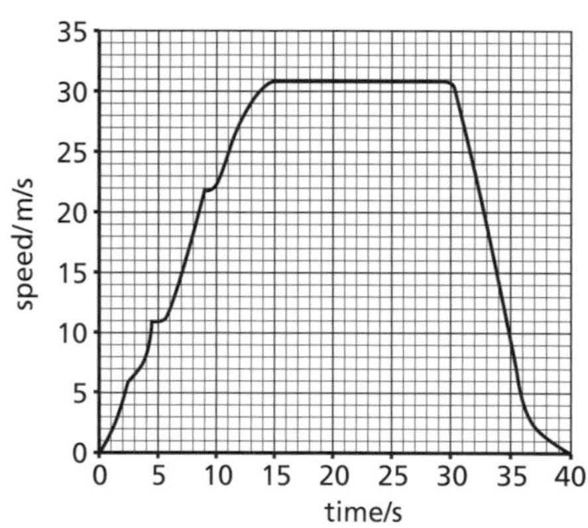

(i) What is the maximum speed of the motorcycle?
(ii) Whilst accelerating, the motorcycle changes gear three times. State *one* of the speeds at which the gear is changed.
(iii) For how long is the motorcycle slowing down?

b On another occasion, the motorcycle is made to increase its speed at a constant rate for 10 s. The speed/time graph for this is shown below.

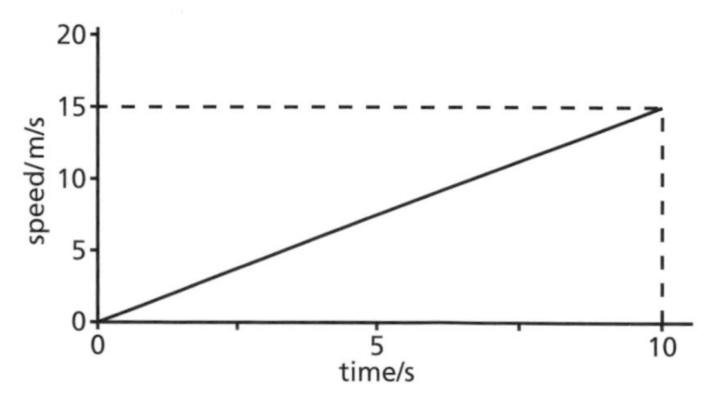

How far does the motorcycle travel in these 10 s?

(*UCLES IGCSE Physics Core, Nov 2000*)

42 The table below gives some data about an accelerating car.

time/s	0	1	2	3	4	6	8	10
speed/m/s	0	5	10	15	19	24	25	25

a Plot the speed/time graph for the motion.
b How far did the car travel during the first 3 s?
c What was the top speed of the car?
d How far would the car travel in 3 s if travelling at its top speed?

(*UCLES IGCSE Physics Core, May 2001*)

43 Which one of the velocity–time graphs **A** to **E** represents an object falling freely in a vacuum?

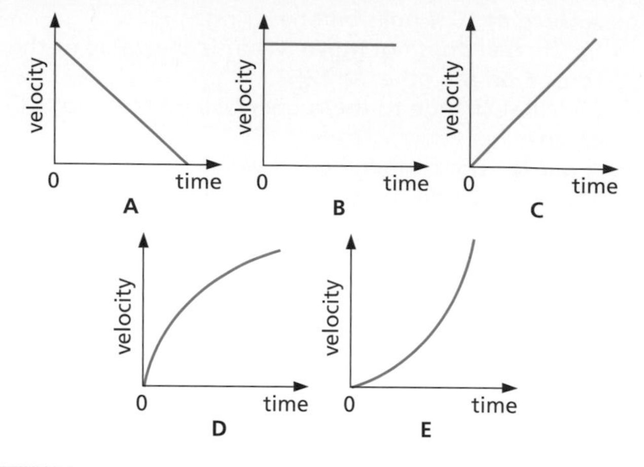

44 The driver of a car does not press his foot down on the brake immediately he sees a hazard ahead. This delay in braking is called the 'thinking time'. For a car travelling at 20 m/s, the car travels 14 m during this 'thinking time'. What was the 'thinking time' of the driver?

(*UCLES IGCSE Physics Core, Nov 1997*)

45 A 3 kg mass falls with its terminal velocity. Which combination **A** to **E** gives its weight, the air resistance and the resultant force acting on it?

	Weight	**Air resistance**	**Resultant force**
A	0.3 N down	zero	zero
B	3 N down	3 N up	3 N up
C	10 N down	10 N up	10 N down
D	30 N down	30 N up	zero
E	300 N down	zero	300 N down

46 A stone of mass 2 kg is dropped from a height of 4 m. Neglecting air resistance, the k.e. of the stone in joules just before it hits the ground is

A 6 **B** 8 **C** 16 **D** 80 **E** 160

47 An object of mass 2 kg is fired vertically upwards with a k.e. of 100 J. Neglecting air resistance, which of the numbers in **A** to **E** below is

a the velocity in m/s with which it is fired,
b the height in m to which it will rise?

A 5 **B** 10 **C** 20 **D** 100 **E** 200

48 An object has k.e. of 10 J at a certain instant. If it is acted on by an opposing force of 5 N, which of the numbers **A** to **E** below is the further distance it travels in metres before coming to rest?

A 2 **B** 5 **C** 10 **D** 20 **E** 50

49 A boy whirls a ball at the end of a string round his head in a horizontal circle, centre O. If he lets go of the string when the ball is at X in the diagram, the ball flies off in the direction

A 1 **B** 2 **C** 3 **D** 4 **E** 5

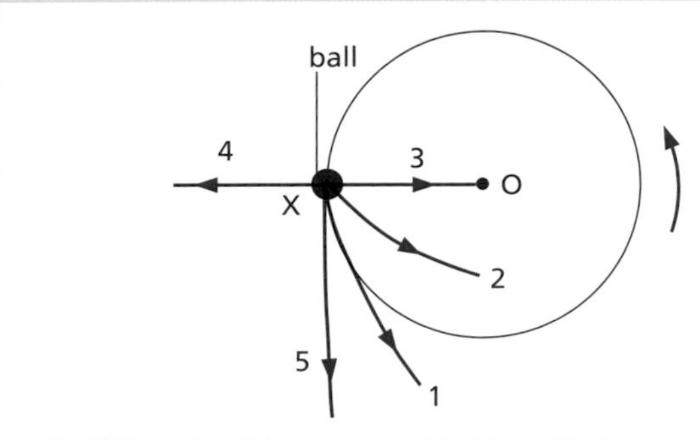

50 The diagram shows a metre rule with a pivot at its centre. A 250 g mass is placed on the 25 cm mark.

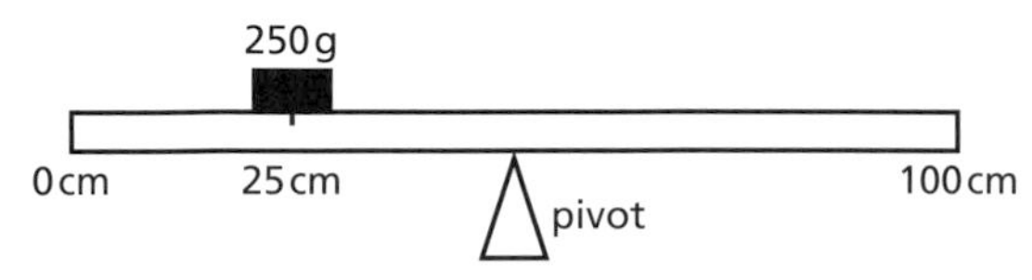

a (i) Calculate the weight of the 250 g mass. Write down the equation that you use and show all your working. (g = 10 N/kg).
(ii) Describe what happens to the metre rule when this mass is positioned as shown in the diagram.
(iii) Calculate, in N cm, the moment of the weight in **a**(i) about the pivot. Write down the equation that you use and show all your working.

b (i) Explain what is meant when the metre rule is said to be in *equilibrium*.
(ii) Describe how you would carry out an experiment to show that there is no net moment on the metre rule when it is in equilibrium.

(*UCLES IGCSE Physical Science Extended, Nov 2000*)

51 The diagram shows a person raising a concrete block from a river bed by using two pulleys.

a As shown in the diagram, the top of the block is

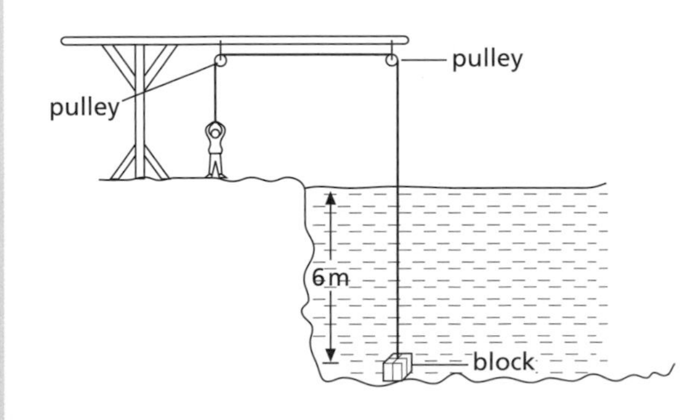

6.0 m below the water surface. The density of water is 1000 kg/m^3 and the acceleration of free fall is 10 m/s^2. Calculate the water pressure acting on the top of the block.

b The block is raised through the water. At one point, the water pressure acting on the top of the block is 4.5×10^4 Pa. The area of the top of the block is 0.015 m^2. Calculate the downward force exerted by the water on the top of the block.

c When the block is clear of the water, it is raised a further 4.0 m. The weight of the block is 550 N. Calculate the work done on the block as it is raised the 4.0 m through the air.

d Some of the energy the person uses to raise the block is converted into heat energy. Sketch the diagram at the start of the question and indicate, using an arrow and the letter H, *two* places where heat is released. For each place, explain why heat is released there.

(*UCLES IGCSE Physics Extended, Nov 2000*)

52 The diagram shows a stone, tied to the end of a piece of string, being rotated in a vertical circle.

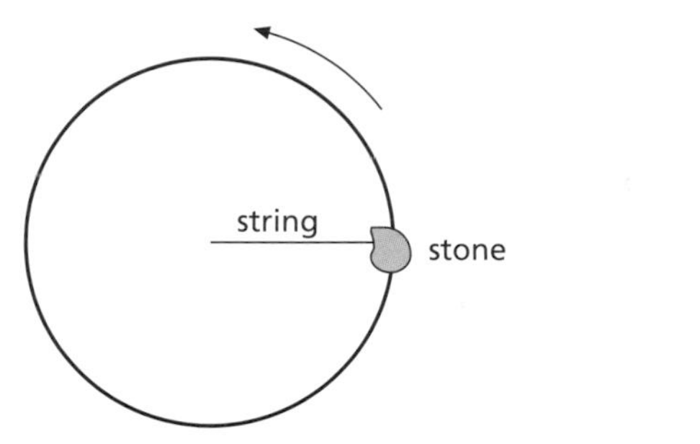

As the stone moves in the circle, its speed remains constant but its velocity is continually changing.

a What is the difference between the terms *speed* and *velocity*?

b Copy the diagram and draw an arrow on the string to show the direction of the force exerted by the string on the stone.

c Describe how you would measure accurately the time that it takes for the stone to make one complete revolution.

d The stone's mass is 105 g. Calculate its weight. State the equation that you use and show all your working (g = 10 N/kg).

e Draw a labelled diagram and describe how you would measure the density of the stone.

(*UCLES IGCSE Physical Science Extended, May/June 2000*)

53

A player kicks a stationary ball of mass 0.4 kg. The ball moves off with a speed of 3 m/s.

a Calculate the kinetic energy of the ball immediately after it has been kicked. Write down the equation that you use and show all your working.

b (i) Write down the maximum potential energy of the ball.
(ii) Calculate the maximum height that the ball could reach. Write down the equation that you use and show all your working. (g = 10 N/kg)
(iii) Explain why it is unlikely that this height will be reached by the ball.

c When the ball has been kicked several times, the temperature of the air in the ball rises. Explain what has happened to the air molecules in the ball.

(*UCLES IGCSE Physical Science Extended, Nov 1999*)

Heat and energy

54 Some smoke is mixed with the air in a glass box. The box is lit brightly from the side and its contents studied from above through a microscope.

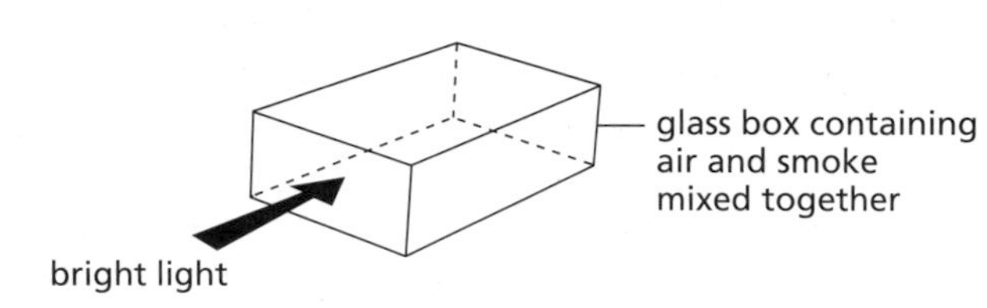

a Bright specks are seen moving in continuous and jerky random movement.
(i) What are the bright specks? (Choose *one*.)

air molecules.
smoke molecules.
smoke particles.

(ii) What is the explanation for the jerky random movement? (Choose *one*.)

The air molecules bombard each other.
The smoke particles bombard each other.
The air molecules bombard the smoke particles.
The air molecules bombard the glass.
The smoke particles bombard the glass.

b The contents of the glass box exert a pressure on the glass walls. Choose *any* of the following sentences which might help explain this pressure.

The air molecules bombard each other.
The smoke particles bombard each other.
The air molecules bombard the smoke particles.
The air molecules bombard the glass.
The smoke particles bombard the glass.

(*UCLES IGCSE Physics Core, May/June 2000*)

55 Which one of the following statements is *not* true?

A Temperature tells us how hot an object is.
B Temperature is measured by a thermometer which uses some property of matter (e.g. the expansion of mercury) that changes continuously with temperature.
C Heat flows naturally from an object at a lower temperature to one at a higher temperature.
D The molecules of an object move faster when its temperature rises.
E Temperature is measured in °C, heat is measured in joules.

56 a Copy the table below and write two different physical properties which may be used to measure temperature. An example has been given to help you.

The change in volume	of	a liquid
	of	
	of	

b When creating a temperature scale, fixed points are needed. Explain what is meant by a *fixed point*.

c Copy the table below and state the upper and lower fixed points used when calibrating a liquid-in-glass thermometer with a centigrade temperature scale.

The upper fixed point is the temperature of	Its value is
The lower fixed point is the temperature of	Its value is

d The diagram shows how the temperature changes with time for a substance as it is heated steadily from a solid to a liquid and then to a gas.

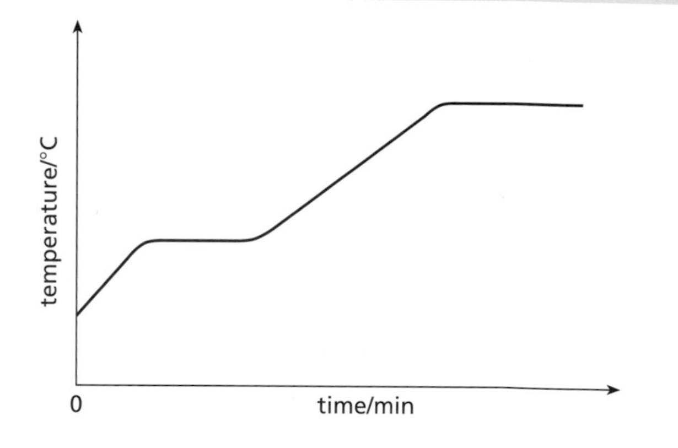

Copy the diagram and
(i) label the melting point and the boiling point of the substance,
(ii) indicate the time when the substance is completely liquid.

(*UCLES IGCSE Physics Core, Nov 1999*)

57 The diagram shows a ring with gear teeth, which is to be fitted to a flywheel.

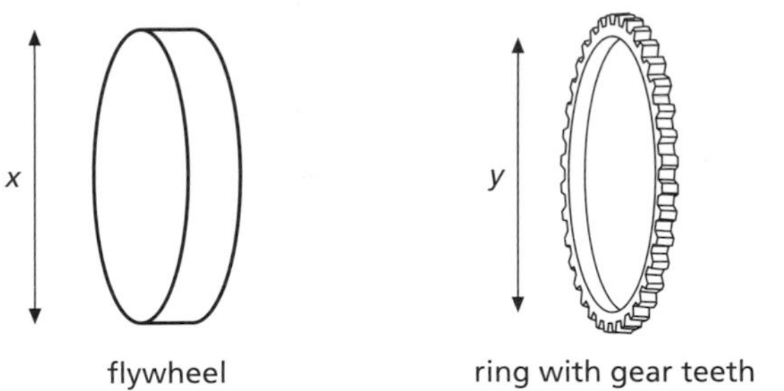

The internal diameter, *y*, of the ring is slightly smaller than the diameter, *x*, of the flywheel. The ring is heated to a high temperature.

a State why the ring will now fit on to the flywheel.
b State what will happen when the ring is allowed to cool.
c Explain why diameter *y* is designed to be smaller than diameter *x*.

(*UCLES IGCSE Physical Science Core, May/June 2000*)

58 The pressure exerted by a gas in a container

1 is due to the molecules of the gas bombarding the walls of the container
2 decreases if the gas is cooled
3 increases if the volume of the container increases.

Which statement(s) is (are) correct?

A 1, 2, 3   **B** 1, 2   **C** 2, 3   **D** 1   **E** 3

59 **a** Complete the following sentence.

'The temperature of a body rises when the ..... energy of its molecules is increased.'

b The diagram gives details about an empty beaker and the same beaker with different substances in it.

	empty beaker	beaker + water	beaker + sand
mass	250 g	500 g	500 g
energy needed to raise temperature by 1 °C	125 J	1175 J	325 J

(i) Which of the arrangements has the highest thermal capacity?
(ii) 1. What is the mass of water?
2. What is the mass of the sand?
3. How much energy is needed to raise the temperature of the water by 1 °C?
4. How much energy is needed to raise the temperature of the sand by 1 °C?
5. Use your answers above to suggest why, on a sunny day, the temperature of the sand on a beach rises faster than the temperature of the sea.

(*UCLES IGCSE Physics Core, Nov 2000*)

60 A student measures the temperature of ice in a beaker. The temperature rises from −5 °C to 0 °C, then remains for several minutes at 0 °C, before rising to room temperature.

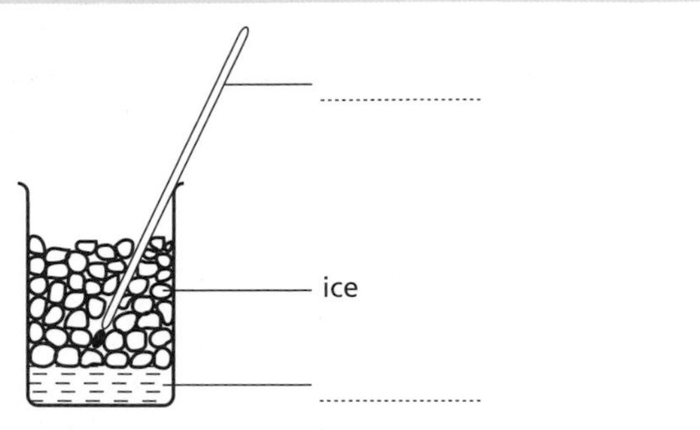

a The diagram below shows the beaker and its contents whilst the temperature remains at 0 °C.
(i) Name the instrument used to measure the temperature.
(ii) Name the liquid in the beaker.
b Use words from the list below to complete the following sentences about this experiment.

boils	**freezes**	**melts**
electrons	**molecules**	**protons**
lose	**need**	

At 0 °C, the ice . . . . . . to form water. The . . . . . . in the ice are held firmly in the crystal structure. These particles . . . . . . energy so that they are able to leave the crystal structure and move about as water.

(*UCLES IGCSE Physical Science Core, May 1999*)

61 A drink is cooled more by ice at 0 °C than by the same mass of water at 0 °C because ice

A floats on the drink
B has a smaller specific heat capacity
C gives out latent heat to the drink as it melts
D absorbs latent heat from the drink to melt
E is a solid.

62 **a** Some students are asked to write down what they know about evaporation of a liquid. Here are their statements, some of which are correct and some incorrect. Which are correct?

A 'Evaporation occurs at any temperature.'
B 'Evaporation only occurs at the boiling point.'
C 'Evaporation occurs where the liquid touches the bottom of the container.'
D 'Evaporation occurs at the surface of the liquid.'
E 'It is the higher energy molecules which escape.'
F 'The molecules gain energy as they escape.'
G 'The liquid temperature **always** rises when evaporation occurs.'
H 'Rapid evaporation produces cooling.'

b Sometimes after shaving, men splash a liquid, called an aftershave, over their faces. This makes their faces feel fresher as the aftershave evaporates.
(i) Which of the statements in part **a** explains why the aftershave, even though it is at room temperature, cools the skin?
(ii) Suggest why the aftershave cools the skin better than water at room temperature.

(*UCLES IGCSE Physics Core, Nov 2000*)

63 Which of the following statements is/are true?

1 In cold weather the wooden handle of a saucepan feels warmer than the metal pan because wood is a better conductor of heat.
2 Convection occurs when there is a change of density in parts of a fluid.
3 Conduction and convection cannot occur in a vacuum.

A 1, 2, 3 **B** 1, 2 **C** 2, 3 **D** 1 **E** 3

64 Which one of the following statements is *not* true?

A Energy from the Sun reaches the Earth by radiation only.
B A dull black surface is a good absorber of radiation.
C A shiny white surface is a good emitter of radiation.
D The best heat insulation is provided by a vacuum.
E A vacuum flask is designed to reduce heat loss or gain by conduction, convection and radiation.

65 A mass of water is put into a beaker made of insulating plastic foam. An equal mass of glycerine is put into an identical insulating beaker. Identical thermometers are then put in each beaker. The thermal capacity of the beaker of water is greater than the thermal capacity of the beaker of glycerine.

The two liquids are then each heated at the same rate by a small immersion heater. They are stirred whilst being heated.

a (i) Describe in detail what you would expect to see happening to the readings of the thermometers during the first minute after the heaters are switched on.
(ii) Explain your answer to **a** (i).

b (i) The heaters are left switched on. Describe in detail what happens to the thermometer readings after a long time. (Assume that the rate of supply of heat from the immersion heaters is always greater than the rate of loss of heat from the heaters.)
(ii) Explain your answer to **b** (i).

(*UCLES IGCSE Physics Core, May 1998*)

66 Electronic components usually generate heat when they are in use but they must not be allowed to become too hot. To take the heat away from a component quickly, it is fixed to a metal 'heat sink'. The diagram shows a section through one such heat sink.

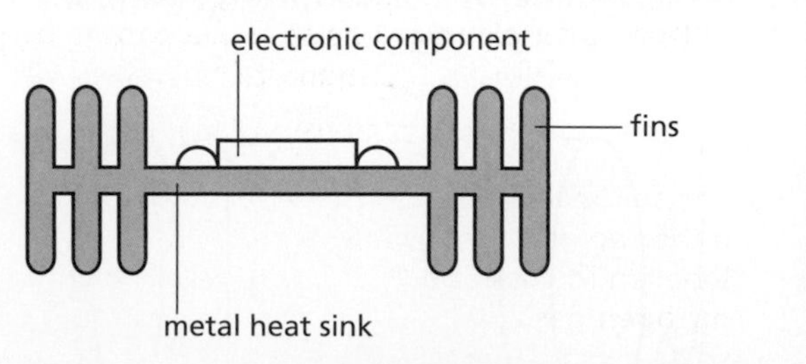

a Suggest why the heat sink is made from a metal, rather than some other material.
b Suggest why the heat sink has fins.

(*UCLES IGCSE Physics Core, Nov 1998*)

67 In an experiment to find the specific latent heat of fusion of ice, an electric heater, of power 200 W, is used. The following readings are taken.

mass of ice at 0 °C, before heating started, 0.54 kg
mass of ice at 0 °C, after 300 s of heating, 0.36 kg

a Calculate a value of the specific latent heat of fusion of ice.
b Explain, in molecular terms, how heat is transferred from the surface of a block of ice to its centre.

(*UCLES IGCSE Physics Extended, May/June 2000*)

68 The diagram shows a piece of apparatus which could be used to find the specific heat capacity of a metal at high temperatures.

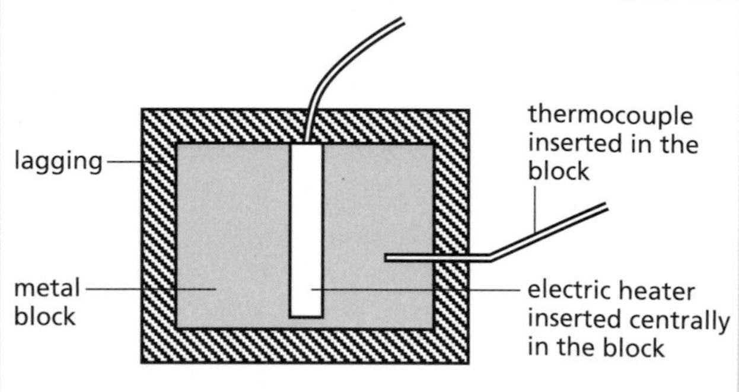

Results from an experiment using the apparatus are recorded as follows:

mass of the metal block, 1.0 kg;
power of the heater, 200 W;
time for which the heater is switched on, 2.5 minutes (150 s);
rise in temperature during this time, from 160 °C to 210 °C.

a Describe the experimental steps which were taken to obtain these results.
b Use the results to calculate an average value for the specific heat capacity of the metal over this temperature range.
c The temperature of the metal was measured by using a thermocouple.
(i) Draw a labelled diagram of a thermocouple being used as a thermometer.
(ii) Describe the action of a thermocouple when measuring a temperature change.

(iii) Suggest two reasons why use of a thermocouple might have an advantage over a mercury-in-glass thermometer.

(*UCLES IGCSE Physics Extended, Nov 1999*)

69 The following diagram shows a student's design for a thermometer. The student stated that the material labelled M could be a copper rod, alcohol or nitrogen gas.

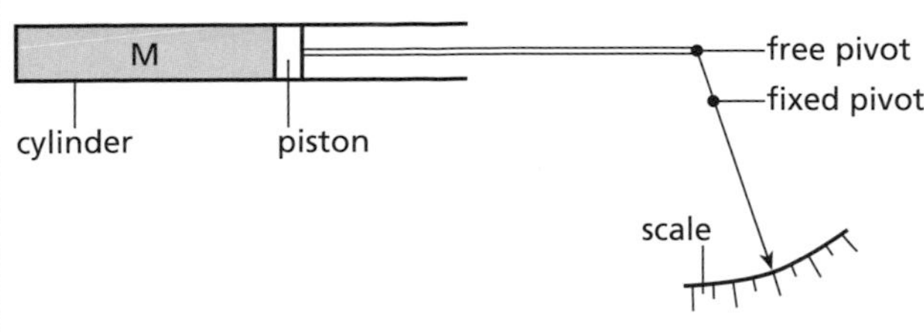

a Explain what is meant by the term *sensitivity of the thermometer*.
b (i) State which of the three suggested materials would give a thermometer of greatest sensitivity.
(ii) Explain your answer.
c (i) State which of the three materials would allow the thermometer to measure the largest range of temperature.
(ii) Explain your answer.
d The student found that the temperature scale of this thermometer was *non-linear*. Explain what this means.

(*UCLES IGCSE Physics Extended, Nov 2000*)

70 a A student determines the specific heat capacity of water. It is found that 15.5 kJ of energy supplied raise the temperature of 0.45 kg of water by 8.2 °C. Calculate the specific heat capacity of water.
b A cylinder, which is closed by a gas-tight moveable piston, contains 0.0060 pm^3 of gas.
The gas has its pressure raised from 2.0×10^5 Pa to 3.5×10^5 Pa, without any change in temperature.
(i) Describe how this could be achieved.
(ii) Calculate the volume when the pressure is 3.5×10^5 Pa.

(*UCLES IGCSE Physics Extended, Nov 2000*)

71 The diagram shows a solar furnace. This focuses radiation onto a container.

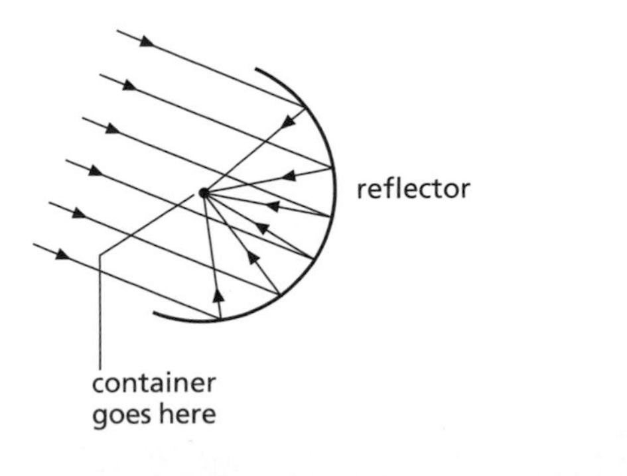

a (i) Suggest the best colour for the surface of the container.
(ii) Explain your choice.
b (i) Suggest a suitable material for the surface of the reflector.
(ii) Explain your choice.

(*UCLES IGCSE Physical Science Extended, May 1999*)

Electricity

72 A charged ebonite rod has negative charges all over its surface. It is held above three small pieces of aluminium foil, one positively charged, one negatively charged and one uncharged. This is shown in the diagram.

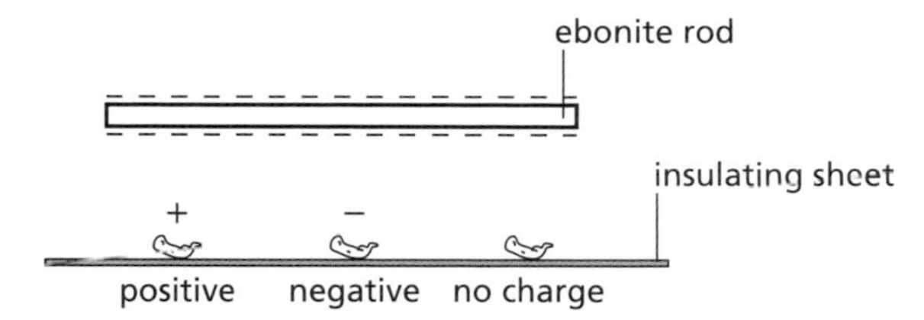

a State which if any of the pieces of aluminium are attracted by the ebonite rod.
b Ebonite is an insulator. What is meant by the term *insulator*?
c Write down the name of another insulating material.

(*UCLES IGCSE Physics Core, May 2001*)

73 For the circuit below calculate
a the total resistance,
b the current in each resistor,
c the p.d. across each resistor.

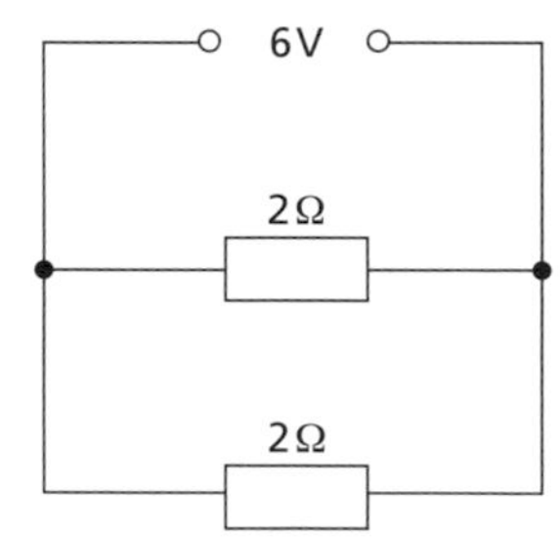

74 Repeat question 73 for the circuit below.

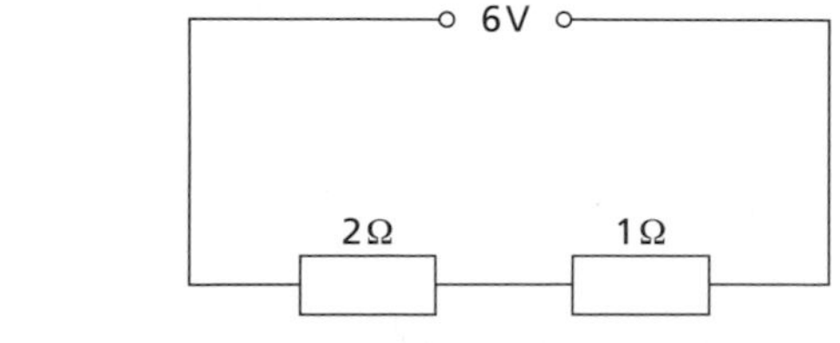

75 An electric kettle for use on a 230 V supply is rated at 3000 W. For safe working, the cable supplying it should be able to carry at least

A 2 A **B** 5 A **C** 10 A **D** 15 A **E** 30 A

76 Which one of the following statements is *not* true?

A In a house circuit lamps are wired in parallel.
B Switches, fuses and circuit breakers should be placed in the neutral wire.
C An electric fire has its earth wire connected to the metal case to prevent the user receiving a shock.
D When connecting a three-core cable to a 13 A three-pin plug the *brown* wire goes to the *live* pin.
E The cost of operating three 100 W lamps for 10 hours at 10 p per unit is 30 p.

77 A laboratory technician wants to make a resistor of value 64 Ω, using some resistance wire. He takes 1.0 m of this wire. The wire is shown in the diagram as AC. He connects up the circuit shown.

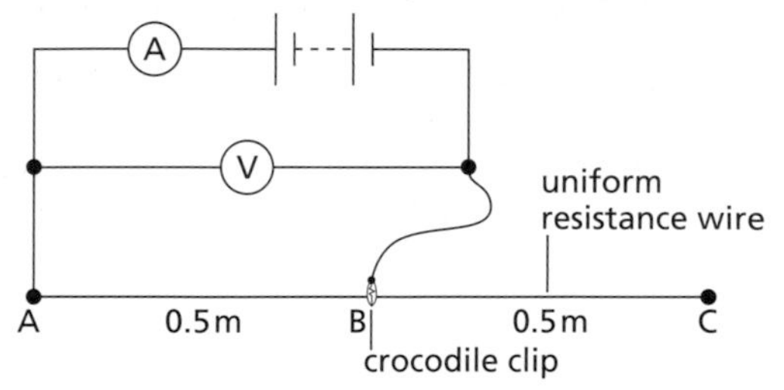

a He connects the crocodile clip at B, which is 0.5 m from A.
Here are the readings he gets.

voltmeter reading	12 V
ammeter reading	1.5 A

Calculate the resistance of wire AB.

b The laboratory technician now connects the crocodile clip to C, to measure the resistance of 1 m of the wire. The wire has constant thickness.
(i) Write the voltmeter reading and ammeter reading he obtains. Ignore the effects of the resistance of the ammeter, voltmeter and battery.
(ii) What is the resistance of wire AC?

c Use your answer to **b** to answer the following questions.
(i) What is the resistance per metre of this wire?
(ii) What length of wire does the laboratory technician need for the 64 Ω resistor?

(*UCLES IGCSE Physics Core, May/June 2000*)

78 Which of the units **A** to **E** could be used to measure
a electric charge,
b electric current,
c p.d.,
d energy,
e power?

A ampere **B** joule **C** volt **D** watt
E coulomb

79 The diagram shows an electric circuit.

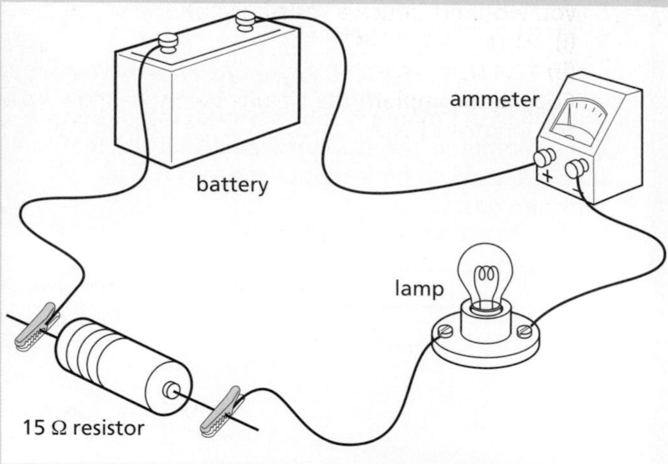

a The lamp lights, but the ammeter needle moves the wrong way. What change should be made so that the ammeter works correctly?
b What does an ammeter measure?
c Draw a circuit diagram of the circuit in the figure, using correct circuit symbols.
d (i) Name the instrument that would be needed to measure the potential difference (p.d.) across the 15 Ω resistor.
(ii) Using the correct symbol, add this instrument to your circuit diagram in **c**, in a position to measure the p.d. across the 15 Ω resistor.
e The potential difference across the 15 Ω resistor is 6 V.
Calculate the current in the resistor.
f Without any further calculation, state the value of the current in the lamp.
g Another 15 Ω resistor is connected in parallel with the 15 Ω resistor that is already in the circuit.
(i) What is the combined resistance of the two 15 Ω resistors in parallel?
30 Ω, 15 Ω, 7.5 Ω or zero?
(ii) State what effect, if any, adding this extra resistor has on the current in the lamp.

(*UCLES IGCSE Physics Core, Nov 2006*)

80 a The diagram shows a coil of thin wire and a lamp connected to a 4V supply.

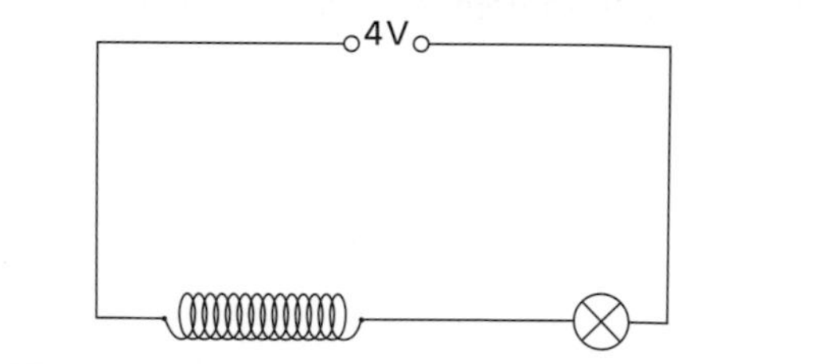

The lamp is marked 1.5 V, 0.6 W. The lamp lights at normal brightness. Calculate
(i) the current in the lamp,
(ii) the resistance of the lamp,
(iii) the charge flowing through the lamp in 20 s.

b The resistance of the coil of wire shown in the circuit in part **a** is 6.2 Ω and its length is 1.0 m. Using only 1.0 m lengths from the same reel of

wire, and without cutting any of them, state how you would produce a resistance of
(i) 3.1 Ω,
(ii) 12.4 Ω.
Copy and complete the circuits below to show how the lengths of wire are connected in each case.

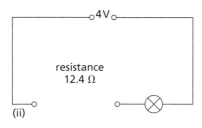

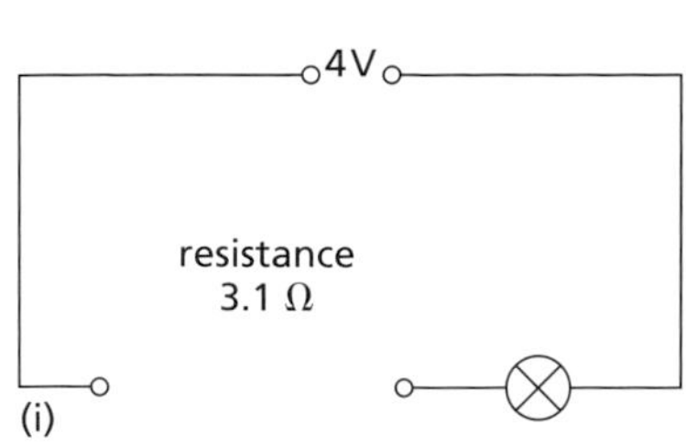

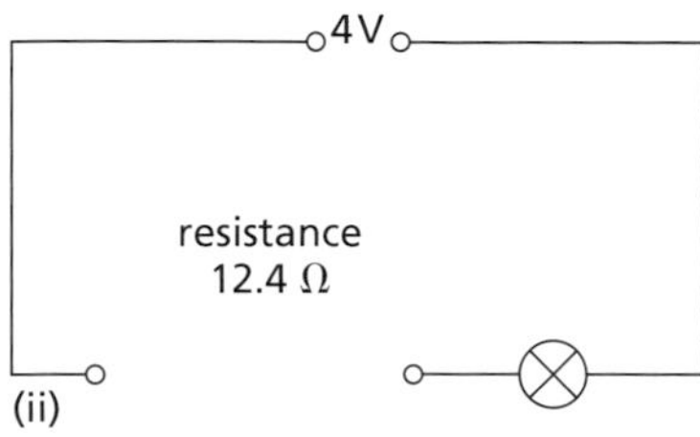

c In a similar circuit to that shown in part **a**, the resistance of the coil is 5.0 Ω and the current through it is 0.6 A. Calculate the heat energy produced in the coil in 20 s.

(*UCLES IGCSE Physics Extended, May 2001*)

81 a The following is a list of devices which may be used in electrical circuits.

capacitor **diode** **microphone**
multimeter **transistor**

Using each device only once, name the device which
(i) allows current to flow in one direction only;
(ii) can store charge;
(iii) can act as an electronic switch;
(iv) can be used to measure voltage;
(v) can act as a sound sensor.

b The diagram below shows a simple electronic bicycle alarm system. There are two switches in the system:

- a key operated switch which the rider turns on when the bicycle is parked;
- a pressure switch fitted in the saddle.

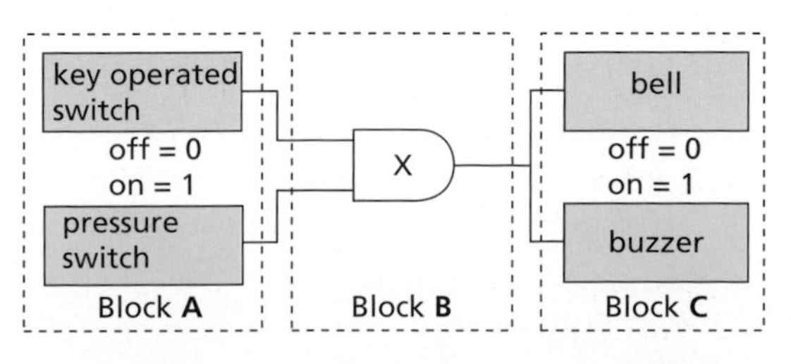

(i) The alarm system consists of three blocks, **A**, **B** and **C**. Use the following words to name each block.

control circuit **output sensors** **input sensors**

(ii) What sort of logic gate is gate X?
(iii) Copy and complete the truth table for the system.

Key operated switch	Pressure switch	Buzzer	Bell
1	1		
1	0		
0	1		
0	0		

(iv) What conditions will cause the alarm to go off?

(*SEG Foundation, Summer 1998*)

82 The diagram shows part of an alarm system used to protect CD players on display in a shop. Each CD player is placed on two pressure switches. When a pressure switch is pressed down it provides a high output (1). The alarm bell sounds if the CD player is lifted up.

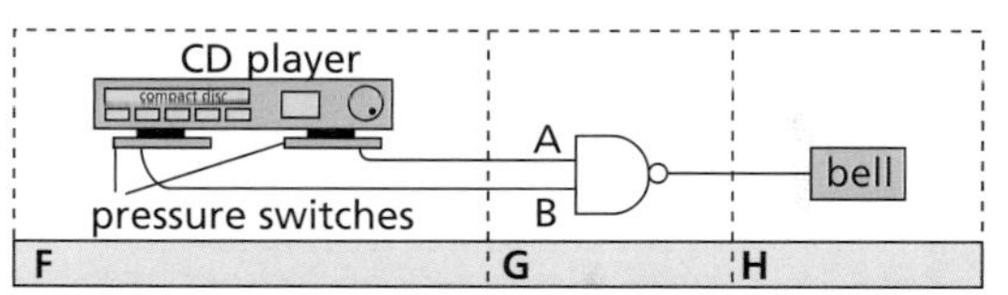

a This electronic alarm system consists of three parts **F**, **G** and **H**. Write the name of each of these parts of the system. Chosoe the *best* word from the list below.

input **output** **processor**
relay **supply** **thermistor**

b The system contains a NAND logic gate.
(i) Write the truth table for this logic gate, started below.

Input A	Input B	Output
0	0	

(ii) The NAND gate can be replaced by a combination of two other logic gates to give the same effect. Copy the diagram below and draw this combination in the box labelled **G**. Label the gate.

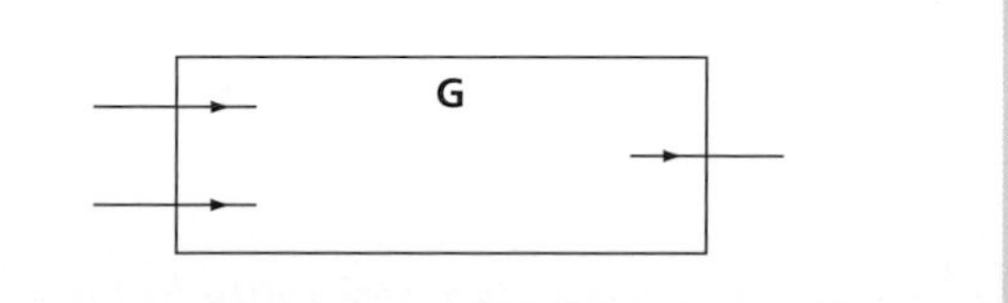

(iii) Somebody suggests replacing the NAND gate with a NOR gate. Describe how this would change the way the system works.

(*OCR Higher, June 1999*)

83 The diagram below shows an electronic system installed in a greenhouse.

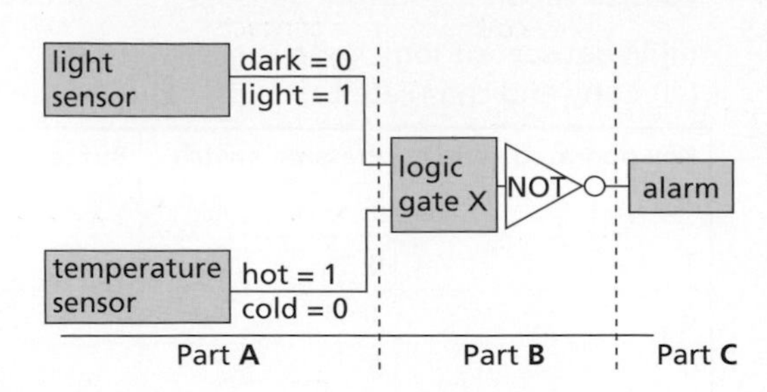

The alarm is designed to operate when there is an output from the NOT gate.

a (i) Which part, **A**, **B** or **C** of the system acts as the **processor**?
(ii) Which part, **A**, **B** or **C** of the system is **controlled** by the processor?

b The alarm should be activated if the temperature in the greenhouse drops during the night. Which type of logic gate should be used at X?

c Name a device suitable for use as
(i) the light sensor,
(ii) the temperature sensor.

d The following graph shows show the resistance of the temperature sensor changes with temperature. Use the graph to find the resistance of the temperature sensor at 33 °C.

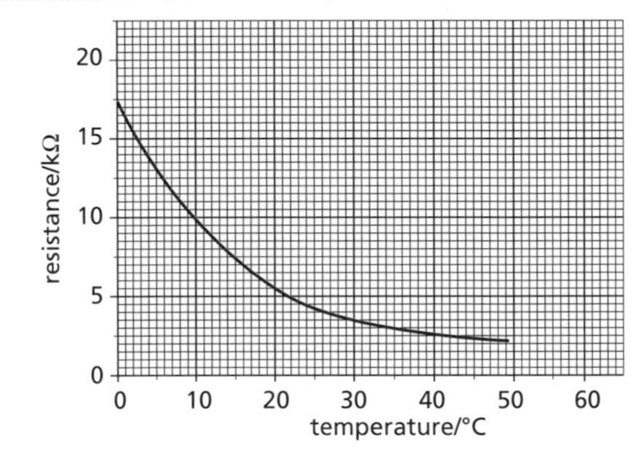

e The temperature sensor is connected in the circuit below.

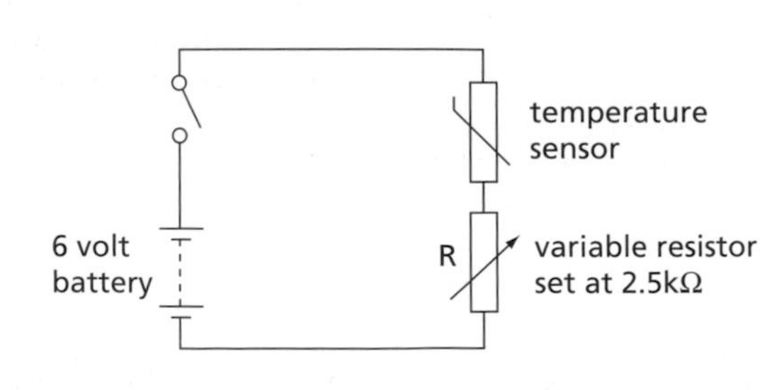

(i) What is the total resistance in the circuit when the temperature is 33 °C?
(ii) Calculate the current in the circuit when the temperature is 33 °C.

f Resistor R is connected to an electronic circuit as shown below.

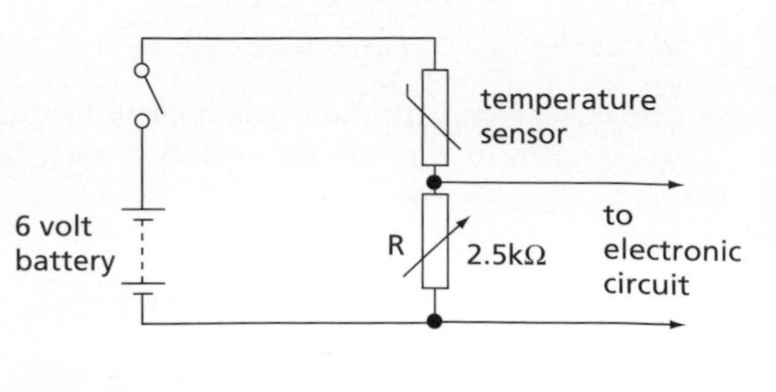

The electronic circuit is switched on when the potential difference across R is 1.4 V or more.
(i) Calculate the output voltage when the temperature is 33 °C.
(ii) Will the electronic circuit be ON when the temperature is 33 °C?
(iii) Calculate the output voltage when the temperature is 0 °C.
(iv) Will the electronic circuit be ON when the temperature is 0 °C?
(v) Suggest a use for this circuit.

(*AQA* (*NEAB*) *Higher, June 1998*)

84 Which one of the following statements about the transistor circuit shown below is *not* true?

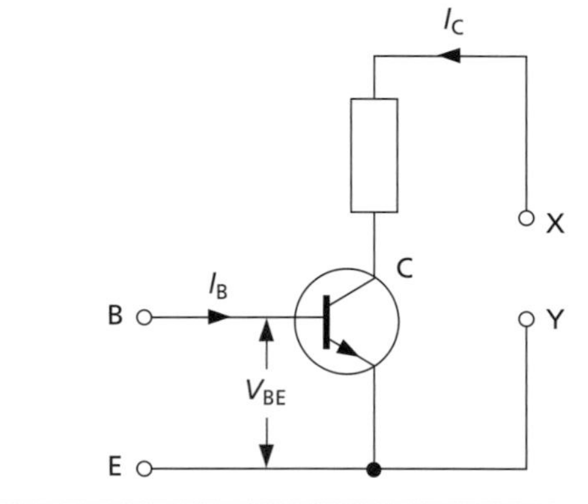

A The collector current I_C is zero until base current I_B flows.
B I_B is zero until the base–emitter p.d. V_{BE} is +0.6 V.
C A small I_B can switch on and control a large I_C.
D When used as an amplifier the input is connected across B and E.
E X must be connected to supply − and Y to the +.

Electromagnetic effects

85 a Two pieces of metal are lying end to end, with a small gap between them. They are not connected to anything else.

In (i) and (ii) below, write out the correct completion of the statement.
(i) If two pieces of metal attract each other, they

must both be magnets.
might both be magnets.
cannot both be magnets.

(ii) If two pieces of metal repel each other, they

must both be magnets.
might both be magnets.
cannot both be magnets.

b The diagram shows a reed switch in the absence of a magnetic field. If a circuit is connected to X and Y, the reed switch is classed as 'normally closed'.

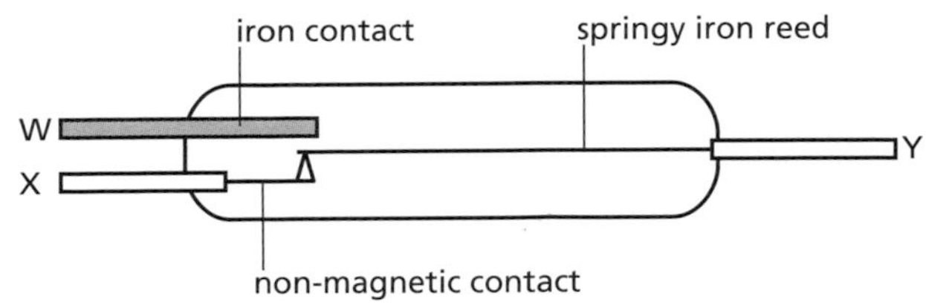

When a magnet is held close to the reed switch, the iron parts become magnetised as shown below.

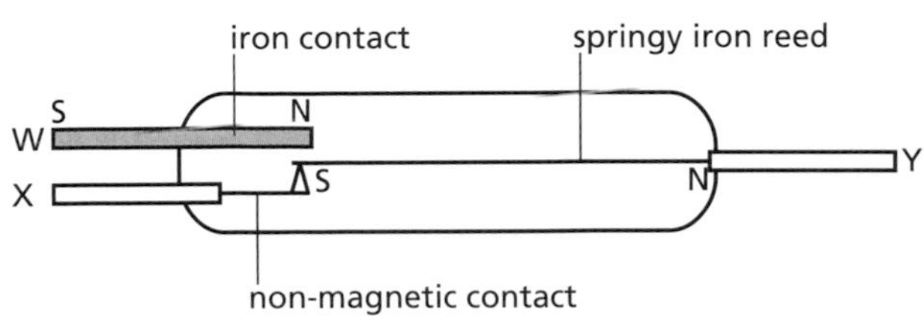

(i) State and explain what happens as a result of this magnetisation.
(ii) Where would you make electrical connections for this to be a normally open switch (i.e. open **unless** the magnet is present).

c A normally closed reed switch is used as part of a burglar alarm. The diagram shows how the reed switch and magnet are positioned, and it also shows the battery and the bell. The bell should ring when the door is open.

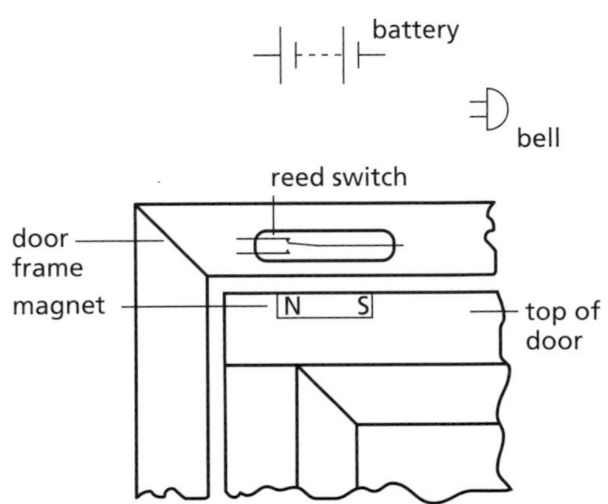

(i) Draw the circuit connecting the reed switch, the battery and the bell.
(ii) Why does the bell **not** ring whilst the door remains closed?
(iii) Why does the bell ring when a burglar opens the door?
(iv) Suggest **one** disadvantage of this simple circuit.

(UCLES IGCSE Physics Core, Nov 1998)

86 The diagram below shows a relay, which is an electromagnetic switch.

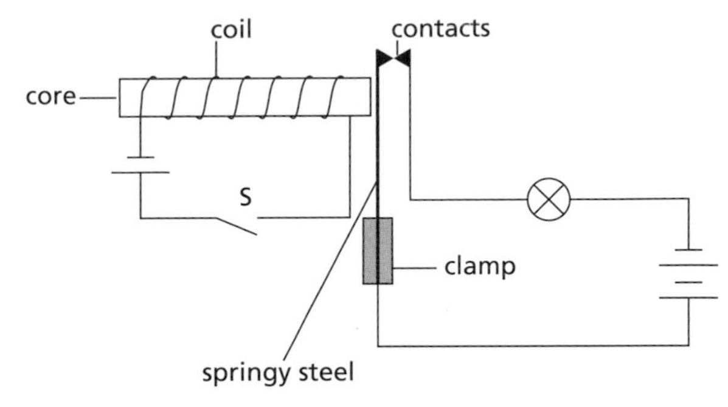

a Name the material that the core is made of.
b When switch **S** is closed, state what happens
(i) in the coil,
(ii) in the core,
(iii) at the contacts,
(iv) at the lamp.

(UCLES IGCSE Physical Science Core, May 1999)

87 Which one of the following statements about the diagram below is *not* true?

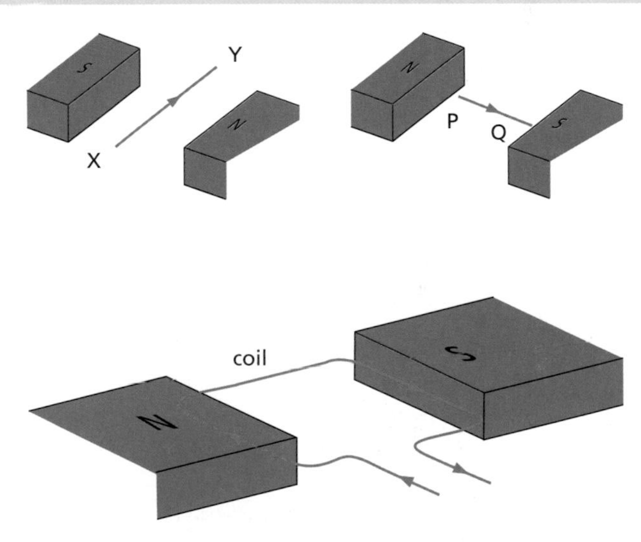

A If a current is passed through the wire XY, a vertically upwards force acts on it.
B If a current is passed through the wire PQ, it does not experience a force.
C If a current is passed through the coil, it rotates clockwise.
D If the coil had more turns and carried a larger current, the turning effect would be greater.
E In a moving-coil loudspeaker a coil moves between the poles of a strong magnet.

88 Which one of the following statements **A**–**E** is *not* true when a magnet is pushed N pole first into a coil as in the diagram overleaf?

A A p.d. is induced in the coil and causes a current through the galvanometer.
B The induced p.d. increases if the magnet is pushed in faster and/or the coil has more turns.
C Mechanical energy is changed to electrical energy.

D The coil tends to move to the right because the induced current makes face X a N pole which is repelled by the N pole of the magnet.
E The effect produced is called electrostatic induction.

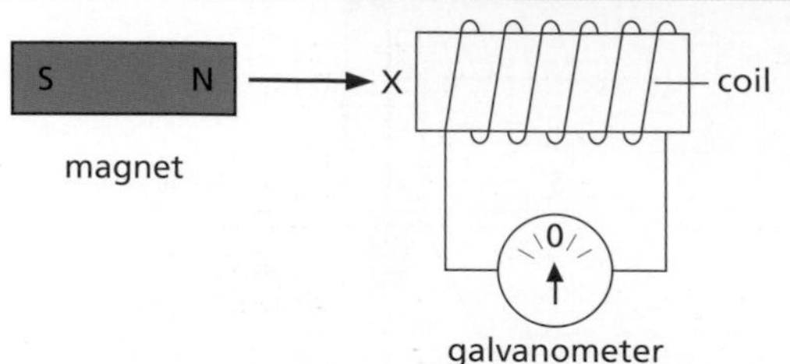

89 a Using words from the list below copy and complete the following sentences about transformers.

alternating	**charges**	**current**	**direct**
electric	**gravitational**	**induces**	**magnetic**
magnetises	**resistance**	**voltage**	

Transformers work using ..... current. As the current changes size and direction, the ..... field from one coil ..... a voltage across the other coil.

b (i) A step-up transformer, with a primary coil of 500 turns, is used to transform 200 V to 1000 V. Calculate the number of turns on the secondary coil. Write down the equation that you use and show all your working.
(ii) Calculate the power in the output circuit of the transformer if it carries a current of 10 A. Show all your working.

c Explain why voltages are usually stepped-up by transformers before power is transmitted over large distances.

d Transformers can have an efficiency of 99%. Explain what is meant by *efficiency*.

(*UCLES IGCSE Physical Science Extended, Nov 1999*)

90 Below is a block diagram of an electrical generating and distribution system.

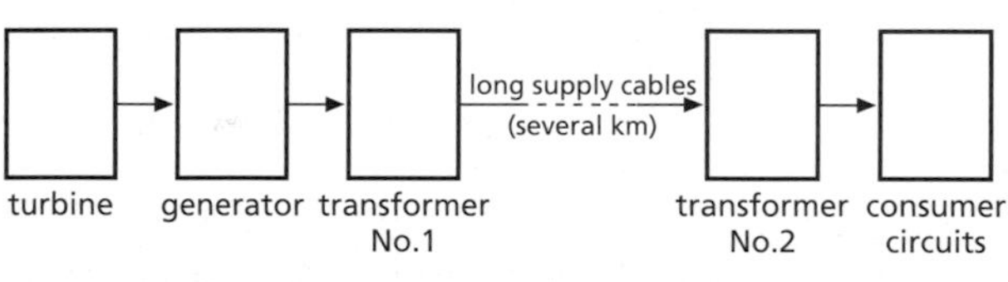

a The generator produces an e.m.f. by a process called electromagnetic induction.
(i) Name two factors and state how they are changed in order to increase the output e.m.f. of the generator.
(ii) Explain what is meant by the statement 'the induced e.m.f. acts in such a direction as to produce effects to oppose the change causing it'.

b (i) The diagram below shows the basic parts of transformer No. 1 which is 100% efficient. Using the information on the diagram, calculate the current in the supply cables.
(ii) Describe the function of transformer No. 2.
(iii) Explain why the use of the two transformers results in a big reduction in power loss in the supply cables.

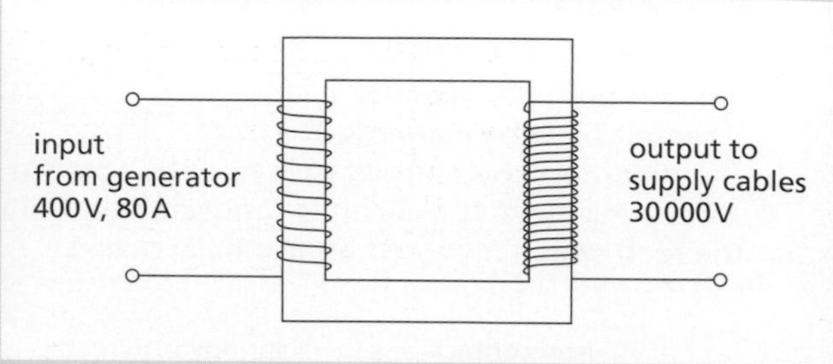

c The diagram below shows one of the consumer circuits with three electrical appliances R, S and T, connected into the circuit.

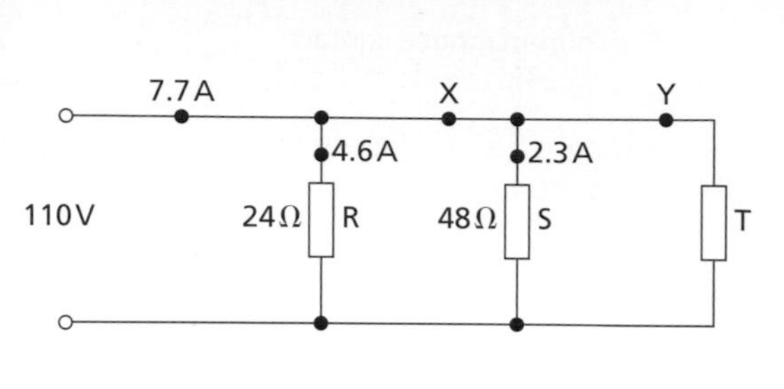

Using the current, voltage and resistance values shown, calculate
(i) the current at point X and at point Y,
(ii) the resistance of appliance T,
(iii) the combined resistance of appliances R and S,
(iv) the power developed in appliance R,
(v) the energy converted by the appliance S in 2 minutes (120 s).

(*UCLES IGCSE Physics Extended, Nov 1999*)

91 The diagram shows a simple electrical generator. By turning the handle, the single coil may be spun between the poles of the magnet.

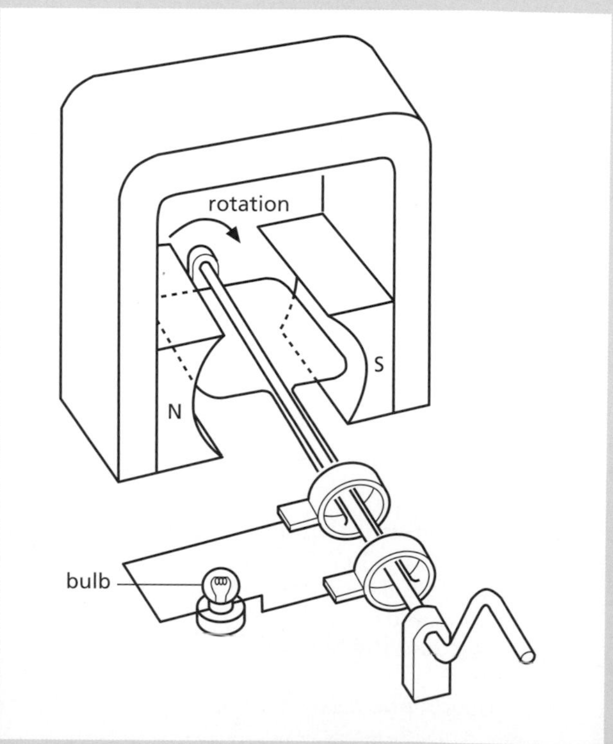

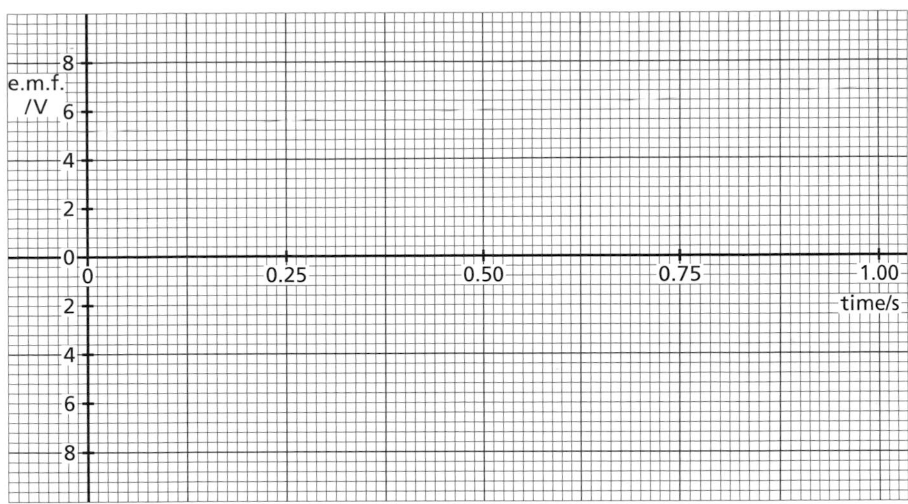

a The handle is turned so that the coil makes two complete revolutions per second. The maximum output is 7 V. On axes like those above, sketch this output over a period of 1 s.

b Explain
(i) how an e.m.f. is induced,
(ii) why the e.m.f. varies in magnitude and direction.

(*UCLES IGCSE Physics Extended, Nov 2000*)

Electrons and atoms

92 Identify the radiations X, Y and Z in the diagram below.

	A	**B**	**C**	**D**	**E**
X	alpha	beta	gamma	gamma	beta
Y	beta	alpha	alpha	beta	gamma
Z	gamma	gamma	beta	alpha	alpha

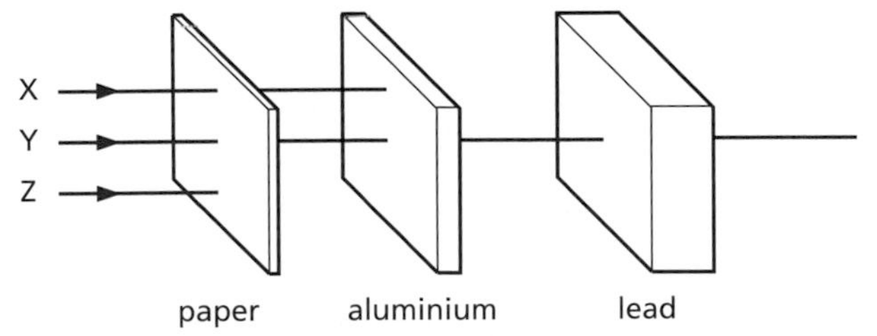

93 The radioactive substance whose decay curve is given in the next column has a half-life in minutes of

A 1
B 2
C 3
D 4
E 5

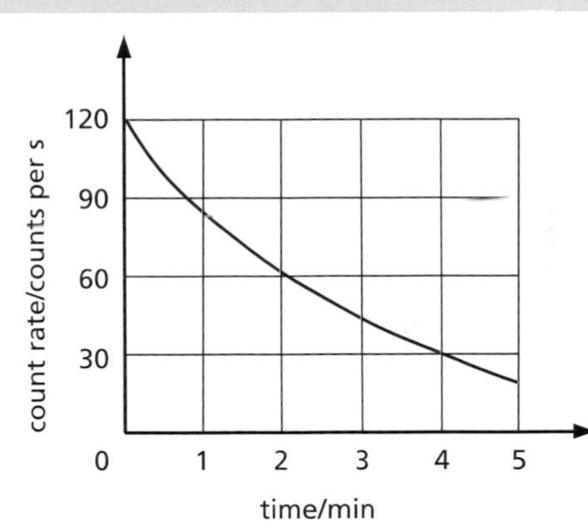

94 A radioactive source which has a half-life of 1 hour gives a count-rate of 100 counts per second at the start of an experiment and 25 counts per second at the end. The time taken by the experiment was, in hours,

A 1 **B** 2 **C** 3 **D** 4 **E** 5

95 Which one of the following statements is *not* true?

A An atom consists of a tiny nucleus surrounded by orbiting electrons.
B The nucleus contains protons and neutrons, called nucleons, in equal numbers.
C A proton has a positive charge, a neutron is uncharged and their mass is about the same.
D An electron has a negative charge of the same size as the charge on a proton but it has a much smaller mass.
E The number of electrons equals the number of protons in a normal atom.

96 A lithium atom has a nucleon (mass) number of 7 and a proton (atomic) number of 3.

1 Its symbol is ${}^{7}_{4}\mathrm{Li}$.
2 It contains 3 protons, 4 neutrons and 3 electrons.
3 One of its isotopes has 3 protons, 3 neutrons and 3 electrons.

Which statement(s) is (are) correct?

A 1, 2, 3 **B** 1, 2 **C** 2, 3 **D** 1 **E** 3

97 Which symbol **A** to **E** is used in equations for nuclear reactions to represent
a an alpha particle,
b a beta particle,
c a neutron,
d an electron?

A $^{0}_{-1}e$ **B** $^{1}_{0}n$ **C** $^{4}_{2}He$ **D** $^{1}_{-1}e$ **E** $^{1}_{1}n$

98 **a** Radon $^{220}_{86}Rn$ decays by emitting an alpha particle to form an element whose symbol is

A $^{216}_{85}At$ **B** $^{216}_{86}Rn$ **C** $^{218}_{84}Po$ **D** $^{216}_{84}Po$ **E** $^{217}_{85}At$

b Thorium $^{234}_{90}Th$ decays by emitting a beta particle to form an element whose symbol is

A $^{235}_{90}Th$ **B** $^{230}_{89}Ac$ **C** $^{234}_{89}Ac$ **D** $^{232}_{88}Ra$ **E** $^{234}_{91}Pa$

99 A special electricity supply passes a current through a resistor R. A cathode ray oscilloscope is connected across R. The arrangement is shown below.

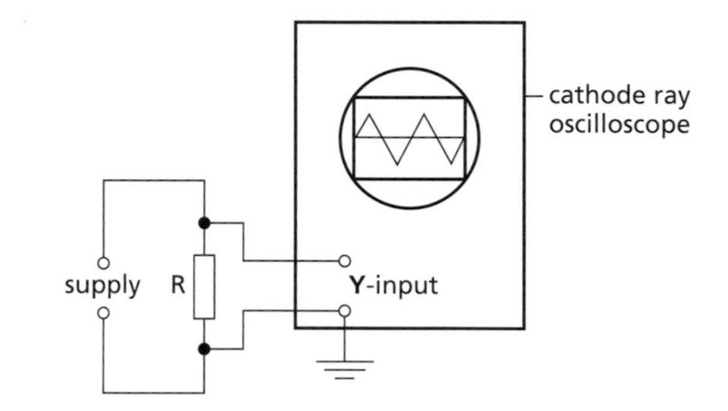

A waveform is seen on the screen of the oscilloscope. This waveform is shown by the broken line on the diagram below.

a Resistor R is changed for another, of half the resistance of R. The electricity supply is adjusted to pass the same current as before. Copy the diagram below and draw the new waveform which is seen on the screen.

b With this new resistor left in the circuit, the frequency of the electricity supply is halved. On another copy of the diagram below, draw the waveform which is now seen on the screen.

c The special electricity supply is now removed and replaced by a battery. On another copy of the diagram below, show what might now be seen on the screen.

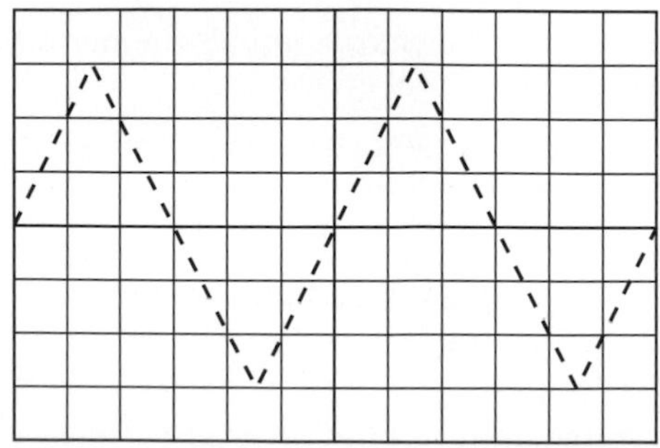

(UCLES IGCSE Physics Core, Nov 1998)

100 The diagram represents a neutral lithium atom.

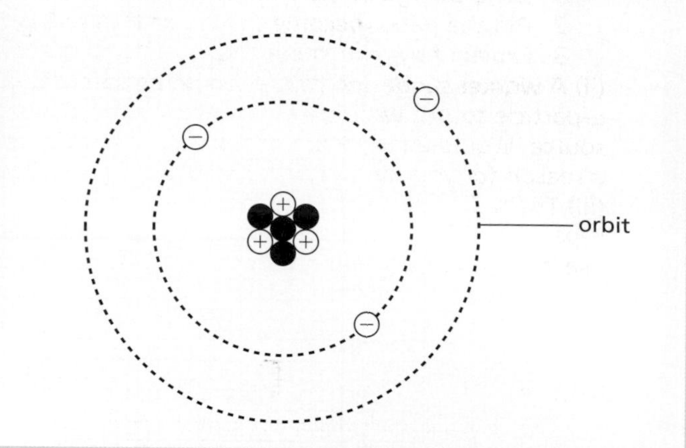

a Use your knowledge of the structure of the atom to identify the particles in the atom.
(i) Name the particles represented by ⊖
(ii) Name the particles represented by ⊕
(iii) Name the particles represented by ●

b Name the group of particles at the centre of the atom.

c Complete the nuclide notation of this lithium atom by writing the appropriate numbers missing in the boxes below.

(UCLES IGCSE Physics Core, May 1998)

101 **a** What do β-particles consist of?

b Put α-particles, β-particles and γ-rays in order of penetrating ability, from most penetrating to least penetrating.

c In a factory which makes paper, the paper passes through a thickness recorder, which consists of a β-particle source and a detector.

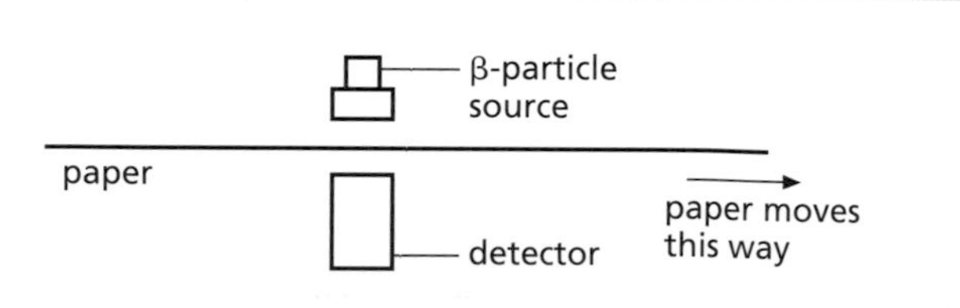

During one production period, the readings, in counts/s from the detector were as follows.

time, in minutes	0	2	4	6	8	10	12	14	16	18	20
reading, in counts/s	90	91	90	88	89	91	171	170	172	169	170

(i) 1. At approximately what time was there a large change in the thickness of the paper?
2. Did the paper become thicker or thinner?
3. Explain how you know this.

(ii) A worker suggested that it would be safer if an α-particle source were used instead of a β-particle source. Would an α-particle source be suitable? Give a reason for your answer.

(iii) The radioactive decay curve for one particular β-particle source which was being considered for use in the thickness recorder is shown below.

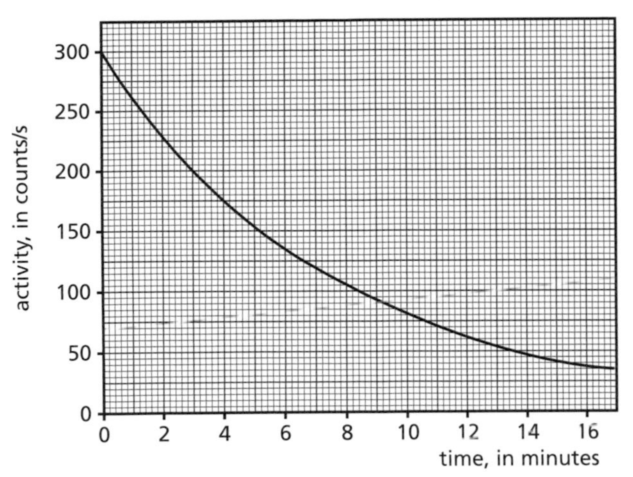

1. What is the half-life of the source?
2. Would this source be suitable for use in the thickness recorder? Give your reasons.

(UCLES IGCSE Physics Core, Nov 1998)

102 The diagrams in parts **a** and **b** below represent the screen of a cathode ray oscilloscope (c.r.o.) when two different power sources are connected to it. The trace is arranged to be in the middle of the screen before each of the sources is connected. The time base is set to 20 ms per division and the y-gain to 2 V per division.

a

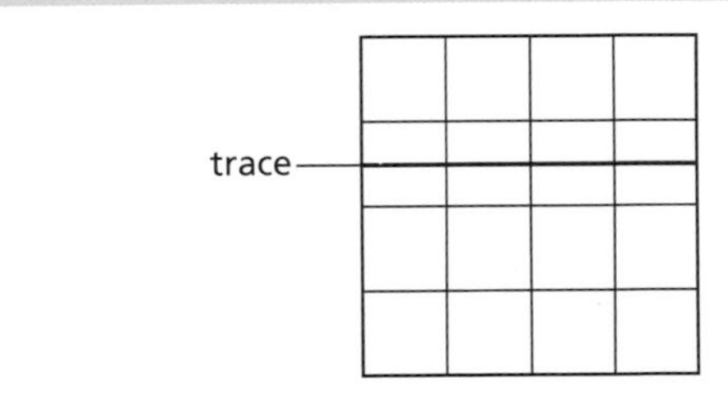

(i) Name the type of voltage.
(ii) Write down the value of the voltage.

b

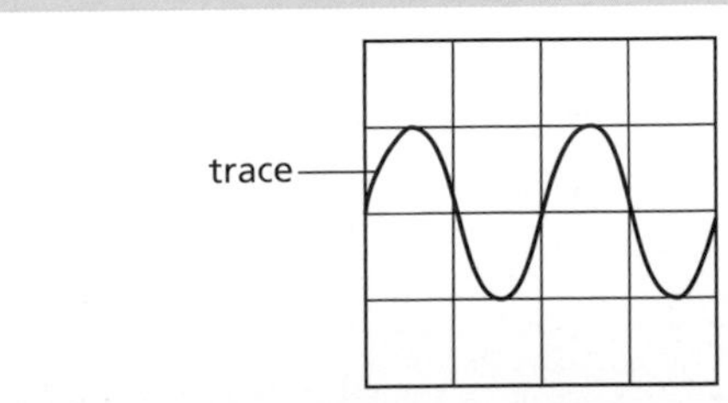

(i) Name the type of voltage.
(ii) Write down the value of the peak voltage (amplitude).
(iii) Write down the time for one cycle (period).

c (i) A single diode is now connected in series with the power source in **b**. Draw the new trace seen.
(ii) Explain why the diode produces this trace.

(UCLES IGCSE Physical Science Extended, May 1999)

103 The diagram shows an uncharged metal plate held in a wooden clamp and stand.

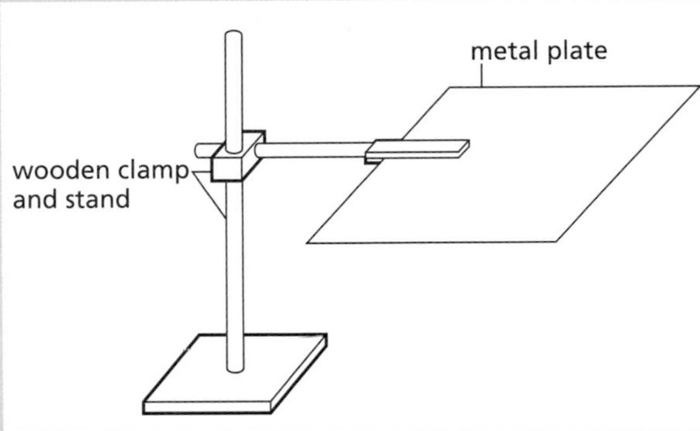

a A polythene rod is charged negatively by rubbing it with a duster. Suggest, in terms of the movement of electrons,
(i) how the polythene becomes negatively charged,
(ii) how the metal plate can be positively charged without the polythene touching the plate.

b A strong α-particle emitting source is brought close to, but not touching, the positively charged metal plate. Explain why the plate rapidly loses its charge.

(UCLES IGCSE Physics Extended, Nov 2000)

104 a A laboratory needs to find a radioactive isotope which will produce very intense ionisation of air. The apparatus is shown below.

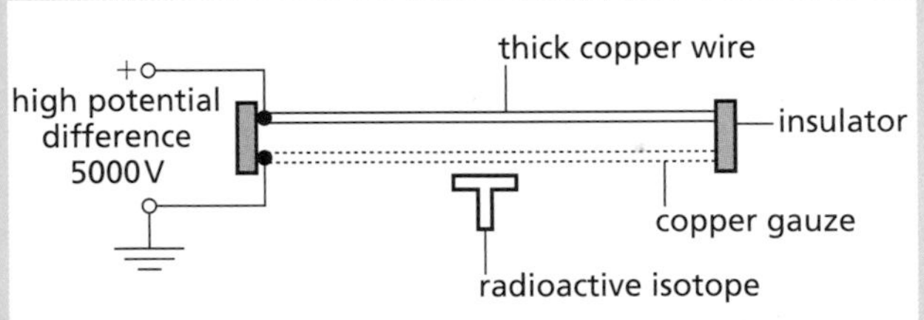

(i) Explain why sparks jump between the gauze and the wire when a radioactive isotope with high ionising properties is brought near the gauze.
(ii) An α-emitting source, a β-emitting source and a γ-emitting source, each of the same activity, are tested. One source gives no sparks at all, the second gives only a few sparks per second and the third many sparks per second. State the relative quantities of ionisation produced by each type of emitter.

b Some of the results of a comparison between α-particles, β-particles and γ-rays are shown in the table below.

	α	β	γ
mass	4 units		
constitution	2 protons +2 neutrons		
charge	+2 units		

Copy and complete the table by filling in the blank boxes.

(UCLES IGCSE Physics Extended, May 1999)

105 a Copy and complete the following table for α-particles. The first answer has been given.

property/ nature	complete this column	
symbol	^4_2He	(hint: it is a helium nucleus)
mass number		(hint: write down the number of proton charges)
charge		
ionisation of gases		(hint: choose from: strong, weak or almost none)
deflection in a magnetic field		(hint: choose from: towards N, towards S or at right angles to the magnetic field lines)
deflection in an electric field		(hint: choose from: towards +ve, towards −ve or no deflection)

b The diagram shows the paths of α-particles scattered by the nuclei of metal atoms in thin foils.

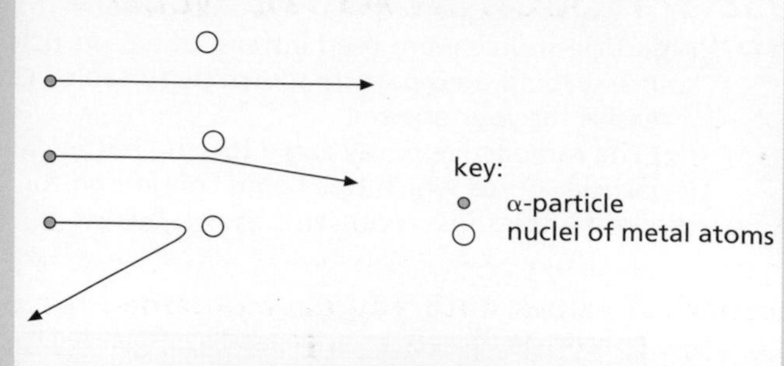

Explain what can be deduced from the paths shown about
(i) the mass of the nucleus of a metal atom compared to the mass of an α-particle,
(ii) the charge on the nucleus of a metal atom,
(iii) the volume occupied by a metal atom compared to its nucleus.

(UCLES IGCSE Physics Extended, May 2001)

Mathematics for physics

USE THIS SECTION AS THE NEED ARISES

Solving physics problems

When tackling physics problems using mathematical equations it is suggested that you **do not substitute numerical values until you have obtained the expression in symbols which gives the answer**. That is, work in symbols until you have solved the problem and only then insert the numbers in the expression to get the final result.

This has two advantages. First, it reduces the chance of errors in the arithmetic (and in copying down). Second, you write less since a symbol is usually a single letter whereas a numerical value is often a string of figures.

Adopting this 'symbolic' procedure frequently requires you to change round an equation first. The next two sections and the questions that follow them are intended to give you practice in doing this and then substituting numerical values to get the answer.

Equations – type 1

In the equation $x=a/b$, the subject is x. To change it we **multiply or divide both sides** of the equation by the same quantity.

To change the subject to a

We have

$$x = \frac{a}{b}$$

If we multiply both sides by b the equation will still be true.

$$\therefore \quad x \times b = \frac{a}{b} \times b$$

The b's on the right-hand side cancel

$$\therefore \quad b \times x = \frac{a}{\not b} \times \not b = a$$

$$\therefore \quad a = b \times x$$

To change the subject to b

We have

$$x = \frac{a}{b}$$

Multiplying both sides by b as before, we get

$$a = b \times x$$

Dividing both sides by x:

$$\frac{a}{x} = \frac{b \times x}{x} = \frac{b \times \not x}{\not x} = b$$

$$\therefore \quad b = \frac{a}{x}$$

Note that the **reciprocal** of x is $\frac{1}{x}$.

Can you show that

$$\frac{1}{x} = \frac{b}{a}?$$

Now try the following questions using these ideas.

Questions

1 What is the value of x if
a $2x=6$ **b** $3x=15$ **c** $3x=8$
d $\frac{x}{2}=10$ **e** $\frac{x}{3}=4$ **f** $\frac{2x}{3}=4$
g $\frac{4}{x}=2$ **h** $\frac{9}{x}=3$ **i** $\frac{x}{6}=\frac{4}{3}$

2 Change the subject to
a f in $v=f\lambda$ **b** λ in $v=f\lambda$
c I in $V=IR$ **d** R in $V=IR$
e m in $d=\frac{m}{V}$ **f** V in $d=\frac{m}{V}$
g s in $v=\frac{s}{t}$ **h** t in $v=\frac{s}{t}$

3 Change the subject to
a I^2 in $P=I^2R$ **b** I in $P=I^2R$
c a in $s=\frac{1}{2}at^2$ **d** t^2 in $s=\frac{1}{2}at^2$
e t in $s=\frac{1}{2}at^2$ **f** v in $\frac{1}{2}mv^2=mgh$
g y in $\lambda=\frac{ay}{D}$ **h** ρ in $R=\frac{\rho l}{A}$

4 By replacing (substituting) find the value of $v=f\lambda$ if
a $f=5$ and $\lambda=2$ **b** $f=3.4$ and $\lambda=10$
c $f=1/4$ and $\lambda=8/3$ **d** $f=3/5$ and $\lambda=1/6$
e $f=100$ and $\lambda=0.1$ **f** $f=3\times10^5$ and $\lambda=10^3$

5 By changing the subject and replacing find
a f in $v=f\lambda$ if $v=3.0\times10^8$ and $\lambda=1.5\times10^3$
b h in $p=10hd$ if $p=10^5$ and $d=10^3$
c a in $n=a/b$ if $n=4/3$ and $b=6$
d b in $n=a/b$ if $n=1.5$ and $a=3.0\times10^8$
e F in $p = F/A$ if $p=100$ and $A=0.2$
f s in $v=s/t$ if $v=1500$ and $t=0.2$

Equations – type 2

To change the subject in the equation $x=a+by$ we **add or subtract the same quantity from each side**. We may also have to divide or multiply as in type 1. Suppose we wish to change the subject to y in

$$x=a+by$$

Subtracting a from both sides,

$$x-a=a+by-a=by$$

Dividing both sides by b,

$$\frac{x-a}{b}=\frac{by}{b}=y$$

$$\therefore \quad y=\frac{x-a}{b}$$

Questions

6 What is the value of x if
a $x+1=5$ b $2x+3=7$ c $x-2=3$
d $2(x-3)=10$ e $\frac{x}{2}-\frac{1}{3}=0$ f $\frac{x}{3}+\frac{1}{4}=0$
g $2x+\frac{5}{3}=6$ h $7-\frac{x}{4}=11$ i $\frac{3}{x}+2=5$

7 By changing the subject and replacing, find the value of a in $v=u+at$ if
a $v=20$, $u=10$ and $t=2$
b $v=50$, $u=20$ and $t=0.5$
c $v=5/0.2$, $u=2/0.2$ and $t=0.2$

8 Change the subject in $v^2=u^2+2as$ to a.

Proportion (or variation)

One of the most important mathematical operations in physics is finding the relation between two sets of measurements.

a) Direct proportion

Suppose that in an experiment two sets of readings are obtained for the quantities x and y as in Table M1 (units omitted).

Table M1

x	1	2	3	4
y	2	4	6	8

We see that when x is doubled, y doubles; when x is trebled, y trebles; when x is halved, y halves; and so on. There is a one-to-one correspondence between each value of x and the corresponding value of y.

We say that y is **directly proportional** to x, or y **varies directly** as x. In symbols

$$y \propto x$$

Also, the **ratio** of one to the other, e.g. y to x, is always the same, i.e. it has a constant value which in this case is 2. Hence

$$\frac{y}{x}=\text{a constant}=2$$

The constant, called the **constant of proportionality** or **variation**, is given a symbol, e.g. k, and the relation (or law) between y and x is then summed up by the equation

$$\frac{y}{x}=k \quad \text{or} \quad y=kx$$

Notes

1 In practice, because of inevitable experimental errors, the readings seldom show the relation so clearly as here.

2 If instead of using numerical values for x and y we use letters, e.g. x_1, x_2, x_3, etc., and y_1, y_2, y_3, etc., then we can also say

$$\frac{y_1}{x_1}=\frac{y_2}{x_2}=\frac{y_3}{x_3}=\ldots\ldots=k$$

or $$y_1=kx_1,\ y_2=kx_2,\ y_3=kx_3, \ldots\ldots$$

b) Inverse proportion

Two sets of readings for the quantities p and V are given in Table M2 (units omitted).

Table M2

p	3	4	6	12
V	4	3	2	1

There is again a one-to-one correspondence between each value of p and the corresponding value of V but when p is doubled V is halved; when p is trebled, V has one-third its previous value; and so on.

We say that V is **inversely proportional** to p, or V **varies inversely** as p, i.e.

$$V \propto \frac{1}{p}$$

Also, the **product** $p \times V$ is always the same (=12 in this case) and we write

$$V=\frac{k}{p} \quad \text{or} \quad pV=k$$

where k is the constant of proportionality or variation and equals 12 in this case.

Using letters for values of p and V we can also say

$$p_1V_1=p_2V_2=p_3V_3=\ldots\ldots=k$$

Graphs

Another useful way of finding the relation between two quantities is by a graph.

a) Straight line graphs

When the readings in Table M1 are used to plot a graph of y against x, a **continuous** line joining the points is **a straight line passing through the origin**, Figure M1. Such a graph shows there is direct proportionality between the quantities plotted, i.e. $y \propto x$. But note that the line must go through the origin.

A graph of p against V using the readings in Table M2 is a curve, Figure M2. However if we plot p against $1/V$ (Table M3) (or V against $1/p$) we get a straight line through the origin, showing that $p \propto 1/V$, Figure M3 (or $V \propto 1/p$).

Table M3

p	V	$1/V$
3	4	0.25
4	3	0.33
6	2	0.50
12	1	1.00

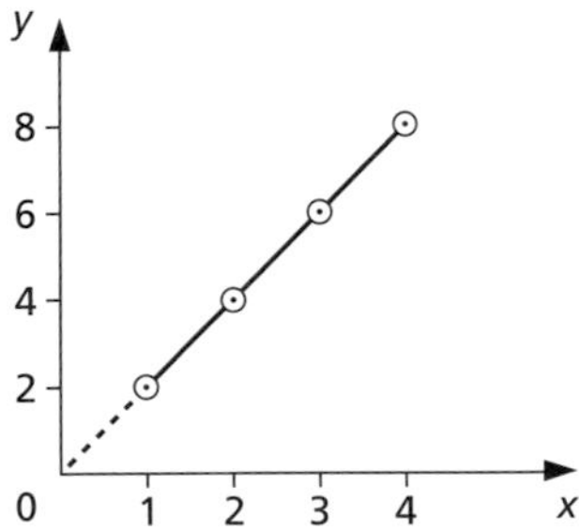

Figure M1

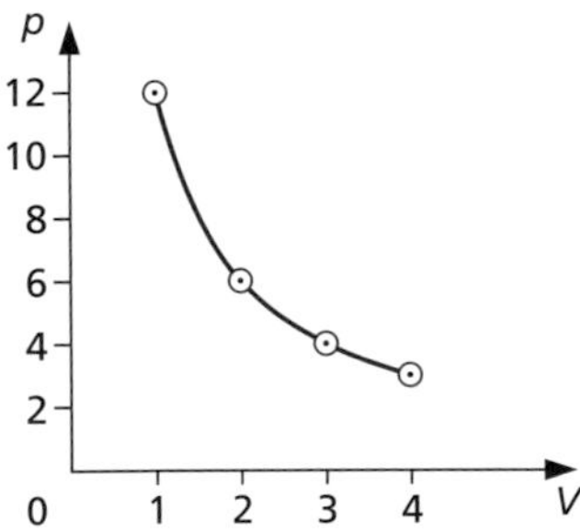

Figure M2

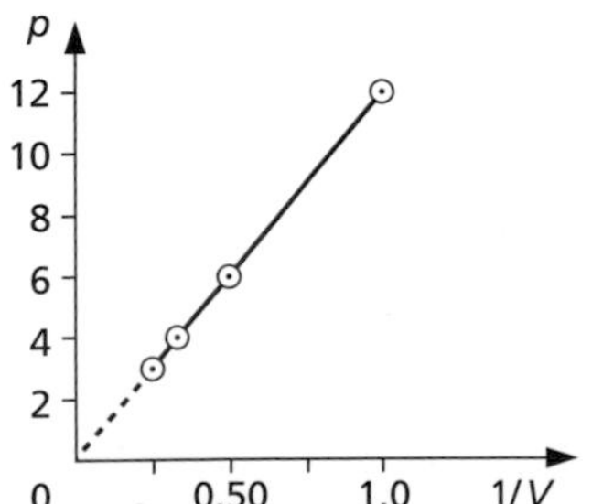

Figure M3

b) Slope or gradient

The slope or gradient of a straight line graph equals the constant of proportionality. In Figure M1, the slope is $y/x = 2$; in Figure M3 it is $p/(1/V) = 12$.

In practice points plotted from actual measurements may not lie exactly on a straight line due to experimental errors. The 'best straight line' is then drawn 'through' them so that they are equally distributed about it. This automatically averages the results. Any points that are well off the line stand out and may be investigated further.

c) Variables

As we have seen, graphs are used to show the relationship between two physical quantities. In an experiment to investigate how potential difference, V, varies with the current, I, a graph can be drawn of V/V values plotted against the values of I/A. This will reveal how the potential difference depends upon the current (see Figure M4).

In the experiment there are two **variables**. The quantity I is varied and the value for V is dependent upon the value for I. So V is called the **dependent variable** and I is called the **independent variable**.

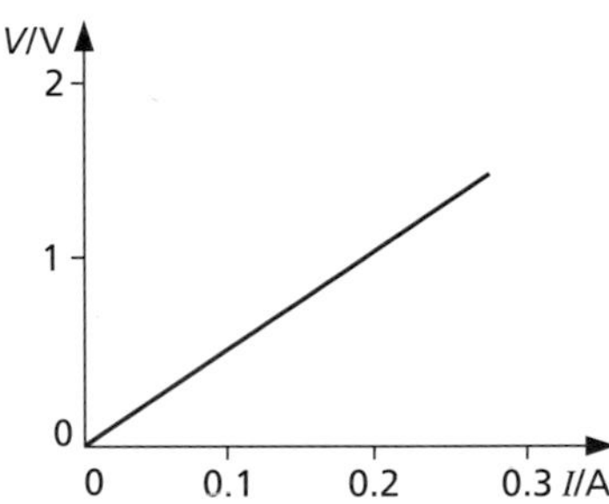

Figure M4

Note that in Figure M4 each axis is labelled with the quantity *and* the unit. Also note that there is a scale along each axis. The statement V/V *against* I/A means that V/V, the dependent variable, is plotted along the y-axis and the independent variable I is plotted along the x-axis (see Figure M5).

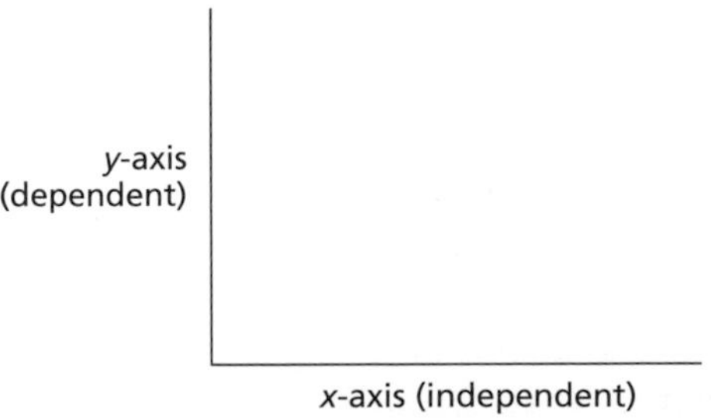

Figure M5

d) Practical points

(i) The axes should be labelled giving the quantities being plotted and their units, e.g. I/A meaning current in amperes.
(ii) If possible the origin of both scales should be on the paper and the scales chosen so that the points are spread out along the graph. It is good practice to draw a large graph.
(iii) The scale should be easy to use. A scale based on multiples of 10 or 5 is ideal. Do not use a scale based on a multiple of 3; such scales are very difficult to use.
(iv) Mark the points ⊙ or ×.

Questions

9 In an experiment different masses were hung from the end of a spring held in a stand and the extensions produced were as shown below.

Mass/g	100	150	200	300	350	500	600
Extension/cm	1.9	3.1	4.0	6.1	6.9	10.0	12.2

a Plot a graph of **extension** along the vertical (y) axis against **mass** along the horizontal (x) axis.
b What is the relation between **extension** and **mass**? Give a reason for your answer.

10 Pairs of readings of the quantities m and v are given below.

m	0.25	1.5	2.5	3.5
v	20	40	56	72

a Plot a graph of m along the vertical axis and v along the horizontal axis.
b Is m directly proportional to v? Explain your answer.
c Use the graph to find v when $m = 1$.

11 The distances s (in metres) travelled by a car at various times t (in seconds) are shown below.

s/m	0	2	8	18	32	50
t/s	0	1	2	3	4	5

Draw graphs of
a s against t,
b s against t^2.
What can you conclude?

Further experimental investigations

Stretching of a rubber band

Set up the equipment as shown in Figure 12.3 (p. 55) but replace the spring with a thick rubber band. Draw up a table in which to record stretching force/N, scale reading/mm, and total extension/mm. Take readings for increasing loads on the hanger.

Plot a graph with stretching force/N along the x-axis and extension/mm along the y-axis. Draw the best straight line through your points; are your results consistent with Hooke's law for all loads?

(If weights and a hanger are not available you could use coins (all similar) in a paper cup instead; in this case the stretching force would be proportional to the number of coins used.)

Toppling

The stability of a body can be investigated using a 1 litre drinks carton or can as shown in Figure E1a. When the carton is tilted so that the centre of mass moves outside the base, the carton will topple over.

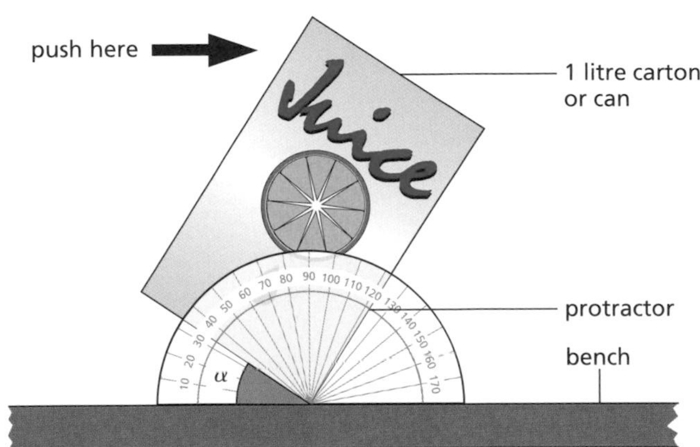

a

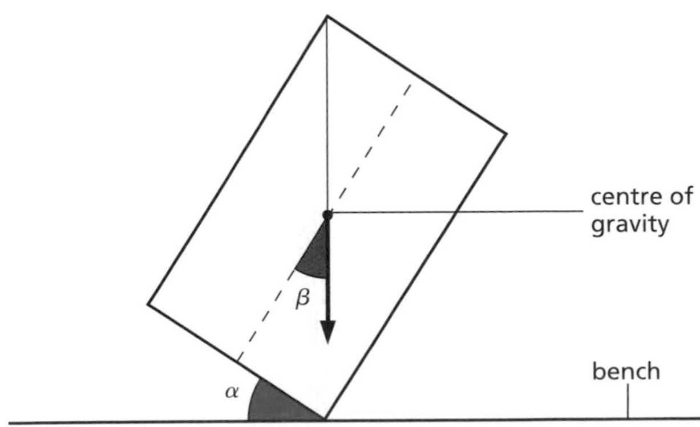

b

Figure E1

(i) Attach a protractor to the bench with Blu-tack. Fill a carton with water and gently push it at the top so that it tilts. Measure the maximum angle, α, that the carton can be tilted through without toppling; repeat your measurement several times and obtain an average value for α.

(ii) Draw a full-size diagram of the face of the carton; mark the centre of mass on the face and measure the angle β between the long side and a diagonal as shown in Figure E1b; how do your values for α and β compare?

(iii) Repeat part (i) with the carton half full, a quarter full and empty. Draw up a table of your results as shown below.

Liquid volume/litres	$\alpha_1/°$	$\alpha_2/°$	$\alpha_3/°$	Average $\alpha/°$
1.0				
0.5				
0.25				
0.0				
0.5 (frozen)				

Where is the centre of mass of an empty carton?

Plot a graph with volume/litres on the y-axis and $\alpha/°$ on the x-axis. What angle of topple would you expect if the carton was one third full of water? How does changing the position of the centre of mass affect the stability of the carton?

(iv) Put a half-full carton in the freezer; when the water is fully frozen repeat part (i); add your results to the table.

Will the carton be more or less stable when the water has melted? How are the centre of mass and the angle of topple changed by freezing the water?

(v) Turn a full carton on its side and repeat steps (i) and (ii). Is the carton more or less stable than when upright? Explain why.

Summarize the factors that influence the stability of a body.

Cooling and evaporation

For this experiment you will need two heat sensors connected to a datalogger and computer. Use some cotton thread to tie a piece of tissue paper loosely over one of the heat sensors. Insert both heat sensors into a beaker of hot water and wait until they reach a constant temperature.

Experiment 1: With the datalogger running, remove the heat sensors from the water and quickly dry the sensor that is not covered by tissue paper. Hang each sensor on a retort stand and allow each to cool to room temperature. Use the computer to record a graph of temperature (on the y-axis) versus time (on the x-axis) for each sensor – these are 'cooling curves'.

Discuss the general shape of the cooling curves – when do the bulbs cool most rapidly? How do the cooling curves differ for the 'wet' compared with the 'dry' heat sensor? Which sensor reaches the lower temperature – can you explain why?

Experiment 2: Repeat the first experiment but this time hang the sensors in a draught to cool. An artificial draught can be produced by an electric cooling fan. Compare the cooling curves recorded by the computer with those obtained when there was no draught (Experiment 1). Comment on how the rate of cooling and the lowest temperature reached have changed for each sensor and try to explain your results.

From your findings, summarize the factors that affect the rate at which an object cools.

(If dataloggers and computers are not available this experiment could be done with mercury thermometers and a 'team' of students to help record temperatures manually every 15 seconds!)

Variation of the resistance of a wire with length

Several different lengths of resistance wire (constantan SWG 34 is suitable) are needed, in addition to the equipment shown in Figure 36.5 (p. 175).

Cut the following lengths (l) of resistance wire: 20 cm, 40 cm, 60 cm, 80 cm and 100 cm. Wind each wire into a coil, ensuring that adjacent turns do not touch if the wire is not insulated. Set up the circuit shown in Figure 36.5 with the shortest coil in position R. (Set the rheostat near the midway position.) Draw up a table in which to record l, I, V and R for each coil. Determine R ($= V/I$) from your readings; repeat the measurements and calculation of R for each coil.

Draw a graph with average R values on the y-axis and l values on the x-axis. Is it consistent with the relation $R = \rho l/A$ (see p. 178)? Calculate the slope of the graph.

Measure the diameter of the constantan wire with a micrometer screw gauge and determine a value for the resistivity, ρ, of the wire.

Practical test questions

1 In this experiment, you are to determine a quantity called the refractive index of the material of a transparent block.

Carry out the following instructions using a ray trace sheet on the pin board supplied and referring to Figure P1.

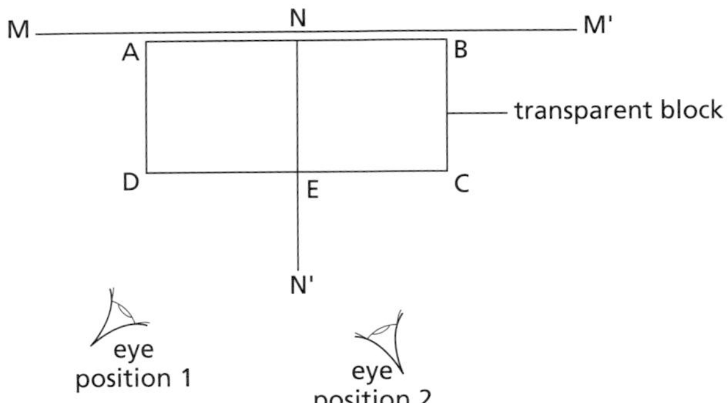

Figure P1

a Place the transparent block flat on the ray trace sheet along the line **MM'** so that the vertical line on the block is at point **N**.
b Draw the outline of the block, **ABCD**, as shown in the diagram.
c Label, with the letter **E**, the point where the line **NN'** cuts the line **CD**.
d With your eye at 'eye position 1' shown in the diagram, look through the block so that you can see the vertical line.
e Push a pin $\mathbf{P_1}$ into the pin board on, or close to, the line **DC**. Place a second pin $\mathbf{P_2}$ between your eye and **the pin** $\mathbf{P_1}$ so that it hides the pin $\mathbf{P_1}$ and the vertical line. Label the positions of $\mathbf{P_1}$ and $\mathbf{P_2}$.
f Remove the pins and the block.
g Draw a line through $\mathbf{P_1}$ and $\mathbf{P_2}$ and continue the line until it meets **NN'**. Label the position **F** where the line meets **NN'**.
h Measure and record the distance **EF**.
i Measure and record the distance **EN**.
j Calculate n, the refractive index of the material of the block using the equation

$$n = \frac{\mathbf{EN}}{\mathbf{EF}}$$

k With your eye at 'eye position 2' shown in Figure P1, look through the block so that you can see the image of the vertical line.
l Repeat the procedure in **e–j**.
m Calculate the average value of the refractive index from the two values that you have obtained.

You will need:
(i) Two optics pins.
(ii) Pin board, 300 mm × 210 mm or larger. A cork mat would be suitable.
(iii) 300 mm rule.
(iv) Transparent glass or Perspex block, approximately 20 × 65 × 115 mm.

Note:
The transparent block is to have a strip of paper taped to one of the longer sides. The strip of paper is to have a thin straight line drawn across it on both sides. The paper is to be positioned so that the line is at the mid-point of the long side of the block and vertical, as shown in Figure P2. The line must be clearly visible when viewed **through** the block.

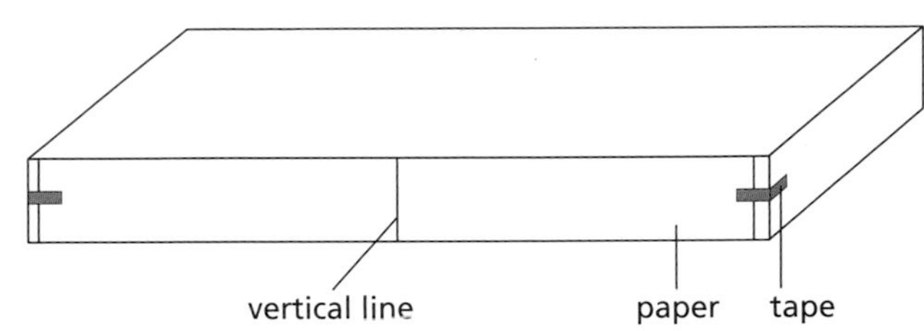

Figure P2

The ray trace sheet should be marked as shown in Figure P3 (suitably sized).

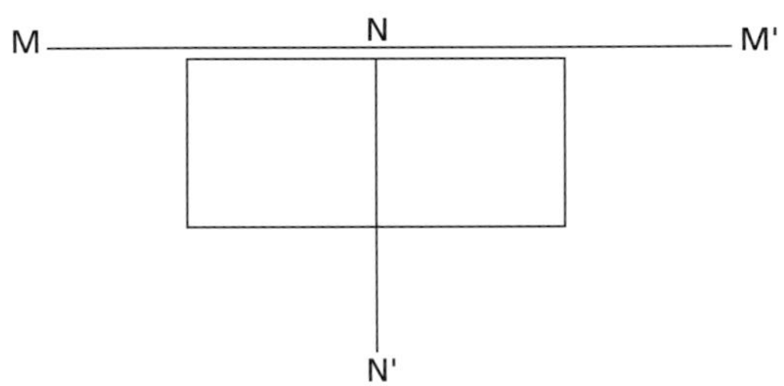

Figure P3

(*UCLES IGCSE Physics Practical Test, May 2001*)

2 In this experiment you are to find the density of a sample of Plasticine.

Carry out the following instructions, referring to Figures P4 and P5.

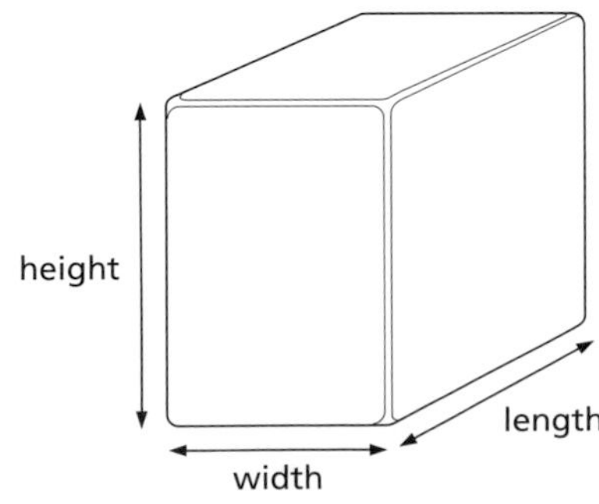

Figure P4

a Using a rule, measure the length, height and width of the block of Plasticine supplied.

b Calculate the volume of the block, using the following equation:

$$\text{volume} = \text{length} \times \text{height} \times \text{width}$$

c Place the metre rule on the pivot so that the 50 cm mark is directly over the pivot, as shown in Figure P5.

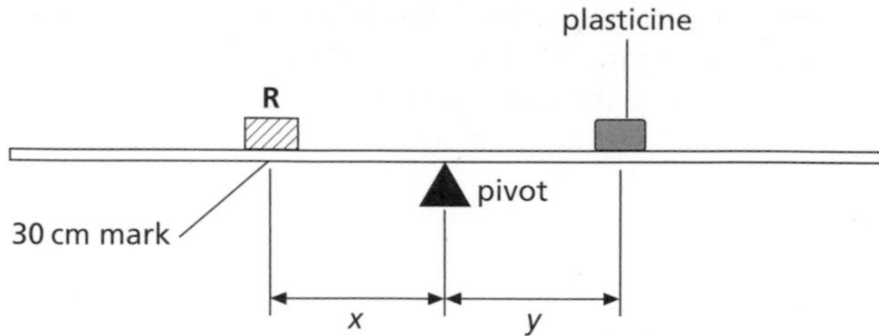

Figure P5

Place the load **R** on the metre rule so that its centre is above the 30 cm mark. Record x, the distance from the 30 cm mark to the pivot. (The first value should be $x = 20.0$ cm.)

Place the Plasticine on the metre rule and move it until the rule balances (do not change the position of **R** or of the pivot). Measure and record y, the distance between the pivot and the centre of mass of the Plasticine.

d Calculate m, the mass of the Plasticine, using the equation:

$$m = Rx/y \qquad \text{where } R = 60\,\text{g}$$

e Repeat steps **c** and **d** using different values of x to obtain a total of three sets of results.

f Calculate and record the average value for m.

g Calculate the density of the Plasticine using the equation:

$$\text{density} = \text{mass/volume}$$

h By means of clearly labelled diagrams and a brief description, explain how you could find the volume of some Plasticine if you were supplied with a measuring cylinder and water. (You are **not** required to carry this out.)

You will need:
(i) Plasticine mass 50 g ± 5 g, moulded into a rectangular shape.
(ii) Metre rule (see note below).
(iii) 300 mm or 500 mm rule.
(iv) Triangular block to act as a pivot for the metre rule. This block is to stand on the bench and should be 3 to 5 cm high.
(v) A 60g mass, labelled **R** (see note 2).

Note:
1. The metre rule should balance on the pivot when the 50 cm mark is directly over the pivot. It may be necessary to tape a small mass (e.g. Plasticine) to one end of the rule to achieve this.
2. The 60 g mass should be suitable for placing on a metre rule in a balancing experiment.

(*UCLES IGCSE Physics Practical Test, May 2000*)

3 In this experiment you are to investigate the effect on a thermometer of blackening the bulb.

Two thermometers, **A** and **B** (with a blackened bulb), will have been set up for you. You should not change the position of the thermometers or the lamp. Each one is 1 cm from the lamp. Do not remove the screen from its position in front of the lamp – it is to prevent you being dazzled as you take temperature readings.

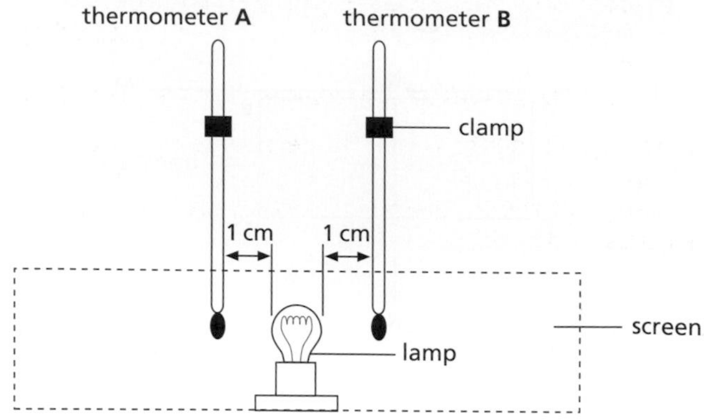

Figure P6

Carry out the following instructions referring to Figure P6. Record all temperatures to the nearest 1°C in a table like the one below.

a Measure room temperature with each of the thermometers and record the two values in the table for time 0 s.

b Switch on the lamp. Record the temperature shown on each thermometer after 30 seconds.

c Continue recording the temperature readings every 30 seconds for a total of 180 seconds.

d Suggest a conclusion for this experiment, justifying your conclusion by reference to your table of readings.

e Using the readings for thermometer **B** only, plot the graph of temperature/°C (y-axis) against time/s (x-axis). Draw the best-fit curve.

Time/s	Thermometer **A** temperature/°C	Thermometer **B** temperature/°C
0		
30		
60		
90		
120		
150		
180		

You will need:
(i) A thermometer, −10 to 110 °C, labelled **A**.
(ii) A thermometer, −10 to 110 °C (see note 1 below), labelled **B**.
(iii) 12 V, 24 W lamp (a car headlamp bulb is suitable).
(iv) Power supply and leads for the lamp.
(v) 2 clamps, 2 bosses, 2 stands.
(vi) Screen (stiff card or hardboard).
(vii) Stopclock or stopwatch.

Notes:

1. This thermometer is to have a blackened bulb. Powder paint, emulsion paint or poster paint are suitable for this. The two thermometers should otherwise be as similar as possible. Thermometer **B** should give a significantly greater temperature rise compared to **A** when the lamp is switched on.
2. The apparatus is to be set up as shown in Figure P6. The screen must be of suitable size and positioned so that the candidate can take readings from the thermometers without being dazzled by the lamp.

(*UCLES IGCSE Physics Practical Test, May 2000*)

4 In this experiment, you are to determine the power dissipated in lamps.

Carry out the following instructions, referring to Figures P7 and P8.

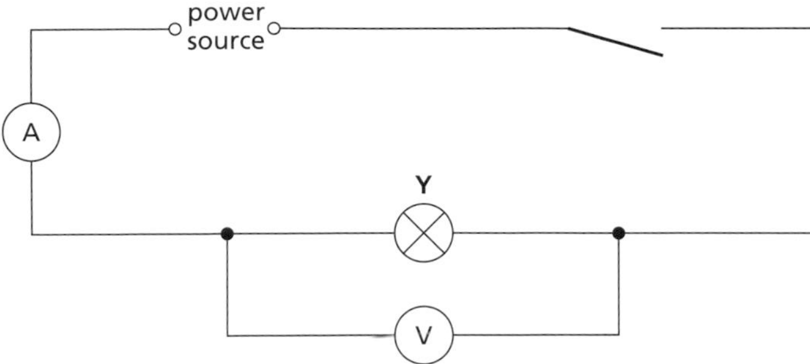

Figure P7

The circuit shown in Figure P7 will have been set up for you, using lamp **Y**.

a Switch on. Measure and record in a table like the one below the current *I* through the lamp and the p.d. *V* across the lamp. Switch off.

b Calculate the power *P* of the lamp, using the equation

$$P = VI$$

Record this value of *P* in your table.

c Replace lamp **Y** with lamp **Z**. Repeat the procedure in **a** and **b**, recording the corresponding values of *I*, *V* and *P* in your table.

d Complete the column headings in your table by inserting the appropriate units in each of the *I*, *V* and *P* columns.

Lamps	*I*/	*V*/	*P*/
Y			
Z			
Y and **Z** in series			
Y and **Z** in parallel			

e Set up the circuit with lamps **Y** and **Z** connected in series, as shown in Figure P8.

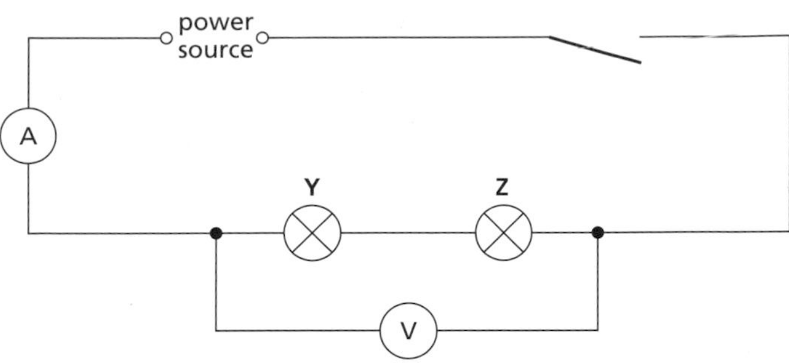

Figure P8

f Switch on. Measure and record the current *I* and the p.d. *V* across both lamps. Switch off.

g Calculate the power *P* of the two lamps when connected in series, using the equation

$$P = VI$$

Record this value of *P* in your table.

h Draw a circuit diagram showing
- the two lamps connected in parallel,
- the voltmeter connected to measure the p.d. across the lamps,
- the ammeter connected to measure the total current passing through the two lamps.

i Set up the circuit described in **h**.

j Switch on. Measure and record the current *I* and the p.d. *V*.

k Calculate the power *P* of the two lamps using the equation

$$P = VI$$

Record this value of *P* in your table.

l If the power source for each of the circuits with two lamps were a dry cell and you left the circuits switched on until the cells completely ran down, which circuit would stop working first: the series circuit or the parallel circuit? Justify your answer by reference to your results.

You will need:
(i) Voltage source of approximately 1.5–2.5 V.
(ii) Two 2.5 V, 0.2 A lamps (one labelled '**Y**', the other labelled '**Z**') in suitable holders.
(iii) Ammeter capable of reading up to 1.0 A with minimum precision of 0.05 A.
(iv) Voltmeter capable of measuring the supply p.d. with a minimum precision of 0.1 V.
(v) 8 connecting leads.
(vi) Switch. (The switch can be an integral part of the power supply. In this case only 7 connecting leads are required.)

(*UCLES IGCSE Physics Practical Test, May 2001*)

Alternative to practical test questions

1 The IGCSE class is investigating reflection in a plane mirror. Figure P9 shows a ray diagram that a student is constructing.

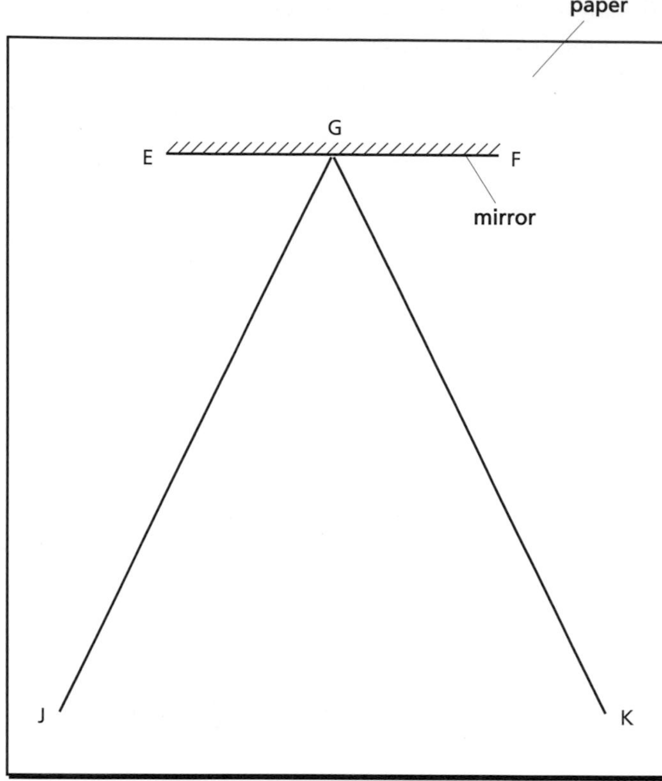

Figure P9

a (i) Copy the diagram and draw a normal GH to line EF.
(ii) Mark a point A on line GJ so that the distance AG is 11.5 cm.
(iii) Measure the angle of incidence *i* between line GJ and the normal.

b The student pushes two pins into the paper on line GJ, one at point A, and the other at a point B nearer to the mirror. He views the images of the pins from the direction indicated in the diagram. He then pushes in two pins on line GK between his eye and the mirror so that these two pins and the images of the pins on line GJ appear exactly one behind the other.
(i) On your diagram, mark suitable positions for the pins on lines GJ and GK. Label the marks with letters B, C and D.
(ii) To obtain an accurate result for this experiment, would you view the tops, bases or central parts of the pins when lining them up? Give a reason for your answer.

(*UCLES IGCSE Physics Alternative to Practical, Nov 2006*)

2 The IGCSE class is investigating the swing of a loaded metre rule. The arrangement of the apparatus is shown in Figure P10.

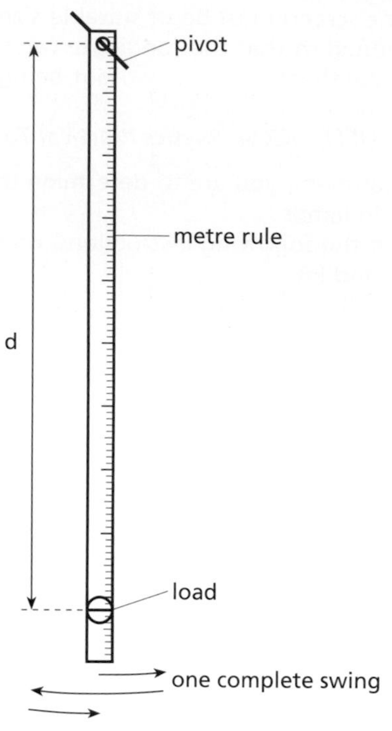

Figure P10

The load is attached to the metre rule so that its centre is 90.0 cm from the pivot. The rule is displaced a small distance to one side and allowed to swing. The time *t* taken for 10 complete swings is recorded. This is repeated using different values of the distance *d*. The readings are shown in the table.

d/	*t*/	*T*/
90.0	18.35	
85.0	17.87	
80.0	17.53	
75.0	17.06	
70.0	16.72	

a Copy the table and complete the column headings.

b Calculate the period *T* for each value of *d*. The period is the time taken for one complete swing. Enter the values in your table.

c Plot a graph of *T*/s (*y*-axis) against *d*/cm (*x*-axis). Start the *x*-axis at $d = 70.0$ cm and the *y*-axis at a suitable value of *T*/s to make best use of the graph paper.

d A student suggests that *T* is proportional to *d*. State whether or not the results support this suggestion and give a reason for your answer.

e Explain why the student takes the time for 10 swings and then calculates the time for one swing (the period), rather than just measuring the time for one swing.

(*UCLES IGCSE Physics Alternative to Practical, Nov 2006*)

3 Some students were asked to carry out a simple experiment to compare different heat insulation materials.

a One student measured the temperature of hot water in insulated beakers (all the same size), waited for 5 minutes for the water to cool and then measured the temperatures again. Figure P11 shows how one tudent recorded the results.

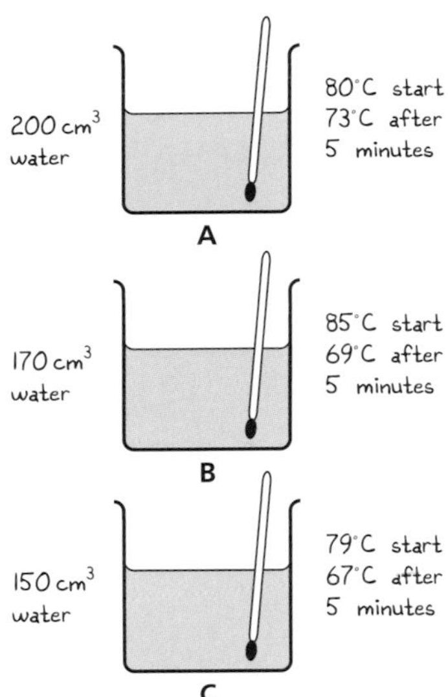

Figure P11

(i) Calculate the temperature fall for each beaker, **A**, **B** and **C**.

If you had **only** these results and no information about the way these results had been obtained, which beaker would **appear** to be the best insulated?

(ii) Suggest a simple, practical way to overcome the problem of heat loss, by evaporation and convection, from the surface of the water in the beaker.

(iii) Look at Figure P11 again. Suggest **one** further improvement that you would make to improve the reliability of the experiment.

b Another student carried out a similar experiment, with proper control of the variables, and took temperature readings every 4 minutes. Room temperature during the experiment was 19 °C. He plotted a graph of temperature against time to show the cooling of the water in each beaker. Figure P12 shows the graph obtained.

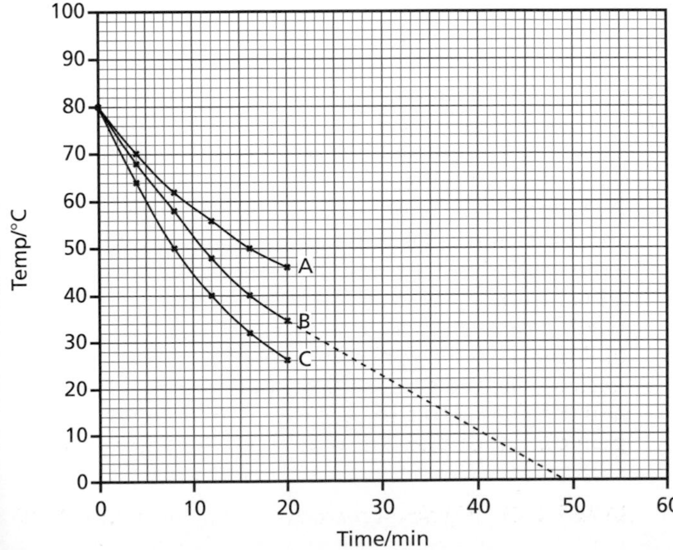

Figure P12

(i) From the graph, which beaker, **A**, **B** or **C**, was best insulated?

(ii) The student extended graph line **B** with a dotted line as shown. Explain why this does **not** show a realistic continuation of the cooling of the water.

(iii) Copy the graph and extend line **A** to show a realistic result up to 60 minutes.

(UCLES IGCSE Physics Alternative to Practical, May 2001)

4 Figure P13 shows a circuit in which three resistors are connected to a d.c. power supply.

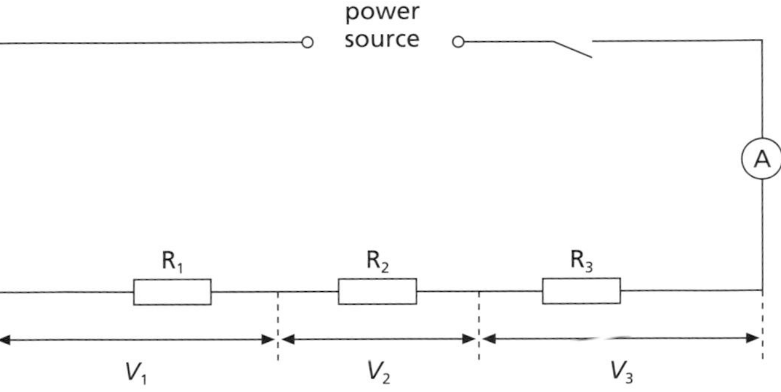

Figure P13

A 0–1 V voltmeter was used to measure the potential differences V_1, V_2 and V_3. Figure P14 represents the face of the voltmeter when reading these values.

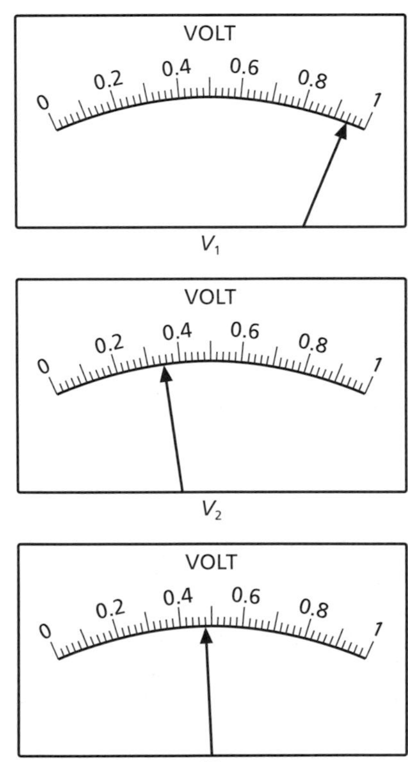

Figure P14

a (i) Record the reading of each potential difference, shown in Figure P14, in a copy of the table below.

potential difference	V/V
V_1	
V_2	
V_3	

(ii) Using the values in your table, predict the voltmeter reading when a 0–5 V voltmeter is connected across all three resistors together.
(iii) The current, I, in the circuit is 0.35 A. Using the values in your table and the equation $R = V/I$, calculate the resistances R_1, R_2 and R_3.

b Draw a circuit diagram showing the same components as in Figure P13 but with
- the three resistors in parallel,
- a voltmeter connected to record the potential difference across all three resistors,
- the ammeter connected to record the current through R_1 only.

(*UCLES IGCSE Physics Alternative to Practical, May 2001*)

5 When investigating the magnetic field due to a bar magnet, a student places the magnet on a sheet of paper as shown in Figure P15. The edge of the paper is placed so that it is parallel to the direction of the Earth's magnetic field. The bar magnet is then placed as shown so that it is at right angles to the direction of the Earth's magnetic field. (In Figure P15, the lines OX and OY are perpendicular to each other.) A small plotting compass is used to investigate the magnetic field.

a It is found that there are positions where the small magnet in the **plotting compass** points so that it is parallel to the line OX. Some of these positions are located and are labelled A, B, C, D, E, F, G and H, as shown on Figure P15. The positions shown also lie on straight lines that come from the centre of the bar magnet.

Describe how you would locate the position labelled A. Your answer should explain
(i) what you would do to help you judge when the small magnet in the plotting compass is parallel to OX,
(ii) how you would ensure that the small magnet of the plotting compass is not sticking,
(iii) what you would do so as to mark the point A on the radial line,
(iv) how you would avoid making a parallax error when locating the point A.

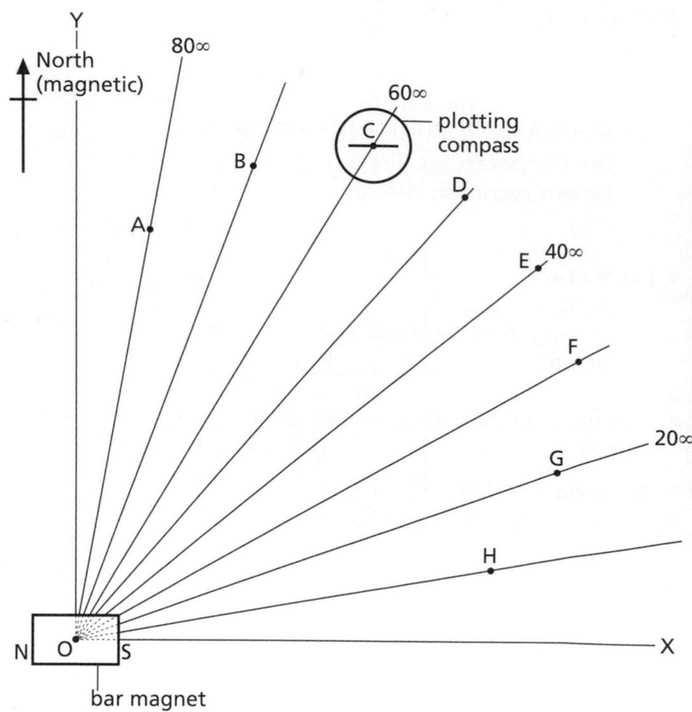

Figure P15

b The plotting compass is at point C as shown in Figure P15.
(i) Indicate which end of the small magnet of the plotting compass is a North pole.
(ii) The compass is at point C. It is then moved along the radial line so that it is closer to the bar magnet. Describe and explain what happens to the small magnet of the plotting compass.

(*UCLES IGCSE Physics Alternative to Practical, May 2000*)

Answers

Higher level questions are marked with *.

Light and sight

1 Light rays

1 Larger, less bright
2 a 4 images b Brighter but blurred
3 C
4 Before; sound travels slower than light

2 Reflection of light

1 a 40° c 40°, 50°, 50° d Parallel
2 A
3 Top half

3 Plane mirrors

1 B
2 D
3 4 m towards mirror

4 Refraction of light

3 250 000 km/s
4 C

5 Total internal reflection

1 a Angle of incidence = 0
b Angle of incidence > critical angle
3 Periscope, binoculars

6 Lenses

1 Parallel
2 a Converging
c Image 9 cm from lens, 3 cm high
3 Distance from lens:
a > 2F b 2F c between F and 2F
d < F
4 Towards
5 a 4 cm
b 8 cm behind lens, virtual, $m = 2$

Additional questions

1 B
2 b parallel to the ray incident at X
4 a Ray passes into air and is refracted away from the normal
b Total internal reflection occurs in water
5 E
6 A
8 C

9* b (i) 14 cm (ii) 10 cm (iii) 4 cm
c Same size, virtual, upright, laterally inverted
10* a (i) 40°
(ii) increase angle of incidence at Q
b 1.556
11* a (iii) 45°, 30°
b 1.41
12* A: converging $f = 10$ cm
B: converging $f = 5$ cm

Waves and sound

7 Mechanical waves

1 a 1 cm b 1 Hz c 1 cm/s
2 A, C
3 a Speed of ripple depends on depth of water
b AB since ripples travel more slowly towards it, therefore water shallower in this direction
4 a Trough
b (i) 3.5 mm (ii) 17 mm/s
(iii) 5 Hz

8 Electromagnetic radiation

1 a 0.7 μm b 0.4 μm
2 a B b D
3 a Ultraviolet
b Microwaves
c Gamma rays
d Infrared
e Infrared/microwaves
f X-rays
4 a 3 m b 2×10^{-4} s

9 Sound waves

1 1650 m (about 1 mile)
2 a 2 × 160 = 320 m/s
b 240/(3/4) = 320 m/s
c 320 m
3 a Reflection, refraction, diffraction, interference
b Vibrations are perpendicular to rather than along the direction of travel of the wave; longitudinal
4* b 132 Hz
c 198 m
5 b (i) 1.0 m (ii) 2.0 m

Additional questions

1 a 30/6 = 5 cm b 4 Hz c $v = f\lambda$
d 20 cm/s
3* (i) X down (ii) Y up (iii) Z up
4 E
5 a (i) A and G
(ii) C and B
b Number of complete waves generated per second
c 5 Hz
6* a 20°
b Angle of incidence is greater than critical angle
c (i) 41.8°
(ii) angle of incidence is less than critical angle
d (i) single frequency
(ii) 2×10^8 m/s
(iii) $v_{air} = 1.50\ v_{glass}$
7 B
8* a Direction of travel of the wave
b (i) position of air molecules at a given instant
(ii) rarefaction
(iii) compression
c Oscillates about R in the same direction as that in which the wave is travelling
d 2.3 cm

Matter and measurements

10 Measurements

1 a 10 b 40 c 5 d 67 e 1000
2 a 3.00 b 5.50 c 8.70 d 0.43
e 0.1
3 a 1×10^5; 3.5×10^3; 4.28×10^8; 5.04×10^2; 2.7056×10^4
b 1000; 2 000 000; 69 000; 134; 1 000 000 000
4 a 1×10^{-3}; 7×10^{-5}; 1×10^{-7}; 5×10^{-5}
b 5×10^{-1}; 8.4×10^{-2}; 3.6×10^{-4}; 1.04×10^{-3}
5 10 mm
6 a two b three c four d two
7 24 cm^3
8 40 cm^3; 5
9 80
10 a 250 cm^3 b 72 cm^3
11 a 53.3 mm b 95.8 mm
12 a 2.31 mm b 14.97 mm

11 Density

1 a (i) 0.5 g (ii) 1 g (iii) 5 g
b (i) 10 g/cm^3 (ii) 3 kg/m^3
c (i) 2.0 cm^3 (ii) 5.0 cm^3
2 a 8.0 g/cm^3 b 8.0×10^3 kg/m^3
3 15 000 kg
4 130 kg
5 1.1 g/cm^3

12 Weight and stretching

1 a 1 N b 50 N c 0.50 N
2 a 120 N b 20 N
3 a 2000 N/m b 50 N/m
4 A

Additional questions

1 a Metre, kilogram, second
b Different number of significant figures
c (i) πr^2 (ii) $\frac{4}{3}\pi r^3$ (iii) $\pi r^2 h$

2 a 0.46 kg
b (i) 15 cm^3
(ii) 515 cm^3
(iii) 0.89 g/cm^3
3 a C b C
c 30 minutes d 6 minutes
4 b 8 cm c 2 cm
5 a 8.5 cm
b 19.0 cm
c 10.5 cm
6* c (i) 6 mm (ii) 13 mm

Forces and resources

13 Moments and levers

1 E
2 (i) C (ii) A (iii) B
3* a (i) 1.5 N (iv) 0.75 N
b (i) increases
(ii) position of books, strength of rope

14 Centres of mass

2 a B b A c C
3 Tips to right
4* a No resultant force or moment
b Take moments about centre of rule
c 0.5 N, downwards
5* a Narrow track and short wheel base; passengers raise centre of mass
b 6000 N m
c Centre of mass raised so that weight lies outside wheel base at smaller angles of tilt

15 Adding forces

1 40 N
2 50 N
3 25 N
4 50 N at an angle of 53° to the 30 N force

16 Energy transfer

1 a Electrical to sound
b Sound to electrical
c k.e. to p.e.
d Electrical to light (and heat)
e Chemical to electrical to light and heat
2 A chemical; B heat; C k.e.; D electrical
3 180 J
4 1.5×10^5 J
5 a 150 J b 150 J c 10 W
6 500 W
7 a $(300/1000) \times 100 = 30\%$
b Heat c Warms surroundings

17 Energy sources

1 a 2% b Water
c Cannot be used up d Solar, wind
e All energy ends up as heat which is difficult to use and there is only a limited supply of non-renewable sources
2 Renewable, non-polluting (i.e. no CO_2, SO_2 or dangerous waste), low initial building cost of station to house energy converters, low running costs, high energy density, reliable, allows output to be readily adjusted to varying energy demands
4 b 0.25

18 Pressure and liquid pressure

1 a (i) 25 Pa (ii) 0.50 Pa
(iii) 100 Pa
b 30 N
2 a 100 Pa b 200 N
3 a A liquid is nearly incompressible
b A liquid transfers the pressure applied to it
4 1 150 000 Pa (1.15×10^6 Pa) (ignoring air pressure)
5 a (i) D
(ii) has largest area in contact with floor
b AB, BC, CD, area – *decrease*
mass, weight – *stay the same*
density, pressure – *increase*
6 a vacuum
b atmospheric pressure
c 740 mmHg
d becomes less; atmospheric pressure lower

Additional questions

1 a A yes, B yes, C no
b Convert to type B
2 a Turning effect
b (i) to balance beam and lift load
(ii) smaller than
3 a (i) W_1 (ii) towards pivot
b Between W and W_1 but nearer to W
4 a 7 N b 13 N
5 a Electricity transferred to k.e. and heat
b Electricity transferred to heat
c Electricity transferred to sound
6 3.5 kW
7 a (i) BC (ii) p.e. increased
b (i) greater than
(ii) time taken decreased
8 a (i) potential
(ii) kinetic
(iii) turbine; electrical
b As steam in coal-fired power station; in tidal power stations
9* a 20 000 N/m^2
b 0.0025 m^2
c p.e. converted to k.e. (and heat)
10 E
11 B
12 a (ii) pressure
b (i) 5 cm (ii) pressure difference
(iii) gas pressure is less than air pressure
(iv) height difference would decrease
(v) height difference would be greater

Motion and energy

19 Speed, velocity and acceleration

1 a 20 m/s b 6.25 m/s
2 a (i) Ahmed
(ii) longest time
b (i) speed = distance/time
(ii) 8 m/s
3 a 15 m/s b 900 m
4 2 m/s^2
5 a Uniform acceleration
b 75 cm/s^2
6 50 s
7 a 6 m/s b 14 m/s
8 4 s
9 a 1 s
b (i) 10 $cm/tentick^2$
(ii) 50 cm/s per tentick
(iii) 250 cm/s^2
c 0

20 Graphs and equations

1 a 60 km b 5 hours c 12 km/h
d 2 e $1\frac{1}{2}$ hours
f 60 km/$3\frac{1}{2}$ h = 17 km/h
g Steepest line: EF
2 a 100 m b 20 m/s c Slows down
3 a 5/4 m/s^2 b (i) 10 m (ii) 45 m
c 22 s
4 a OP accelerating, PQ greater acceleration, QR constant speed, RS deceleration
b O and S c 6 m/s
d 70 s e find area under OPQRS

21 Falling bodies

1 a (i) 10 m/s (ii) 20 m/s
(iii) 30 m/s (iv) 50 m/s
b (i) 5 m (ii) 20 m (iii) 45 m
(iv) 125 m
2 3 s; 45 m

22 Force and acceleration

1 D
2 20 N
3 a 5000 N b 15 m/s^2
4 a 4 m/s^2 b 2 N
5 a 0.5 m/s^2 b 2.5 m/s c 25 m
6 a 1000 N b 160 N
7 a 5000 N b 20 000 N; 40 m/s^2
8 a (i) weight (ii) air resistance
b Falls at constant velocity (terminal velocity)
10 The rock *moves to the left* and *accelerates*

23 Kinetic and potential energy

1 a 2 J b 160 J c 100 000 = 10^5 J
2 a 20 m/s b (i) 150 J (ii) 300 J
3 a 1.8 J b 1.8 J c 6 m/s d 1.25 J
e 5 m/s
4 3.5×10^9 W = 3500 MW

5* a 70 J
b 12 m/s
c some p.e. lost as heat and sound
6* a 7 m
b (i) fatigue, poor eyesight
(ii) A increases; B none; C increases
c Less friction
d (i) -6.25 m/s^2 (ii) 5625 N
(iii) 72 m

24 Circular motion

1 Force is greater than string can bear
2 a Sideways friction between tyres and road
b (i) larger (ii) smaller
(iii) larger
3 Slicks allow greater speed in dry conditions but in wet conditions treads provide frictional force to prevent skidding
4 5000 s (83 min)

Additional questions

1 a OA, BC: accelerating;
DE: decelerating;
AB, CD: uniform velocity
b OA: $a = +80$ km/h^2;
AB: $v = 80$ km/h;
BC: $a = +40$ km/h^2;
CD: $v = 100$ km/h;
DE: $a = -200$ km/h^2
c OA 40 km; AB 160 km;
BC (5 + 40) = 45 km;
CD 100 km; DE 25 km
d 370 km
e 74 km/h
2 a Uniform velocity
b 600 m
c 20 m/s
3 A
4 E
5 a 36 minutes
b Mr B arrives 6 minutes after Mr A
6* b (i) 4.5 s
(ii) 1. -10 m/s^2 2. 9 s
(iii) -45 m/s
7 a B, less air resistance
b (i) X caused by gravity
(ii) Y caused by air resistance
(iii) speed of sky diver will go up
(iv) X stays the same
c Y increases
8* a (i) constant acceleration
(ii) constant velocity
b 20 s
d (i) yes
(ii) no
e 120 m
9 b (i) constant acceleration at 3.5 m/s^2
(ii) 112 m
c Kinetic energy → heat energy
10* a (i) 12 000 J
(ii) 11 m/s
11* c 8 m d $s \propto v^2$
e Driver slows down and thinking time is less
f Braking distance would be less

Heat and energy

25 Molecules

1 B
2 a Air is readily compressed
b Steel is not easily compressed

26 Thermometers

1 a 1530 °C b 19 °C c 0 °C
d -12 °C e 37 °C
2 C
3 a Property must change continuously with temperature
b Volume of a liquid, resistance, pressure of a gas
c (i) platinum resistance
(ii) thermocouple (iii) alcohol

27 Expansion of solids, liquids and gases

2 Aluminium

28 The gas laws

1 a level down in tube
b pressure increases with temperature
c slow, affected by changes in air pressure, small temperature range, no scale
2* a (i) random (ii) hit and rebound
b (i) increase (ii) increase
c (i) random, fast moving in gas; vibration in solid
(ii) long way apart in gas, very close in solid
3 a 15 cm^3 b 6 cm^3
4* b (i) P proportional to $1/V$
c 196

29 Specific heat capacity

1 15 000 J, 15 000 J/°C
2 A = 2000 J/(kg °C); B = 200 J/(kg °C); C = 1000 J/(kg °C)
3 a Steel b 56 000 J
4 Specific heat capacity of jam is higher than that of pastry so it cools more slowly

30 Specific latent heat

1 a 3400 J b 6800 J
2 a $5 \times 340 + 5 \times 4.2 \times 50 = 2750$ J
b 1700 J
3 680 s
4 a 0 °C b 45 g
5 a 9200 J b 25 100 J
6 157 g
7 a Ice has a high specific latent heat of fusion
b Water has a high specific latent heat of vaporization
8 Heat drawn from the water when it evaporates
9 Heat drawn from the milk when the water evaporates
10* a (i) to measure heat supplied from the surroundings
(ii) 440 J/g
(iii) not all the heat goes to melting the ice
b (i) no change
(ii) increase p.e. of molecules
11* a (i) liquid (ii) liquid
b (i) 440 °C
c (i) constant (ii) p.e. increases

31 Conduction and convection

3 a If small amounts of hot water are to be drawn off frequently it may not be necessary to heat the whole tank
b If large amounts of hot water are needed it will be necessary to heat the whole tank

32 Radiation

1 a (i) radiation (ii) convection
(iii) conduction
b (i) black
(ii) black surfaces are good absorbers of radiation; shiny surfaces are poor absorbers of radiation
4* a Hot air rises at A so cooler air is drawn in at C
b (i) radiation
(ii) black surfaces are good emitters of radiation
5* a (i) vacuum
(ii) no medium is present to transfer heat by conduction or convection
b (i) silvered inner surfaces
(ii) radiation reflected

Additional questions

1 a (i) D (ii) B
b Collisions with gas molecules
c (i) faster (ii) increases
2* b Each division on the scale is the same size and corresponds to the same change in temperature
c Reduce the diameter of the capillary
d Quicker response time, larger temperature range
3 B
4 A
5 a Pressure arises from the collision of gas molecules with the sides of the container
b (i) speed increases
(ii) pressure increases as collisions with the walls are more frequent
(iii) they will absorb radiation, heat up and may explode
6 1200 J
7* a 5000 J/(kg °C)

b (ii) more heat is needed to raise the temperature of the water in the pool when heat is lost to its surroundings
8* a (i) 20 °C; 15 °C
(ii) more heat lost at higher temperatures
b 4.2 J/g °C
9 Metal is a better conductor of heat than rubber
11 a Convection – hot air rises
b Conducton – carpet is a good insulator
c Radiation – electromagnetic waves can travel in a vacuum
d Conduction – snow is a good insulator; radiation – white snow reflects heat back into the hole; rate of cooling is reduced if person curls up so that her surface area/volume ratio is smaller
12* a (i) conduction
(ii) free electrons carry energy from one atom to the next
b Four surfaces face one heat source at equal distances for the same time; thermometer behind each surface

Electricity

33 Static electricity
1 D
2 Electrons are transferred from the cloth to the polythene
3 a (i) so particles separate
(ii) to attract paint and help it stick
b To discharge aircraft

34 Electric current
1 a 5 C b 50 C c 1500 C
2 a 5 A b 0.5 A c 2 A
4 B
5 C
6 All read 0.25 A

35 Potential difference
1 a 12 J b 60 J c 240 J
2 a 6 V
b (i) 2 J (ii) 6 J
3 B
4 b Very bright
c Normal brightness
d No light
e Brighter than normal
f Normal brightness
5 a 6 V b 360 J
6 $x = 18, y = 2, z = 8$

36 Resistance
1 3 Ω
2 20 V
3 C
4 A = 3 V; B = 3 V; C = 6 V
5 2 Ω
6 a 15 Ω b 1.5 Ω
7 D
8* b (ii) 2 A (iii) 5 Ω
c (i) increases

37 Capacitors
2 a (i) maximum (ii) zero
b (i) maximum (ii) zero

38 Electric power
1 a 100 J b 500 J c 6000 J
2 a 24 W b 3 J/s
3 C
4 2.99 kW
5 Fuse is in live wire in **a** but not in **b**
7 a 3 A b 13 A c 13 A
8 40 p
9* b (ii) 500 ms

39 Electronic systems
1 b L_1 lights, L_2 does not
c L_1 and L_2 light
d L_1 lights, L_2 does not
2 a $V_1 = V_2 = 3$ V
b $V_1 = 1$ $V_2 = 5$ V
c $V_1 = 4$ V, $V_2 = 2$ V
4* a A: resistor, B: LDR C: transistor D: lamp
b C
c resistance of LDR low in light, high in dark: when $V_{be} > 0.6$ V, transistor conducts and lamp lights

40 Digital electronics
1 A AND B OR C NAND D NOR
2 A OR B NOT C NAND D NOR E AND
3 a (i) NOT (ii) $\begin{smallmatrix}01\\10\end{smallmatrix}$
b Switch 0; gate output high
5 a (i) OR (ii) 0, 1, 1, 1
b (i) AND gates
(ii) power off or door open or water level low
6 a (i) thermistor
b (ii) analogue; smooth
c (i) 700 Ω (ii) 2100 Ω (iii) 1 V
(iv) output remains high as input to NOT is less than 1 V
7* a P diode; Q transistor; R_1 LDR
b (i) 4.8 V
(ii) output from the NOT gate is low
c 0.06 V

Additional questions
1 C
2 b (i) repel
(ii) positively charged glass, uncharged copper
3 C
4 B
5 a (ii) 2 Ω
6 B
7 a 2 Ω b 2/3 A c 1/3 A
8 4/3 Ω; 2 A
9 a 60 Ω
b (i) p.d./current
(ii) 0.1 A
c 4 V d 4 V
10* b (i) 1.6 A (ii) 2.5 Ω
c Slope increases
11* 30 Ω
12 a (i) 2 kW (ii) 60 W (iii) 850 W
b 4 A
13 a Ceiling-mounted pull-cord switch
b Remote contact
14 a A
b B
15 a Electric charge
b Voltage rises slowly to a maximum value
c Timer to control the length of time a lamp stays on
16* a 100 joules of electrical energy is transferred to other forms of energy each second; 220 joules of electrical energy is transferred to other forms of energy when 1 coulomb passes through the lamp
b $P = IV$; $I = 100/220 = 0.45$ A
c $V = IR$; $R = 220/0.45 = 484\ \Omega$
d (ii) diode
17* b (i) low, OFF
(ii) low, OFF
c (ii) no change

Electromagnetic effects

41 Magnetic fields
1 C
3 a (i) S–N; N–S (ii) magnet 2
b (i) becomes magnetized
(ii) easily magnetized
(iii) to slide easily
(iv) steel paper clip (becomes magnetized and is attracted to fixed iron tip; aluminium is not magnetized)

42 Electromagnets
1 a N b east
2 S
3 a (i) 4.5 N
(ii) increase number of turns of wire
b (i) 1 A (ii) no

43 Electric motors
1 E
2 Clockwise
3 E

44 Electric meters
3 a 0–5 V, 0–10 V
b 0.1 V c 0–5 V
d Above the 4
e Parallax error introduced

45 Generators
1 a (i) zero (ii) negative

46 Transformers
2 B

3 **a** 24 **b** 1.9 A

4* **a** (i) increase number of turns on the secondary coil
(ii) core is divided into sheets which are insulated from each other
b (i) voltage can be easily changed
(ii) used to step voltages up or down
(iii) to avoid contact with high voltage
c (i) $N_s/N_p = V_s/V_p$ (ii) 4000
d (i) resistance of cable decreases
(ii) radiation

Additional questions

1 **c** (i) L, N, O
(ii) proportional

2 **a** To complete the circuits to the battery negative
b One contains the starter switch and relay coil; the other contains the relay contacts and starter motor
c Carries much larger current to starter motor
d Allows wires to starter switch to be thin since they only carry the small current needed to energize the relay

4 B

7* **a** (i) magnetic field formed in the coil constantly changes
(ii) changing magnetic field is transferred to the secondary coil
(iii) e.m.f. induced by changing magnetic field cutting secondary coil
b $N_s > N_p$
c magnetic field not transferred to secondary coil
d 0.2 A

Electrons and atoms

47 Electrons

1 **a** A −ve, B +ve
b Down

2 **a** B
b (iii) change value of high voltage
(iv) no beam, electrons repelled by anode
c (i) +12 V (ii) −12 V

48 Radioactivity

1 **a** α **b** γ **c** β **d** γ
e α **f** α **g** β^- **h** γ

2 **a** Cosmic rays, radon gas
b Lift with forceps, keep away from eyes
c 120, 60, 30, 15, 0, 0
e (i) absorbed (ii) γ

3 25 minutes

4 D

5* **a** (i) β, γ (ii) γ
d (i) ¼ (ii) 9000 million years

6* **a** Electron (β^--particle)
b Approx. 6000 years
c Cloth about 800 years old; possible

49 Atomic structure

1 **a** (i) 84 (ii) 210
(iii) 84 (iv) 210 − 84 = 126
b $^{206}_{82}Pb$

2 **a** 92
b outside nucleus
c 146
d nucleus
e reduces to 90

3* **a** (i) B (ii) more electrons
b (i) A and C
(ii) same number of electrons and protons
c (i) ● (ii) ○ (iii) ×
d ●●
○○

Additional questions

1 **a** (i) electron beam strikes fluorescent screen
(ii) focus (iii) time base
(iv) increasing voltage applied to *X* plates
b (i) Y input
(ii) trace moves horizontally; trace moves vertically

2 **a** electrons
b A
c (i) D
(ii) show where electrons hit screen
d causes deflection
e so electrons not stopped by collisions with gas molecules

3 **a** (i) AH (ii) GF (iii) BE

4* **a** (i) top plate positive, lower plate negative
(ii) electrons attracted to positive
b (i) time base applied to X plates
(ii) varying voltage applied to Y plates

5 **a** 1.6×10^{-16} J **b** 1.9×10^7 m/s

6 **c** (i) 3 μg

7 **a** (i) 6 protons, 6 neutrons
(ii) 6 protons, 8 neutrons, $N > Z$
(iii) 6 protons
b (ii) $N = Z$
(iii) $^{12}_{6}C$, $N = Z$
c (i) number of protons increases to 7; number of neutrons decreases to 7
(ii) electron

8 **a** (i) time taken for activity to fall to half original value
(ii) radiation received from surroundings
b (i) 55 (ii) 1. 27.5 2. 52.5
(iii) 15 min
(iv) there is always a background count
(v) curve to left of one shown, flattening out at 25 counts/min

9* **a** (i) Electrons are knocked from gas atoms and ions formed
(ii) α-particles have a much greater mass and charge, are slower moving and produce more ion pairs/cm

10* **b** (i) α
(ii) $^{216}_{84}Po + ^{4}_{2}He$
(iii) radon decays
(iv) radioactive radon gas can be breathed in and may lodge in the lungs

Revision questions

1 **a** 60° **b** 30°

2 **a** Refraction **b** POQ
c Towards **d** 40°
e 90 − 65 = 25°

3 C

4 **a** Refraction **b** (ii) 90 − 39 = 51°
c (ii) total internal reflection since angle of incidence > critical angle

5 **c** (i) 2 m
(ii) 2 m away from mirror

6* **b** virtual, inverted, same size as object
c ray strikes glass normally
d 2×10^8 m/s
e $i > c$ so total internal reflection occurs

7 **a** Dispersion
b (i) red
(ii) violet

8 B

9* **a** (ii) virtual, upright, same size, same distance from mirror

10 C

11 E

12 **a** Become circular **b, c** No change

13 **a** Sound **b** Infrared **c** X-rays
d (i) sound cannot travel in absence of air; electromagnetic waves can travel in a vacuum
(ii) radio

14 **a** (i) infrared
(ii) X-rays
b (i) radio (ii) γ-rays

15 **a** Longitudinal
b (i) compression
(ii) rarefaction

16 **a** Sound travels more slowly than light
b (i) sound is reflected from the mountain
(ii) sound has travelled further
c 1320 m

17 A
18 D
19 B
20 a (i) 6 (ii) 15 cm
(iii) circular; constant depth since the wavelength is constant
(iv) wood bobs up and down
b (i) water waves are transverse waves; sound waves are longitudinal waves
21* b 5×10^{14} Hz; $v = f\lambda$
c (i) no change (ii) reduced
22 E
23 A
24 a Weight
b Mass of the potatoes
c Packing density of smaller potatoes is higher
25 C
26 a Density = mass/volume
b 9 g/cm^3
c No
d Brass
e Aluminium
27* a limit of proportionality
b force proportional to extension
c OQ extension proportional to force;
QR extension/unit force greater
d 4.0 N/mm
28* a Yes, 1 mm = 0.001 m
b E
29 c 8000 N
30 A
31 B
32 a Same
b Power of A greater
c (i) chemical
(ii) food
d Potential
e p.e. → k.e. → heat, sound
33 a force, perpendicular distance from pivot
b (i) force, moment
(ii) $F_1 + F_2 + W$ (iii) F
34 a (i) 10 N (ii) stretched
b (i) 120 N (ii) up
(iii) use more blocks
35* a 1800 N b (i) 4500 J
(ii) 1800 W
36 a Chemical, kinetic
b Work done/time taken
c Increases; p.e. increases
37 C
38 D
39 A
40 0.4 cm/s
41 a (i) 31 m/s
(ii) 6 m/s, 11 m/s or 22 m/s
(iii) 10 s
b 75 m
42 b 22.5 m c 25 m/s d 75 m
43 C
44 0.7 s
45 D
46 D
47 a B b A
48 A
49 E
50* a (i) 2.5 N
(ii) tips anticlockwise
(iii) 62.5 N cm
51* a 6×10^4 Pa
b 675 Pa
c 2200 J
52* d Weight = mg $= \frac{105}{1000} \times 10$
$= 1.05$ N
53* a k.e. = $\frac{1}{2}mv^2 = \frac{1}{2} \times 0.4 \times 3^2$
$= 1.8$ J
b (i) 1.8 J
(ii) 0.45 m
(iii) air resistance slows ball
c Move faster
54 a (i) smoke particles
(ii) air molecules bombard the smoke particles
b Air molecules bombard the glass
55 C
57 a y increases as ring expands
b y decreases as ring contracts
c To ensure a tight fit
58 B
59 a Kinetic
b (i) beaker + water
(ii) 1. 250 g 2. 250 g
3. 1050 J 4. 200 J
5. thermal capacity of sand is less than that of water
60 a (i) thermometer
(ii) water
b melts; molecules; need
61 D
62 a A, D, E, H
b (i) H
(ii) alcohol is more volatile (changes to vapour more easily)
63 C
64 C
65 a (i) temperature rises faster in glycerine than in water
(ii) thermal capacity of water is higher so for a given amount of heat supplied the temperature reached is lower
b (i) rise to a maximum temperature and then remains constant while the liquid boils
(ii) heat supplied is used to vaporize the liquid
66 a Metals are good conductors of heat
b Fins have a large surface area/volume ratio so have a high rate of cooling
67* 333×10^3 J/kg
68* b 600 J/(kg °C)
69* a Amount of movement of the pointer for a given change in temperature
b (i) nitrogen gas
(ii) would expand most for a given temperature change
c (i) nitrogen
(ii) remains a gas over a wide range in temperature
d Equal divisions on the scale did not represent equal changes in temperature
70* a 4200 J/(kg °C)
b (i) compress gas slowly with piston
(ii) 0.0034 m^3
71* a (i) black
(ii) good absorber of radiation
b (i) polished metal such as aluminium
(ii) good reflector of radiation
72 a Positively and uncharged aluminium are attracted
b an insulator does not conduct electricity
c Polythene
73 a 1 Ω
b 3 A
c 6 V
74 a 3 Ω
b 2 A
c 4 V across 2 Ω; 2 V across 1 Ω
75 D
76 B
77 a 8 Ω
b (i) 12 V, 0.75 A (ii) 16 Ω
c (i) 16 Ω (ii) 4 m
78 a E b A c C d B e D
79 a interchange connections on ammeter or battery
b current
d (i) voltmeter
e 0.4 A
f 0.4 A
g (i) 7.5 Ω
(ii) increases
80*a (i) 0.4 A (ii) 3.75 Ω (iii) 8 C
b (i) connect two lengths in parallel
(ii) connect two lengths in series
c 36 J
81 a (i) diode
(ii) capacitor
(iii) transistor
(iv) multimeter
(v) microphone
b(i) A input sensor; B control circuit; C output sensor
(ii) AND
(iii) 1 1 1 1
1 0 0 0
0 1 0 0
0 0 0 0
(iv) key on and saddle pressed

82* **a** F input; G processor; H output
b (i) 0 0 1
0 1 1
1 0 1
1 1 0
(ii) AND + NOT
(iii) 0 0 1
0 1 0
1 0 0
1 1 0
Alarm sounds only if both pressure switches released

83* **a** (i) B (ii) C
b OR
c (i) LDR (ii) thermistor
d 3 kΩ
e (i) 5.5 kΩ (ii) 1.09 mA
f (i) 2.7 V (ii) yes (iii) 0.75 V
(iv) no

84 E

85 **a** (i) *might* both be magnets
(ii) *must* both be magnets
b (i) reed becomes magnetized and is attracted to magnet W so that switch opens
(ii) WY
c (ii) switch is open
(iii) iron reed demagnetizes and switch closes
(iv) bell stops ringing when burglar shuts door again

86 **a** Soft iron
b (i) current flows
(ii) becomes magnetized
(iii) open
(iv) lamp does not light

87 C

88 E

89*a alternating, magnetic, induces
b (i) 2500; $V_s/V_p = N_s/N_p$
(ii) 10^4 W

90*b (i) 1.07 A
(ii) step-down
c (i) 3.1 A, 0.8 A (ii) 138 Ω
(iii) 16 Ω (iv) 506 W
(v) 30 360 J

92 E

93 B

94 B

95 B

96 C (symbol is $^7_3$Li)

97 **a** C **b** A **c** B **d** A

98 **a** D **b** E

100 **a** (i) electrons
(ii) protons
(iii) neutrons
b Nucleus
c $^7_3$Li

101 **a** Electrons
b γ most penetrating, then β, and α are least penetrating
c (i) 1.12 minutes
2. Thinner
3. count rate increased
(ii) no; α-particles do not pass through a thick sheet of paper
(iii) 1. 5.3 minutes
2. no; count rate falls too quickly because the half-life is short

102*a (i) d.c. (ii) 1 V
b (i) a.c. (ii) 2 V
(iii) 40 ms

103*a (i) electrons transferred from duster to polythene
(ii) bring polythene close to lower side of metal plate so that positive charge is induced on the lower surface and negative charge on the upper surface; touching the upper surface will allow negative charge (electrons) to flow to earth leaving the plate positively charged
b α-particles ionize air; negative electrons are attracted to positively charged plate and neutralize it

104*a (i) air becomes ionized
(ii) α-particles most ionizing, then β-particles, and γ-rays are least ionizing
b β-particles have the mass of an electron (≈ 0 units) and are electrons with charge of −1 units; γ-rays have no mass and are photons with no charge

105*a 4 units; +2 units; strong; at right angles to the magnetic field lines; towards −ve
b (i) high
(ii) high and +ve
(iii) large

Mathematics for physics

1 **a** 3 **b** 5 **c** 8/3 **d** 20 **e** 12
f 6 **g** 2 **h** 3 **i** 8

2 **a** $f = v/\lambda$ **b** $\lambda = v/f$ **c** $I = V/R$
d $R = V/I$ **e** $m = d \times V$
f $V = m/d$ **g** $s = vt$ **h** $t = s/v$

3 **a** $I^2 = P/R$ **b** $I = \sqrt{(P/R)}$
c $a = 2s/t^2$ **d** $t^2 = 2s/a$
e $t = \sqrt{(2s/a)}$ **f** $v = \sqrt{(2gh)}$
g $y = D\lambda/a$ **h** $\rho = AR/l$

4 **a** 10 **b** 34 **c** 2/3 **d** 1/10 **e** 10
f 3×10^8

5 **a** 2.0×10^5 **b** 10 **c** 8
d 2.0×10^8 **e** 20 **f** 300

6 **a** 4 **b** 2 **c** 5 **d** 8 **e** 2/3
f −3/4 **g** 13/6 **h** −16 **i** 1

7 $a = (v - u)/t$ **a** 5 **b** 60 **c** 75

8 $a = (v^2 - u^2)/2s$

9 **b** Extension ∝ mass because the graph is a straight line through the origin

10 **b** No: graph is a straight line but does not pass through the origin
c 32

11 **a** is a curve
b is a straight line through the origin, therefore $s \propto t^2$ or s/t^2 = a constant = 2

Alternative to practical test questions

1 **a** (iii) 27° **b** bases, pins may not be vertical

2 **a** cm; s; s
b 1.835; 1.787; 1.753; 1.706; 1.672
d no; graph is not a straight line through origin
e greater accuracy

3 **a** (i) A 7 °C; B 16 °C; C 12 °C; beaker A
(ii) cover the beaker with a piece of card
(iii) have water at the same start temperature in each beaker
b (i) A
(ii) cooling curves are not linear; the water will not cool so far below room temperature

4 **a** (i) V_1 = 0.94 V, V_2 = 0.35 V, V_3 = 0.48 V
(ii) 1.77 V
(iii) R_1 = 2.7 Ω, R_2 = 1.0 Ω, R_3 = 1.4 Ω

5 **b** (i) right-hand end is North pole
(ii) N pole of compass rotates clockwise towards S pole of magnet because the magnetic field becomes stronger closer to the magnet

Index

Photo acknowledgements

p.x *t* Philippe Plailly/Science Photo Library, *b* Space Telescope Science Institute/NASA/Science Photo Library; **p.xi** *tl* Mauro Fermarellio/Science Photo Library, *cl* PurestockX/photolibrary.com, *tr* NASA/Science Photo Library, *br* Martin Bond/Science Photo Library; Christine Boyd; **p.1** © Brian Kennett; **p.2** *l* Alexander Tsiaras/Science Photo Library, *r* © Tom Tracy Photography/Alamy; **p.6** © Owen Franken/Corbis; **p.8** © Colin Underhill/Alamy; **p.9** © Phil Schermeister/Corbis; **p.12** © Cn Boon/Alamy; **p.13** Alfred Pasieka/Science Photo Library; **p.16** *l* Last Resort, *cr* CNRI/Science Photo Library, *br* © vario images GmbH & Co.KG/Alamy; **p.18** *t* © S.T. Yiap Selection/Alamy, *b* Last Resort; **p.25** Adrienne Hart-Davis/Science Photo Library; **p.28** *all* Andrew Lambert/Science Photo Library; **p.29** *t* Andrew Lambert/Science Photo Library, *b* HR Wallingford Ltd; **p.30** © Bruce Coleman Inc./Alamy; **p.33** *t* US Geological Survey/Science Photo Library, *b* Mohamad Zaid/Rex Features; **p.34** Unilab (a Philip Harris Education brand) Image supplied by Unilab Science www.philipharris.co.uk; **p.35** Westinghouse; **p.36** Jonathan Watts/Science Photo Library; **p.38** Andrew Drysale/Rex Features; **p.39** *l* Science Photo Library, *r* © Corbis. All Rights Reserved; **p.43** Lawrence Livermore Laboratory/Science Photo Library; **p.44** *both* NASA/Science Photo Library; **p.47** © David J. Green – studio/Alamy; **p.48** John Walmsley/Education Photo; **p.52** © Israel images/Alamy; **p.54** *both* Ross Land/Getty Images; **p.59** Rex Features; **p.64** Kerstgens/SIPA Press/Rex Features; **p.65** *t* BBSRC/Silsoe Research Institute, *b* TransBus International; **p.68** © Arnulf Husmo/ Getty Images; **p.71** Ray Fairall/Photoreporters/Rex Features; **p.72** *t* © Richard Cummins/Corbis, *tc* Glen Dimplex Heating Ltd, *bc* © Alt-6/Alamy, *b* Scottish & Southern energy plc; **p.77** *tl* © Alex Bartel/Science Photo Library, *cl* Aurora Vehicle Association Inc., *br* © ALSTOM; **p.78** *cl*, *bl* Mark Edwards/Still Pictures, *br* © ALSTOM; **p.80** AP Photo/Ben Margot; **p.85** © Esa Hiltula/Alamy; **p.86** © image100/Corbis; **p.91** © David Stoecklein/Corbis; **p.92** © Images-USA/Alamy; **p.94** Andrew Lambert/Science Photo Library; **p.100** © Images&Stories/Alamy; **p.102** PSSC Physics © 1965, Education Development Center, Inc.; D.C Health & Company; **p.104** © Arbortech Industries Limited; **p.107** Agence DPPI/Rex Features; **p.109** Charles Ommanney/Rex Features; **p.110** © Volker Moehrke/zefa/Corbis; **p.111** TRL Ltd/Science Photo Library; **p.113** © Albaimages/Alamy; **p.114** Rex Features; **p.115** Photo ESA; **p.121** © Don B. Stevenson/Alamy; **p.122** Dr Linda Stannard, UCT/Science Photo Library; **p.124** *t* Claude Nuridsany & Marie Perennou/Science Photo Library, *b* Last Resort; **p.126** Photo by Pete Mouginis-Mark; **p.129** *l* Photo courtesy of BOC Limited, *r* © Chris Mattison/Alamy; **p.131** © Martyn F. Chillmaid/Science Photo Library; **p.135** Philippe Plailly/Eurelios/Science Photo Library; **p.143** © Mark Sykes/Alamy; **p.149** *l both* Rockwool Ltd; *r* © Fogstock LLC/SuperStock; **p.150** *both* © sciencephotos/Alamy; **p.151** © Don B. Stevenson/ Alamy; **p.153** Gail Goodger/Hutchison; **p.155** James R. Sheppard; **p.159** Richard R. Hansen/Science Photo Library; **p.160** Keith Kent/Science Photo Library; **p.162** © Martyn F. Chillmaid/Science Photo Library; **p.171** Andrew Lambert/Science Photo Library; **p.175** *both* RS Components; **p.180** *t* RS Components, *b* Andrew Lambert/Science Photo Library; **p.187** RS Components; **p.188** Siemens Metering Limited; **p.191** *t* © iStockphoto.com/216Photo, *b* AJ Photo/Science Photo Library; **p.192** *both* RS Components; **p.193** Andrew Lambert/Science Photo Library; **p.194** *both* Andrew Lambert/Science Photo Library; **p.196** © Martyn F. Chillmaid/Science Photo Library; **p.200** Unilab (a Philip Harris Education brand) www.philip harris.co.uk; **p.201** *l* © Claude Charlier/Science Photo Library, *r* James King-Holmes/Science Photo Library; **p.202** Moviestore Collection; **p.209** Rafael Macia/photolibrary.com; **p.212** *both* Andrew Lambert/Science Photo Library; **p.216** Alex Bartel/Science Photo Library; **p.221** Elu Power Tools; **p.225** *l* RS Components, *r* Unilab (a Philip Harris education brand) Image supplied by Unilab Science, Ashby de la Zouch www.philipharris.co.uk, **p.229** *both* © ALSTOM; **p.233** ALSTOM T & D Transformers Ltd; **p.239** CERN/Photo Science Library; **p.243** Andrew Lambert/Science Photo Library; **p.248** *t* The Royal Society, Plate 16, Fig 1 from CTR Wilson , Proc. Roy. Soc. Lond. A104, pp.1–24 (1923), *b* The Royal Society, Plate 16, Fig 1 from CTR Wilson , Proc. Roy. Soc. Lond. A104, pp.1–24 (1923); **p.249** Lawrence Berkeley Laboratory/Science Photo Library; **p.250** Lippke; **p.251** © University Museum of Cultural Heritage – University of Oslo, Norway (photo Eirik Irgens Johnsen); **p.258** *both* Courtesy of the Physics Department, University of Surrey.